7th Biennial Conference on Classical and Quantum Relativistic Dynamics of Particles and Fields 2010

AA003000

(IARD 2010)

Journal of Physics: Conference Series Volume 330

Hualien, Taiwan
30 May – 1 June 2010

Editors:

L. P. Horwitz
Martin C. Land
Da-Shin Lee
Bei-Lok Hu
Tepper Gill

ISBN: 978-1-61839-353-1
ISSN: 1742-6588

7th Biennial Conference on Classical and Quantum Relativistic Dynamics of Particles and Fields 2010

(IARD 2010)

Journal of Physics: Conference Series Volume 330

Hualien, Taiwan
30 May - 1 June 2010

ISBN: 978-1-61839-353-1
ISSN: 1742-6588

IOP Institute of Physics

Printed from e-media with permission by:

Curran Associates, Inc.
57 Morehouse Lane
Red Hook, NY 12571

Some format issues inherent in the e-media version may also appear in this print version.

Copyright© (2010) by the Institute of Physics
All rights reserved.

Printed by Curran Associates, Inc. (2011)

For permission requests, please contact the Institute of Physics
at the address below.

Institute of Physics
Dirac House, Temple Back
Bristol BS1 6BE UK

Phone: 44 1 17 929 7481
Fax: 44 1 17 920 0979

techtracking@iop.org

Additional copies of this publication are available from:

Curran Associates, Inc.
57 Morehouse Lane
Red Hook, NY 12571 USA
Phone: 845-758-0400
Fax: 845-758-2634
Email: curran@proceedings.com
Web: www.proceedings.com

TABLE OF CONTENTS

PREFACE

Although the subject of relativistic dynamics has been explored from both classical and quantum mechanical points of view since the work of Einstein and Dirac, its most striking development has been in the framework of quantum field theory. The very accurate calculations of spectral and scattering properties, for example, of the anamolous magnetic moment of the electron and the Lamb shift in quantum electrodynamics, and many qualitative features of the strong and electroweak interactions, demonstrate the very great power of description achieved in this framework. Yet, many fundamental questions remain to be clarified, such as the structure of classical realtivistic dynamical theories on the level of Hamilton and Lagrange in Minkowski space as well as on the curved manifolds of general relativity. There moreover remains the important question of the covariant classical description of systems at high energy for which particle production effects are not large, such as discussed in Synge's book, *The Relativistic Gas*, and in Balescu's book on relativistic statistical mechanics. In recent years, the study of high energy plasmas and heavy ion collisions has emphasized the importance of developing the techniques of relativistic mechanics.

The results of Linder *et al* (*Phys. Rev. Lett.* **95** 0040401 (2005)) as well as the more recent work of Palacios *et al* (*Phys. Rev. Lett.* **103** 253001 (2009)) and others, have shown that there must be a quantum theory with coherence in time. Such a theory, manifestly covariant under the transformations of special relativity with an invariant evolution parameter, such as that of Stueckelberg (*Helv. Phys. Acta* **14** 322, 588 (1941); **15** 23 (1942); see also R P Feynman *Phys. Rev.* **80** 4401 and J S Schwinger *Phys. Rev.* **82** 664 (1951)) could provide a suitable basis for the study of such questions, as well as many others for which the application of the standard methods of quantum field theory are difficult to manage, involving, in particular, local properties of spacetime structure.

The scope of this series of conferences is, however, much wider. There have been recent develpments in the understanding of the quantum properties of spacetime, the application of quantum field theory and statistical quantum field theory to problems in relativistic dynamics, as well as new techniques in general relativity; some of these topics have been discussed in the IARD 2010 conference, and which will be reported in these Proceedings.

It was for this purpose, to bring together researchers from a wide variety of fields, such as particle physics, astrophysics, cosmology, heavy ion collisions, plasma research, and mathematical physics, with a common interest in relativistic dynamics, that this Association was founded.

The International Association for Relativistic Dynamics was organized at its first meeting as an informal session of seminars among researchers with common interest in February 1998 in Houston, Texas, with John R Fanchi as president.

The second meeting took place, in 2000, at Bar Ilan University in Ramat Gan, Israel, the third, in 2002, at Howard University in Washington, DC, and the fourth, on 12–19 June 2004, in Saas Fee, Switzerland. In 2006, the meeting took place at the University of

Connecticut campus in Storrs, Connecticut, and the sixth meeting, in Thessaloniki, Greece. The seventh meeting, took place at the National Dong Hwa University in Hulien, Taiwan from 30 May to 1 June 2010.

This meeting forms the basis for the Proceedings that are recorded in this volume of *Journal of Physics: Conference Series*. Along with the work of some of the founding members of the Association, we were fortunate to have lecturers from application areas that provided strong challenges for further developments in quantum field theory, statistical quantum field theory and its potential applications to relativistic quantum information theory, cosmological problems, and in the dynamics of systems described in the framework of general relativity.

The opening session of IARD 2010 was held jointly with the closing seesion of the RQI-N workshop on relativistic quantum information that took place from 28–30 May. This joint meeting emphasized the importance of including dynamical models in relativistic quantum information theory, and of utilizing the perspective of quantum information in extracting results with strong implications for application in relativistic dynamics.

Topics discussed at the conference and reported in this volume included investigations into problems in general relativity, relations between quantum field theory, cosmology and, in its statistical aspects, to the extraction of classical attributes of macroscopic quantum systems. There was also a very fundamental study by David R Finkelstein, of the stucture of spacetime itself, posing the possibility that the spacetime manifold emerges from an underlying quantum complex, composed of simplices with spin 1/2 and Fermi statistics, resulting in the regularization of the Standard Model and perhaps a regularized structure for quantum gravity.

H T Cho and B L Hu study the vacuum expectation value of the stress energy tensor of a minimally coupled massless scalar field and its role as a source in the Einstein–Langevin equations of quantum gravity, governing the induced metric of fluctuations above the mean field dynamics of the semiclassical theory. C H Chou, B L Hu and Y Subasi study macroscopic quantum phenomena from the point of view of correlations, coupling and criticality, and explain how a macroscopic quantum system may, in this way, acquire classical attributes but still retain some quantum features. S Y Lin discusses a connection with quantum information science as one of the consequences of his work on local projective measurements on relativistic fields.

In the field of cosmology, F H Ho and J M Nester study Poincaré gauge theory with a metric compatible connection to an independent dynamics associated with torsion and curvature. They find a propagating 0^+ mode that could account for accelerated expansion. They discuss, in particular, a model in the Bianchi class A, and present a Lagrangian and a typical dynamical evolution. J T Hsiang, C H Wu, L H Ford and K W Ng review investigations of the effects of a quantum stress tensor of a conformal field on inflationary cosmology. They find that the quantum stress tensor fluctuations lead to effects that can depend upon the total expansion factor during inflation, which may contribute to a non-scale invariant and non-Gaussian component to the primordial spectrum of perturbations,and may be observable.

In the framework of quantum field theory, A N Kvinikhidze and B Blankleider show that a relativistic quantum mechanics emerges from light frame quantum field theory, and that in the case of baryon-like conservation, these theories are equivalent. With T Skawronski, they show in a second paper the power of gauging for several body problems, and demonstrate how this idea can be applied to the study of parton distributions, two nucleon currents in cutoff quantum field theory, and in a potential model for πN scattering. C M Chen and J R Sun study a holographic dual of the Reissner–Nordström black hole in a quantum gravity description from the perspective of the AdS/CFT correspondence.

On a fundamental level, somewhat related to the ideas of Finkelstein, A Gersten and A Moalem discuss the factorization of the d'Alembertian in a 4×4 representation of 'relativistic quaternions' to find an interpretation of Maxwell's equations; with an 8×8 factorization, they obtain spin two fields as in gravitation. They discuss a general method for obtaining field equations for zero mass particles and arbitrary spin.

M Pavsic has developed a generalization of the theory of Stueckelberg, mentioned above, applicable to general relativity. He finds a source of the world time τ in $M_{2,4}$, achieving a 5D metric tensor and a resolution of the 'problem of time' in this framework. In a basic investigation of the structure of the theory of special relativity, closely related to the original work of Minkowski and previous work of M Pavsic, Z Oziewicz has proposed a groupoid (non-group) covariance, defining the electric and magnetic fields as tensor on the 4D spacetime. P O'Hara, using a linearized metric of general relativity had previoulsy found that on the geodesics, one finds the Dirac equation. In the paper in this volume he studies the result on arbitrary curves, and proposes equations of motion.

In the application of Stueckelberg's theory to the covariant harmonic oscillator problem (such as the model studied by Feynman, Kislinger and Ravndal (R P Feynman, M Kislinger and F Ravndal *Phys. Rev.* D **3** 2706 (1971)), it has long been known (R Arshansky and L P Horwitz *J. Math. Phys.* **30** 66, 213 (1989)) that there are exact solutions providing the non-relativistic spectrum up to relativistic corrections, with no 'ghosts', in terms of variables separated in terms of the (relative) spacelike invariant radial coordinate, angular and hyperbolic angular coordinates. However, no ladder representation for annihilation-creation operators was obtained. M Land has made considerable progress in this direction in his paper on harmonic oscillator states by studying the problem in two and three dimensions, with symmetries O(1), O(3) and O(2,1). He finds that for all $s \neq 0$ solutions, the SU(n) symmetry of the general oscillator Hamiltonian discussed by Bars (I Bars *Phys. Rev.* D **79** 045009 (2009)) is spontaneously broken by the ground state, and he demonstrates the connection of this symmetry breaking to non-separability into one dimensional Cartesian solutions. In a second paper, Land discusses the 5D (based on the gauge fields that leave the Stueckelberg–Schrödinger equation locally gauge invariant) Abraham–Lorentz–Dirac equations of self-interaction of a relativistic charged particle; he shows that a statistical interpretation of the contributions of the event current to the measured currents can lead to an effective regularization of the theory, in which pre-acceleration of the event by future values of the fields is not present.

On a phenomenological level, the work of S Bai, Z Cao, W B Han, C Y Lin, H J Yo and J P Yu studies a simulation of the two and three black hole configurations, developing new and powerful methods for these important problems. N Ben-Amotz studies the possibiliy of describing gravitation phenomenologically in terms of an exponential model. In a second paper, he studies the measured radial expansion rate of the Universe using the Einstein addition formula for velocity. He finds a modified Hubble law which explains the Olber's paradox (as the linear Hubble law does also), and for which the existence of dark energy becomes unnecessary for explaining existing data on dependence of luminosity as a function of redshift for type Ia supernovas.

We thank the Scientific Advisory Committee for their invaluable guidance and advice:

Stephen Adler (Institute for Advanced Study)
Itzhak Bars (University of Southern California)
Gordon Baym (University of Illinois)
Jacob Bekenstein (Hebrew University)
Fred Cooper (Los Alamos National Laboratory)
Bei-Lok Hu (University of Maryland)
Werner Israel (University of Victoria)
E V Shuryak (Brookhaven National Laboratory)
L S Shulman (Clarkson University)
William Unruh (University of British Columbia)

The organizers express their gratitude to the academic sponsors for their support and hospitality:

National Science Council (Taiwan)
National Center for Theoretical Sciences (Taiwan)
National Dong Hwa University (Taiwan)
Institute of Physics, Academia Sinica (Taiwan)

Finally, we thank the participants who contributed through their lectures, personal discussions, and these papers, to the advancement of the subject and our understanding.

For the Editors and Organizing Committee,

L P Horwitz (Tel-Aviv University, Bar Ilan University), Editor-in-Chief
Martin C Land (Hadassah College), IARD President
Da-Shin Lee (National Dong Hwa University), Chairman of the Local Organizing Committee
Bei-Lok Hu (University of Maryland)
Tepper Gill (Howard University)

7th Biennial Conf. on Classical and Quantum Relativistic Dynamics of Particles and Fields IOP Publishing
Journal of Physics: Conference Series **330** (2011) 012001 doi:10.1088/1742-6596/330/1/012001

Quantum simplicial dynamics

David Finkelstein
Georgia Institute of Technology

Abstract. Present-day quantum field theory can be regularized by a decomposition into quantum simplices. This replaces the infinite-dimensional Hilbert space by a high-dimensional spinor space and singular canonical Lie groups by regular spin groups. It radically changes the uncertainty principle for small distances. Gaugeons, including the gravitational, are represented as bound fermion-pairs, and space-time curvature as a singular organized limit of quantum non-commutativity.
Keywords: Quantum logic, quantum set theory, quantum gravity, quantum topology, simplicial quantization.

1. Simplicial quantization

1.1. Quantization and de-radicalization

One way to regularize the Standard Model and general relativity while preserving their agreement with empirical particle spectra and selection rules is to analysis them into finite quantum simplicial elements. This might be a physical regularization, not a formal one.

A quantum theory represents its unmixed system input and outtake processes by rays in a space of what Heisenberg called "probability vectors". A "probability-amplitude vector" is more accurate; a "ψ vector" or simply "a ψ" is quicker. If the ψ space is finite-dimensional, the operators of the theory have only finite spectra, and divergences do not occur. Such a quantum theory is called *regular* [5].

Call a Lie algebra or group regular or singular according to whether its Cartan-Killing form is so. Semisimple Lie algebras are regular; canonical Lie algebras are singular.

Quantum theories today use infinite-dimensional ψ's for unitary representations of singular or non-compact Lie algebras.

The singular Lie algebras include the canonical/Bose-Einstein Lie algebra

$$\mathsf{h}(M) = \mathrm{alg}(x^\mu, y_\mu, i) : [x^\mu, y_{\mu'}] = \delta^{\mu'}_{\mu}, \quad i \le \mu, \mu' \le M, \tag{1}$$

of classical space-time coordinates and differentiators, ordinary or gauge-covariant; of quantum coordinates and their canonical momenta in general; of the boson fields in particular; and the functional algebras of gauge theory, including the diff of general covariance.

The non-compact Lie groups include the Lorentz group. Its Hilbert space representations are singular but its spinor representations are regular.

The radicals of the singular algebras indicate how singular they are. Canonical quantization reduced the radical dimensions of some Lie algebras of classical physics from ∞ to 1 by decontraction. Here the remaining singular Lie algebras are further reformed into simple subalgebras of $\mathsf{spin}(N, N)$ for some large finite N determined by the number of quantum events in the system history. This eliminates the radicals.

The limit $N \to \infty$ is not a contraction in the usual sense. It goes by discrete steps instead of a homotopy. Yet in a weak sense still to be made precise, the step from N to $N+1$ is small when N is large. If N is large enough, for example, the difference between systems of N and $N+1$ quanta will be experimentally unresolvable by present means. Let us term a succession of small steps toward regularity a *reformation* whether they are discrete steps or infinitesimal.

Quantizations have required "painful renunciations", as Bohr put it. Now the ideas of space-time, Hilbert space, and a fundamental theory are renounced. A typed and graded spinor space $\mathbb{S}$ replaces and elaborates the usual Hilbert space. Its indefinite metric has a familiar physical interpretation, essentially given by Dirac (§1.2).

Classical systems with ancestrally finite sets for their sample spaces (= phase spaces) are intrinsically finite in their properties (§1.2) . Finite sets are also simplices, whose vertices are their elements and whose points are the convex statistical mixtures of their vertices. Quantum simplices replace the classical statistical mixture by quantum superposition (§1.2).

Vertices of a simplex may themselves be simplices of a deeper level. Call a simplex a *complex* when we wish to emphasize that it is not only finite but ancestrally finite, finite all the way down.

The quantum complex of §1.2 is regular, with its ψ's in a spinor space, finite dimensional by definition, and seems rich enough to represent any finite quantum structure and any of the experimental transformation groups of quantum physics as closely as necessary. It seems closer to experiment to use ψ's in a spinor than in a Hilbert space.

From a logic of yes-or-no questions about one individual one can construct a logic of how-many questions about a quantity of such individuals. Such a construction is called *quantification* after William Hamilton, ca. 1850. The higher-order algebra of sets (or classes) is a quantification theory. It transforms a theory with sample space S into a quantified theory with sample space 2^S, the power set of S.

As von Neumann pointed out, when Heisenberg replaced the commutative algebra of dynamical variables by his non-commutative algebra of dynamical variables, he effectively revised the first-order logic of physical systems; for predicates are merely two-valued dynamical variables. Quantum theory replaced Boolean logic by a projective logic.

This destroyed classical quantification theory at its Boolean base and necessitated a quantum replacement. Several quantum quantifications were swiftly constructed; but from scratch, as if classical quantification had never existed. Classical quantifications include such assemblies as combinations and permutations, sets and sequences. Quantum sets were instead called Fermi-Dirac ensembles. Quantum sequences were called Maxwell-Boltzmann ensembles. These quantifications convert each one-quantum ψ into an annihilation/creation (or A/C) operator $\widehat{\psi}$, acting on many-quantum ψ vectors, and generating an algebra that determines the statistics.

Call a classical sequence $(e, e, \ldots, e)$ of any number of equal elements, a *sib*. This construct was too useless to merit a name before the advent of the quantum theory. A quantum sib, however, is a Bose-Einstein ensemble, whose every ψ is a superposition of tensor powers like $\vec{e} \otimes \ldots \otimes \vec{e}$, where $\vec{\ }$ indicates a ψ vector.

All the classical modes of quantification mentioned are easily expressed in terms of power sets, but the converse is not true. Moreover, the only regular quantification theory among thm is the quantum set, the Fermi-Dirac ensemble, whose ψ space is a Grassmann algebra. This suggests that if we choose to start with just one kind of quantification, even as a toy theory, we ought to try the Fermi-Dirac kind first. Let us do so.

1.2. The spinor space $\mathbb{S}$

Since space-time belongs to a deeper logical type than the fields that are functions on it, their joint quantization calls for a quantum theory of several levels.

Journal of Physics: Conference Series **330** (2011) 012001 doi:10.1088/1742-6596/330/1/012001

Unitization (Glaserfeld's term) is the bracing operation

$$\iota : a \mapsto \{a\} \equiv \iota a \equiv \overline{a}. \tag{2}$$

Unlimited iteration of the classical operations 1, ι and $\vee$ generates an infinite set $\widehat{\mathbb{S}}$ of classical (finite) complexes.

Correspondingly, the ψ space $\mathbb{S}$ of the generic quantum complex is finitely generated by $+$ and $\mathbb{R}$ for quantum superposition, the unitization operator or unitizor ι, subject to linearity, and the Grassmann product $\vee$.

Iterated unitization is standard in classical thought; Peano used it to generate the natural numbers. It already occurs to a limited extent at several junctures in quantum physics today; as when we treat helium nuclei as point particles. There ι is a phenomenological description of a more complex process of dynamical binding of four fermions. Likely ι is phenomenological here too.

Write the set of all finite subsets of a set s as $S = \bigvee s$ or 2^s. Let us call the set s the *logarithm* of 2^s. For example, $\widehat{\mathbb{S}}$ is its own logarithm: $\widehat{\mathbb{S}} = 2^{\widehat{\mathbb{S}}}$.

. As usual, use the same symbol for corresponding classical and quantum constructs when context makes it clear which is meant. When necessary, designate a quantum correspondent of x by $\vec{x}$. For example, write the Grassmann or exterior algebra of a vector space $\mathcal{V}$ as $\bigvee\mathcal{V} \equiv \mathbf{2}^{\mathcal{V}}$. This is the vector space spanned by Grassmann products of finitely many vectors in $\mathcal{V}$.

Like $\widehat{\mathbb{S}}$, $\mathbb{S}$ is its own logarithm, $\mathbb{S} = \bigvee\mathbb{S} = 2^{\mathbb{S}}$, and is doubly graded, by Grassmann grade g and Quine type T:

$$\mathbb{S} = \bigoplus_{g,T} \mathbb{S}^{(g)}[T] \tag{3}$$

The spinors of $\mathbb{S}$ are used as ψ's for quantum complexes.

These spinors carry many physical variables besides spin. For clarity, call a quantum complex a *plexon*, and call a ψ vector for a plexon a *plexor*. Spinors are special cases.

Since simplicial quantum theory replaces the Hilbert space of canonical quantum theory by a Grassmann algebra $\mathbb{S}[E]$, it is a graded (or "super-") quantum theory of an extreme kind. In the terms of Bryce DeWitt, plexons are all soul and no body.

In nature, all elementary fermions seem to have spin 1/2. The systems represented in the Grassmann algebra $\mathbb{S}$ are fermions. Therefore give each type $\mathbb{S}$ the structure of a spinor space, not a Hilbert space (§2.3). The Cartan theory of spinors enables us to assign a spinor structure to every Grassmann algebra $\mathbb{S}[T]$. The underlying quadratic space is the direct sum

$$\mathcal{W}[T] := \mathrm{Bi}\,\mathbb{S}^{(1)}[T] := \mathbb{S}^{(1)}[T] \oplus \mathrm{Dual}\,\mathbb{S}^{(1)}[T]. \tag{4}$$

The quadratic space underlying the spinor space $\mathbb{S}$ is the bipolar space of (14).

In homological algebra, a *complex*—a collection of simplices— is usually represented as a sum of simplices. But in $\mathbb{S}$, addition is mere quantum superposition, not collection. It does not increase the number of simplices but merely changes the possibility for one simplex. One assembles plexons into a plexon of higher type according to quantum logic; not by adding them but by multiplying their unitizations. The vertices of the resulting complex are the unitizations of the simplices of the type below. A complex is a simplex of simplices, not a sum of simplices.

1.3. Index conventions

$\mathbb{S}[T]$:= the subspace of $\mathbb{S}$ consisting of all polyadics of type T.

$\mathrm{Bi}\,\mathbb{S}[T] := \mathbb{S}^{(1)}[T] \oplus \mathrm{Dual}\,\mathbb{S}^{(1)}[T]$.

1_t := a typical polyadic basis spinor in the Grassmann algebra $\mathbb{S}[T]$, indexed by a lower-case version of T, the symbol for the type.

$1_{\bar{t}}$:= a typical monadic basis vector in the first-grade subspace $\mathbb{S}^{(1)}[T+1]$:

$$1_{\bar{t}} := \overline{1_t} := \iota(1_t) =: 1_{(2^t)}. \tag{5}$$

$1_{\ddot{t}}$:= a typical basis vector for the bispace $\mathrm{Bi}\,\mathbb{S}[T+1]$.

In the classical basis t is the serial number of 1_t. t and $\bar{t}$ take on $\mathrm{hexp}\,T$ values. $\ddot{t}$ takes on $2\,\mathrm{hexp}\,T$ values.

1.4. Bar codes

An algebraic *bar code* is easier to read and write than a graph of a multidimensional complex. Construct a bar code from a formula for the complex in the symbols 1, ι, and $\vee$, by omitting the symbols 1 and $\vee$ as usual, writing ιx as the bar symbol $\bar{x}$, ordering factors by their serial numbers, and write superpositions as usual. $\mathbb{S}$ is the vector space of quantum bar codes (Table 1).

Table 1. Bar codes by rank r and serial number n

6	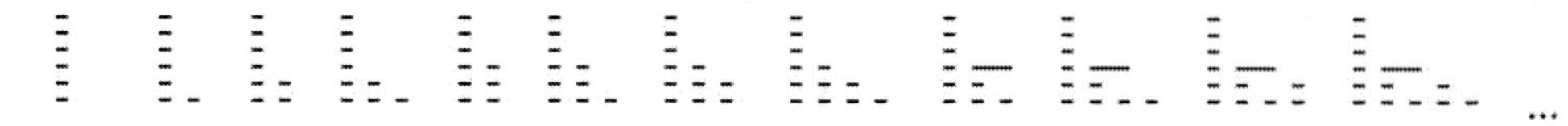 … **hexp 6** …
5	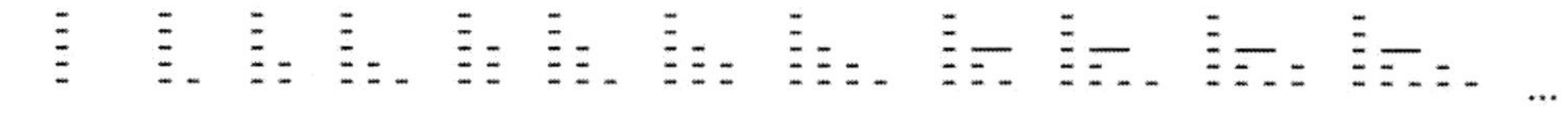 … **hexp 5** …
4	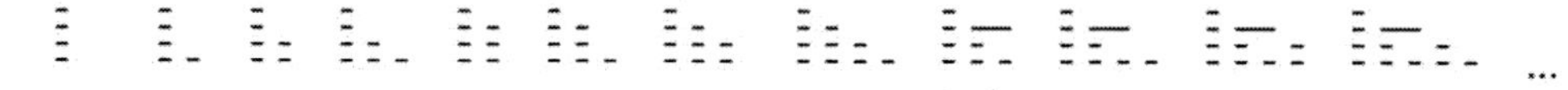 … 16 17 18 19 20 21 22 23 24 25 26 27 …
3	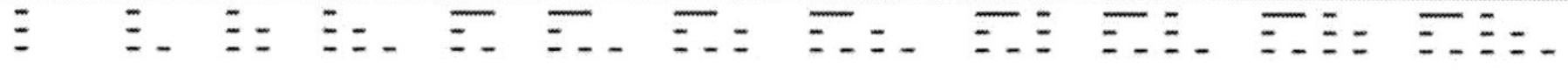 4 5 6 7 8 9 10 11 12 13 14 15
2	 2 3
1	 1
0	0
r	$\mathbf{1}_n$ n

Each bar code describes an oriented simplicial complex, whose vertices are its factors, oriented by the order of their multiplication. They must be read from the bottom up. Which bars stand for the same vertex is determined by what lies below them and not by what lies above them. The bar codes constructed without quantum superposition, using only $\vee$ and ι, form a classical basis of $\mathbb{S}$. Their signs are fixed by serially ordering their monadic factors.

A g-adic is a simplex of exactly g vertices. Each vertex of a simplex is a segment of its bar code covered by a single bar, as long as necessary, representing a monadic.

The rank of a polyadic is the height in bars of its highest monadic factor.

A simplicial quantum theory of type T is one whose plexors have type $\mathbb{S}[T] \subset \mathbb{S}$.

Assume that there exist:

(i) a cellular plexon type $\mathbb{S}[C]$ that supports the Lie algebra of spin and the unitary charges.
(ii) an event type $\mathbb{S}[E]$ that supports the orbital variables as well.
(iii) a field type $\mathbb{S}[F] = \mathbb{S}[E+1]$ that supports field variables as well.

Let us roughly estimate the event type $\mathbb{S}[E]$ whose dimension N is large enough to pass for the Hilbert space of a fermion. Provisionally, $C = 3$, $\mathrm{Dim}\,\mathbb{S}[C] = \mathrm{hexp}\,2 = 16$ might suffice for the spin and charge of one cell. Then $E = 4$ would imply $N \leq \mathrm{hexp}\,4 = (2^{16})$, far too small for a quasi-continuum of events; while $E = 5$ implies $N \leq \mathrm{hexp}\,5 = 2^{64\mathrm{K}}$ (where $\mathrm{K} := 2^{10}$), far more than required.

The Poincaré group can be approximated within present experimental error, though non-uniformly, by a subgroup of $\mathrm{SO}(\mathbb{S}[5])$ but not $\mathrm{SO}(\mathbb{S}[4])$. The Standard Model groups seem to be faithfully represented in $\mathbb{S}[5]$ up to experimental accuracy.

Therefore assume tentatively that $C = 3$, $E = 5$, and $F = 6$.

2. Space-time quantizations

2.1. Yang quantum space

Most physical theories today assume an absolute space-time, with many reference frames related by a relativity group. To simplify the Poincare-Heisenberg Lie algebra

$$\mathsf{ph}(4) = \mathrm{alg}(x^m, p_m, L_{m'm}, i) \tag{6}$$

of special relativistic mechanics, Yang dropped this assumption and adopted the decontraction $\mathsf{ph}(4) \nleftarrow \mathsf{so}(5,1)$ [34]. This relativizes the split into position space and momentum space, and quantizes the imaginary i of the commutation relations. He retained a complex Hilbert space representation of infinity dimensions, however, with its separate central i.

Many reformations decontract the canonical Lie algebra $\mathsf{h}(1)$ to the sphere Lie algebra $\mathsf{so}(3)$:

$$\mathsf{so}(3): \quad [p,q] = r, \quad [q,r] = \lambda p, \quad [r,p] = \lambda q, \quad \lambda \to 0, \tag{7}$$

replacing the canonical i by a variable r, and the singular Killing form by a regular one with negative-definite Killing form. Slightly different reformations de-contract $\mathsf{h}(1)$ to the hyperboloid Lie algebra:

$$\mathsf{so}(2,1): \quad [p,q] = r, \quad [q,r] = \lambda p, \quad [r,p] = -\lambda q, \quad \lambda \to 0, \tag{8}$$

with indefinite Killing form. Another decontraction of the canonical Lie algebra $\mathsf{h}(1) := \mathrm{alg}(q,p,i)$ of (1) acts on a spinor space instead of a Hilbert space, representing $\mathsf{so}(3)$ by dyadic spin operators in a Clifford algebra $\mathbb{S}[2] \otimes \mathrm{Dual}\,\mathbb{S}[2]$, of type 2:

$$p = \gamma_{23}, \quad q = \lambda\gamma_{31}, \quad r = \lambda\gamma_{12}, \quad \lambda \to 0. \tag{9}$$

A unification of position and momentum occurs in other gauge theories in quantum space-time [16].

2.2. Feynman quantum space

Feynman proposed a spinorial space-time quantization that in retrospect fits into the neutral spinor space $\mathbb{S}$, not a Hilbert space $\mathcal{H}$. He wrote the coordinates as sums of many commuting Dirac matrix-vectors:

$$x^m \sim \gamma^m(1) \oplus \ldots \oplus \gamma^m(N). \tag{10}$$

Journal of Physics: Conference Series **330** (2011) 012001 doi:10.1088/1742-6596/330/1/012001

This was undoubtedly suggested by the relativistic proper-time equation of motion

$$\frac{dx^m}{d\tau} = \gamma^m \tag{11}$$

for a Dirac particle. These Dirac spin operators $\gamma^m(n)$ belong to a quantum element of space-time. (10) concerns an assembly of N such elements with a Palev statistics, which had not been discovered yet. An implied quantum unit of time X is omitted from (10).

Feynman's proposal must be extended to represent momentum and i. This can be done within the same Dirac algebra, as noted by Marks [17]. For the Standard Model fermions, $i\gamma^{4321}$ is ± 1. This suggests that at a deeper level γ^{4321} performs some of the functions of i. If we set

$$\begin{aligned} p_m &\sim \gamma_{4321}(1)\gamma_m(1) \oplus \ldots \oplus \gamma_{4321}(N)\gamma_4(N), \\ i &\sim \frac{\gamma_{4321}(1) \oplus \ldots \oplus \gamma_{4321}(N)}{N}, \end{aligned} \tag{12}$$

then the commutator of δx and δp has a γ^{4321} where the canonical theory has i. (12) omits a quantum unit of energy

$$\mathsf{P} = \frac{\hbar}{\mathsf{NX}}, \tag{13}$$

required to balance the units. (12) defines a simplicial quantum space. In the limit of classical space-time, $\mathsf{X} \to 0$, $\mathsf{P} \to 0$, and $\mathsf{N} \to \infty$. The spin now serves as an element of space-time and momentum-energy, as well as angular momentum $\delta L_{m'm} = \gamma_{m'm}/2$. Soon its simplex will be enlarged to carry the unitary charges as well, and to describe entire events. When it is necessary in order to reduce confusion, call a spin with this enlarged physical interpretation and algebra, a quantum simplex, or *plexon*.

The space-time coordinates, Lorentz angular momenta, momenta, and i of a plexon of the cellular type correspond 1-1 to $4+6+4+1 = 15$ traceless 4×4 Dirac matrices in the real Dirac spinor space $4\mathbb{R}$ of the plexon, soon to be enlarged. The commutator Lie algebra of $\mathrm{Cliff}(3\mathbb{R} \ominus 1\mathbb{R})$ is $\mathsf{sl}(4) \cong \mathsf{so}(3,3)$ instead of the Yang $\mathsf{so}(5,1)$. The Feynman quantum space has an automorphism $\gamma^m \mapsto \gamma^{4321}\gamma^m$ of the Clifford algebra of polynomials in the γ^m; it exchanges space-time with momentum-energy.

The Feynman quantum space is not represented in Hilbert space here but in a spinor space $\mathbb{S}[E]$, and so it is regular. If the extra two coordinates of $\mathsf{so}(3,3)$ are frozen out in the condensation that produces i, as assumed, their peculiar timelike signature seems to lead to no contradictions with macroscopic experience.

Each type $\mathbb{S}[T]$ is a spinor space for a quadratic vector space $\mathcal{W}[T]$. The vector-wise unitization $\iota'\mathbb{S}[T] = \mathbb{S}^{(1)}[T+1]$ is the null semivector space for the next spinor space, of type $T+1$, depending on context. This context-dependence can be eliminated merely by explicitly tagging spinors and semivectors differently, so it seems harmless.

As spinor space, the spin group Spin$[T]$ doubly covers the orthogonal group SO$[T]$ of the *bipolar space*

$$\mathcal{W}[T] := \mathrm{Dual}\,\mathbb{S}^{(1)}[T] \oplus \mathbb{S}^{(1)[T]} =: \mathrm{Bi}\,\mathbb{S}^{(1)}[T] \tag{14}$$

with neutral *bipolar norm* $b : \mathcal{W} \to \mathrm{Dual}\,\mathcal{W}$ given by

$$\begin{aligned} \forall Q \in \mathbb{S},\ \forall Q' \in \mathrm{Dual}\,\mathbb{S},\ W &:= Q + Q' : \\ bW \circ W := \|W\| = \|Q + Q'\| &:= Q' \circ Q. \end{aligned} \tag{15}$$

The bipolar space is also called the quantum space [21]. A basis I_s for monadics in $\mathbb{S}$ defines a basis $I_{\bar{s}}$ for the bipolar vectors of $\mathcal{W}$. These in turn define a basis $\gamma_{\bar{s}}$ for the first grade of the

Journal of Physics: Conference Series **330** (2011) 012001 doi:10.1088/1742-6596/330/1/012001

Clifford algebra over $\mathcal{W}$. Grassmann left-multiplication defines the first grade generator γ_s of the type-T Clifford algebra Cliff $\mathcal{W}^{(1)}[T]$:

$$\forall\psi \in \mathbb{S}[T], I_s \in \mathcal{W}[T-1] : \gamma_s\psi := I_s \vee \psi. \tag{16}$$

And the dual basis elements γ^s are dually defined by Grassmann left-differentiation with respect to I_s.

The Dirac spin operators γ^m come from a low type C of this Clifford algebra. The Fermi-Dirac annihilation/creation operators $\psi(x)$ belong to a high type E of the same Clifford algebra.

The indefinite metric b of $\mathcal{W}$ gives rise to that of special relativity, which distinguishes the forbidden spacelike directions from the allowed timelike ones.

2.3. Neutral spinor form

Any spinor space has a bilinear *spinor form*, that is invariant under its spin group Spin$[T]$. Cartan called it C; in electron theory it is often designated by β. For an irreducible spinor space, β is unique up to a numerical factor.

The addition of a quantum A to a system is equivalent to the removal of a quantum $B = \text{Dual}\,A$ from the system. The anti-particle $\overline{A}$ of a particle A with energy E is a dual particle with energy $-E$.

Consider a basis $\gamma^{m\pm}$ of Clifford algebra generators that are anti-commuting square roots of ± 1. Let $\Pi_\pm$ be the product of the operators $\gamma^{m\pm}$ that have square (say) ± 1. Let 1_s be a spinor basis in which the γ_{m+} are symmetric and the γ_- are skew-symmetric. Let $E = \sum 1_s \otimes 1_s$ be the Euclidean quadratic form that is diagonal in the basis 1_s. Then two spinor forms, possibly proportional to each other, are

$$\beta_\pm = \Pi_\pm E \tag{17}$$

Write $\beta[T]$ for the spinor form of $\mathbb{S}[T]$. Then $\beta[2]$ is skew-symmetric and $\beta[T]$ is neutral and symmetric for $T > 2$.

The physical interpretation of the indefinite metric is borrowed from Dirac's electron theory: The neutral spinor metric β distinguishes input operations from outtake operations (§3.1).

The normed space $\mathcal{W}[2] = \mathbb{S}[2] \oplus \text{Dual}\,\mathbb{S}[2]$ of type 2 is a neutral quadratic space of eight dimensions. It completes the first cycle of the Bott periodicity. Its reduced spinor and dual-spinor spaces are also of eight dimensions, and are related to $\mathcal{W}[2]$ by triality.

One may choose any type C of $\mathbb{S}$ to be the ψ space for an ancestral cell, and analyze any spinor of the event type $E > C$ as a complex of clones of the cell C. Then any transformation $L : \mathbb{S}[C] \to \mathbb{S}[C]$ induces a transformation $\sum_C^E L : \mathbb{S}[T] \to \mathbb{S}[T]$ for all $T > C$, the *cumulant* of L.

If X is any basic polyadic in $\mathbb{S}[E]$, a product of basic monads, then by masking all bars above those of type C one exhibits a collection of polyads of type C, whose transforms under L are well defined. To transform the complex X, one transforms each of these type-C polyads in X in turn and unmasks. The sum of all these transforms is $\sum_C^{C+1} L$. This is extended to general polyadics by linearity.

In this way any representation of a Lie algebra on the ancestral cell extends to one on the entire complex, its cumulant. $\mathbb{S}$ is a set theory with null foundation. This construction re-interprets it as a set theory with foundation represented by $\mathbb{S}[C]$.

For example, the cell space $\mathbb{S}[2]$ can be identified with real Dirac spinor space. Then all higher-type spinors of $\mathbb{S}[T]$, $T > 2$, represent simplices composed of spins 1/2 by iterated Fermi quantification. The Dirac spin vector γ^m for the generic fermion simplex is part of the spin vector γ^c of its ancestral cell. The total spin angular momentum $J_{c'c}$ is the cumulant over the system complex, of the cell spin angular momenta $\gamma_{c'c}$ of its constituent cell simplices.

This builds in the spin-statistics correlation as an identity.

Journal of Physics: Conference Series **330** (2011) 012001 doi:10.1088/1742-6596/330/1/012001

The orthogonal group SO(4, 4) of $\mathcal{W}[2]$ does not include Yang SO(5, 1) but it includes the simplicial quantum $\mathsf{sl}(4) = \mathsf{spin}(3,3)$, which in turn includes Lorentz $\mathsf{spin}(3,1)$.

The two extra timelike dimensions of $\mathsf{spin}(3,3)$ are supposed to freeze out in the condensation of i.

Classical space-time is supposed to emerge as a statistically smoothed approximation to an organized form of such a quantum cellular complex.

Isospin and color, the unitary charges, must commute with spin and orbital angular momentum. They may be represented by operators on the extra 12 spinor dimensions of the ancestral spin cell.

2.4. Quantization as quantification

Let the generic one-quantum ψ be expanded in a monadic basis $\{1_q\}$ with numerical amplitudes ψ^q:

$$\psi = \sum_q \psi^q 1_q = \psi^q 1_q. \tag{18}$$

"Second quantization" replaces the numerical coefficients ψ^q by operators $\widehat{\psi}^q$. This is often said to misconstrue what is not a quantization at all, since it introduces no quantum constant or homotopy parameter, but a mere quantification, going from one quantum to many, replacing the ψ vector 1_q, not its numerical coefficient, by an annihilation/creation operator $\widehat{1}_q$.

But the difference between $\widehat{\psi}^q$ and $\widehat{1}_q$ is a mere duality. This suggests deeper connections between quantization and quantification going both ways:

1. Quantum quantification is also a second quantization in a generalized sense. In a certain singular organized limit, in which the number of quanta $N \to \infty$, a quantum field theory becomes a classical field theory. Were the quantum field theory a true quantization, the return to a classical field theory would be a contraction, a homotopy. In fact the limit $N \to \infty$ can be regarded as a generalized contraction, one that proceeds by small discrete steps instead of by a homotopy. In this extended sense, a quantification is also a quantization. The small quantum constant that it introduces is $1/N$. In an organized limit $1/N \to 0$ the field operators are centralized.

2. Canonical quantization entails a quantification with Bose statistics. To see this, break canonical quantization down into two steps:

(i) *Atomize*. Select and decontract a Lie algebra a of operators, to become the variables of the quantum atomic element.

(ii) *Quantify*. Form the variables of a quantity of quantum atoms as the polynomials in a and limits thereof.

The elements of $\widehat{a}$ represent operations that put in and take out quanta of excitation, such as phonons. For the harmonic oscillator, for example, a is spanned by the elements q and p, and is the ψ space of one phonon. The polynomials in p and q, and their limits, correspond to symmetric tensors over a. The quantized harmonic oscillator is then a bosonic assembly of phonons, a quantified phonon.

The point is that a regular quantum theory requires a quantization based on a regular statistics like the Fermi or Palev statistics. Bose statistics does not work.

The corresponding steps of $\mathbb{S}$ simplicial quantization are:

(i) *Atomize*. Select a graded Lie algebra a of basic system variables and decontract it into a subalgebra $\widehat{a}$ of $\mathbb{S}[E] \subset \mathbb{S}$ of simplicial quantum events.

(ii) *Quantify*. Form the Grassmann algebra in $\mathbb{S}[E]$ generated by $\widehat{a}$.

This is a Fermi quantification.

The simplicial quantum theory will use at least six successive quantifications. During the contraction of a simplicial quantum theory to a canonical classical or quantum theory, moreover, the type T increases beyond any finite bound.

The Grassmann functor $\bigvee$ quantifies, converting a one-quantum ψ space to a many-quantum one, and also gauges, cloning one cell Lie algebra into many isomorphs. Any coordinate system (x^1, x^2, x^3, x^4) orders space-time events into a sequence of sequences of sequences of sequences, implying that events have Maxwell-Boltzmann statistics. No such quanta are found in nature. Fermions and bosons are found, and bosons are singular, so let us assemble space-time as a quantum simplex of simplices of ... of simplices.

Cartan constructed spinors as elements of the exterior algebra over a null "semivector" space. Chevalley noted that Grassmann algebras describe simplicial complexes. Combining their insights, each quantum simplicial cell with n vertices, is described by a spinor, which we have called a plexor to avoid spacial connotations, and is therefore a spin, or plexon.

The plexor has 2^n components, one for each possible simplicial face. Further structure resides in the deeper levels, covered by one or more bars, successive unitizations. In the singular continuum limit, $\mathsf{X} \to 0$, $\mathsf{N} \to \infty$.

In $\mathbb{S}$, Grassmann grade counts simplicial quantum events, Quine type counts nested ι's, and the basic type-E dynamical operators $J^A[E] \in \mathsf{so}[E]$, being angular momentum operators of an orthogonal group $\mathrm{SO}[E]$, count angular momentum in units of the roots of the Lie algebra $\mathsf{so}[E]$. $\mathsf{so}[C]$ respects the neutral quadratic form $\mathsf{b}[C-1]$, which contracts to the Minkowski metrical form.

The action of $\mathsf{so}(3,3)$ on type $C+1$ is induced by its action on the ancestral cell of type C. Let us call this familiar summation process and its iterates *cumulation*, and write it as $J[C+1, C] = \Sigma J[C]$. The action on type E is then the cumulant $J[E, C] = \Sigma^{E-C} J[C]$.

3. Simplicial quantum events

At first Einstein described a space-time point or event operationally as a smallest possible occurrence, like the collision of two small hard bodies. It is oddly anachronistic to take collisions of classical macroscopic bodies to be elementary today. Later, events were redefined by their radar coordinates but this concept is still classical, since an electromagnetic wave is actually an unanalyzed photon beam. There seems to be no experimental evidence for an elementary process that defines only a space-time location, as general relativity assumes. In standard theories, nevertheless, fermion events are represented against a classical Minkowski space-time background, by a ψ vector in an infinite-dimensional complex Hilbert space. Both the Minkowski and Hilbert spaces, having singular groups, must be singular contractions of regular structures closer to experiment. An operational quantum construct of event is still needed.

A single photon is more elementary than a radar signal. The most elementary quantum events to which we have experimental access today are elementary fermion input/outtake operations. Let us take these as the elementary events of the next physics, and regard the various classical event constructs as contracted or truncated descriptions.

These more physical events have coordinates besides position in space-time, forming a non-commutative algebra. According to the Standard Model, fermionic event today carries one hypercharge variable y, three isospin variables τ^k, four space-time position variables x^m, four momentum-energy variables p_m, four spin variables γ^m, eight color charges χ^c, a ternary generation index g, and a binary variable distinguishing input from outtake operations. Let us tentatively omit the generation g for reasons given in §5.

A simplicial quantum theory represents a fermion event by a monadic in a subalgebra $\mathbb{S}[E] \subset \mathbb{S}$ of some finite type E; call this the *event type*. The first-grade event subalgebra $\mathbb{S}^{(1)}[E]$ replaces Hilbert space as ψ space to support the algebra $\mathrm{Alg}[E]$ of single-event operators. Higher grades

Journal of Physics: Conference Series **330** (2011) 012001 doi:10.1088/1742-6596/330/1/012001

represent composites of several fermions. The fermion field operators of the Standard Model form a singular Clifford algebra. Those of a simplicial theory are regular Clifford monadics that operate on $\mathbb{S}[E]$. They are elementary in that they are not products of non-trivial factors, but they generally have a great many ancestral elements of lower type, since experimental events have high orbital quantum numbers.

The differentiators ∂_m of canonical quantum theory refer to two levels: An infinite-dimensional orbital level of the variables x^m, p_m and a four-dimensional tangent-space level with the index m. The group Diff acts on the field and respects the tangent spaces. Both levels are customarily built into the kinematics by restricting the system to differentiable fields and their limits.

In a corresponding class of simplicial quantum theories, monadics $\widehat{1}_c$ of a lower type C are used as annihilation/creation operators for the foundational elements of which events are composed, and the group of the simplicial quantum theory is required to respect the partition of the system into cells.

This restriction could be a consequence of the dynamics, not the kinematics. Non-differentiable or cell-breaking transformations may merely require too much energy. The cell level C might be determined by the action function of the event level, and the kinematics could have the larger group $\mathsf{spin}(\mathbb{S}[E])$ of the event space. This amounts to assuming that the foundation ψ space is one-dimensional, and should be kept in mind.

Each g-adic in $\mathbb{S}^{(g)}[E]$ is a ψ for a complex of g events with Fermi statistics. The singular Lie algebra of Standard Model fermion variables is to be a contraction of a semisimple subalgebra of simplicial quantum event operators in $\mathrm{Alg}[E]$.

3.1. Bipolar quantum spaces

Dirac compared a negative probability to a bank overdraft [7]. Let us expand this somewhat terse theory of the indefinite metric.

Two frames may share common light cones, and each may draw its future time axis as a vertical arrow, but may still disagree on the distinction between past and future, so that a space-like ray r that appears to begin at the origin in one frame may appear to end there in another:

Frame F Frame F' (19)

If the space-time metric is a dynamical variable, then we cannot restrict our attention to timelike vectors, and an output from the origin can become an input to the origin under this change of frame. Let us take this to indicate that when differential locality breaks down, the distinction between input and output may become relative, like the distinction between past and future.

In general, an operation may introduce an excitation in one circumstance and eliminate one in another. The distinction between input and outtake is then relative to the choice of medium or vacuum.

Let the difference of two normed spaces designate their direct sum provided with the difference norm:

$$\forall Q \in \mathbb{S},\ \forall Q' \in \mathbb{S}' :\quad \|Q \ominus Q'\| := \|Q\| - \|Q'\|. \tag{20}$$

Minkoski vectors belong neither to the space $\mathbb{S}$, nor to Dual $3Q$, but to their *bipolar* direct sum

$$\mathrm{Bi}\,\mathbb{S} := \mathbb{S} \ominus \mathrm{Dual}\,\mathbb{S}. \tag{21}$$

Journal of Physics: Conference Series **330** (2011) 012001 doi:10.1088/1742-6596/330/1/012001

Each experimental frame now splits the bipolar space $\mathbb{S}[E]$ into two unipolar subspaces of positive-definite and negative-definite norms respectively. Let us take flux directed from the experimenter into the experiment to be positive. Kets have positive norms and bras have negative norms.

The spinor norm thus gives probability flux rather than net probability. The Lorentz group has regular finite-dimensional isometric representations in the orthogonal group of Bi $\mathbb{S}$.

The ambient spin complex is not random but highly organized locally, into something like a crystal dome whose cell is (say) SO(3, 3) invariant. The dome also supports the particle spectrum, sharp bands of highly coherent transmission.

Each simple Lie algebra of a physical theory has a Killing form with important physical meaning. In the original Yang space, it defines a Minkowski metric on the energy-momentum vector, giving the squared rest-energy. In this way the metrical structure of these quantum space-times derives from their quantum structure.

3.2. Simple quantum theories

Nowadays a physical theory involves several Lie or near-Lie algebras, some graded. Let us call a system simple or semisimple if all these algebras are simple or semisimple, respectively, and otherwise *compound*; one bad algebra spoils the barrel. Since the commutative coordinate algebra of a classical system is compound, only a quantum system can be simple. Since neither $\mathsf{sl}(\infty)$ nor $\mathsf{su}(\infty)$ is a Lie algebra, let alone a simple one, only a quantum system with a finite-dimensional ket space can be simple.

To recover the singular canonical quantum theory (1) from a simple one like (7), one must not only take a singular commutative limit $h, \mathsf{X} \to 0$ but must also freeze the degrees of freedom in the quantized imaginary variable $r = [p, q]$, so that r can be treated as a constant $\mathsf{N}i$. This freezing can result from a self-organization of many plexons, like that of the spins of a ferromagnet. Let us refer to such a construction briefly as a *singular organized limit*, and write, for example, $\widetilde{q} \nrightarrow iq$, $\widetilde{p} \nrightarrow ip$, $\widetilde{r}/4N \nrightarrow i$. The symbol $\widehat{x}$ designates a real decontraction of a canonical (classical or quantum) construct $x \nleftarrow \widehat{x}$. The symbol $\widetilde{x}$ indicates a real decontraction of an imaginary canonical construct $ix \nleftarrow \widetilde{x}$.

The lost Hilbert space, canonical commutation relations, and Bose statistics must return in a singular organized limit.

3.3. Evolution of simplicial quantum theory

Heisenberg effectively introduced a quantum space for phase space in 1924, and suggested quantizing space-time in 1930 as a way to eliminate the remaining infinities. Julian Wess describes how this idea passed from Heisenberg to Peierls, to Pauli, to Oppenheimer, to Snyder [27], to Yang C. N. [34], who made simplified the Lie algebra of the orbital variables $x^\mu, p_\mu, L_{\mu'\mu}$ and i but still represented it in the singular algebra $\mathsf{su}(\infty)$ of Hilbert space.

Feynman considered the quantum space (10) in about 1941, before undertaking the Lamb shift.

Independently Penrose R. [20] quantized the Euclidean 2-sphere by representing its points as the directions of the sums of many Pauli spins with Bose-Einstein statistics.

On still another track to quantum space-time, Segal [23] made a suggestion already implicit in Yang's note: that physics evolves toward simple Lie algebras by homotopies that reduce Lie-algebra radicals. Segal's argument was Darwinian: A compound Lie algebra has a singular Killing form, making it labile. A semisimple Lie algebra is stable in this respect. As measurements of the structure constants improve, therefore, a compound Lie algebra has survival probability 0 relative to its semisimple neighbors, which outnumber it by ∞ to 1. Segal wrote this before the Golden Age of Gauge, but it applies to gauge groups as well. Singular Lie algebras are indeed singular cases; almost all Lie algebras are semisimple.

This Darwinian argument for stabilization must be used with discretion. For example, within the universe of linear spaces with bilinear products the Jacobi identity is as unstable as commutativity, and it is not proposed here to stabilize it. In addition, evolution by small random changes is slow. Simple organizations not only compete but also cooperate, and so form complex organizations faster than random changes can. The fact that relativity and quantum theory are de-radicalizations is sufficient reason to try another.

Gerstenhaber, influenced by Segal, described homologically a rich terrain of Lie algebras connected by homotopies, such as *contractions* [14], that carry groups out of stable valleys of simplicity, to ridges between the valleys, and up to singular peaks [12].

According to the simplicity principle, physics is flowing glacially down the simplicity gradient to a valley in Gerstenhaberland. The Galileo Lie algebra, for example, is on a ridge between the valleys of Lorentz $\mathsf{so}(3,1)$ and the orthogonal group $\mathsf{so}(4)$. Similarly, the Poincaré-canonical Lie algebra $\mathrm{alg}(x^m, p_m, L_{m'm}, i)$ is on a ridge between $\mathsf{so}(3,3)$ and $\mathsf{so}(5,1)$.

Group contraction and deformation quantization were also introduced by others [14, 2].

Vilela Mendez, inspired by Gerstenhaber, rediscovered the Yang SO(5, 1) group [31], representing it in Hilbert space, and proposed high-energy physical consequences. The Yang group was rediscovered several times since then.

Galiautdinov [11], Shiri-Garakani [26], and others have studied theories with the Yang SO(5, 1) invariance in $\mathbb{S}$. Baugh [1] represented the Yang $\mathsf{so}(5,1)$ Lie algebra in $\mathsf{sl}(6\mathbb{R})$.

Now decontraction algorithms are being intensively studied [24].

4. Simplicial quantum relativity

Assume that the statistics Lie algebra of simplicial events and the Lie algebra of the ψ space are simple [19]. Represent $\mathsf{spin}(3,3)$ in $\mathsf{spin}(N,N)$, whose indefinite metric leads to a *bipolar relativity* (§3.1).

Compare the quantized orbital variables of the simple spaces mentioned, with $ch\mathsf{PX}$ units and dimensionless orbital variables. $\widehat{x}$ is a quantized x, $\widetilde{x}$ is a quantized ix, $\delta\widehat{q}$ is the contribution of one cell to a sum $\widehat{q} = \sum \delta\widehat{q}$ over many; and $1 \le k \le 3$, $1 \le m \le 4$:

Feynman [8]	$\delta\widehat{x}^m \sim \gamma^m$	—	
Yang [34]	$\widetilde{x}^m \sim i(g^{mn}\eta^5\partial_n - \eta^m\partial_5)$	$\widehat{p}_m \sim i(\eta_6\partial_m - \eta_m\partial_6)$	(22)
Penrose [20]	$\delta\widehat{x}^k \sim \sigma^k$	—	
Simplicial —	$\delta\widetilde{x}^m \sim \gamma^{4321}\gamma^m$	$\delta\widetilde{p}_m \sim \gamma_m$	

The proposals of Penrose and Feynman maintain an absolute distinction between space-time and energy-momentum. For group simplicity Yang relativized this distinction within a relativity Lie algebra $\mathsf{so}(5,1)$, but represented this within the singular Lie algebra $\mathsf{su}(\infty)$ of a singular Hilbert space $L^{(2)}(6\mathbb{R},\mathbb{C})$. This respects both reciprocity and parity. For the sake of regularity, the simplicial quantum theory represents both time and energy within a spin representation of Lorentz $\mathsf{spin}(3,1)$ instead. This also introduces fewer variables. Furthermore it violates parity and reciprocity, in better accord with experiment. The factor γ^{4321} has been shifted from δp to δx to make the gaugeon dynamics easier.

Unitary charges of the Standard Model kind are short transverse struts to the dome. The cellular plexons of the ambient dome therefore have a higher dimension than classical space-time, are spins of a higher orthogonal group. Dirac spins are short longitudinal struts. Space-time coordinate lines are longitudinal and long. If they include transverse struts, their unitary charges cancel within a few cells. The quantized imaginary $\mathring{\imath}$—where $^\circ$ stands for "operator"—is macroscopic and composed of longitudinal struts, and so can be classed as longitudinal, but it is polarized and frozen. There are now three non-zero vacuum expectation values to account for:

Journal of Physics: Conference Series **330** (2011) 012001 doi:10.1088/1742-6596/330/1/012001

g. Gravity's $g_{\mu\nu}$, which breaks space-time $\mathsf{sl}(4\mathbb{R})$.

h. Higgs h_i, which breaks electroweak $\mathsf{su}(2)$.

i. The i in the Heisenberg equation and the canonical commutation relations, which breaks complex-plane $\mathsf{sl}(2\mathbb{R})$ and time reversal $\mathbb{Z}_2 \subset \mathrm{SL}(2\mathbb{R})$.

Let us represent the dome and its excitations within $\mathbb{S}$ using as few ι types as possible.

When i becomes a variable $\mathring{\imath}$ and leaves the center of the algebra, a real quantum theory arises of the Stückelberg kind [29]. This also recalls the Hestenes theory in which $i = \gamma^{4321}$ [13], and the quaternionic i that was proposed as a mass-generating Higgs field [30].

The canonical relations work in a segment of the spectrum of $-\mathring{\imath}^2$ so near to its maximum value 1 as to be indistinguishable from it in atomic experiments. Yet this narrow band must have a multiplicity that passes today for infinite. For but one example, the band

$$1 - \mathsf{N}^{-1/2} < -\mathring{\imath}^2 \leq 1 \tag{23}$$

is both narrow and populous enough, with multiplicity $\mathrm{O}(\sqrt{N}) \to \infty$.

In the singular limit $\mathring{\imath} \to i$, $E \to 0$, and (supposedly) $(\mathring{\imath})^2 \to -1$, the classical space-time and canonical commutation relations are to emerge. We must suppose that

$$\mathsf{XPN} = \hbar, \quad \mathsf{N} \gg 1, \quad \hbar/\mathsf{X} \gg 1\,\mathrm{TeV}, \quad \frac{\hbar}{\mathsf{NX}} \approx 0, \tag{24}$$

in the sense that $\hbar/\mathsf{NX}$ is presently not resolvable from 0.

To achieve this one may hypothesize an organization akin to polarization that centralizes ("superselects") the variable $\mathring{\imath}$ and contracts it to the imaginary unit of complex quantum theory: $\mathring{\imath} \nrightarrow i$. The Heisenberg relation $[x^m, p_m] = \hbar i$ returns in a singular limit of many cells. This means zero-point estimates based on canonical quantum theory are gross overestimates in the cell domain, including the ones that call for infinite renormalization and energy density.

The coordinates, momenta, and $\mathring{\imath}$ are now cumulated spin variables. To make all components of a cumulated spin small is not impossible in the way that making position and momentum coordinates small is supposed to be. It may merely require a meltdown of $\mathring{\imath}$.

4.1. Reciprocity and locality

In canonical classical and quantum theories and all their regularizations considered here, the kinematics obeys reciprocity while experience seems to flout it, especially by its locality. The locality principle permits fields to couple only when they are evaluated at the same position, but different momenta (wave numbers) generally couple.

In the simplicial quantum space (22), reciprocity is the transformation

$$R : \gamma^m \mapsto \gamma^{4321}\gamma^m \tag{25}$$

R is an automorphism of the Dirac Clifford algebra. The simplicial event space has reciprocity. Locality presently breaks it. In the simplicial quantum theory, however, a simplicial "ultra-locality" is tautological: Two simplices γ^A, γ^B grade-commute unless they are neighbors, share a vertex γ^a, $a \in A, B$.

4.2. Palev statistics

Bose-Einstein statistics is a defining though singular part of canonical quantization. Fermi-Dirac statistics generalizes from Lie algebras to graded Lie algebras, such as Grassmann algebras, in order to deal with variables without classical correspondents, like fermion creators and spins.

Journal of Physics: Conference Series **330** (2011) 012001 doi:10.1088/1742-6596/330/1/012001

A *Palev statistics* (of class a) is one whose one-quantum annihilation/creation operators generate a semisimple graded Lie algebra a. Fermion pairs are not exactly bosons, only quasi-bosons, but they are exact Palevons. They can simulate bosons as polarized spins can simulate an oscillator by precessing about the polarization axis.

Most attention has been given to the simple ungraded case; but Stoilov and Van der Jeugt find that the Palev statistics of class $\mathsf{su}(1|5)$ is especially relevant to the Standard Model [28].

In pure Grassmann simplicial quantum theories, like the present study, one must represent empirical bosons as event dyads. These obey a Palev statistics, with fermionic cores. Presumably this fermion core must show up in high-energy photon-photon collisions. Arguments against composite photons or gravitons are well known, but are based on Bose-Einstein statistics and the canonical commutation relations, and therefore must break down in exactly the domain of interest.

With this generalized concept of statistics, any quantum number can be regarded as a number of quanta. This is familiar for the linear harmonic oscillator, where the quantum is the phonon of excitation.

For instance, the quanta of the dipole rotator in three dimensions, with Lie algebra (7)–call them rotatons, since the term "roton" is preempted—have a statistics that is neither Bose-Einstein nor Fermi-Dirac, but Palev of the $\mathsf{spin}(3) = A_1 = B_1$ class. The individual kinds of Palevons correspond to root vectors of the Palevon Lie algebra. The statistics of the Palevon is determined by an irreducible representation of the Palev Lie algebra. Where the canonical commutation relations have just one practical representation, each Palev statistics has infinitely many. The $\mathsf{so}(3)$ Palev statistics is isomorphic to the rotator (7), and has only one kind of quantum and its dual. The Bose-Einstein statistics arises from the representation $D(N)$ of (7) when N is allowed to approach infinity with organization that polarizes and centralizes one component, say r. The corresponding physical theory, however, has a finite N that must be measured.

The same $\mathsf{so}(3)$ class of Palev statistics includes the spinor representation $D(1/2)$. Its representation in $\mathbb{S}$ requires spinors of four real components and is then given by

$$p = \gamma^{32}/2, \quad q = \gamma^{21}/2, \quad r = \gamma^{13}/2. \tag{26}$$

A spin $1/2$ is thus a composite of up to 2 rotatons with Palev statistics. It is necessary to see how Palev statistics describes a condensation resembling the Bose-Einstein one. This requires many Palevons, $N \gg 1$, in at least two classes, one to be filled and the other to be emptied.

A monadic of any type embraces monadics of the type below. As a result, lower-type operators $L[T]$ induce higher-type ones $L[T+1] = \Sigma L[T]$. In particular, quantum spin operators of the ancestral cell induce orbital angular momentum operators of the event. The cumulant $\Sigma^n L[T]$ represents the operator $L \in \mathrm{Alg}(\mathbb{S}[T])$ in the algebra $\mathrm{Alg}(\mathbb{S}[T+n])$ of every higher type $T+n$.

Let us apply the cumulation process to the Lie algebra of the simplicial quantum space-time of (22). Let $\gamma^{\ddot{b}}$ be four real Dirac spin operators acting on spinors 1_c of a cellular level $\mathbb{S}[C]$, with Minkowskian metric form

$$g^{\ddot{b}'\ddot{b}} = \frac{1}{2}[\gamma^{\ddot{b}'}, \gamma^{\ddot{b}}]. \tag{27}$$

The ancestral group generators $\delta J_{c'c}[C]$ of the cellular spin are

$$\begin{array}{rcl}
\delta\widetilde{x}^{\ddot{b}} & = & \mathsf{X}\,\gamma^{\ddot{b}}, \quad \ddot{b} = 1, 2, 3, 4, \\
\delta\widetilde{p}_{\ddot{b}} & = & \mathsf{P}\,\gamma^{4321}\gamma_{\ddot{b}}, \\
\delta\,\mathring{i} & = & \mathsf{N}^{-1}\,\gamma^{4321}, \\
\delta\widetilde{L}_{\ddot{b}'\ddot{b}} & = & h\gamma^{4321}\gamma_{\ddot{b}'\ddot{b}}, \quad \text{as} \\
\mathrm{h}(4) & \nleftarrow & \mathsf{so}(3,3) \nleftarrow \mathsf{sl}(6)
\end{array} \tag{28}$$

Journal of Physics: Conference Series **330** (2011) 012001 doi:10.1088/1742-6596/330/1/012001

The simplicially quantized imaginary $\mathring{\imath}$ is normalized to unit magnitude with a factor N^{-1}. To form macroscopic monad coordinates, we must cumulate these atomic cell variables at least twice, to reach at least type 6, with dimensionality $\operatorname{hexp} 6 = 2^{64\mathrm{K}}$, more than ample for a quasi-continuum.

Consider six 8×8 $\gamma^{\ddot{b}}$ of the cellular type C associated with $\mathsf{spin}(3,3)$. Choose a frame with metric $\ddot{b} = \operatorname{diag}(1,1,1,-1,-1,-1)$ and label the γ's as $\gamma^1,\ldots,\gamma^4,\gamma^X,\gamma^Y$, with spontaneous polarization of γ^{YX}: $\mathring{\imath} = \Sigma^2\gamma^{YX}[C] \nrightarrow i$. Define the Dirac γ^m as the remaining four first-grade elements of $\gamma^c[C]$. They define the Minkowski space-time by cumulation and the singular limit.

4.3. The Umklapp problem

When Heisenberg proposed to quantize space-time, Pauli pointed out the Umklapp problem [33]. Feynman raised it for his quantized space-time. Discreteness of the space coordinates with spacing X,

$$x = n\mathsf{X}, \quad (n \in \mathbb{Z}), \tag{29}$$

ordinarily implies periodicity of the momenta with period $2\pi\mathsf{P} = 2\pi\hbar/\mathsf{X}$:

$$p \cong p + 2\pi n\mathsf{P}. \tag{30}$$

Then the maximum momentum $p = \pi n\mathsf{P}$ in one direction is indistinguishable from the maximum momentum $-p$ in the reverse direction. Phonons in crystals flip their momenta by this process, where the missing momentum is taken up by the crystal; but not photons in the vacuum.

Simplicial quantization creates no Umklapp problem. The spectrum of the momentum $\mathsf{P}L_{mY}$ is still bounded by $\pm\mathsf{NP}$, and is still discrete and uniformly spaced between these bounds, as in a crystal, but it is not uniform in spectral multiplicity, and $L \doteq \mathsf{N}$ is not the same eigenvector as $L \doteq -\mathsf{N}$. None of the variables $L_{c'c}$ is periodic.

Furthermore, to attain a maximum value for L_{mY} requires polarizing all N spins in the sum along the mY direction. Since L_{mY} does not commute with L_{YX}, this disrupts the dome polarization of the L_{XY} that produces $\mathring{\imath}$, and so must be difficult, perhaps unfeasible.

5. Simplicial quantum gauge

Gauge fields are associated with non-commutative fermion momentum variables. A canonical gauge field theory with a semisimple gauge Lie algebra g splits the total momentum-energy p_m into an invariant kinetic part π_m and a residual potential part Γ_m:

$$p_m = \pi_m + \mathsf{P}\Gamma_m \tag{31}$$

p_m is integrable ($[p_m, p_{m'}] = 0$) but not gauge invariant. π_m is gauge invariant but not integrable. P is a constant with dimensions of energy-momentum ($c = 1$).

Simplicial quantization also produces non-commutative momentum components. Since the ancestral atoms of momentum do not commute, neither do their cumulants, the simplicial quantum momenta, the reformations of the singular infinitesimal translations canonical quantum theories. Some of this quantum non-commutativity survives into general relativity as part of the curvature, perhaps including a cosmological constant contribution from the ground complex, a gravitational energy-momentum ("dark energy"). We can be certain that this contribution is finite, and reasonably sure that it is much smaller than the canonical commutation relations permit, since the mean magnitude of the zero-point energy of a simplicial quantum field oscillator is less than its canonical correspondent by a cosmologically large numerical factor.

Therefore it is natural to ask whether gauge fields are simplicial quantum effects expressed in the canonical quantum limit; much as Poisson Brackets are canonical quantum effects expressed in the classical limit. Let us call such a gauge theory simplicial.

Journal of Physics: Conference Series **330** (2011) 012001 doi:10.1088/1742-6596/330/1/012001

In a more canonical theory, the gauge vector fields are "internal" parts of the gravitational field [6]. In a simplicial theory, the gravitational field is an external self-organization of many plexons.

In the simplicial quantum gauge theory, the momentum-energy $\widetilde{p}$, positional coordinates $\widetilde{x}$, and quantized imaginary $\mathring{\imath}$ are generators of $\mathsf{sl}(6) \sim \mathsf{spin}(3,3)$, represented as polyadic fermion processes.

Since gaugeons do not obey Fermi statistics they cannot have monadic ψ's. The simplest possibility is dyadic, as if they were fermion pairs. Dyadic plexons obey a Palev statistics, and the gauge group is their Palev Lie algebra.

De Broglie suggested that photons are neutrino pairs, a special case. In the canonical quantum theory this seems untenable. According to the Heisenberg uncertainty relations, when two identical fermions are close in their position coordinates they must be far apart in their momenta on the average. The forces binding them would have to be large for high relative momenta. There seem to be no such forces. On the contrary, asymptotic freedom seems to be observed, as well as inferred by Gross and Wilczek from non-abelian quantum gauge theory.

The simplicial quantum gauge theory reopens this classic question by weakening the Heisenberg uncertainty principle enormously, allowing small values of both relative position and relative momentum at once when the organization of $\mathring{\imath}$ is highly broken. In the case of a definite gauge metric and an irreducible group, the conservation of the Casimir operator still blocks the possibility that p, q, r can all be small. But the simplicial quantum metric is indefinite; its Casimir operator can be constant even when its terms decrease.

There are other indications for dyadic simplicial gaugeons:

1 *Spin.* The spins and the statistics add up correctly. This is what motivated De Broglie to propose the fermion-pair theory of the photon and Feynman to consider a fermion-quartet for the graviton. The graviton is the noblest particle, in the way that helium is the noblest gas and the alpha particle a noblest (that is, most magical) nucleus. It would be natural to suppose that, like them, it is composed of four fermions that have saturated each other's valences. Here it is supposed that it and the photon are composed of two fermions that have saturated each other's valences.

The Minkowski index m of the gaugeon vector $\Gamma_{m\alpha}(x)$, where α indexes a basis for the gauge Lie algebra, contributes $\hbar$ to the gaugeon spin angular momentum, as usual. The two fermions in the dyad are at slightly different quantum positions and can orbit about each other. The index m indicates a relative orbital angular momentum of $1\hbar$.

The Lie algebra index α contributes another unit spin for the graviton, where the unitary spins are saturated and α indexes the gauge Lie algebra $\mathsf{spin}(3,1)$; and not for the unitary gaugeons, where the fermion spin valences are saturated. Thus all the gaugeons, including the graviton, can be fermion-dual-fermion pairs, plus higher-order terms of more complex structure.

Every fermion has spin 1/2 and Fermi-Dirac statistics, and some have color or isospin. It is well-known how to represent fermions in Hilbert space. Then the fermion basis ψ's are tensor products of more elementary ψ's supporting orbital variables, spin, isospin, color, and generation number. These are its valences. Since they are distinguishable, there is no physical change if Grassmann products are used instead of tensor products. Then the whole construction is readily carried out in $\mathbb{S}$; this adds no insight as yet.

As for the gaugeons: The generic gaugeon has as semisimple gauge group the semi-simple product of all the simple gauge groups of the fermions. If it is a pair composed of just one kind of fermion, the fermion must have everything, so it is a quark. The isospin gaugeons are pairs of quarks in which all the color valences are saturated; the color gaugeons saturate their isospin valences; gravitons saturate both, leaving only their spins open.

The pair theory of gaugeons has a generational problem, however. There are three generations g_1, g_2, g_3 of fermions, and nine possible pairs $(g_i g_j)$, but only one generation of gaugeons is seen.

Journal of Physics: Conference Series **330** (2011) 012001 doi:10.1088/1742-6596/330/1/012001

In one resolution of this conflict, one quark generation g_1 would be more elementary than the other two, and gaugeons are pairs $g_1 g_1$ of quarks of that generation. The other two generations q_2, g_3 of fermions may then be block from pairing by their structures. In that case, however,they would couple to gaugeons more weakly than the generation g_1. This would disagree with the Standard Model.

In a resolution that saves the Standard Model, only the symmetric superposition $g_1 g_1 + g_2 g_2 + g_3 g_3$ is a stable gaugeon. This gaugeon couples to all generations alike.

2 *Connection.* A fermion dyad $\psi \vee \widetilde{\psi}$ (where the tilde indicates the dual particle) indeed transports a third fermion from ψ' to ψ as a gauge connection should, according to

$$\psi'' = (\psi \vee \widetilde{\psi}') \circ \psi'. \tag{32}$$

3 *Coupling* If the gaugeon is a fermion pair, the Dirac action of the canonical gauge theory musy be a contraction of a polyadic simplicial action. Let us de-contract it:

5.1. Simplicial gauge dynamics

The Dirac fermion-gaugeon coupling density A_{D} is an invariant trilinear form in annihilation/creation operators $\psi(x)$ for a fermion, $D_m(x)$ for a gaugeon, and $\overline{\psi}$ for a dual fermion:

$$A_{\mathrm{D}}(x) = i\overline{\psi}(x)\gamma^m(x)D_m(x)\psi(x) = \succ\!\!\bullet\!\!=, \tag{33}$$

in which ">" designates the two fermions and "=" the gaugeon. All four factors are spin operators with invisible spinor indices, and are multiplied accordingly. Schematically speaking, A_{D} couples a vector $\overline{\psi}\gamma^m\psi$ of the fermion pair to a vector D_m of the gaugeon at infinitesimally separated space-time points; it is a V-V coupling. Clearly one can form S-S, V-V, T-T, A-A, P-P couplings, and S-P and V-A couplings, as in early theories of beta decay.

The system history ψ is the time-ordered exponential

$$\Omega = \mathrm{Texp} \int (dx) A_{\mathrm{D}}(x). \tag{34}$$

A_{D} is required to be skew-hermitian so that Ω is unitary.

(27) defines a Clifford algebra only when the gravitational metric $g^{m'm}(x)$ is treated as a constant quadratic form. When quantum gravity comes into the action, this space-time Clifford algebra breaks down. On the other hand, there is no reason to assume that the Fermi-Dirac Clifford algebra also breaks down, so it is still reasonable to represent quantum gravitational processes with spinors in $\mathbb{S}$.

Let us infer from the theories of the other gaugeons that the simplicial gravitational gaugeon too is described by a simplicial correspondent $\widehat{D}$ of the differentiator D_m, with curvature field $[\widehat{D}_{m'}, \widehat{D}_m] = \widehat{R}_{m'm}$; not (say) by the simplicial correspondent of the Dirac spin vector g^m, since (27) breaks down.

Suppose that the gaugeons in D_m are bound fermionic pairs, associated with two proximate events rather than one, as discussed in §4.2 and §5. Then the Dirac action (33), which couples two fermions and a gaugeon, is actually a surrogate action for a four-fermion ("Fermi") coupling, effective when two of the fermions are bound; a tetradic

$$A^{(4)} \sim \Gamma^{e'''}{}_{e''}{}^{e'}{}_{e}\, \overline{\psi}_{e'''}\overline{\psi}^{e''}\overline{\psi}_{e'}\overline{\psi}^{e} \tag{35}$$

in which (Γ) is an invariant coupling tensor with real components.

For a physical theory with this tetradic coupling to be possible, the time development should be an orthogonal operator and its generator $A^{(4)}$ must be skew-symmetric with respect to the spinor metric form β:

$$(\beta A^{(4)})^{\mathsf{T}} = -\beta A^{(4)}, \tag{36}$$

where $(\ldots)^{\mathsf{T}}$ is the transpose of $(\ldots)$. This holds for four-dimensional spinors of $\mathsf{spin}(3,1)$, whose β is skew-symmetric, if (Γ) has the symmetry property

$$\Gamma^{e'''}{}_{e''}{}^{e'}{}_{e} = \Gamma^{e'}{}_{e}{}^{e'''}{}_{e''}. \tag{37}$$

The simplest-looking tetradic action

$$A^{\rm SS} \sim (\overline{\psi} \circ \psi)(\overline{\psi} \circ \psi) \tag{38}$$

also has this property; but violates locality grossly, since every event in one pair is coupled to every event in the other. In the Dirac action, only infinitesimally separated events are coupled.

The simplicial action closest to the Dirac action is the V-V coupling

$$A^{\rm VV} := (\overline{\psi}\gamma^{\ddot{b}}\psi)(\overline{\psi}\gamma_{\ddot{b}}\psi) \in \text{Cliff}\,\mathcal{W}[E] \tag{39}$$

Each factor is a simplicial tensor with invisible indices, whose structure is defined so that it corresponds to that of the Dirac action. Since $\gamma^{\ddot{b}}$ puts in or takes out a single vertex, it couples only adjacent plexons, and this may suffice for the experimental locality. A more explicit formulation is reserved for the computational stage.

Turn from fermions to gaugeons. The usual gaugeon action of the Standard Model has the form

$$A_{\rm G} \sim [D_{m'}, D_m][D^{m'}, D^m]. \tag{40}$$

This becomes an eight-fermion coupling, effective when all eight fermions are bound into four pairs. A corresponding plexon action is octadic:

$$\widehat{A}^{(8)} \sim \Gamma^{e^8}{}_{e^7}\ldots{}^{e'}{}_{e}\, \overline{\psi}_{e^8}\psi^{e^7}\ldots\overline{\psi}_{e'}\psi^{e} \tag{41}$$

in which (Γ) is an invariant coupling tensor with real components.

It is impossible to repress the speculation that the octadic gaugeon action $A^{(8)}$ is an effective action of second order in the tetradic action, operative when eight fermions bind into four pairs. In the continuum-based theory such speculation may be idle, due to infinities, but in the simplicial theory the coefficients, including the fine-structure constant, are mathematically well-defined, and probably computable for sufficiently small-scale experiments.

5.2. Space-time curvature as quantum effect

The classical momentum components ∂_m commute with each other, but the gauge-covariant momentum components $i\hbar D_m$ do not, nor do the simplicial quantum components $\widehat{p}_m$. This must contribute to space-time curvature and the gauge tensor fields. It is parsimonious to conjecture that the classical non-commutativity of the gauge differentiator, like that of the Poisson Bracket, is entirely a vestige of the quantum non-commutativity of the pair annihilation/creation operators.

The simplicial quantum curvature is the quantum commutator:

$$\begin{aligned} K_{\ddot{b}'\ddot{b}} &= \mathsf{P}^2[J_{Y\ddot{b}'}, J_{Y\ddot{b}}] \\ &= -\mathsf{P}^2 J_{\ddot{b}'\ddot{b}}, \end{aligned} \tag{42}$$

in a frame where the form b is in Sylvester normal form with $b_{YY} = -1$. K includes the gravitational curvature R and the unitary gauge curvatures as non-unified terms in a sum. In the simplicial quantum theory the commutators $[J_{e'''e''}[E], J_{e'e}[E]]$ are related to other components of the same generating tensor J by the structure tensor c of $\mathsf{spin}[E]$. Thus the infinity of differential concomitants of the gauge manifold, including the curvature, are replaced in the simplicial quantum theory by the finitely many components of the generating tensor $J[E]$ of type E.

6. Discussion

Spinors belong to a Grassmann algebra (Cartan); a Grassmann algebra represents a simplicial complex (Chevalley). It has long been suspected [4, 18, 15] that quantum gauge theory might be a singular limit of a simplicial complex theory, and that the gauge charges are vestiges of quantum Burgers(-Volterra) vectors for defects of various kinds. Then the manifold-based gauge theory is not fundamental but a high-level consequence of the underlying cellular dynamics of the complex.

The simplicial quantum theory presented here represents the quantum system as a simplicial quantum complex and its gaugeons, including the graviton, as fermion-dual-fermion pairs. Differential transport is a a singular continuum limit of annihilation reactions of the form

$$(\psi \vee \psi'^{\beta}) \vee \psi'' \rightarrow \psi, \tag{43}$$

where a pair effects a quantum transport by replacing the fermion ψ'' outside the pair by the fermion ψ inside the pair. The classical principle of gauge invariance is not basic but emerges from the simplicial quantum structure of the deeper levels as a smoothing approximation.

New quantum constants enter: a large integer N limiting the number of events, and quantum units X and E of time and energy. The usual imaginary i becomes a quantum spin operator, which has to be frozen and centralized by a spontaneous organization in the vacuum. The bras and kets belong to a space $\mathbb{S}$, a real Grassmann algebra over itself, not only graded but also typed in the sense of Quine. Its first-grade elements are spinors. Its neutral spinor metric assigns positive norms to kets, negative to bras.

This paper is a palimpsest. Between its lines can dimly be made out remnants of an earlier work called the space-time code [9, 10]. Instead of quantizing the Standard Model, working from the top down, the space-time code attempted to quantize space-time from the bottom up.

The weakened uncertainty relations for small distances and momenta make it possible that gaugeons, including gravitons, are fermion pairs, as their transformation properties suggest.

A plausible fermion-gaugeon action is then a tetradic (35) in fermion annihilation/creation operators, summed over the cells of the complex. Additional action terms like the Maxwell, Hilbert-Einstein, and Yang-Mills actions for the separate gaugeon fields are not necessary if they are effective actions for organizations derived from (35). Then the gauge coupling constants are order parameters of this organization.

The road from dynamics to particle spectrum and cross-sections is not yet entirely computational. Existing quantum concepts of binding and scattering rest on unstable classical foundations in Minkowski space-time and idealizations like stationary or plane ψ waves. They too must now be transferred to the more stable quantum foundations of the quantum complex, finite in both the large and the small.

Acknowledgments

Roger Penrose and Richard Feynman graciously showed me unpublished spin quantizations of space or space-time that were most important for this work. I am obliged for other helpful communications to Otto B. Bassler, James Baugh, Walter L. Bloom, Jr., Dustin Burns, David

Edwards, Shlomit Ritz Finkelstein, Andrei Galiautdinov, Josef Maria Jauch, Dennis Marks, Zbigniew Oziewicz, David Speiser, Tchavdar Palev, Heinrich Saller, Stephen Selesnick, Abraham Sternlieb, Sarang Shah, Frank (Tony) Smith, and Carl-Friedrich von Weizsäcker.

References

[1] Baugh J, 2002 Regular Quantum Dynamics. Ph. D. Thesis, School of Physics, Georgia Institute of Technology ().

[2] Bayen F, Flato M, Fronsdal C, Lichnerowicz A, and Sternheimer D 1977 Quantum mechanics as a deformation of classical mechanics. *Letters in Mathematical Physics* 1:521–530

[3] Blatt J M and Weisskopf V 1952 *Theoretical Nuclear Physics*. John Wiley, New York, 1952.

[4] Bohm D 1962 A proposed topological formulation of the quantum theory In Good I J *The Scientist Speculates: An Anthology Of Partly-Baked Ideas*, 302-314, Heinmann, London

[5] Bopp F and Haag R 1950 Über die Möglichkeit von Spinmodellen. *Zeitschrift für Naturforschung* 5a:644

[6] Connes A 1994 *Non-commutative geometry*. Academic Press, San Diego

[7] Dirac P A M 1974 *Spinors in Hilbert Space*. Plenum, New York

[8] Feynman R P Personal communication ca 1961. Feynman left this line of thought to work on the Lamb shift.

[9] Finkelstein D 1969 Space-time code *Physical Review* 184:1261

[10] Finkelstein D 1996 *Quantum Relativity* Heidelberg: Springer

[11] Galiautdinov A A and Finkelstein D R 2002 Chronon corrections to the Dirac equation. *Journal of Mathematical Physics* **43**, 4741

[12] Gerstenhhaber M 1964 *Annals of Mathematics* 32:472

[13] Hestenes D 1966 *Space-Time Algebra* Gordon & Breach, New York

[14] Inönü E and Wigner E P 1953 *Proceedings of the National Academy of Science* 39:510–524

[15] Kleinert H 1989 *Gauge Fields in Condensed Matter*, Vol I, II World Scientific, Singapore, Available online.

[16] Madore J , Schraml S , Schupp P , and Wess J 2000 Gauge theory on noncommutative spaces. *Eur. Phys. J.* C16:161167, arXiv:hep-th/0001203.

[17] Dennis Marks Private communication 2008

[18] Mermin N D 1971 The topological theory of defects in ordered media *Reviews of Modern Physics* 51:591

[19] Palev T D 1977 Lie algebraical aspects of the quantum statistics. Unitary quantization (A-quantization). Joint Institute for Nuclear Research Preprint JINR E17-10550. Dubna hep-th/9705032.

[20] Penrose R 1971 Angular momentum: an approach to combinatorial space-time. In Bastin T (ed.), *Quantum Theory and Beyond*, 151–180. Cambridge Penrose kindly shared much of this seminal work with me ca. 1960.

[21] Saller H 2006 *Operational quantum theory I. Nonrelativistic structures*. Springer, New York

[22] Saller H 2006 *Operational quantum theory II. Relativistic structures*. Springer, New York

[23] Segal I E 1951 A class of operator algebras which are determined by groups. Duke Mathematical Journal 18:221–265 Especially §6A.

[24] Salom I and Sijacki D 2011 Generalization of the Gell-Mann Decontraction Formula for sl(n,?) and su(n) Algebras. *International Journal of Geometric Methods in Modern Physics* 08: 395

[25] Selesnick S A 2003 *Quanta, Logic and Spacetime*. World Scientific

[26] Shiri-Garakani, M and Finkelstein D R Finite quantum theory of the harmonic oscillator. quant-ph/0411203. Based on Ph.D. thesis of Mohsen Shiri-Garakani.

[27] Snyder H P 1947 Quantized space-time. *Physical Review* 71:38

[28] Stoilova N I and Van der Jeugt J 2005 All fundamental fermions fit inside one su(1—5) irreducible representation. *International Journal of Theoretical Physics* 44:1157-1165

[29] Stückelberg E C G 1960 Quantum theory in real Hilbert space. *Helvetica Physica Acta* 33:727–752

[30] Tavel M , Finkelstein D and Schiminovich S 1965 Weak and electromagnetic interactions in quaternion quantum mechanics, *Bulletin of the American Physical Society* 9:435

[31] Vilela Mendes R 1994 Deformations, stable theories and fundamental constants. *Journal of Physics* A 27:8091

[32] C F von Weizsaecker Komplementaritat und Logik I–II. *Naturwissenschaften* 42:521–529, 545–555. 1955

[33] Wess J 2005 Gauge theories on noncommutative spacetime treated by the Seiberg-Witten method. In *Quantum Field Theory and Noncommutative Geometry*; edited by U. Carow- Watamura, Y. Maeda, and S. Watamura, Springer, Berlin

[34] Yang C N 1947 On Quantized Space-Time. *Physical Review* 72:874,

Journal of Physics: Conference Series **330** (2011) 012002 doi:10.1088/1742-6596/330/1/012002

Correlations of the stress-energy tensor in AdS spaces via the generalized zeta-function method

H. T. Cho[1] **and B. L. Hu**[2]

[1] Department of Physics, Tamkang University, Tamsui, New Taipei City, TAIWAN

[2] Maryland Center for Fundamental Physics, Department of Physics, University of Maryland, College Park, Maryland 20742-4111, USA

E-mail: [1]`htcho@mail.tku.edu.tw` ; [2]`blhu@umd.edu`

Abstract. We calculate the vacuum expectation values of the stress-energy bitensor of a minimally coupled massless scalar field in anti-de Sitter (AdS) spaces. These correlators, also known as the noise kernel, act as sources in the Einstein-Langevin equations of stochastic gravity [1, 2] which govern the induced metric fluctuations beyond the mean-field dynamics described by the semiclassical Einstein equations of semiclassical gravity. Because the AdS spaces are maximally symmetric the eigenmodes have analytic expressions which facilitate the computation of the zeta-function [3, 4]. Upon taking the second functional variation of the generalized zeta function introduced in [5] we obtain the correlators of the stress tensor. Both the short and the long geodesic distance limits of the correlators are presented.

1. Introduction

After the seminal work of Hawking [6] it is realized that with quantum effects black holes can evaporate. Black holes behave like thermal objects and can be in equilibrium with a thermal gas at the Hawking temperature. However, this equilibrium state is unstable to the absorption of the thermal gas by the black hole in an asymptotically flat spacetime [7]. In order to obtain an equilibrium state one has to put the black hole in an ad hoc cavity with some appropriate boundary condition [8]. The situation with black holes in an anti-de Sitter (AdS) spacetime is rather different as explored by Hawking and Page [9]. Above the critical temperature there are two masses of the black hole which can be in equilibrium with the thermal gas. The larger mass black hole is actually stable. This phase transition is now known as the Hawking-Page transition.

The recent interest in this Hawking-Page transition is mainly due to the AdS/CFT correspondence [10]. This transition is conjectured to be related to the confinement phenomenon in QCD under this correspondence. Moreover, the quasinormal modes of the AdS black hole perturbations are supposed to give information on the thermal states of the conformal theory at the boundary [11].

To study the Hawking-Page transition in a dynamical manner, the corresponding fluctuations of the quantum fields on the background spacetime must be taken into account. A systematic way to do so is by the theory of stochastic gravity [1, 2]. To the lowest order the fluctuations are represented by a noise kernel which is the vacuum expectation value of the correlation of the stress energy tensor. This is one of the reasons why we are interested in such a consideration here. For the AdS black hole case, one should deal with the stress tensor of a thermal gas in the

Journal of Physics: Conference Series **330** (2011) 012002 doi:10.1088/1742-6596/330/1/012002

Hartle-Hawking vacuum state. However, before taking such an endeavor we would like to get a grip on the problem by calculating the case with the scalar field in AdS spacetime.

The stress energy tensor correlators have been obtained before, notably, by Osborn and Shore [12] for maximally symmetric spacetimes which include the de Sitter and anti-de Sitter spaces via conformal field theory techniques. They have considered both the scalar and the fermion cases. However, they are only interested in the conformally coupled fields and have made extensive use of the traceless condition of the correlator. In [13] the authors have also calculated this quantity for quantum scalar fields in de Sitter spacetime. Their results can be analytically extended to AdS spaces and vice versa.

As this quantity is of basic importance it is useful to derive these expressions with a different method. The zeta function method we adopt here was introduced by Dowker and Critchley [3] and Hawking [4]. It is one of the most elegant methods used for the regularization of the stress energy tensor for quantum fields in Riemannian (not pseudo-Riemannian) spaces. One can use the zeta function to construct the regularized effective action and upon taking its first functional variation obtain the regularized stress energy tensor. (For a systematic exposition, see [14]). This method is generalized by Phillips and Hu [5] to calculate the noise kernels of quantum scalar fields in $S^1 \times R^1$ spaces (useful for finite temperature theory and Casimir energy density considerations) and for an Einstein universe, by way of the second order variation of the regularized effective action (see also [15]). This method is used in our present calculation of the noise kernel for AdS spacetimes in N-dimensions.

In the next section, we shall describe the generalized zeta-function method of Phillips and Hu in some details as applied to a scalar field. Then in Section III, the correlators of the stress-energy tensors in AdS^N are calculated. The small and large geodesic distance expansions of the correlators are given in Section IV. The results are then compared with those in [13]. Finally conclusions and discussions are presented in Section V.

2. Scalar field in spacetimes admitting an Euclidean section

To make the discussions general we first consider a massive m scalar field ϕ coupled to an N-dimensional Euclideanized space (with contravariant metric $g^{\mu\nu}(x)$, determinant g and scalar curvature R) with coupling constant ξ described by the action

$$S[\phi] = \frac{1}{2}\int d^N x\sqrt{g(x)}\phi(x)H\phi(x), \tag{1}$$

where H is the quadratic operator

$$H = -g^{\mu\nu}\nabla_\mu\nabla_\nu + m^2 + \xi R, \tag{2}$$

and R is the scalar curvature. The effective action defined by $W = \ln\mathcal{Z}$ is related to the generating functional $\mathcal{Z}$ by

$$\mathcal{Z} = \int \mathcal{D}\phi\, e^{-S[\phi]}. \tag{3}$$

The expectation value of the stress-energy tensor can be obtained by taking the functional derivative of the effective action

$$\langle T_{\mu\nu}\rangle = -\frac{2}{\sqrt{g(x)}}\frac{\delta W}{\delta g^{\mu\nu}(x)}. \tag{4}$$

This formal expression is divergent at coincident limit and some procedure of regularization must be implemented. Here we adopt the procedure of ζ-function regularization as shown by

Dowker and Critchley [3] and Hawking [4]. As shown by Phillips and Hu [5] the fluctuation of the stress-energy tensor is obtained by taking another derivatives of the regularized effective action W,

$$\begin{aligned}\Delta T^2_{\mu\nu\alpha'\beta'}(x,x') &\equiv \langle T_{\mu\nu}(x)T_{\alpha'\beta'}(x')\rangle - \langle T_{\mu\nu}(x)\rangle\langle T_{\alpha'\beta'}(x')\rangle \\ &= \frac{4}{\sqrt{g(x)g(x')}}\frac{\delta^2 W}{\delta g^{\mu\nu}(x)\delta g^{\alpha'\beta'}(x')}\end{aligned} \tag{5}$$

Note that although the regularized expectation value of the stress-energy tensor is dependent on one spacetime point, the fluctuation of the stress energy is a bitensor defined at two separate spacetime points through the two functional derivatives taken with respect to these two separate spacetime points.

Conventionally one define the ζ-function of an operator H,

$$\zeta_H(s) = \sum_n \left(\frac{\mu}{\lambda_n}\right)^s = \mathrm{Tr}\left(\frac{\mu}{H}\right)^s \tag{6}$$

where λ_n is the eigenvalues of H and μ represents the renormalization scale. The ζ-function regularized effective action of the operator H is

$$W_R = \frac{1}{2}\frac{d\zeta}{ds}\bigg|_{s\to 0}. \tag{7}$$

Using the proper-time method [3] one can write the ζ-function as

$$\zeta_H(s) = \frac{\mu^s}{\Gamma(s)}\int_0^\infty dt\, t^{s-1}\mathrm{Tr}e^{-tH} \tag{8}$$

$$W_R = \frac{1}{2}\frac{d}{ds}\left[\frac{\mu^s}{\Gamma(s)}\int_0^\infty dt\, t^{s-1}\mathrm{Tr}e^{-tH}\right]_{s\to 0} \tag{9}$$

Taking the first variation of the ζ-function,

$$\begin{aligned}\delta\zeta_H &= -\frac{\mu^s}{\Gamma(s)}\int_0^\infty dt\, t^s \mathrm{Tr}\left(\delta H e^{-tH}\right) \\ &= -\frac{\mu^s}{\Gamma(s)}\int_0^\infty dt\, t^s \sum_n e^{-t\lambda_n}\langle n\,|\delta H|\,n\rangle\end{aligned} \tag{10}$$

we obtain the regularized expectation value of the stress-energy tensor which is given by

$$\begin{aligned}\langle T_{\mu\nu}(x)\rangle &= \frac{1}{2}\frac{d}{ds}\left[-\frac{\mu^s}{\Gamma(s)}\int_0^\infty dt\, t^s\sum_n e^{-t\lambda_n}\left(-\frac{2}{\sqrt{g(x)}}\langle n|\frac{\delta H}{\delta g^{\mu\nu}(x)}|n\rangle\right)\right]_{s\to 0} \\ &= -\frac{1}{2}\frac{d}{ds}\left\{\frac{\mu^s}{\Gamma(s)}\int_0^\infty dt\, t^s\sum_n e^{-t\lambda_n}\, T_{\mu\nu}\left[\phi_n(x),\phi_n^*(x)\right]\right\}_{s\to 0}\end{aligned} \tag{11}$$

where

$$\begin{aligned}T_{\mu\nu}\left[\phi_n(x),\phi^*_{n'}(x)\right] &\equiv -\frac{2}{\sqrt{g(x)}}\langle n'|\frac{\delta H}{\delta g^{\mu\nu}(x)}|n\rangle \\ &= -\frac{2}{\sqrt{g(x)}}\int d^N x' \sqrt{g(x')}\phi^*_{n'}(x')\left[\frac{\delta H}{\delta g^{\mu\nu}(x)}\phi_n(x')\right]\end{aligned} \tag{12}$$

and $\phi_n(x)$ is a eigenfunction of the operator H, namely,

$$H\phi_n = \lambda_n \phi_n \tag{13}$$

where λ_n are the eigenvalues corresponding to the eigenfunctions ϕ_n. We then have

$$\begin{aligned} T_{\mu\nu}\left[\phi_n(x), \phi^*_{n'}(x)\right] &= -\left(\partial_\mu \phi^*_{n'} \partial_\nu \phi_n + \partial_\nu \phi^*_{n'} \partial_\mu \phi_n\right) + g_{\mu\nu}\left(g^{\alpha\beta}\partial_\alpha \phi^*_{n'} \partial_\beta \phi_n + \phi^*_{n'} g^{\alpha\beta} \nabla_\alpha \nabla_\beta \phi_n\right) \\ &\quad -2\xi\left[g_{\mu\nu} g^{\alpha\beta} \nabla_\alpha \nabla_\beta (\phi^*_{n'}\phi_n) - \nabla_\mu \nabla_\nu (\phi^*_{n'}\phi_n) + R_{\mu\nu}\phi^*_{n'}\phi_n\right] \end{aligned} \tag{14}$$

In the approach of Phillips and Hu [5], through the use of the Schwinger method [16], the second variation of the ζ-function can be written as

$$\begin{aligned} \delta_2\delta_1\zeta_H &= \frac{\mu^s}{2\Gamma(s)} \int_0^\infty du \int_0^\infty dv (u+v)^s (uv)^\nu \\ &\quad \left\{ \mathrm{Tr}\left[(\delta_1 H) e^{-uH} (\delta_2 H) e^{-vH}\right] + \mathrm{Tr}\left[(\delta_2 H) e^{-uH} (\delta_1 H) e^{-vH}\right]\right\} \end{aligned} \tag{15}$$

and

$$\begin{aligned} \Delta T^2_{\mu\nu\alpha'\beta'}(x,x') &= \frac{1}{2}\frac{d}{ds}\left\{ \frac{\mu^s}{\Gamma(s)} \int_0^\infty du \int_0^\infty dv (u+v)^s (uv)^\nu \sum_{n,n'} e^{-u\lambda_n - v\lambda_{n'}} \right. \\ &\quad \left. T_{\mu\nu}\left[\phi_n(x), \phi^*_{n'}(x)\right] T_{\alpha'\beta'}\left[\phi_{n'}(x'), \phi^*_n(x')\right] \right\}_{s,\nu\to 0} . \end{aligned} \tag{16}$$

Note that in this Phillips-Hu prescription [5], an additional regularization factor $(uv)^\nu$ has been introduced. This is because the authors were interested in the fluctuations of the stress-energy tensor, that is, in the coincident limit of $\Delta T^2_{\mu\nu\alpha'\beta'}(x,x')$ where under this limit further divergences occur which call for an additional regularization factor. (See also [15]). However, our present purpose is focused on getting the correlators with two points separated, i.e., in the non-coincident case. Hence, apart from the fact that the expression in Eq. (16) is more symmetric with this factor, the keeping of this factor above is actually a matter of convenience. Here we can first take the $s \to 0$ limit without spoiling the regularization and the expression in Eq. (16) will become

$$\begin{aligned} \Delta T^2_{\mu\nu\alpha'\beta'}(x,x') &= \frac{1}{2}\int_0^\infty du \int_0^\infty dv (uv)^\nu \sum_{n,n'} e^{-u\lambda_n - v\lambda_{n'}} \\ &\quad T_{\mu\nu}\left[\phi_n(x), \phi^*_{n'}(x)\right] T_{\alpha'\beta'}\left[\phi_{n'}(x'), \phi^*_n(x')\right]\Big|_{\nu\to 0} . \end{aligned} \tag{17}$$

We shall see from the following calculations that with this expression the integrations over u and v effectively separate. The calculations are therefore simplified considerably.

3. Correlations of the stress-energy tensor

The Euclideanized N-dim AdS space or the hyperbolic space H^N can be described by the metric

$$ds^2 = d\sigma^2 + a^2 \sinh^2\left(\frac{\sigma}{a}\right) d\Omega^2_{N-1} \tag{18}$$

where σ is the geodesic distance, a the radius of the AdS^N space, and $d\Omega^2_{N-1}$ the metric for the $(N-1)$-sphere. The eigenfunctions $\phi_{\kappa lm}$ obey the equations,

$$\left(-g^{\mu\nu}\nabla_\mu\nabla_\nu - \frac{\rho_N}{a^2}\right)\phi_{\kappa lm} = \left(\frac{\kappa^2}{a^2}\right)\phi_{\kappa lm} \tag{19}$$

where $\rho_N = (N-1)/2$, and are given by

$$\begin{aligned}\phi_{\kappa lm} &= c_l(\kappa)\left(\sinh\frac{\sigma}{a}\right)^{1-\frac{N}{2}} P^{1-l-\frac{N}{2}}_{-\frac{1}{2}+i\kappa}\left(\cosh\frac{\sigma}{a}\right) Y_{lm}(\Omega) \\ &\equiv f_{\kappa l}(\sigma) Y_{lm}(\Omega) \end{aligned} \tag{20}$$

where $P^\mu_\nu(x)$ is the associated Legendre function. $Y_{lm}(\Omega)$ are the hyperspherical harmonics obeying the addition theorem,

$$\sum_m Y^*_{lm}(\Omega)Y_{lm}(\Omega') = \frac{(2l+N-2)\Gamma\left(\frac{N-2}{2}\right)}{4\pi^{\frac{N}{2}}} C_l^{\frac{N-2}{2}}(\Omega\cdot\Omega') \tag{21}$$

where C^n_l is the Gegenbauer polynomial.

The normalization constant is given by

$$c_l(\kappa) = \frac{|\Gamma(i\kappa+\rho_N)|}{|\Gamma(i\kappa)|} \tag{22}$$

Later we shall need the normalization constant for $l=0$. For odd N,

$$|c_0(\kappa)|^2 = \frac{1}{a^N}\prod_{j=0}^{\rho_N}\left(\kappa^2+j^2\right) = \frac{1}{a^N}\sum_{n=1}^{\rho_N} c_{2n}\kappa^{2n} \tag{23}$$

and for even N,

$$|c_0(\kappa)|^2 = \frac{1}{a^N}(\kappa\tanh\pi\kappa)\prod_{j=\frac{1}{2}}^{\rho_N}\left(\kappa^2+j^2\right) = \frac{1}{a^N}\tanh\pi\kappa\sum_{n=0}^{\rho_N-\frac{1}{2}} c_{2n+1}\kappa^{2n+1} \tag{24}$$

In both cases we have turned it into a finite sum.

In a maximally symmetric space like the hyperbolic space H^N, any bitensor can be expressed in terms of a set of basic bitensors [17]. The first basic bitensor is the bi-scalar function $\tau(x,x')$ which is the geodesic distance between x and x'. Using the covariant derivative one could define

$$n_\mu = \nabla_\mu\tau(x,x') \quad ; \quad n_{\alpha'} = \nabla_{\alpha'}\tau(x,x') \tag{25}$$

where $n_\mu(x,x')$ is a vector at x and a scalar at x', while $n_{\alpha'}(x,x')$ is a scalar at x and a vector at x'. Next, we have the parallel propagator $g_\mu{}^{\alpha'}(x,x')$ which parallel transports any vector v^μ from x to x'. The transported vector is $v^\mu g_\mu{}^{\alpha'}(x,x')$. It is easy to see that $n_\mu(x,x') = -g_\mu{}^{\alpha'} n_{\alpha'}(x,x')$. $\tau(x,x')$, $n_\mu(x,x')$, $n_{\alpha'}(x,x')$ and $g_\mu{}^{\alpha'}(x,x')$ constitute this set of basic bitensors. All other bitensors can be expressed in terms of them with coefficients depending only on the geodesic distance between the two points. For example,

$$\nabla^\mu n_\nu = A(\tau)\left(g^\mu{}_\nu - n^\mu n_\nu\right) \tag{26}$$

$$\nabla^\mu n_{\alpha'} = B(\tau)\left(g^\mu{}_{\alpha'} + n^\mu n_{\alpha'}\right) \tag{27}$$

$$\nabla^\mu g_{\nu\alpha'} = -(A(\tau)+B(\tau))\left(g^\mu{}_\nu n_{\alpha'} + g^\mu{}_{\alpha'} n_\nu\right) \tag{28}$$

where $A(\tau)$ and $B(\tau)$ are function of τ only. For hyperbolic H^N spaces, $A = \coth(\tau/a)/a$ and $B = -\mathrm{csch}(\tau/a)/a$.

Since $\Delta T^2_{\mu\nu\alpha'\beta'}(x,x')$ is a symmetric bitensor, one could express it in terms of these basic bitensors. Taking the symmetries of the indices into account, we have

$$\begin{aligned}\Delta T^2_{\mu\nu\alpha'\beta'}(x,x') &= C_1(\tau)n_\mu n_\nu n_{\alpha'}n_{\beta'} + C_2(\tau)\left(n_\mu n_\nu g_{\alpha'\beta'} + g_{\mu\nu}n_{\alpha'}n_{\beta'}\right)\\ &\quad +C_3(\tau)\left(n_\mu g_{\nu\alpha'}n_{\beta'} + n_\nu g_{\mu\alpha'}n_{\beta'} + n_\mu g_{\nu\beta'}n_{\alpha'} + n_\nu g_{\mu\beta'}n_{\alpha'}\right)\\ &\quad +C_4(\tau)\left(g_{\mu\alpha'}g_{\nu\beta'} + g_{\mu\beta'}g_{\nu\alpha'}\right) + C_5(\tau)g_{\mu\nu}g_{\alpha'\beta'} \end{aligned} \tag{29}$$

Using the derivatives in Eqs. (26) to (28), the conservation conditions $\nabla^\mu \Delta T^2_{\mu\nu\alpha'\beta'}(x,x') = 0$ can be expressed as three equations on the coefficients,

$$\begin{aligned}&\frac{dC_1}{d\tau} + \frac{dC_2}{d\tau} - 2\frac{dC_3}{d\tau} + (N-1)AC_1 + 2BC_2 - 2\left((N-2)A + NB\right)C_3 = 0\\ &\frac{dC_2}{d\tau} + \frac{dC_5}{d\tau} + (N-1)AC_2 + 2BC_3 - 2(A+B)C_4 = 0\\ &\frac{dC_3}{d\tau} - \frac{dC_4}{d\tau} + BC_2 + NAC_3 - N(A+B)C_4 = 0\end{aligned} \tag{30}$$

Moreover, the traceless condition $g^{\mu\nu}\Delta T^2_{\mu\nu\alpha'\beta'}(x,x') = 0$ can be written as

$$\begin{aligned}C_1 + NC_2 - 4C_3 &= 0\\ C_2 + 2C_4 + NC_5 &= 0\end{aligned} \tag{31}$$

To obtain the five coefficients $C_1(\tau)$ to $C_5(\tau)$ on H^N, we just need to evaluate five components of $\Delta T^2_{\mu\nu\alpha'\beta'}(x,x')$. Here we shall choose $\Delta T^2_{\sigma\sigma\sigma'\sigma'}(x,x')$, $\Delta T^2_{\sigma\sigma\theta'\theta'}(x,x')$, $\Delta T^2_{\sigma\theta\sigma'\theta'}(x,x')$, $\Delta T^2_{\theta\theta\phi_1'\phi_1'}(x,x')$ and $\Delta T^2_{\theta\phi_1\theta'\phi_1'}(x,x')$. Since H^N is homogeneous, the correlator should only depend on the geodesic distance between the two points. Hence, it is possible to simplify the consideration by appropriately choosing x and x'. First, we shall set x and x' to have the same angular coordinates $\Omega' \to \Omega$. After that we shall take the limit $\sigma' \to 0$. In effect we shall take x' to be at the origin. In addition, in order to keep the consideration simple, we shall only work on the massless minimally coupled case.

As $\Omega' \to \Omega$, various bitensors can be simplified as follows.

$$\tau(x,x') = |\sigma - \sigma'|, \tag{32}$$

$$n_\mu(x,x') = \delta_{\mu\sigma}, \tag{33}$$

$$n_{\alpha'}(x,x') = -\delta_{\alpha'\sigma'}, \tag{34}$$

$$g_{\sigma\sigma'}(x,x') = 1, \tag{35}$$

$$g_{\theta\theta'}(x,x') = a^2\sinh\left(\frac{\sigma}{a}\right)\sinh\left(\frac{\sigma'}{a}\right), \tag{36}$$

$$g_{\phi_1\phi_1'}(x,x') = a^2\sinh\left(\frac{\sigma}{a}\right)\sinh\left(\frac{\sigma'}{a}\right)\sin^2\theta, \tag{37}$$

and so on. Note that the non-diagonal elements of $g_{\mu\alpha'}$ vanish in this limit. Using this result the relationship between the various components of $\Delta T^2_{\mu\nu\alpha'\beta'}(x,x')$ and the coefficients C_i also simplify. Then we have, as $\Omega' \to \Omega$,

$$\Delta T^2_{\sigma\sigma\sigma'\sigma'}(x,x') = C_1 + 2C_2 - 4C_3 + 2C_4 + C_5, \tag{38}$$

$$\Delta T^2_{\sigma\sigma\theta'\theta'}(x,x') = a^2\sinh^2\left(\frac{\sigma'}{a}\right)(C_2 + C_5), \tag{39}$$

$$\Delta T^2_{\sigma\theta\sigma'\theta'}(x,x') = -a^2\sinh\left(\frac{\sigma}{a}\right)\sinh\left(\frac{\sigma'}{a}\right)(C_3 - C_4), \tag{40}$$

$$\Delta T^2_{\theta\theta\phi_1'\phi_1'}(x,x') = a^4 \sinh^2\left(\frac{\sigma}{a}\right)\sinh^2\left(\frac{\sigma'}{a}\right)\sin^2\theta C_5, \tag{41}$$

$$\Delta T^2_{\theta\phi_1\theta'\phi_1'}(x,x') = a^4 \sinh^2\left(\frac{\sigma}{a}\right)\sinh^2\left(\frac{\sigma'}{a}\right)\sin^2\theta C_4, \tag{42}$$

The evaluation of the various components of the correlator can be further simplified if we take $\sigma' \to 0$ as only terms with low values of l and l' will contribute.

To see how the procedure goes, we consider $\Delta T^2_{\theta\phi_1\theta'\phi_1'}(x,x')$ in some details. Using the prescription in Eq. (17),

$$\begin{aligned}
&\Delta T^2_{\theta\phi_1\theta'\phi_1'}(x,x') \\
&= \frac{1}{2}\int_0^\infty du u^\nu \int_0^\infty dv v^\nu \int_0^\infty d\kappa \int_0^\infty d\kappa' \sum_{lml'm'} e^{-\frac{u}{a^2}\left(\kappa^2+\frac{(N-1)^2}{4}\right)} e^{-\frac{v}{a^2}\left(\kappa^2+\frac{(N-1)^2}{4}\right)} \\
&\quad T_{\theta\phi_1}[\phi_{\kappa lm}(x), \phi^*_{\kappa' l'm'}(x)] T_{\theta'\phi_1'}[\phi_{\kappa' l'm'}(x'), \phi^*_{\kappa lm}(x')]
\end{aligned} \tag{43}$$

To anticipate that we shall take $\sigma' \to 0$, we only need to consider the first few values of l and l'. For example, when $l = l' = 1$, we have

$$\begin{aligned}
T_{\theta\phi_1}[\phi_{\kappa 1m}(x), \phi_{\kappa' 1m'}(x)] &= -f_{\kappa 1}(\sigma) f_{\kappa' 1}(\sigma) [\partial_\theta Y^*_{1m'}(\Omega)\partial_{\phi_1} Y_{1m}(\Omega) + \partial_{\phi_1} Y^*_{1m'}(\Omega)\partial_\theta Y_{1m}(\Omega) \\
&\quad -2\xi \nabla_\theta \nabla_{\phi_1}(Y^*_{1m'}(\Omega) Y_{1m}(\Omega))]
\end{aligned} \tag{44}$$

The summations over m and m' can be carried out using the addition theorem in Eq. (21). Moreover, in the limit $\Omega' \to \Omega$, we have

$$\partial_\theta \partial_{\theta'}(\Omega\cdot\Omega')\big|_{\Omega'\to\Omega} = 1 \tag{45}$$

$$\partial_{\phi_1}\partial_{\phi_1'}(\Omega\cdot\Omega')\big|_{\Omega'\to\Omega} = \sin^2\theta \tag{46}$$

$$\partial_{\phi_1}\partial_{\phi_1'}(\Omega\cdot\Omega')^2\big|_{\Omega'\to\Omega} = 2\sin^2\theta \tag{47}$$

$$\partial_\theta\partial_{\phi_1}\partial_{\phi_1'}(\Omega\cdot\Omega')^2\big|_{\Omega'\to\Omega} = 2\sin\theta\cos\theta \tag{48}$$

$$\partial_\theta\partial_{\theta'}\partial_{\phi_1}\partial_{\phi_1'}(\Omega\cdot\Omega')^2\big|_{\Omega'\to\Omega} = 2 \tag{49}$$

and the summations can be simplified to

$$\begin{aligned}
&\sum_{mm'} T_{\theta\phi_1}[\phi_{\kappa 1m}(x), \phi^*_{\kappa' 1m'}(x)] T_{\theta'\phi_1'}[\phi_{\kappa' 1m'}(x'), \phi^*_{\kappa 1m}(x')]\Big|_{\Omega'\to\Omega} \\
&= \frac{2\Gamma^2\left(\frac{N}{2}+1\right)}{\pi^N}\sin^2\theta\,[f_{\kappa 1}(\sigma) f_{\kappa' 1}(\sigma) f_{\kappa 1}(\sigma') f_{\kappa' 1}(\sigma')]
\end{aligned} \tag{50}$$

Furthermore, as $\sigma' \to 0$,

$$\begin{aligned}
&\sum_{mm'} T_{\theta\phi_1}[\phi_{\kappa 1m}(x), \phi^*_{\kappa' 1m'}(x)] T_{\theta'\phi_1'}[\phi_{\kappa' 1m'}(x'), \phi^*_{\kappa 1m}(x')]\Big|_{\Omega'\to\Omega,\sigma'\to 0} \\
&= \frac{2}{(2\pi)^N}\sinh^2\left(\frac{\sigma'}{a}\right)\sin^2\theta\,[c_1(\kappa) f_{\kappa 1}(\sigma) c_1(\kappa') f_{\kappa' 1}(\sigma)]
\end{aligned} \tag{51}$$

where we see that $\sinh^2(\sigma'/a)$ is the leading behavior as $\sigma' \to 0$. This behavior conforms with that of $\Delta T^2_{\theta\phi_1\theta'\phi'_1}(x,x')$ in Eq. (42) and that is the only term with this leading behavior. Comparing with Eq.(42), we can extract the coefficient C_4.

$$C_4(\sigma) = \frac{1}{(2\pi)^N a^4 \left(\sinh\frac{\sigma}{a}\right)^2}\left(I_1^{(0)}\right)^2 \tag{52}$$

where we have defined the integral,

$$I_l^{(i)} = \int_0^\infty du u^\nu \int_0^\infty d\kappa \kappa^{2i} e^{-\frac{u}{a^2}\left(\kappa^2+\frac{(N-1)^2}{4}\right)} c_l(\kappa) f_{\kappa l}(\sigma) \tag{53}$$

After some lengthy calculations similar to that above, we have

$$\begin{aligned} C_1 &= -\frac{1}{(2\pi)^N a^4}\left[\frac{(N-1)^2 a^2}{4}\left(\partial_\sigma I_0^{(0)}\right)^2 + a^2\left(\partial_\sigma I_0^{(0)}\right)\left(\partial_\sigma I_0^{(1)}\right)\right] \\ &+ \frac{1}{(2\pi)^N a^4}\left[2a^2\left(\partial_\sigma I_1^{(0)}\right)^2 + I_1^{(0)} I_1^{(1)} - \left(\frac{2a}{\sinh\frac{\sigma}{a}}\partial_\sigma - \frac{(N-1)^2}{4} - \frac{2}{\left(\sinh\frac{\sigma}{a}\right)^2}\right)\left(I_1^{(0)}\right)^2\right] \end{aligned} \tag{54}$$

$$\begin{aligned} C_2 &= \frac{1}{(2\pi)^N a^4}\left[\frac{(N-1)^2 a^2}{4}\left(\partial_\sigma I_0^{(0)}\right)^2 + a^2\left(\partial_\sigma I_0^{(0)}\right)\left(\partial_\sigma I_0^{(1)}\right)\right] \\ &+ \frac{1}{(2\pi)^N a^4}\left[\frac{1}{\left(\sinh\frac{\sigma}{a}\right)^2}\left(I_1^{(0)}\right)^2 - a^2\left(\partial_\sigma I_1^{(0)}\right)^2\right] \end{aligned} \tag{55}$$

$$C_3 = -\frac{1}{2(2\pi)^N a^4 \left(\sinh\frac{\sigma}{a}\right)^2}\left[a\left(\sinh\frac{\sigma}{a}\right)\partial_\sigma - 2\right]\left(I_1^{(0)}\right)^2 \tag{56}$$

$$C_4 = \frac{1}{(2\pi)^N a^4 \left(\sinh\frac{\sigma}{a}\right)^2}\left(I_1^{(0)}\right)^2 \tag{57}$$

$$\begin{aligned} C_5 &= \frac{1}{2(2\pi)^N a^4}\left[-a^2\left(\partial_\sigma I_0^{(0)}\right)\left(\partial_\sigma I_0^{(2)}\right) - \frac{a^2(N-1)^2}{4}\left(\partial_\sigma I_0^{(0)}\right)^2\right. \\ &\left. + \left(I_0^{(1)}\right)^2 + \frac{(N-1)^2}{2}\left(I_0^{(0)} I_0^{(1)}\right) + \frac{(N-1)^4}{16}\left(I_0^{(0)}\right)^2\right] \\ &+ \frac{1}{2(2\pi)^N a^4}\left[a^2\left(\partial_\sigma I_1^{(0)}\right)^2 - I_1^{(0)} I_1^{(1)} + \left(-\frac{(N-1)^2}{4} + \frac{N-5}{\left(\sinh\frac{\sigma}{a}\right)^2}\right)\left(I_1^{(0)}\right)^2\right] \end{aligned} \tag{58}$$

Although we have succeeded in expressing the coefficients in terms of products of integrals involving the associated Legendre functions, the integrals cannot be simplified to known functions. In this form it is hard to check the corresponding conservation conditions. Therefore, we shall consider in the next section the asymptotic behaviors of the above coefficients in the small and large geodesic distance limits. In so doing we could have a better understanding of the correlators and we could also check the conservation and the traceless conditions explicitly for various dimensions.

4. Small and large geodesic distance limits in AdS for different dimensions

In this section we explore the small and the large geodesic distance limits of the coefficients C_1 to C_5 in Euclidean AdS^N for arbitrary N. As we have mentioned above, it is possible to check the conservation and the traceless conditions explicitly for various dimensions in the small distance limit. In addition, the small distance limit also indicates the divergent behavior of the correlators in the coincident limit as $\sigma \to 0$. This type of expansion has been explored in [12] in relation to the generalization of the c-theorem [18] from two to higher dimensions. A general discussion has also been given in [12] on the stress-energy correlators of conformally coupled scalar and fermion fields in constant curvature spaces.

For the small distance limit, we need to consider the integral $I_l^{(i)}(N)$ in Eq. (53) in more detail. First we shall use the integral representation of the Legendre function,

$$P^{1-l-\frac{N}{2}}_{-\frac{1}{2}+i\kappa}\left(\cosh\frac{\sigma}{a}\right) = \frac{\sqrt{2}\,(\sinh\frac{\sigma}{a})^{1-l-\frac{N}{2}}}{\sqrt{\pi}\Gamma\left(-\frac{1}{2}+l+\frac{N}{2}\right)}\left(\frac{\sigma}{a}\right)\int_0^1 dt\,\cos\left(\frac{\kappa\sigma t}{a}\right)\left(\cosh\frac{\sigma}{a}-\cosh\frac{\sigma t}{a}\right)^{l+\frac{N}{2}-\frac{3}{2}} \tag{59}$$

Since the normalization constant is given by

$$|c_l(\kappa)|^2 = \left[\kappa^2+(\rho_N+l-1)^2\right]\left[\kappa^2+(\rho_N+l-2)^2\right]\cdots\left[\kappa^2+(\rho_N)^2\right]|c_0(\kappa)|^2 \tag{60}$$

and for even dimensions, $|c_0(\kappa)|^2$ is given by Eq. (24), we see that if we want to extract the leading contribution of the integrals as $\sigma \to 0$, we need to deal with integrals of the general form

$$\begin{aligned}
&\int_0^1 dt\left(\cosh\frac{\sigma}{a}-\cosh\frac{\sigma t}{a}\right)^{l+\frac{N}{2}-\frac{3}{2}}\int_0^\infty du\,u^\nu\int_0^\infty d\kappa\,\kappa^{2n+2i+1}\left(\cos\frac{\kappa\sigma t}{a}\right)e^{-\frac{u}{a^2}\left(\kappa^2+\frac{(N-1)^2}{4}\right)}\\
=&\int_0^1 dt\left(\cosh\frac{\sigma}{a}-\cosh\frac{\sigma t}{a}\right)^{l+\frac{N}{2}-\frac{3}{2}}\int_0^\infty d\kappa\frac{a^{2(\nu+1)}\Gamma(\nu+1)\kappa^{2n+2i+1}\left(\cos\frac{\kappa\sigma t}{a}\right)}{\left(\kappa^2+\frac{(N-1)^2}{4}\right)^{\nu+1}}
\end{aligned} \tag{61}$$

where n is an integer. This is finite for sufficiently large values of ν. In this analytic continuation procedure, we can represent κ^{2n+2i} by derivatives on the cosine function. Then the above integral becomes

$$\int_0^1 dt\left(\cosh\frac{\sigma}{a}-\cosh\frac{\sigma t}{a}\right)^{l+\frac{N}{2}-\frac{3}{2}}(-1)^{n+i}\left(\frac{\partial}{\partial\eta}\right)^{2n+2i}\int_0^\infty d\kappa\left[\frac{a^2\kappa\cos\eta\kappa}{\kappa^2+\frac{(N-1)^2}{4}}\right] \tag{62}$$

where $\eta = \sigma t/a$. Since the integration over κ is now finite even when $\nu \to 0$, we have taken that limit.

To see how it goes explicitly, we consider the case with $l = 1$, $N = 4$ and $n = i = 1$. Then, expanding in powers of σ/a,

$$\begin{aligned}
&\left(\cosh\frac{\sigma}{a}-\cosh\frac{\sigma t}{a}\right)^{3/2}\left(\frac{\partial}{\partial\eta}\right)^4\int_0^\infty d\kappa\left(\frac{a^2\kappa\cos\eta\kappa}{\kappa^2+\frac{9}{4}}\right)\\
=&\ a^2\left(\frac{a}{\sigma}\right)\left(\frac{3}{\sqrt{2}}\right)(1-t^2)^{3/2}t^{-4}+a^2\left(\frac{\sigma}{a}\right)\left(\frac{3}{8\sqrt{2}}\right)(1-t^2)^{3/2}\left[t^{-4}+4t^{-2}\right]\\
&+a^2\left(\frac{\sigma}{a}\right)^3\left(\frac{1}{640\sqrt{2}}\right)(1-t^2)^{3/2}\left\{13t^{-4}+108t^{-2}+103-810\left[2\gamma+\ln\left[\frac{9t}{4}\left(\frac{\sigma}{a}\right)^2\right]\right]\right\}
\end{aligned} \tag{63}$$

where we have put $a^2b = 9/4$ for $N = 4$. Using the formula,

$$\int_0^1 dt\,t^\alpha(1-t^2)^\beta = \frac{\Gamma\left(\frac{1+\alpha}{2}\right)\Gamma(1+\beta)}{2\Gamma\left(\frac{3}{2}+\frac{\alpha}{2}+\beta\right)} \tag{64}$$

Journal of Physics: Conference Series **330** (2011) 012002 doi:10.1088/1742-6596/330/1/012002

the integration over t can be performed giving

$$\begin{aligned}&\int_0^1 dt\left(\cosh\frac{\sigma}{a}-\cosh\frac{\sigma t}{a}\right)^{3/2}\left(\frac{\partial}{\partial\eta}\right)^4\int_0^\infty d\kappa\left(\frac{\kappa\cos\eta\kappa}{\kappa^2+\frac{9}{4}}\right)\\ =\;& a^2\left(\frac{a}{\sigma}\right)\left(\frac{3\pi}{2\sqrt{2}}\right)-a^2\left(\frac{\sigma}{a}\right)\left(\frac{30\pi}{32\sqrt{2}}\right)-a^2\left(\frac{\sigma}{a}\right)^3\left(\frac{\pi}{10240\sqrt{2}}\right)\times\\ &\left\{883-1215\left[3-4\gamma-4\ln\left(\frac{3}{4}\right)-4\ln\left(\frac{\sigma}{a}\right)\right]\right\}\end{aligned} \tag{65}$$

Since the coefficients C_1 to C_5 are expressed in terms of the integrals $I_l^{(i)}(N)$, the expansion above can be used to obtain the corresponding small geodesic distance limits of the coefficients. For $N=4$,

$$\begin{aligned}C_1(4) &= \frac{8}{\pi^4\sigma^8}\left(1-\frac{7\sigma^2}{12a^2}+\cdots\right)\\ C_2(4) &= -\frac{2}{\pi^4\sigma^8}\left(1+\frac{19\sigma^2}{6a^2}+\cdots\right)\\ C_3(4) &= \frac{1}{\pi^4\sigma^8}\left(1-\frac{23\sigma^2}{24a^2}+\cdots\right)\\ C_4(4) &= \frac{1}{4\pi^4\sigma^8}\left(1-\frac{4\sigma^2}{3a^2}+\cdots\right)\\ C_5(4) &= \frac{1}{\pi^4\sigma^8}\left(1-\frac{5\sigma^2}{24a^2}+\cdots\right)\end{aligned} \tag{66}$$

One can then check that the conservation conditions in Eq. (30) are satisfied to the order indicated.

With the expressions in the small distance limit of the coefficients C's, it is possible to make some comparison with our results to that in [13]. For simplicity, we just consider the leading behavior of these coefficients. In [13], these coefficients, named P, Q, R, S, and T there, are expressed in terms of the Wightman function (in de Sitter spacetime),

$$G(\sigma)=c_{m,N}F\left(h_+,h_-;\frac{N}{2};\frac{1+\cos\sigma/a}{2}\right) \tag{67}$$

where $F(\alpha,\beta;\gamma;z)$ is the hypergeometric function and

$$h_\pm=\frac{1}{2}\left[(N-1)\pm\sqrt{(N-1)^2-4m^2a^2}\right]\quad;\quad c_{m,N}=\frac{\Gamma(h_+)\Gamma(h_-)}{(4\pi a)^{N/2}\Gamma(N/2)} \tag{68}$$

In the small geodesic distance limit, $\sigma\to 0$,

$$G(\sigma)=\frac{\Gamma\left(\frac{N}{2}-1\right)}{(4\pi)^{N/2}}\left(\frac{2}{\sigma}\right)^{N-2}+\cdots. \tag{69}$$

The leading behavior of the coefficients are then

$$C_1(N) = \frac{N^2\Gamma^2(N/2)}{2\pi^N\sigma^{2N}}+\cdots$$

$$\begin{aligned}
C_2(N) &= -\frac{N(N-2)\Gamma^2(N/2)}{4\pi^N\sigma^{2N}}+\cdots \\
C_3(N) &= \frac{N\Gamma^2(N/2)}{4\pi^N\sigma^{2N}}+\cdots \\
C_4(N) &= \frac{\Gamma^2(N/2)}{4\pi^N\sigma^{2N}}+\cdots \\
C_5(N) &= \frac{(N^2-N-4)\Gamma^2(N/2)}{8\pi^N\sigma^{2N}}+\cdots
\end{aligned} \tag{70}$$

These agree with our results in Eq. (66) when one puts $N=4$. It is also easy to show that the coefficients in Eqs. (55) to (58) indeed have the same leading behavior as above.

To consider the large geodesic distance limit for the various coefficients, we go back to the integral $I_l^{(i)}$ in Eq. (53). First,

$$f_{\kappa l}(\sigma)=c_l(\kappa)\left(\sinh\frac{\sigma}{a}\right)^{1-\frac{N}{2}}P^{1-l-\frac{N}{2}}_{-\frac{1}{2}+i\kappa}\left(\cosh\frac{\sigma}{a}\right) \tag{71}$$

and from the asymptotic behavior as $\sigma\to\infty$ of the associated Legendre function, we have

$$P^{1-l-\frac{N}{2}}_{-\frac{1}{2}+i\kappa}\left(\cosh\frac{\sigma}{a}\right)\sim e^{-\frac{\sigma}{2a}}e^{i\frac{\kappa\sigma}{a}} \tag{72}$$

The integrations over u and κ will only give powers of σ. Hence, the leading asymptotic behavior of $I_l^{(i)}$, independently of the parameters l and i, is

$$I_l^{(i)}\sim e^{-(N-1)\sigma/2a} \tag{73}$$

Using this result we can estimate the asymptotic behavior of the various coefficients in Eqs. (55) to (58),

$$\begin{aligned}
C_1, C_2, C_5 &\sim e^{-(N-1)\sigma/a} \\
C_3 &\sim e^{-N\sigma/a} \\
C_4 &\sim e^{-(N+1)\sigma/a}
\end{aligned} \tag{74}$$

This behavior is in fact independent of the mass of the field. We can see that an intrinsic infrared cutoff scale is provided by the radius a of the background spacetime.

This large geodesic distance limit can also be compared with the result in [13]. Since there the authors considered the de Sitter spacetime, one could analytic continue their result by taking

$$Z=\cos\left(\frac{\sigma}{a}\right)\to\cosh\left(\frac{\sigma}{a}\right)\sim e^{\sigma/a} \tag{75}$$

Then our result in Eq. (74) indeed agrees with that in [13].

5. Conclusions and Discussions

In this paper we have calculated the correlators of the stress-energy tensor, or the noise kernel in the theory of stochastic gravity in AdS^N spacetimes. The method we have used is the generalized zeta-function regularization procedure devised in [5]. The zeta-function regularization method was originally used to deal mainly with one loop effective action and the corresponding operators associated with it, notably the expectation value of the stress-energy tensor. The method introduced in [5] enables one to deal also with operator correlation functions. To understand the correlations more closely and also to facilitate further applications, we have developed their small and long geodesic distance limits. Particularly in the small geodesic distance limit, we could verify explicitly that the correlators satisfy the conservation equation.

We have compared our results with that in [13]. The agreement shows that the zeta-function method used here is an useful alternative to evaluate correlators of operators. We are confident that this method can be used in other applications which are also of significant interest to us. As discussed in the Introduction, we would like ultimately to explore the physics in the AdS-black hole spacetimes. In order to do that we need to extend our consideration to first the finite temperature case, and then to the black hole case. These cases will be considered in our subsequent works.

Acknowledgments

HTC is supported in part by the National Science Council of the Republic of China under the Grant NSC 99-2112-M-032-003-MY3, and by the National Center for Theoretical Sciences. BLH is supported in part by the National Science Foundation under grants PHY-0601550 and PHY-0801368 to the University of Maryland.

References

[1] Hu B L and Verdaguer E 2008 *Living Rev. Rel.* **11** 3.
[2] BHu L and Verdaguer E 2003 *Class. Quant. Grav.* **20** R1.
[3] Dowker J S and Critchley R 1976 *Phys. Rev. D* **13** 224
Dowker J S and Critchley R 1976 *Phys. Rev. D* **13** 3224.
[4] Hawking S W 1977 *Commun. Math. Phys.* **55** 133.
[5] Phillips N G and Hu B L 1997 *Phys. Rev. D* **55** 6123.
[6] Hawking S W 1975 *Commum. Math. Phys.* **43** 199.
[7] Hawking S W 1976 *Phys. Rev. D* **13** 191.
[8] York J W 1985 *Phys. Rev. D* **31** 775.
[9] Hawking S W and Page D N 1983 *Commun. Math. Phys.* **87** 577.
[10] Witten E 1988 *Adv. Theor. Math. Phys.* **2** 505.
[11] Horowitz G T and Hubeny V E 2000 *Phys. Rev. D* **62** 024027.
[12] Osborn H and Shore G M 2000 *Nucl. Phys. B* **571** 287.
[13] Guillem Perez-Nadal Albert Roura and Enric Verdaguer 2010 *JCAP* **05** 036.
[14] Elizalde E 1994 *Zeta Function Regularization Techniques with Applications* (World Scientific Singapore).
Elizalde E 1995 *Ten Physical Applications of Spectral Zeta Functions* (Springer-Verlag Berlin).
[15] Cognola G Elizalde E and Zerbini S 2002 *Phys. Rev. D* **65** 085031.
[16] Schwinger J 1951 *Phys. Rev.* **82** 664.
[17] Allen B and Jacobson T 1986 *Commun. Math. Phys.* **103** 669.
[18] Zamolodchikov A B 1986 *JETP Lett.* **43** 730.

Journal of Physics: Conference Series **330** (2011) 012003 doi:10.1088/1742-6596/330/1/012003

Macroscopic Quantum Phenomena from the Correlation, Coupling and Criticality Perspectives

C H Chou[1], B L Hu[2] and Y Subaşı[2]

[1]Department of Physics, National Cheng Kung University, Tainan, Taiwan 701 and National Center for Theoretical Sciences (South), Tainan, Taiwan 701

[2] Maryland Center for Fundamental Physics and Joint Quantum Institute, University of Maryland, College Park, Maryland 20742, USA

E-mail: blhu@umd.edu

Abstract.

In this sequel paper we explore how macroscopic quantum phenomena can be measured or understood from the behavior of quantum correlations which exist in a quantum system of many particles or components and how the interaction strengths change with energy or scale, under ordinary situations and when the system is near its critical point. We use the nPI (master) effective action related to the Boltzmann-BBGKY / Schwinger-Dyson hierarchy of equations as a tool for systemizing the contributions of higher order correlation functions to the dynamics of lower order correlation functions. Together with the large N expansion discussed in our first paper [1] we explore 1) the conditions whereby an H-theorem is obtained, which can be viewed as a signifier of the emergence of macroscopic behavior in the system. We give two more examples from past work: 2) the nonequilibrium dynamics of N atoms in an optical lattice under the large $\mathcal{N}$ (field components), $2PI$ and second order perturbative expansions, illustrating how N and $\mathcal{N}$ enter in these three aspects of quantum correlations, coherence and coupling strength. 3) the behavior of an interacting quantum system near its critical point, the effects of quantum and thermal fluctuations and the conditions under which the system manifests infrared dimensional reduction. We also discuss how the effective field theory concept bears on macroscopic quantum phenomena: the running of the coupling parameters with energy or scale imparts a dynamical-dependent and an interaction-sensitive definition of 'macroscopia'.

1. Key points: correlation, interaction strength, effective scales and dimensions

We continue our investigations begun in an earlier paper [1] (MQP1) into the key features of macroscopic quantum phenomena (MQP), seeking to identify and assemble the necessary ingredients toward the construction of a viable theoretical framework. Not only does this new field bear on the foundational issues of both quantum and statistical mechanics it also sees a widening range of manifestations and applications. (For background please refer to references in [1]). We want to examine the conditions or criteria whereby a macroscopic quantum system may take on classical attributes, and, more interestingly, that it keeps some of its quantum features. Our first paper focused on the large N approximation, highlighting the subtle yet important difference between results obtained from a leading-order large-N approximation, namely, the mean field theory which is quantum in nature, from a corresponding classical theory. Here we focus on the key features of correlations and couplings in a quantum system of many particles or components, especially its behavior near the critical point, to investigate which factors contribute to, or determine a system as being macroscopic, and under what conditions would quantum features persist or dominate in that limit.

Before delving into these main subjects, it may be useful for the purpose of orientation to refresh what is already known in quantum statistical mechanics of a) how the thermodynamic properties of a quantum system depend on N, the number of particles present, and b) how qualitative changes arise in a many-body quantum system, such as the formation of a Bose-Einstein condensate (BEC), one of the most commonly cited examples (the other being quantized fluxes in Josephson junctions superconductivity) of MQP. We want to point out the qualitative differences in the meanings of 'quantumness' in these familiar cases before we launch our investigations into the other quantum features we are more interested in, namely, quantum correlations, fluctuations, coherence and entanglement, and how they may show up or even persist at a macroscopic scale.

1.1. N dependence in quantum statistical mechanics

Note from the start that one of the two basic premises of quantum statistical mechanics (QSM) [2], namely, the assumption of *random phase approximation*, precludes quantum coherence considerations. All descriptions are in terms of probabilities, not amplitudes. The other basic premise, that the system has *equal a priori probability* to be found in any of its accessible states, refers to the equilibrium condition. Note, however, that decoherence and quantum to classical transition [3, 4] are fundamentally nonequilibrium processes.

The 'quantum' in QSM refers to effects due to **spin-statistics**, in the difference between bosons and fermions, and **distinguishability**: different combinatorics in distinguishable (classical feature) versus identical (quantum feature) particles, which underlie the Gibbs paradox [5]. Thus the answer to a) above is simple: the dependence of the partition function Z on N is explicit, and the thermodynamic limit is well known. With the correct counting of distinguishable versus indistinguishable particles included in the partition function, the difference between quantum and classical features in this restricted sense is also well accounted for.

QSM describes finite temperature equilibrium processes for a quantum (in the above restricted sense) system in or near equilibrium, such as in the ordinary (thermal) critical phenomena mediated by energy fluctuations in a canonical ensemble (temperature), or numbers / species fluctuations in a grand canonical ensemble (chemical potential) description, but cannot address issues related to quantum coherence and entanglement, the central issues in quantum information. To the extent entanglement is a useful marker for quantum phase transition QSM cannot address issues of quantum noise or fluctuation phenomena.

BEC is the well-known process where one can pinpoint the number N_c of a system of N bosonic atoms undergoing a transition where the salient features of the system undergo marked qualitative changes on the large scale, exemplifying MQP. There, the critical number N_c and temperature T_c which define the phase transition also depend on the dimensionality. (E.g., a two-dimensional Bose gas undergoes transition into vortex-antivortex pairs – see e.g., [6]). BEC is certainly a quantum entity, carrying macroscopic quantum coherence described by the N-atom wave function as many exquisite experiments carried out in the last decade have witnessed. Yet its pivotal feature, its 'quantumness', originates from the spin-statistics properties (bosonic vs fermionic condensate) and the critical phenomenon which describes the formation of a condensate is a classical, not a quantum phase transition. This addresses query b) above.

Making the distinction between 'quantumness' arising from particle spin-statistics as in quantum statistical mechanics and that of large scale quantum coherence as in BEC offers a way to discern what would be called MQP in the modern sense of the word: usually it is the latter, not the former, which is the deciding factor. To name two examples: 1) Can the behavior of a collection of quantum gas be regarded as MQP? In solid state physics one treats the conduction electrons in a metal as a degenerate Fermi gas because the room temperature is much lower

Journal of Physics: Conference Series **330** (2011) 012003 doi:10.1088/1742-6596/330/1/012003

than the Fermi temperature of the metal. The electron's de Broglie thermal wavelength is much larger than the atomic spacing. But one would not regard the thermal properties of metals as MQP. Similar situations occur in the white dwarfs and neutron stars where the degenerate pressure of electrons or neutrons respectively balances the gravitation attraction to keep the stars in relative stable configurations. One does not refer to the existence of white dwarfs or neutron stars as examples of MQP. 2) Can one view collective excitations as MQP? Phonons are the quantized long wavelength modes of lattice vibrations. Like waves all such collective variables involve a large number of atoms and can span a large spatial extent. But one would not call these collective phenomena as MQP. However for a large population of bosonic gas accumulating in the ground state at an extremely low temperature forming a Bose-Einstein condensate(BEC), we do regard it as a MQP. The crucial difference is the large scale quantum coherence established in such systems. Similarly, for superconductors: the Cooper pairs existing at large spatial separation (compared to the extent of electron wave functions) are quasiparticles with full quantum coherence [1]

1.2. Scaling and interaction, RG running and criticality

In parallel to explicating the (rather restrictive) meaning of 'quantumness' used widely in QSM mentioned above here we wish to explore the meaning of 'macroscopia' in like spirit. Macroscopia in the most direct way conjures large numbers: many particles or components and/or large sizes or scales. Paper I focused on the 'many' aspect, here we focus on the 'scale' aspect. If one goes by the scale of a quantum system alone without other considerations such as the interactions (or coupling strength) amongst its underlying constituents then the quantum behavior of a macro object such as the Universe would be similar to that of a micro one, such as the internal degrees of freedom of an atom. In fact, both obey a harmonic oscillator equation of motion, albeit the former (for the wave function of the universe) has a negative spring constant. The scale is usually measured by the inverse mass of the relevant process. The physically relevant scale (or mass) is different from the bare scale (or mass) by a ratio given by a renormalization constant which captures the relevant interactions involved. Interaction is measured by the coupling constants which vary ('run') with energy according to the renormalization group (RG) equations. The infrared behavior of an interacting quantum system at or near its critical point such as the universality class it belongs to, is of special significance. We will review briefly these familiar concepts below as the notion of 'macroscopic' for quantum systems near criticality is determined by its infrared behavior and the interaction strength of its constituents.

1.2.1. Scales, interaction strength and effective theories Let us examine the calculation of the energy levels of a hydrogen atom as an example. In a standard textbook treatment one uses the Schroedinger equation for an electron moving in the static Coulomb field of the proton. To a good approximation, the only properties of a proton which are relevant to this problem are its mass and charge. The knowledge of the quark structure of the proton is not necessary to compute the energy levels of the hydrogen atom.

The required knowledge of the proton depends on how accurate one asks for the energy levels. A more detailed calculation, taking hyperfine splitting into account, for instance, requires the knowledge of the spin of the proton and the value of the magnetic dipole moment. An even more accurate calculation requires the knowledge of the proton charge radius and details of the proton structure.

[1] Another more practical angle or functional criterion towards this demarkation is to ask whether one could use such a system to perform quantum information processing, where quantum coherence and entanglement are essential. Cold atoms in an optical lattice, atom assembly, as well as superconducting flux qubits are the well-known viable examples. Note, however, there are proposed QIP schemes such as the NMR based ones, which do not invoke quantum entanglement directly, yet could achieve speed-ups over classical computation.

Journal of Physics: Conference Series **330** (2011) 012003 doi:10.1088/1742-6596/330/1/012003

The typical length scale characteristic of the hydrogen atom is the Bohr radius $r_0 = 1/(m_e\alpha)$ (in units where $\hbar = c = 1$). The typical momentum scale is $1/r_0 = m_e\alpha$, the typical energy scale characteristic of the hydrogen atom is $\sim m_e\alpha^2$, and the typical time scale is $1/(m_e\alpha^2)$. We can get a quantitative estimate of the errors caused by the neglected interactions.

In an effective theory description the relevant interactions also depend on the question being asked. For the hydrogen atom the energy spectrum can be computed to the accuracy $(m_e\alpha/M_w)^2$ while ignoring the weak interactions. But if one is interested in atomic parity violation, the weak interactions are the leading contributions because the electromagnetic and strong interactions conserve parity. The effect of atomic parity violation will be very small because the weak scale is much larger than the atomic scale.

As one approaches shorter and shorter distances (or higher and higher energies), one needs to invoke the relevant physical processes at that scale. The high energy processes affecting low energy physics appear in physical parameters measured at the low energy regime. This is captured nicely by effective field theories, the essence of which we will discuss in a later section.

1.2.2. RG running and criticality In an interacting quantum system the variation of interaction strength with energy is described by the RG equations which are derived from the renormalization of the coupling constants (for ultraviolet divergence). Assuming that the system scales homogeneously this enables one to obtain the infrared properties of a quantum system from its ultraviolet behavior. At the critical point the correlation functions of such a system usually diverge: the system 'feels', so to speak, all the way to infinity. Let us consider how this interesting physical situation bears on macroscopic quantum phenomena.

Critical phenomena involves many different length scales. Consider a physical system going through a continuous phase transition under some temperature change: the correlation length diverges as the temperature approaches the critical temperature, and fluctuations at all length scales are involved. When the temperature lowers across the critical point, the system's order parameter will develop a nonzero value, where the relevant degrees of freedom act in a correlated manner. In this sense the system's behavior near its critical point indeed manifests in a macroscopic scale.

Note traditional treatment of critical phenomena is classical in that the external parameters, temperature, magnetic field, etc, are classical. But it could also be quantum in nature, the system being driven by quantum noise or vacuum fluctuations. Quantum phase transition is brought about by the change of the coupling parameters when the system is at zero temperature. It results from the energy competition between different ground states as a function of the coupling parameter, and quantum entanglement in the system has been shown to provide a useful measure of such transitions. We hope to address this issue in a following paper when we turn our attention to the role of quantum entanglement in the description of MQP.

Naively one would expect that at the critical point as the correlation length goes to infinity, the system would not depend on the microscopic details such as the lattice spacing in the Ising model, or the interaction amongst the microscopic constituents. But it does, and that makes it more interesting. It is this particular aspect of MQP which fascinates us, namely, what and how known microscopic details determine a quantum system's macroscopic attributes, or, even more challenging, how to decipher a system's microscopic features from the macroscopic phenomena. An example is the dialectic relation between emergent and quantum gravity pertaining to the micro and macro structures of spacetime [7].

On this point it is perhaps worth making the following observation: systems in a critical state can exist for a long time to certain classes of observers with specially interesting physical consequences, and many qualitative features in the ensuing phase depends on what came out of the transition. An example is inflationary cosmology where it is believed that the universe underwent a first (old) or second (new) order transition at the GUT ($10^{14}GeV$) scale, with a

Journal of Physics: Conference Series **330** (2011) 012003 doi:10.1088/1742-6596/330/1/012003

duration of 68 e-folding time to be able to account for the entropy content (10^{80} photons) in the present universe. During the exponential expansion the universe appears to be static for a local observer, its scale factor described by a de-Sitter -Einstein solution. In fact, in such an epoch between the metastable false vacuum and reheating to the true vacuum, one can use the language of critical phenomena to describe its dynamics, replacing the evolution by scaling (see, e.g., [8]). The microscopic features of the inflaton field – the quantum fluctuations which existed in the early universe – are magnified during this phase transition into galaxies of today. The effect of inflation is like a giant zoom lens at work [9] and the universe we observe today is in this sense a truly macroscopic quantum phenomenon [10]. The exponential red-shifting of outgoing waves from a black hole which accounts for the thermal feature of Hawking radiation has a similar nature, acting like a microscope in a pictorial description [11].

1.3. Correlation in relation to fluctuations and coherence. Organization of this paper

Unlike the situations of large N where an expansion in inverse N yields meaningful physical results, e.g., the leading order corresponds to a semiclassical limit, quantum correlations are less direct or transparent in their physical meanings. Correlation is related to fluctuations (as one form of fluctuation-dissipation relation indicates) and at the quantum level, quantum correlation can be a measure of quantum coherence (as the correlation history approach to quantum decoherence shows [12]. It is also related to but distinct from quantum entanglement (see work of Cirac [13]) It would be very useful to identify the conditions whereby these three properties could offer a measure (no matter how tenuous) of the macroscopic features of a quantum system and how they signify the quantum (albeit residual) nature of a macroscopic system. For this we use three examples to explore these aspects. Part I of this paper consisting of Sections 2 and 3 discusses the relation between the loop, large N and nPI (n-particle irreducible) expansions, the first two signifying the appearance of quantum and macro features and how quantum correlations is placed in regard to them. In Sec. 2 we use the existence of a H theorem at the next-to-leading-order in the quantum mechanical $O(N)$ model to demarcate which order of large N expansion will entropy generation first ensue and thus can be (approximately) regarded as acquiring some macroscopic attributes. In Sec. 3 we provide a concrete example of the above connections in the Bose-Hubbard model for N atoms in an optical lattice. The nonequilibrium dynamics of this system is described by the equations of motion obtained from the closed-time-path (CTP) two particle irreducible ($2PI$) effective action. We compare results from a large $\mathcal{N}$ expansion where $\mathcal{N}$ is the number of fields (see Eq. (3.5) below) under second order coupling with the exact (numerical) solution to see how quantum correlation plays out against $\mathcal{N}$, the number of components of the N body wave function. Although $\mathcal{N}$ or N are not exactly indicators of the macroscopic, they offer a probe into the interplay between these three aspects: Correlation, fluctuations and coupling strength. Part II of this essay discusses the effective field theory (Sec. 4) and effective infrared dimensional reduction (Sec. 5). Though we didn't dwell on this point as we believe it is quite well known, we want to point out the fact that the critical behavior of interacting quantum systems can provide a sense of macroscopia quite different from the simplistic depictions (referring to the size of the system or how many components it possesses). Key to this is the fact that coupling strengths can vary ('run') with energy scale according to the RG equations. In Sec. 4 we describe effective field theory, where this key notion well adopted in critical phenomena studies is put to good use in that the low energy phenomenology (or long wavelength behavior) is effectively independent of or largely insensitive to the high energy processes. In Sec. 5 we discuss the conditions where an interacting quantum system near the critical point may take on a lower-dimensional appearance. This effective infrared dimensional reduction is shown by an eigenvalue analysis: wherever the eigen-spectrum of the invariant operator of the order parameter field possesses a band structure with a gap separating the lowest mode or band from the higher sector, the infrared behavior of this system behaves effectively as

Journal of Physics: Conference Series **330** (2011) 012003 doi:10.1088/1742-6596/330/1/012003

that of a lower dimension. These examples serve to show the intricacy of even a simple notion such as 'macroscopic' for a quantum system, that it is invariably tethered with the interaction strength, enters into the correlation lengths and undergoes qualitative changes when the system is close to the condition of criticality. The characterization of quantum macroscopic phenomena by quantum entanglement is left for a later paper.

Part I: Correlation hierarchy, nPI, NLO-large N and H-theorem

2. Correlation entropy and H theorem at NLO

To prepare for our discussions on how quantum correlations play out in the quantum - macro issues we first introduce the nPI effective action with n=2 as example, which is a special case of the so-called master effective action [14] (MEA), with the specific model of a quantum mechanical $O(N)$ model (QMON) which we have used in our first paper (MQP1) to illustrate the large N expansion. We then recount (see [15, 16]) how nPI is related to the loop expansion on the one hand, where one customarily takes as a measure of quantum features, and the large N expansion on the other, where one may regard as signifying macroscopic features. (These latter two relations were expounded in MQP1.) We then discuss the implications of the existence of a H theorem proven in [17] using the QMON model. In the next section we will use the example of the nonequilibrium dynamics of BEC atoms to illustrate how this threefold-relation amongst $1/N$, $2PI$ and expansion in the (second order) coupling strength pertain to quantum and macro issues.

We paraphrase the results from [17] where a correlation entropy for an interacting quantum field is proposed, and a proof of the existence of a H-theorem for the quantum mechanical $O(N)$ model [18, 19] is provided. For the former, Calzetta and Hu followed the paradigm of Boltzmann-BBGKY [20] and proposed a correlation entropy (of the nth order) for an interacting quantum field [21, 22] obtained by 'slaving' the higher (n+1 th) order correlation functions in the Schwinger-Dyson system of equations (See [15, 16] and below). They then derived the closed time path (CTP) [23] two particle irreducible ($2PI$) [24] effective action (EA) for this model up to the next-to-leading order (NLO) in $1/N$ [17] and used the CTP $2PI$ EA to prove an H-theorem for the correlation entropy of a quantum mechanical $O(N)$ model at the NLO level. We begin with an introduction to the correlation hierarchy.

2.1. Boltzmann-BBGKY hierarchy and Schwinger-Dyson equations

As is well known in nonequilibrium statistical mechanics, truncation of the BBGKY hierarchy at a finite nth order yields a closed system [2]. The system of equations describing the n-particle distribution functions are time reversal invariant. When a causal factorization condition such as Boltzmann's molecular chaos assumption is imposed – that is, assuming that the n+1 th correlation function can be factorized into a product of n th correlation functions initially but not finally (after the collisions) – time-irreversibility appears and an H-theorem is obtained. This type of *coarse graining* of the hierarchy i.e., truncation plus causal factorization into a set of coupled equations for the n-particle distribution functions is called 'slaving' in the language of [15, 16]. Slaving of the n+1 -particle distribution function renders an otherwise closed system (of the n+1 th order correlation functions) an effectively open system (of nth order correlation functions) and ushers in the appearance of dissipative dynamics [22]. Noise and fluctuations [25, 26] should also appear, as required by the fluctuation-dissipation relation (FDR) [27], now

[2] The truncation of the correlation hierarchy is not just an arbitrary mathematical procedure, it reflects the fact that in realistic conditions, measurement are of finite accuracy associated with limited resolution of the instruments. Thus the relevant physical degrees of freedom are often limited to the lower end of the hierarchy, namely, the mean field and two-point function.

Journal of Physics: Conference Series **330** (2011) 012003 doi:10.1088/1742-6596/330/1/012003

manifesting not just for an open system near equilibrium (as depicted in linear response theory), but for an effectively open system [15, 16].

A familiar example is the Boltzmann equation for dilute gases. At sufficiently low density the description of a molecular gas by a truncated BBGKY hierarchy with only the one-particle and two-particle distribution functions may be justified. Because of the low density one may further assume that before the collisions the colliding partners are uncorrelated, i.e., the two-particle distribution function factorizes into a product of single-particle distribution functions. This is the physical basis for Boltzmann's imposition of the molecular chaos hypothesis. Of course the colliding partners become correlated after interaction. This assumed distinction between initial and final conditions is the origin of the macroscopic arrow of time, the appearance of dissipation and foe entropy generation in the Boltzmann paradigm. FDR calls for a rightful place for fluctuations in the Boltzmann equation, as demonstrated in the derivation of a stochastic Boltzmann equation [28, 29].

There are other approximation schemes that allow for the derivation of the kinetic equations from the correlation hierarchy [20]. To see the relation between correlation and interaction in this context of kinetic theory consider a system in which particles interact via a potential of the form $V(r_i, r_j) = \lambda v(r_i, r_j)$. For weakly coupled systems, i.e. $\lambda \ll 1$, the hierarchy of equations can be written for a certain order of λ.

To $O(\lambda)$ the equation for the one-particle distribution function, namely, the Vlasov equation, is closed, i.e., it does not involve any two-particle distribution function. We emphasize that in this case truncation alone *not* followed by slaving is sufficient to obtain a closed equation for the one-particle distribution function. There is no collision term for short-ranged interaction, but interactions between particles do exist represented by an average potential of all other particles present. Vlasov equation is useful for describing the kinetics of systems containing many particles interacting under long range forces. Recall we have seen it in our first paper as the equation obtained from the leading order large N expansion in the O(N) model. Vlasov equation is obtained as a mean field theory with reversible dynamics and no H-theorem.

To $O(\lambda^2)$ the set of equations for the one-particle and two-particle distribution functions is also closed. By slaving the two-particle distribution function to the one-particle distribution function the Landau equation is obtained. Similar to the Boltzmann equation, Landau equation also has a collision term accounting for the collisions between individual particles at $O(\lambda^2)$. Landau equation describes irreversible dynamics and its solutions satisfy the H-theorem. What is different from the Boltzmann equation is the form of the collision integral. This difference can be traced back to the absence of a term in the equation for the two-particle distribution function at $O(\lambda^2)$ compared to the low density approximation.

The situation in quantum kinetic field theory is completely analogous. By quantum kinetic field theory (see, e.g., [21, 22]), we are referring to the hierarchy of coupled equations for the relativistic Wigner function and its higher-correlation analogs, obtainable from the variation of the master effective action [15, 16] whose variation yields the Schwinger-Dyson equations. This is a quantum analogue of the BBGKY hierarchy, expressed in a representation convenient for distinguishing between microscopic (quantum field-theoretic) and macroscopic (transport and relaxation) phenomena [3].

One may choose to work with a truncation of the hierarchy of the Wigner function and its higher correlation analogs, or one may instead perform a slaving of, for example, the Wigner-transformed four-point function, which leads (within the context of perturbation theory) directly to the relativistic Boltzmann equation [22] and the usual H-theorem. The truncation and subsequent slaving of the hierarchy within quantum kinetic field theory can be carried out at

[3] It should be pointed out that in order to *identify* the relativistic Wigner function with a distribution function for quasiparticles, one must show that the density matrix has *decohered*, and this is neither guaranteed nor required by the existence of a separation of macroscopic and microscopic time scales [30].

Journal of Physics: Conference Series **330** (2011) 012003 doi:10.1088/1742-6596/330/1/012003

any desired order, as dictated by the initial conditions and relevant interactions.

2.2. Entropy from slaving the higher correlations: the CTP 2PI EA

A general procedure has been presented for obtaining coupled equations for correlation functions at any order l in the correlation hierarchy, which involves a truncation of the *master effective action* at a finite order in the loop expansion [15, 16]. By working with an l loop-order truncation of the master effective action, one obtains a closed, time-reversal invariant set of coupled equations for the first $l+1$ correlation functions, $\hat{\phi}, G, C_3, \ldots, C_{l+1}$. In general, the equation of motion for the highest order correlation function will be linear, and thus can be formally solved using Green's function methods. The existence of a unique solution depends on supplying causal boundary conditions. When the resulting solution for the highest correlation function is back-substituted into the evolution equations for the other lower-order correlation functions, the resulting dynamics is *not* time-reversal invariant, and generically dissipative. Thus, as was described before, with the slaving of the higher-order (Wigner-transformed) correlation function in quantum kinetic field theory, we have rendered a closed system (the truncated equations for correlation functions) into an *effectively open system*. In addition to dissipation, one expects that an effectively open system will manifest noise/fluctuations, as shown in [15, 16] for the case of the slaving of the four-point function to the two-point function in the symmetry-unbroken $\lambda\Phi^4$ field theory. Thus a framework exists for exploring irreversibility and fluctuations within the context of a unitary quantum field theory, using the truncation and slaving of the correlation hierarchy. The effectively open system framework is useful for precisely those situations, where a separation of macroscopic and microscopic time scales (which would permit an effective kinetic theory description) does *not* exist, such as is encountered in the thermalization issue.

As a particular coarse graining measure the slaving of higher correlation functions to lower-order correlation functions within a particular truncation of the correlation hierarchy has several important benefits. It can be implemented in a truly nonperturbative fashion. This necessitates a nonperturbative resummation of daisy graphs, which can be incorporated in the truncation/slaving of the correlation hierarchy in a natural way.

In [16] the authors are interested in the growth of entropy due to the coarse graining of the *correlation hierarchy* by slaving a higher correlation function. The simplest nonperturbative truncation of the Schwinger-Dyson equations for the $\lambda\Phi^4$ field theory which contains the time-dependent Hartree-Fock approximation is the two-loop truncation of the master effective action, in which only the mean field $\hat{\phi}$, the two-point function G, the three-point function C_3 are dynamical. All higher order correlation functions obey algebraic constraints, and can thus be expressed in terms of the three dynamical correlation functions.

While this truncation of the Schwinger-Dyson equations is well-defined and could in principle be solved, it is disadvantageous because, as stated above, without some coarse graining, the system will not manifest irreversibility and will not equilibrate. Therefore we slave the three-point function to the mean field and two-point function, and thus arrive at an effectively open system. In principle, a systematic analysis of the coarse-grained dynamics of the mean field and two-point function should include stochasticity [15, 16].

The $2PI$ formalism is also suitable for addressing this question because, provided an auxiliary field is cleverly introduced, the $2PI$ CTP effective action can be found in closed form at each order in $1/N$ [17, 31]. We now use the familiar quantum mechanical $O(N)$ model to illustrate how to derive the $2PI$ effective action and then explore the conditions for the existence of an H-theorem. This model was used in our first paper [1] for the discussion of large N and macroscopia.

2.3. $O(N)$ λX^4 theory

As we recall the system dynamics is described by the Hamiltonian with variables X_A and their conjugate momenta P^A, where A, B are the $O(N)$ group indices, with

$$H = \frac{1}{2}\left\{P^B P^B + M^2 X_B X_B + \frac{\lambda}{4N}(X_B X_B)^2\right\} \tag{2.1}$$

The classical action

$$S = \int dt\, \frac{1}{2}\left\{\dot{X}_B \dot{X}_B - M^2 X_B X_B - \frac{\lambda}{4N}(X_B X_B)^2\right\} \tag{2.2}$$

where M_0^2 and λ_0 are the mass parameter and coupling constant. We rescale $X_B \equiv \sqrt{N} x_B$

$$S = N\int dt\, \frac{1}{2}\left\{\dot{x}_B \dot{x}_B - M^2 x_B x_B - \frac{\lambda}{4}(x_B x_B)^2\right\} \tag{2.3}$$

Discarding a constant term, we may rewrite the classical action as

$$S = N\int dt\, \frac{1}{2}\left\{\dot{x}_B \dot{x}_B - \left[\frac{M^2}{\sqrt{\lambda}} + \frac{\sqrt{\lambda}}{2} x_B x_B\right]^2\right\} \tag{2.4}$$

To set up the $1/N$ resummation scheme, it is customary to introduce the auxiliary field χ, writing

$$S = \frac{N}{2}\int\left\{\dot{x}_B \dot{x}_B - \left[\frac{M^2}{\sqrt{\lambda}} + \frac{\sqrt{\lambda}}{2} x_B x_B\right]^2 + \left[\frac{M^2-\chi}{\sqrt{\lambda}} + \frac{\sqrt{\lambda}}{2} x_B x_B\right]^2\right\} \tag{2.5}$$

whence

$$S = N\int dt\, \left\{\frac{1}{2}\dot{x}_B \dot{x}_B - \chi\left[\frac{M^2}{\lambda} + \frac{x_B x_B}{2}\right] + \frac{1}{2\lambda}\chi^2\right\}. \tag{2.6}$$

From now on, we consider χ and x_B as fundamental variables on equal footing.

Because of the $O(N)$ symmetry, the symmetric point must be a solution of the equations of motion. For simplicity, we shall assume we are within this symmetric phase, and treat x_B as a quantum fluctuation. We also split the auxiliary field $\chi = \bar{\chi} + \tilde{\chi}$ into a background field $\bar{\chi}$ and a fluctuation field $\tilde{\chi}$. The action becomes

$$S = S_0 + S_1 + S_2 + S_3 \tag{2.7}$$

S_0 is just the classical action evaluated at $x_B = 0$, $\chi = \bar{\chi}$:

$$S_0 = \frac{N}{\lambda}\int dt\, \left\{\frac{1}{2}\bar{\chi}^2 - M^2\bar{\chi}\right\} \tag{2.8}$$

S_1 contains terms linear in $\tilde{\chi}$ and can be set to zero by a suitable choice of the background field $\bar{\chi}$:

$$S_1 = \frac{N}{\lambda}\int dt\, \left\{\bar{\chi} - M^2\right\}\tilde{\chi} \tag{2.9}$$

S_2 contains the quadratic terms and yields the tree - level inverse propagators,

$$S_2 = N\int dt\, \left\{\frac{1}{2}\dot{x}_B \dot{x}_B - \frac{\bar{\chi}}{2} x_B x_B + \frac{1}{2\lambda}\tilde{\chi}^2\right\} \tag{2.10}$$

Journal of Physics: Conference Series **330** (2011) 012003 doi:10.1088/1742-6596/330/1/012003

Finally S_3 contains the bare vertex

$$S_3 = \left(\frac{-N}{2}\right) \int d^d x \, \{\tilde{\chi} x_B x_B\} . \tag{2.11}$$

To write the $2PI\ CTP\ EA$ we double the degrees of freedom, incorporating a branch label $a = 1, 2$ (for simplicity, if not explicitly written, we assume that the label a includes the time branch, i.e., $x^{Aa} \equiv x^{Aa}(t_a)$). We also introduce propagators $G^{Aa,Bb}$ for the path ordered expectation values

$$G^{Aa,Bb} = \left\langle x^{Aa} x^{Bb} \right\rangle \tag{2.12}$$

and F^{ab} for

$$F^{ab} = \left\langle \tilde{\chi}^a \tilde{\chi}^b \right\rangle \tag{2.13}$$

Because of symmetry, it is not necessary to introduce a mixed propagator, for $\left\langle \tilde{\chi}^a x^{Bb} \right\rangle \equiv 0$. The $2PI\ CTP\ EA$ reads

$$\begin{aligned} \Gamma &= S_0\left[\bar{\chi}^1\right] - S_0\left[\bar{\chi}^2\right] \\ &+ \frac{1}{2} \int dudv \left\{ D_{Aa,Bb}(u,v) G^{Aa,Bb}(u,v) + \frac{N}{\lambda_0} c_{ab} \delta(u,v) F^{ab}(u,v) \right\} \\ &- \frac{i\hbar}{2} \left[\mathrm{Tr}\ \ln G + \mathrm{Tr}\ \ln F \right] + \Gamma_Q \end{aligned} \tag{2.14}$$

where, if the position variable is explicit, $c_{11} = -c_{22} = 1$, $c_{12} = c_{21} = 0$,

$$D_{Aa,Bb}(u,v) = N \delta_{AB} \left[c_{ab} \partial_x^2 - c_{abc} \bar{\chi}^c \right] \delta(u,v), \tag{2.15}$$

and $c_{abc} = 1$ when all entries are 1, $c_{abc} = -1$ when all entries are 2, and $c_{abc} = 0$ otherwise. When we use the compressed notation, it is understood that $c_{ab} \equiv c_{ab}\delta(t_a, t_b)$ and $c_{abc} \equiv c_{abc}\delta(t_a, t_b)\delta(t_a, t_c)$. Γ_Q is the sum of all $2PI$ vacuum bubbles with cubic vertices from S_3 and propagators $G^{Aa,Bb}$ and F^{ab}. Observe that Γ_Q is independent of $\bar{\chi}^c$.

Taking variations of the $2PI\ CTP\ EA$ and identifying $\bar{\chi}^1 = \bar{\chi}^2 = \bar{\chi}$, we find the equations of motion

$$\frac{N}{2} \delta_{AB} D_{ab} - \frac{i\hbar}{2} \left[G^{-1} \right]_{Aa,Bb} + \frac{1}{2} \Pi_{Aa,Bb} = 0 \tag{2.16}$$

$$\frac{N}{2\lambda} c_{ab} - \frac{i\hbar}{2} \left[F^{-1} \right]_{ab} + \frac{1}{2} \Pi_{ab} = 0 \tag{2.17}$$

$$\frac{N}{\lambda} \left\{ \bar{\chi}(t) - M^2 \right\} - \frac{N}{2} \delta_{AB} G^{A1,B1}(t,t) = 0 \tag{2.18}$$

where $D_{ab}(u,v) = c_{ab}\left[\partial_x^2 - \bar{\chi}(u)\right] \delta(u,v)$,

$$\Pi_{Aa,Bb} = 2 \frac{\delta \Gamma_Q}{\delta G^{Aa,Bb}}; \qquad \Pi_{ab} = 2 \frac{\delta \Gamma_Q}{\delta F^{ab}} \tag{2.19}$$

We shall seek a solution with the structure

$$G^{Aa,Bb} = \frac{\hbar}{N} \delta^{AB} G^{ab}(u,v) \tag{2.20}$$

Journal of Physics: Conference Series **330** (2011) 012003 doi:10.1088/1742-6596/330/1/012003

which is consistent with vanishing Noether charges. Then it is convenient to write

$$F^{ab} = \frac{\hbar}{N}H^{ab}; \qquad \Pi_{Aa,Bb} = \delta_{AB}P_{ab}; \qquad \Pi_{ab}(x,y) = NQ_{ab}(x,y) \tag{2.21}$$

The equations become

$$D_{ab} - \mathrm{i}\left[G^{-1}\right]_{ab} + \frac{1}{N}P_{ab} = 0 \tag{2.22}$$

$$\frac{1}{\lambda}c_{ab} - \mathrm{i}\left[H^{-1}\right]_{ab} + Q_{ab} = 0 \tag{2.23}$$

$$\frac{1}{\lambda}\left\{\bar{\chi}(t) - M^2\right\} - \frac{\hbar}{2}G^{11}(t,t) = 0 \tag{2.24}$$

Observe that

$$P_{ab} = \frac{2}{\hbar}\frac{\delta\Gamma_Q}{\delta G^{ab}}; \qquad Q_{ab} = \frac{2}{\hbar}\frac{\delta\Gamma_Q}{\delta H^{ab}} \tag{2.25}$$

These are the exact equations we must solve. The successive $1/N$ approximations amount to different constitutive relations expressing P_{ab} and Q_{ab} in terms of the propagators.

For power counting in orders of N, observe that in any given Feynman graph each vertex contributes a power of N, each internal line a power of N^{-1}, and each trace over group indices another power of N. We have both G and H internal lines, but the G lines only appear in closed loops. On each loop, the number of vertices equals the number of G lines, so there only remains one power of N from the single trace over group labels. Therefore the overall power of the graph is the number of G loops minus the number of H lines. Now, since we only consider $2PI$ graphs, there is a minimum number of H lines for a given number of G loops. For example, if there are two G loops, they must be connected by no less than 3 H lines, and so this graph cannot be higher than $NNLO$. A graph with 3 G loops cannot have less than 5 H lines, and so on.

We conclude that Γ_Q vanishes at LO, and therefore $P_{ab} = Q_{ab} = 0$. There is only one NLO graph, consisting of a single G loop and a single H line. This graph leads to

$$\Gamma_Q^{NLO} = (-\mathrm{i}\hbar)\left(-\frac{1}{2}\right)\left(-\frac{N}{2Z_0\hbar}\right)^2 2N\left(\frac{\hbar}{N}\right)^3 c_{abc}c_{def}\int dudv\, H^{ad}(u,v)\,G^{be}(u,v)\,G^{cf}(u,v) \tag{2.26}$$

Therefore, we get

$$P_{ab} = \mathrm{i}\hbar c_{acd}c_{bef}H^{ce}G^{df} \tag{2.27}$$

$$Q_{ab} = \frac{\mathrm{i}\hbar}{2}c_{acd}c_{bef}G^{ce}G^{df} \tag{2.28}$$

The $2PI$ EA yields equations of motion for the (arbitrary time) two-point functions of the theory. Given a solution of these equations, in principle we may find the expectation values of a large family of composite operators at any given time. Suppose we adopt a coarse-grained description where we choose a certain number of these expectation values as the relevant variables to describe the system. Then there will be a single density matrix which has maximum von Neumann entropy with respect to the class of states reproducing the preferred expectation values. This maximum entropy density matrix is the reduced density matrix for the system, and its entropy its correlation entropy. The H theorem is the statement that the correlation entropy grows in time, when the correlations themselves are evolved using the equations derived from the $2PI$ EA truncated to some order in the $1/N$ expansion. For details in the proof refer to [17].

2.4. *H theorem at NLOLN and its implications*

It is clear that if we could solve the exact evolution, the H theorem would be manifest: Given $\rho(t_0)$ at the initial time t_0, solve the exact Liouville equation up to a time t_1. Let $\bar{\rho}(t_1)$ be the result. We extract the new expectation values from $\bar{\rho}(t_1)$ and use them to construct the new maximum entropy density matrix $\rho(t_1)$. Then $S[\rho(t_1)] \geq S[\bar{\rho}(t_1)]$, by definition, and $S[\bar{\rho}(t_1)] = S[\rho(t_0)]$, because the exact evolution is unitary, thus the H theorem.

What we need to determine is whether, given the approximate dynamics for the expectation values provided by the $1/N$ scheme, the H theorem still holds. From our first paper (MQP1) in this series [1] we know that to LO it does not because LO large N yields a mean field theory whose equation of motion, the Vlasov equation, is unitary. Thus there is no net entropy production and no H-theorem. In such a case, if one works with a Fock representation, for boson fields, the number of particles increases with time and can be used as a measure of field entropy, what Hu and Pavon [32] called an intrinsic measure of field entropy. (See also [33].) However, this should not be confused with the correlation entropy of the Boltzmann kind under study for which the H-theorem is defined.

Of course, we expect better approximations to be closer to the exact dynamics, therefore requiring less external control of (or rather, tampering with) the system. This reduces the entropy loss to the environment. Physically, whenever some higher order correlations are ignored, we expect to obtain an H theorem because such a description of the system is incomplete. This is the physical origin of *correlation entropy*, as explained in the Introduction. In order to have the different correlations evolving according to the $2PI$ $1/N$ equations appropriate to the desired order, instead of the accurate description from the full Schwinger - Dyson hierarchy, we must keep nudging it to conform to this artificially created condition due to our inability to comprehend the complete picture. This is the source of a new kind of noise arising from ignoring the higher order correlation functions, called correlation noise, which, when incorporated in the Schwinger-Dyson hierarchy, yields the (nth-correlation order) stochastic SD equations, as explained in [16].

2.5. *2PI EA, H-theorem, quantum-classical and micro-macro*

We can now ask the question: How does quantum correlations get registered in the $2PI$ EA and how does the existence of H-theorem at the NLO expansion pertain to the quantum-Macro issue?

First, how do the quantum features show up in the $2PI$ effective action? Only equations of evolution for the first two moments can be obtained from the $2PI$ EA. It is known that [34] the equations of motion for the first two moments are identical for classical and quantum systems. Quantum corrections arise from the way the higher order moments are treated implicitly in the effective action approach. Because of this implicit dependence the $2PI$ EA approach is not easy to work with in addressing the differences of quantum and classical dynamics. One could consider going to higher orders, at least 3PI, and see if the third moment equations possess the terms absent in classical dynamics at a given order in large N.

The micro-Macro (m-M) demarkation issue is as difficult to tackle, if not more, as the quantum-classical transition issue. Going from micro to macro requires coarse graining. QMON model is a "democratic" system [4], hence the coarse graining should be "democratic" as well. The $2PI$ approach provides us with a reduced description in terms of the two lowest order Green's functions, which we shall refer to as the relevant variables. Higher order Green's functions are referred to as irrelevant variables. If one keeps track only of the relevant variables information about irrelevant variables is lost, by choice. If these variables are coupled and one wishes to see

[4] Democratic' is a figurative description of a system of identical particles and 'autocratic'refers to a system containing a distinct member from the rest. This corresponds to the Boltzmann versus Langevin paradigms describing an effectively open (by correlation order, which depends on the level of precision of observation) versus an open (by ab initio designation) system. See, [35].

how the relevant variables (our system) is affected by the irrelevant variables (its environment) one needs to take into account the back-action of the irrelevant variables. This back-action included reduced system (in terms of e.g., a reduced density matrix) would show dissipative dynamics and a H-theorem will obtain for the correlation entropy.

The question is whether the H-theorem holds for the dynamics obtained at a certain order of the $1/N$ expansion. There is no H-theorem at the leading order. This can be understood by realizing that the LO is equivalent to Gaussian approximation, and in this approximation correlations of order higher than two have no dynamics. As a result the mean field and the two point function contain all the available partial information because of the approximation imposed. The fact that H-theorem holds at NLO shows that the NLO approximation is capable of capturing the back-action from the irrelevant variables which engenders dissipation in the dynamics of the relevant variables. Very coarsely, in the large N perspective the NLO may be viewed as providing us a witness to the micro-macro transition. Note this does not mean that NLO large N is a good approximation in general. Actually we know from Mihaila et. al. [18] that it fails at long times although the H-theorem holds at long times, albeit only shown for a vacuum initial state [17]. The existence of an H-theorem, hence irreversibility, is a way of arguing that we have gone beyond the micro-domain.

In the QMON model N plays a dual role. On the one hand N is the number of degrees of freedom. On the other hand N appears in the denominator of the interaction term in the system Hamiltonian 2.1. The large N limit is usually associated with the number of components $N \to \infty$ only, however for the QMON model it is the weak coupling limit as well. If we would like to make an analogy to the classical BBGKY hierarchy of section 2.1, large N approximation would have the same effect as the weak coupling approximation. The fact that there is no H-theorem at LO but there is an H-theorem at NLO is analogous to the fact that Vlasov equation doesn't lead to entropy production, whereas the Landau equation does so.

Entropy increase is a signifier of irreversibility and irreversibility is usually associated with macroscopia. LO approximation gives reversible dynamics, while at NLO irreversibility emerges. This fact appears to be in contradiction with the association of irreversibility with macroscopia, since LO theory is successful in describing systems with more components N than NLO theory. The apparent paradox can be resolved by appreciating the dual role of N in the QMON model. The above example shows the interplay between system-size and interaction strength in the emergence of macroscopic qualities in a system. The system-size alone is not enough to qualify the system as macroscopic.

The work of Mihaila et. al. [18] reveals a critical value of N below which the NLO approximation does poorly ($N_c = 18.6$ for the parameters they chose for the simulations). As was discussed in MQP1 this is a way of quantifying when N is actually large enough for a certain order to yield good results. This is a stronger condition than the validity of H-theorem (which holds at NLO for any N), since it requires the dynamics to follow the exact one, rather than just one condition of nondecreasing entropy. We may combine these two results into the following interpretation: The H-theorem is necessary for a satisfactory reduced description of a macroscopic system. It holds at NLO. NLO is a reasonable approximation for the system dynamics for $N > N_c$. Hence for $N > N_c$ the reduced description is consistent.

3. Model study: $2PI$, $NLOLN$, loop expansions in nonequilibrium BEC dynamics

We begin with a description of the well-known Bose-Hubbard model and state our goal in this illustrated example. We introduce the $2PI$ generating functional to construct the $2PI$ effective action (EA) $\Gamma[\phi, G]$ and Green's functions. We perform a perturbative expansions on Γ_2 and define the various approximation schemes. We then derive the equation of motion from the CTP $2PI$ EA and discuss the results under each approximation scheme, starting with the Hartree-Fock-Bogoliubov (HFB) approximation followed by second order expansions which includes the $1/\mathcal{N}$ expansion and then the full second order expansion. The results from these approximations are compared with the exact numerical answers so as to exhibit the effects of quantum correlations, the coupling strength and how the results depend on increasing N.

The dynamics of an ultracold bosonic gas in an optical lattice can be described by a Bose-Hubbard model where the system parameters are controlled by laser light. For a one dimensional lattice the Hamiltonian is:

$$\hat{H} = -J\sum_i(\hat{\Phi}_i^\dagger \hat{\Phi}_{i+1} + \hat{\Phi}_{i+1}^\dagger \hat{\Phi}_i) + \sum_i \epsilon_i \hat{\Phi}_i^\dagger \hat{\Phi}_i + \frac{1}{2}U\sum_i \hat{\Phi}_i^\dagger \hat{\Phi}_i^\dagger \hat{\Phi}_i \hat{\Phi}_i\,, \tag{3.1}$$

where $\hat{\Phi}_i$ and $\hat{\Phi}_i^\dagger$ are the annihilation and creation operators at the site i which obey the canonical commutation relations for bosons. Here, the parameter U denotes the strength of the on-site repulsion of two atoms on the site i; the parameter ε_i denotes the energy offset of each lattice site due to an additional slow varying external potential such as a magnetic trap and J denotes the hopping rate between adjacent sites. Because the next-to-nearest neighbor amplitudes are typically two orders of magnitude smaller, tunneling to them can be neglected. We denote the total number of atoms by N and the number of lattice sites by I. Only a one-dimensional homogeneous lattice with periodic boundary conditions is considered.

A dimensionless parameter that is convenient to describe the different regimes of H is the coupling strength $\lambda \equiv NU/IJ$. Different from a homogeneous system without a lattice where at zero temperature the superfluid fraction is always unity, the presence of the lattice changes the superfluid properties and even at zero temperature, the superfluid fraction decreases with the lattice depth. For strong-coupling strengths [36] $\lambda > \lambda_{crit}$,

$$\lambda_{crit} \sim \frac{2N}{I^2}[2N + I + \sqrt{(2N+I)^2 + I^2}], \tag{3.2}$$

it is known that the ground state undergoes a quantum phase transition from a superfluid to a Mott insulator. In the weakly interacting regime, $\lambda \ll 1$, where tunneling overwhelms the repulsion, to a good approximation quantum fluctuations can be neglected and the properties of the system can be described by replacing the operator on the lattice site i by a classical c number. It can be said that most of the atoms are in the zero quasimomentum state. In the intermediate regime $1 \ll \lambda \ll \lambda_{crit}/2$ the interactions between the bosons can be very strong but the ground state is nevertheless a superfluid. For these interaction parameters a self-consistent HFB-Popov theory gives a good description of the system. However, different from the weak interacting regime where the depletion of the zero quasimomentum state is very small and has little effect on the superfluid properties, in this intermediate regime, depleted atoms spread over the central part of the band and reduces the superfluid fraction. As interactions are further increased the depleted population completely fills the band and cancels the superfluid properties. The system reaches the Mott insulator regime, where atoms are localized at each lattice site and the eigenstates of the system are almost Fock states with vanishing number fluctuations per lattice site. The dynamics in the intermediate regime is the focus of study in [37], where the superfluid properties are important but quantum fluctuations cannot be ignored.

3.1. 2PI EA $\Gamma[\phi, G]$

The first requirement for the study of nonequilibrium processes is a general initial-value formulation depicting the dynamics of interacting quantum fields. The *CTP* or Schwinger-Keldysh effective action formalism [23] serves this purpose. The second requirement is to describe the evolution of the correlation functions and the mean field on an equal footing. The two particle irreducible (*2PI*) formalism [24] where the correlation functions appear also as independent variables serves this purpose. By requiring the generalized (master) *CTP* effective action [14] to be stationary with respect to variations of the correlation functions an infinite set of coupled (Schwinger-Dyson) equations for the correlation functions is obtained which is a quantum analog of the BBGKY hierarchy. The $2PI$ effective action produces two such functions in this hierarchy.

The classical action corresponding to the Bose-Hubbard Hamiltonian (3.1) is given in terms of the complex fields Φ_i and Φ_i^* by

$$\begin{aligned} S[\Phi_i^*, \Phi_i] &= \int dt \sum_i \Big(i\hbar\Phi_i^*(t)\partial_t\Phi_i(t) + J\left(\Phi_i^*(t)\Phi_{i+1}(t) + \Phi_i(t)\Phi_{i+1}^*(t)\right) \\ &\quad - \frac{U}{2}\Phi_i^*(t)\Phi_i^*(t)\Phi_i(t)\Phi_i(t)\Big), \end{aligned} \tag{3.3}$$

where, as before, i denotes the lattice position, J is the hopping rate and U is the interaction strength. We limit the analysis to the case when no external potential is present and include only nearest neighbor hopping. To compactify our notation we introduce $\Phi_i^a (a = 1, 2)$ defined by

$$\Phi_i = \Phi_i^1, \quad \Phi_i^* = \Phi_i^2. \tag{3.4}$$

In terms of these fields the classical action takes the form

$$S[\Phi] = \int dt \sum_i \left(\frac{1}{2} h_{ab}\Phi_i^a(t)\hbar\partial_t\Phi_i^b(t) + J\sigma_{ab}\Phi_{i+1}^a(t)\Phi_i^b(t) - \frac{U}{4\mathcal{N}}(\sigma_{ab}\Phi_i^a(t)\Phi_i^b(t))^2 \right), \tag{3.5}$$

where $\mathcal{N}$ is the number of fields (two in this case), and summation over repeated field indices $a, b = (1, 2)$ is implied. h_{ab} and σ_{ab} are matrices defined as

$$h_{ab} = \mathrm{i}\begin{pmatrix} 0 & -1 \\ 1 & 0 \end{pmatrix} \qquad \sigma_{ab} = \begin{pmatrix} 0 & 1 \\ 1 & 0 \end{pmatrix} \tag{3.6}$$

After second quantization the fields Φ_i^a are promoted to operators. We denote the expectation value of the field operator or mean field by $\phi_i^a(t)$ and the expectation value of the composite field by $G_{ij}^{ab}(t, t')$. Physically, $|\phi_i^a(t)|^2$ is the condensate population and the composite fields determine the fluctuactions around the mean field.

$$\begin{aligned} \phi_i^a(t) &= \langle\Phi_i^a(t)\rangle, & (3.7)\\ \hbar G_{ij}^{ab}(t, t') &= \left\langle T_C\Phi_i^a(t)\Phi_j^b(t')\right\rangle - \langle\Phi_i^a(t)\rangle\left\langle\Phi_j^b(t')\right\rangle. & (3.8) \end{aligned}$$

The brackets denote taking the expectation value with respect to the density matrix and T_C denotes time ordering along a contour C in the complex plane.

All correlation functions of the quantum theory can be obtained from the effective action $\Gamma[\phi, G]$, the two particle irreducible generating functional for Green's functions parametrized by $\phi_i^a(t)$ and the composite field $G_{ij}^{ab}(t, t')$. To get an expression for the effective action we first define the functional $Z[\mathbf{J}, \mathbf{K}]$ [24] as

$$\begin{aligned} Z[\mathbf{J}, \mathbf{K}] &= e^{\mathrm{i}/\hbar W[\mathbf{J},\mathbf{K}]} \qquad (3.9)\\ &= \prod_a \int D\Phi^a \exp\left\{ \frac{\mathrm{i}}{\hbar}\left(S[\Phi] + \int dt \sum_i \mathbf{J}_{ia}(t)\Phi_i^a(t) \right.\right. \\ &\quad \left.\left. + \frac{1}{2}\int dt dt' \sum_{ij} \Phi_i^a(t)\Phi_j^b(t')\mathbf{K}_{ijab}(t, t') \right) \right\}, \end{aligned}$$

where we have introduced the following index lowering convention

$$X_a = \sigma_{ab} X^b. \tag{3.10}$$

The functional integral (3.9) is a sum over classical histories of the field Φ_i^a in the presence of the local source $\mathbf{J}_{ia}$ and the non local source $\mathbf{K}_{ijab}$. The coherent state measure is included in $D\Phi$. The addition of the two-particle source term is what characterizes the $2PI$ formalism.

We define $\Gamma[\phi, G]$ as the double Legendre transform of $W[\mathbf{J}, \mathbf{K}]$ such that

$$\frac{\delta W[\mathbf{J}, \mathbf{K}]}{\delta \mathbf{J}_{ia}(t)} = \phi_i^a(t), \tag{3.11}$$

$$\frac{\delta W[\mathbf{J}, \mathbf{K}]}{\delta \mathbf{K}_{ijab}(t,t')} = \frac{1}{2}[\phi_i^a(t)\phi_i^b(t') + \hbar G_{ij}^{ab}(t,t')]. \tag{3.12}$$

Expressing $\mathbf{J}$ and $\mathbf{K}$ in terms of ϕ and G yields

$$\Gamma[\phi, G] = W[\mathbf{J}, \mathbf{K}] - \int dt \sum_i \mathbf{J}_{ia}(t)\phi_i^a(t) - \frac{1}{2}\int dt dt' \sum_{ij} \phi_i^a(t)\phi_j^b(t')\mathbf{K}_{ijab}(t,t') \tag{3.13}$$
$$-\frac{\hbar}{2}\int dt dt' \sum_{ij} G_{ij}^{ab}(t,t')\mathbf{K}_{ijab}(t,t').$$

From this equation the following identity can be derived:

$$\frac{\delta \Gamma[\phi, G]}{\delta \phi_i^a(t)} = -\left(\mathbf{J}_{ia}(t) + \int dt' \sum_j (\mathbf{K}_{ijad}(t,t'))\phi_j^d(t')\right), \tag{3.14}$$

$$\frac{\delta \Gamma[\phi, G]}{\delta G_{ij}^{ab}(t,t')} = -\frac{\hbar}{2}\mathbf{K}_{ijab}(t,t'). \tag{3.15}$$

In order to get an expression for $\Gamma[\phi, G]$ notice that by using (3.9) for $W[\mathbf{J}, \mathbf{K}]$ and placing it in (3.13) for $\Gamma[\phi, G]$, it can be written as

$$\exp\left(\frac{\mathrm{i}}{\hbar}\Gamma[\phi, G]\right) = \prod_a \int D\Phi^a \exp\left\{\frac{\mathrm{i}}{\hbar}\left(S[\Phi] + \int dt_i\, \mathbf{J}_{ia}(t)\,[\Phi_i^a(t) - \phi_i^a(t)]\right.\right. \tag{3.16}$$
$$\left.\left. + \frac{1}{2}\int dt_i dt_j' \left(\Phi_i^a(t)\mathbf{K}_{ijab}(t,t')\Phi_j^b(t') - \phi_i^a(t)\mathbf{K}_{ijab}(t,t')\phi_j^b(t')\right) - \frac{\hbar}{2}\operatorname{Tr} G\mathbf{K}\right)\right\}$$
$$= \prod_a \int D\Phi^a \exp\left\{\frac{\mathrm{i}}{\hbar}\left(S[\Phi] - \int dt_i\, \frac{\delta \Gamma[\phi, G]}{\delta \phi_i^a(t)}\,[\Phi_i^a(t) - \phi_i^a(t)]\right.\right.$$
$$\left.\left. - \frac{1}{\hbar}\int dt_i dt_j'\, [\Phi_i^a(t) - \phi_i^a(t)]\,\frac{\delta \Gamma[\phi, G]}{\delta G_{ij}^{ab}(t,t')}\left[\Phi_i^b(t') - \phi_i^b(t')\right] + \operatorname{Tr} G\frac{\delta \Gamma[\phi, G]}{\delta G}\right)\right\},$$

where Tr means taking the trace. For simplicity we have denoted $\int dt \sum_i$ by $\int dt_i$. Defining the fluctuation field, $\varphi_i^a = \Phi_i^a - \phi_i^a$, we have

$$\Gamma[\phi, G] - \operatorname{Tr} G\frac{\delta \Gamma[\phi, G]}{\delta G} = -\mathrm{i}\hbar \ln \prod_a \int D\varphi^a \exp\left(\frac{\mathrm{i}}{\hbar} S[\phi, G; \varphi]\right) \tag{3.17}$$

$$S[\phi, G; \varphi] = S[\phi + \varphi] - \int dt_i\, \frac{\delta \Gamma[\phi, G]}{\delta \phi_i^a(t)}\varphi_i^a(t) - \frac{1}{\hbar}\int dt_i dt_j'\, \varphi_i^a(t)\frac{\delta \Gamma[\phi, G]}{\delta G_{ij}^{ab}(t,t')}\varphi_i^b(t'). \tag{3.18}$$

By introducing the classical inverse propagator $iD^{-1}(\phi)$ given by

$$\begin{aligned} \mathrm{i}D_{ijab}(t,t')^{-1} &= \frac{\delta S[\phi]}{\delta\phi_i^a(t)\delta\phi_j^b(t')} \\ &= (\delta_{ij}h_{ab}\partial_t + J(\delta_{i+1j}+\delta_{i-1j})\sigma_{ab})\,\delta(t-t') \\ &\quad -\frac{U}{\mathcal{N}}\left(2\phi_{ia}(t)\phi_{ib}(t)+\sigma_{ab}\phi_i^c(t)\phi_{ic}(t)\right)\delta_{ij}\delta(t-t'), \end{aligned} \tag{3.19}$$

the solution of the functional integro-differential equation (3.17) can be expressed as

$$\Gamma[\phi,G] = S[\phi] + \frac{\mathrm{i}}{2}\,\mathrm{Tr}\ln G^{-1} + \frac{\mathrm{i}}{2}\,\mathrm{Tr}\,D^{-1}(\phi)G + \Gamma_2[\phi,G] + const. \tag{3.20}$$

The quantity $\Gamma_2[\phi,G]$ is conveniently described in terms of the diagrams generated by the interaction terms in $S[\phi+\varphi]$ which are of cubic and higher orders in φ.

$$S_{int}[\phi+\varphi] = -\frac{U}{4\mathcal{N}}\int dt_i\,(\varphi_{ib}(t)\varphi_i^b(t))^2 - \frac{U}{\mathcal{N}}\int dt_i\,\varphi_i^a(t)\phi_{ia}(t)\varphi_i^b(t)\varphi_{ib}(t). \tag{3.21}$$

It consists of all two-particle irreducible vacuum graphs (the diagrams representing these interactions do not become disconnected by cutting two propagator lines) in the theory with propagators set equal to G and vertices determined by the interaction terms in $S[\phi+\varphi]$.

Since physical processes correspond to vanishing sources **J** and **K**, the dynamical equations of motion for the mean field and the propagators are found by using the expression (3.20) in equations (3.14) and (3.15), and setting the right hand side equal to zero. This procedure leads to the following equations:

$$\begin{aligned} h_{ab}\hbar\partial_t\phi_i^b(t) &= -J(\phi_{i+1a}(t)+\phi_{i-1a}(t)) + \frac{U}{\mathcal{N}}(\phi_{id}(t)\phi_i^d(t) + G_{ii\ c}^{\ c}(t,t))\phi_{ia}(t) + \\ &\quad \frac{U}{\mathcal{N}}((G_{iiad}(t,t)+G_{iida}(t,t))\phi_i^d(t) - \frac{\delta\Gamma_2[\phi,G]}{\delta\phi_i^a(t)}, \end{aligned} \tag{3.22}$$

and

$$G_{ijab}^{-1}(t,t') = D_{ijab}(t,t')^{-1} - \Sigma_{ijab}(t,t'), \tag{3.23}$$

$$\Sigma_{ijab}(t,t') \equiv 2\mathrm{i}\frac{\delta\Gamma_2[\phi,G]}{\delta G_{ij}^{ab}(t,t')}. \tag{3.24}$$

Equation (3.23) can be rewritten as a partial differential equation suitable for initial value problems by convolution with G. This differential equation reads explicitly

$$\begin{aligned} h_c^a\hbar\partial_t G_{ij}^{cb}(t,t') &= -J(G_{i+1j}^{ab}(t,t') + G_{i-1j}^{ab}(t,t')) + \frac{U}{\mathcal{N}}(\phi_{id}(t)\phi_i^d(t))G_{ij}^{ab}(t,t') + \\ &\quad \frac{2U}{\mathcal{N}}\phi_i^a(t)G_{ij}^{cb}(t,t')\phi_{ic}(t) + \mathrm{i}\int dt_k''\Sigma_{ikc}^{\ a}(t,t'')G_{kj}^{cb}(t'',t') + \mathrm{i}\delta^{ab}\delta_{ij}\delta(t-t'). \end{aligned} \tag{3.25}$$

The evolution of ϕ^a and G^{ab} is determined by Eqs. (3.22) and (3.25) once $\Gamma_2[\phi,G]$ is specified.

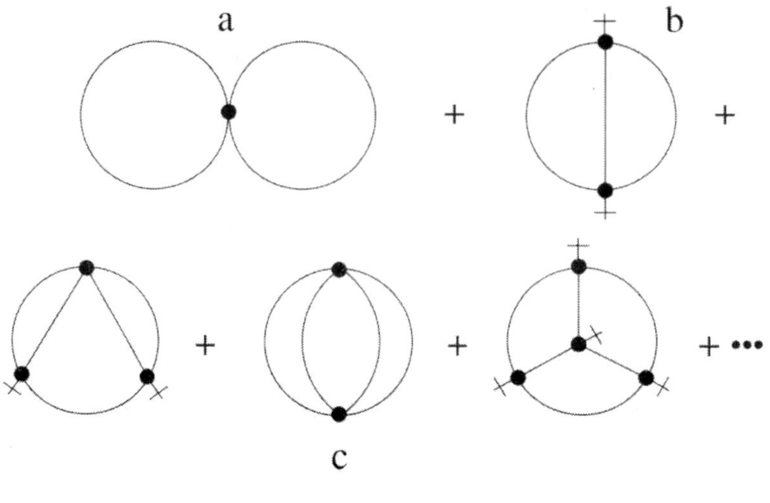

Figure 1. Two-loop (upper row) and three-loop diagrams (lower row) contributing to the effective action. Explicitly, the diagram a is what we call the *double-bubble* , b the *setting-sun* and c the *basket-ball.*

3.2. Perturbative expansion of $\Gamma_2(\phi, G)$ *and approximation schemes*

The diagrammatic expansion of Γ_2 is illustrated in the accompanying Figure where two and three-loop vacuum diagrams are shown.

The dots where four lines meet represent interaction vertices. The expression corresponding to each vacuum diagram should be multiplied by a factor $(-i)^l(i)^{s-2}$ where l is the number of solid lines and s the number of loops the diagram contains.

The action Γ including the full diagrammatic series for Γ_2 gives the full dynamics. It is of course not feasible to obtain an exact expression for Γ_2 in a closed form. Various approximations for the full $2PI$ effective action can be obtained by truncating the diagrammatic expansion for Γ_2. Which approximation is most appropriate depends on the physical problem under consideration:

3.3. The standard approaches

(i) Bogoliubov (One-loop) Approximation:
The simplest approximation consists of discarding Γ_2 altogether. This yields the so called Bogoliubov or one-loop approximation whose limitations have been extensively documented in the literature ([38, 39]).

(ii) Time-dependent Hartree-Fock-Bogoliubov (HFB) Approximation:
A truncation of Γ_2 retaining only the first order diagram in U, i.e., keeping only the *double-bubble* diagram, Fig. *a*, yields equations of motion of ϕ and G which correspond to the time dependent Hartree-Fock-Bogoliubov (HFB) approximation. This approximation violates Goldstone's theorem, but conserves energy and particle number [40, 41, 42]). The HFB equations can also be obtained by using cumulant expansions up to the second order [43] in which all cumulants containing three or four field operators are neglected. The HFB approximation neglects multiple scattering. It can be interpreted as an expansion in terms of Ut/J, (where t is the time of evolution) and is good for the description of short time dynamics or weak interaction strengths.

(i) Second Order expansion:

Journal of Physics: Conference Series **330** (2011) 012003 doi:10.1088/1742-6596/330/1/012003

A truncation retaining diagrams of second order in U containing besides the *double-bubble*, also the *setting-sun* and the *basket-ball*. By including the *setting-sun* and the *Basket-ball* in the approximations we are taking into account two particle scattering processes [44]. Second order terms lead to integro- differential equations which depend on the time history of the system.

(ii) Large-$\mathcal{N}$ approximation
The $1/\mathcal{N}$ expansion is a controlled non-perturbative approximation scheme which can be used to study non-equilibrium quantum field dynamics in the regime of strong interactions[31]. In the large $\mathcal{N}$ approach the field is modeled by $\mathcal{N}$ fields and the quantum field generating functional is expanded in powers of $1/\mathcal{N}$. In this sense the method is a controlled expansion in a small parameter but unlike perturbation theory in the coupling constant, which corresponds to an expansion around the vacuum, the large $\mathcal{N}$ expansion corresponds to an expansion of the theory about a strong quasiclassical field.

In [37] numerical solutions to the equations of motion for a moderate number of atoms and wells up to the second order in the coupling constant U were given. These exact many body solutions are useful for determining the range of validity of the three types of approximations described above, namely, a) the HFB, b) the full second order and c) NLO large $\mathcal{N}$ expansion up to second order in U (the shorthands HFB, 2nd and $1/\mathcal{N}$ respectively are used in the figure). For the effect of the total number of atoms on the dynamics Rey et al have given the numerical solutions for a double well system with fix ratio $UN/J = 4$ and three different values of N: N=20, 40 and 80. The results are given for the evolution of the atomic population per well (Fig. 5), the condensate population per well and total condensate population (Fig. 6) and the quasi-momentum intensities (Fig. 7). (To make the comparisons easier the numerical results obtained for the three different values of N are scaled by dividing them by the total number of atoms. In this way for all the cases we start with an atomic population of magnitude one in the initial populated well.)

In the exact dynamics we see that as the number of atoms is increased the damping effects occur at slower rates. This feature can be noticed in the quantum dynamics of all of the observables depicted in the Figures 5 to 7 of [37]. The decrease of the damping rates as the number of atoms is increased is not surprising because changing the number of atoms alters the quantum coherence properties of the system. As shown in reference [45], the collapse time of the condensate population is approximately given by $t_{coll} \sim \frac{t_{rev}}{\sigma}$ where σ is the variance of the initial atomic distribution and t_{rev} is the revival time which depends on the detailed spectrum for the Hamiltonian. In the kinetic energy dominated regime $t_{rev} \sim hN/J$ (see [46]), thus $t_{coll} \sim \frac{N}{\sigma}$. For our initial conditions the variance is proportional to $\sigma = \sqrt{N}$ so $t_{coll} \sim \sqrt{N}$. Besides damping rates, the qualitative behavior of the exact quantum dynamics is not affected very much as the number of atoms is increased for a fixed UN/J.

The improvement of the $2PI$ approximations as N is increased, as a result of the increase in the initial number of coherent atoms is in fact observed in the plots. Even though the problem of underdamping in the HFB approximation and the overdamping in the second order approaches are not cured, as the number of atoms is increased, we do observe a better matching with the full quantal solution. The $1/\mathcal{N}$ expansion shows the fastest convergence, as can be more easily observed in the quasi-momentum distribution plots, Fig. 7. The better agreement of the $1/\mathcal{N}$ expansion relies on the fact that even though the number of fields is only two in our calculations the $1/\mathcal{N}$ expansion is an expansion about a strong quasiclassical field configuration, a feature we have explained in MQP1.

Journal of Physics: Conference Series **330** (2011) 012003 doi:10.1088/1742-6596/330/1/012003

Part II: Effective Scale and Infrared Dimension

4. Effective field theory, coupling and the effective scale of a quantum system

Effective field theory is a very powerful tool in quantum field theory, and in particular it gives a new point of view about the meaning of renormalization.

Consider a quantum field theory with a characteristic energy scale E_0, and suppose that we are interested in physics at some lower energy scale E. The basic idea of effective field theory is that one can choose a cutoff energy scale Λ at or slightly below E_0, and divide the fields in the path integral into low- and high- frequency parts, $\phi = \phi_L + \phi_H$, where $\phi_L(\phi_H)$ are the field modes with frequencies lower(higher) than Λ. Now integrate over the high frequency fields ϕ_H

$$\int D\phi_H D\phi_L e^{iS(\phi_H,\phi_L)} = \int D\phi_L e^{iS_\Lambda(\phi_L)} \tag{4.1}$$

where $e^{iS_\Lambda(\phi_L)} = \int D\phi_H e^{iS(\phi_H,\phi_L)}$. The $S_\Lambda(\phi_L)$ is called the low-energy or Wilsonian effective action. It's different from the 1PI effective action which are generated from integrating over all frequencies but keeping only 1PI graphs.

One can expand the low-energy effective action in terms of local operators which are consistent with the symmetries of the problem,

$$S_\Lambda(\phi_L) = \int d^D x \sum_i g_i \hat{Q}_i. \tag{4.2}$$

In units of $\hbar = c = 1$, if an operator $\hat{O}_i$ has unit E^{d_i} where d_i is the dimension of operator $\hat{O}_i$ then the parameter g_i has unit E^{D-d_i} where D is the spacetime dimension. In the case of a free scalar field in D dimensions, with action

$$S = \int d^D x \left(\frac{1}{2}\partial_\mu\phi\partial^\mu\phi - \frac{1}{2}m^2\phi^2\right). \tag{4.3}$$

The action S is dimensionless, so the dimension of $\phi(x)$ is determined from the kinetic term to be $[\phi] = \frac{D-2}{2}$ and the dimension of m^2 is $[m^2] = 2$. An operator $\hat{O}_i$ constructed from M ϕ's and N derivatives has dimension $d_i = M(\frac{D-2}{2}) + N$. Define dimensionless couplings $\lambda_i = \Lambda^{d_i - D} g_i$. Now for a process at energy scale E, one can dimensionally estimate the contribution from the i-th term of the effective action

$$\int d^D x g_i \hat{O}_i \sim \lambda_i (\frac{E}{\Lambda})^{d_i - D}. \tag{4.4}$$

If $d_i > D$, this term will become less and less important at lower and lower energies, and is called an irrelevant operator. If $d_i < D$, the operator is more and more important at lower and lower energies and is called a relevant operator. If $d_i = D$ the operator is equally important at all scales and called marginal. In most cases there is only a finite number of relevant and marginal operators so the low energy physics depends only on a finite number of parameters.

The low energy physics depends on the short distance theory only through the relevant and marginal couplings, and possibly through some leading irrelevant couplings if one measures small enough effects.

It should be pointed out that the coupling strength measured will vary with energy(scale). This can easily be seen from the point of view of effective field theory. We could write the path integral as

$$\int_{k \leq \Lambda} D\phi_k e^{iS_\Lambda(\phi_k)} = \int_{k' \leq \Lambda'} D\phi_{k'} e^{iS'_{\Lambda'}(\phi_{k'})}. \tag{4.5}$$

Journal of Physics: Conference Series **330** (2011) 012003 doi:10.1088/1742-6596/330/1/012003

The action $S_\Lambda(\phi_k)$ contains all momentum modes up to some maximum value Λ, whereas the new effective action $S'_{\Lambda'}(\phi_{k'})$ contains momentum modes up to $\Lambda' < \Lambda$. If one take Λ' to be infinitesimally smaller than Λ, $S'_{\Lambda'}$ will be infinitesimally different from S_Λ. This generates a differential equation, the Wilson equation

$$\frac{\partial S_\Lambda}{\partial \Lambda} = F(S_\Lambda), \tag{4.6}$$

where F is some functional of the action. The Wilson equation is the renormalization group flow equation in an infinite dimensional space and it gives the change in action as a function of cutoff. Since

$$S_\Lambda(\phi) = \int d^D x \sum_i g_i \hat{Q}_i, \tag{4.7}$$

the Wilson equation then gives

$$\frac{\partial g_i}{\partial \Lambda} = F(\{g_i\}). \tag{4.8}$$

If one linearizes a solution of the equation, irrelevant operators will correspond to those directions with negative eigenvalues whereas the relevant operators corresponds to those directions with positive eigenvalues.

5. Infrared behavior of quantum fields: finite-size / topology effects

In the above we pointed out that the infrared or large scale behavior of a system depends on many factors: a) Interaction strength: difference between a free theory (Gaussian) and an interacting theory, and for an interacting theory the Gaussian fixed point as different from non-Gaussian ones, relevant versus irrelevant fixed points; b) Coupling strength which runs with energy via the RG equation. All this show the intricacy of the notion of 'large scale' which we need to bear in mind in describing 'macroscopic' quantum behavior. In this section, we demonstrate how the infrared behavior of a quantum system can be altered qualitatively by the geometry and topology of space, or more precisely, the 'finite size' effect. This can be seen from analysis based on rather general arguments, e.g., on whether there exist a band structure in the spectrum of eigenvalues of the invariant operator of the order parameter field. If so there is a dimensional reduction effect where the IR behavior of the system becomes that of a lower dimension. What this conveys is that, for a quantum system near the critical point where its correlation length increases to (near) infinity its large scale or IR behavior or macroscopic features can become sensitive to the size and topology of the underlying spacetime. Cosmological phase transition is an extreme yet valid, even realistic, manifestation of this feature. A more rigorous and complete treatment of this effect is given in Hu and O'Connor [47] from which our present exposition is adapted.

Consider an N-component self-interacting scalar field $\Phi^a (a = 1, \cdots, N)$ on a manifold of dimension D, coupled to the background spacetime with curvature R and coupling constant ξ (conformal coupling for $\xi = 0$ and minimal coupling for $\xi = 1$) described by the action

$$S[\Phi] = \int d^D x \sqrt{g} \left[\frac{1}{2} \Phi^a \nabla^2 \Phi^a + \frac{1}{2} M_\xi^2 \Phi^2 + \frac{\lambda}{4!} \Phi^4 \right], \tag{5.1}$$

where $M_\xi^2 = m^2 + (1 - \xi) R/6$ and $\nabla^2 = -\sqrt{g} \partial_\mu (\sqrt{g} g^{\mu\nu} \partial_\nu)$ is the Laplace-Beltrami operator for scalar fields. For convenience we develop the formalism here in Euclidean space version with signature (+,+,+,+). We begin with a decomposition of the field Φ^a into a background field $\bar{\phi}^a$

Journal of Physics: Conference Series **330** (2011) 012003 doi:10.1088/1742-6596/330/1/012003

and a fluctuation field φ^a , i.e., $\Phi^a = \bar{\phi}^a + \varphi^a$. The background field $\bar{\phi}^a$ is required to satisfy the classical equations of motion with an arbitrary external source. Such a shift eliminates the linear term in the fluctuation field, which is equivalent to performing a Legendre transform. The resultant action is

$$S[\bar{\phi},\varphi] = S[\bar{\phi}] + \int d^D x\sqrt{g}$$
$$\times\left\{\frac{1}{2}\varphi^a\left[\left(\nabla^2 + M_\xi^2 + \frac{\lambda}{6}\bar{\phi}^2\right)\delta^{ab} + \frac{\lambda}{3}\bar{\phi}^a\bar{\phi}^b\right]\varphi^b + \frac{\lambda}{6}\bar{\phi}^a\varphi^a\varphi^2 + \frac{\lambda}{4!}\varphi^4\right\}. \tag{5.2}$$

In the Feynman diagrammatics there are two kinds of vertices: a four point vertex proportional to λ and a three point vertex proportional to $\lambda\bar{\phi}_a(x)$.

The effective action $\Gamma[\bar{\phi}]$ is obtained by functionally integrating over the fluctuation fields:

$$e^{-\Gamma[\bar{\phi}]} = \int [d\varphi] e^{-S[\bar{\phi},\varphi]}. \tag{5.3}$$

The wave operator A^{ab} for the fluctuating field is given by

$$A^{ab} = (\nabla^2 + M_1^2)\frac{\bar{\phi}^a\bar{\phi}^b}{\bar{\phi}^2} + (\nabla^2 + M_2^2)\left(\delta^{ab} - \frac{\bar{\phi}^a\bar{\phi}^b}{\bar{\phi}^2}\right) \tag{5.4}$$

where

$$M_1^2 = M_\xi^2 + \frac{\lambda}{2}\bar{\phi}^2, \quad M_2^2 = M_\xi^2 + \frac{\lambda}{6}\bar{\phi}^2\,. \tag{5.5}$$

Here $\bar{\phi}^a\bar{\phi}^b/\bar{\phi}^2$ and $\delta^{ab} - \bar{\phi}^a\bar{\phi}^b/\bar{\phi}^2$ are orthogonal projectors, the former projects along the direction in the internal space picked out by $\bar{\phi}^a$ and the latter projects into an (N-1)-dimensional subspace orthogonal to the direction of $\bar{\phi}^a$. Note that the operator ∇^2 does not commute with the projectors unless $\bar{\phi}^a$ is a constant.

When the direction in group space picked out by $\bar{\phi}^a$ does not vary from point to point around the manifold (this does not necessarily imply $\bar{\phi}^2$ is a constant) the Green's function for A^{ab} is given by

$$G^{ab} = G_1\frac{\bar{\phi}^a\bar{\phi}^b}{\bar{\phi}^2} + G_2\left(\delta^{ab} - \frac{\bar{\phi}^a\bar{\phi}^b}{\bar{\phi}^2}\right)\,, \tag{5.6}$$

where the G_i $(i = 1, 2)$ are the Green's functions for the operators $\nabla^2 + M_i^2$

The one-loop correction is related to the sum of the logarithms of the determinants of the fluctuation operators. So in this case, by using the projection operators, the one-loop effective action is given by

$$\Gamma[\bar{\phi}] = S[\bar{\phi}] - \frac{1}{2}\operatorname{Tr}\ln G_1 - \frac{N-1}{2}\operatorname{Tr}\ln G_2$$
$$= S[\bar{\phi}] + \frac{1}{2}\sum_l d_l \ln\lambda_{1l} + \frac{N-1}{2}\sum_l d_l\ln\lambda_{2l}\,, \tag{5.7}$$

where $S[\bar{\phi}]$ is the classical action, λ_{il} and d_l are the eigenvalues and degeneracies of the operators $\nabla^2 + M_i^2$.

The determinant of A is formally divergent and needs to be regularized. There are a number of commonly used regularization methods. If the space is Riemannian and has sufficient symmetry so that the spectrum of the invariant operator is known explicitly, then the ζ-function method is probably most convenient. The generalized ζ function is defined by

$$\zeta(\nu) = \sum_{\mathbf{n}}(\mu^{-2}\lambda_{\mathbf{n}})^{-\nu}, \tag{5.8}$$

where λ_n are the eigenvalues of the operator A on the Euclideanized metric obtained by a Wick rotation to imaginary time $\tau = it$. Here a constant mass scale μ is introduced to make the measure $d[\varphi]$ of the functional integral dimensionless. Using the regularization method of Dowker and Critchley, one can express the one-loop effective potential as

$$V^{(1)}(\bar{\phi}) = -\frac{1}{2}\hbar(Vol)^{-1}[\zeta'(0) + \zeta(0)/\nu]. \tag{5.9}$$

The ultraviolet divergence in $V^{(1)}$ can be canceled by the addition of counterterms which is not our concern here. The regularized effective action is useful for the analysis of the infrared behavior of quantum fields in a curved spacetime or spaces of nontrivial topology or finite extent, as shown below. Note, however that one-loop results are insufficient. Leading infrared contributions in higher loop terms need be included by using composite operator techniques.

We learn from the work of [47] that in spacetimes with some compact dimensions the lowest mode of the fluctuation operator has the strongest effect on the symmetry behavior of the system. When the lowest mode is massless it will give the dominant contribution to the effective action. The low energy behavior corresponds to a lower-dimensional system. In what follows we first give a formal derivation of infrared dimensional reduction by examining the result of the decoupling of the higher modes (or bands) in the functional integral for the effective action. We will then give a physical explanation in terms of the correlation lengths and the notion of effective IR dimension (**EIRD**). An alternative way of seeing this problem of dimensional reduction is by spectral analysis. This is applied to direct product spaces with some compact dimensions and to spaces which can be reduced to product spaces.

5.1. Decoupling of the higher modes (or bands)

Let us examine systems where the eigenvalues of the fluctuation operator takes on a band structure. By band structure we mean that the eigenvalues occur in continua with each continuum having a higher lowest eigenvalue than the previous one. This is true for fields on spacetimes with compact sections or for fields with discrete spectrum (e.g. the harmonic oscillator). The procedure is to expand the fields in terms of the band eigenfunctions and convert the functional integral over the fields to an integral over the amplitudes of the individual modes.

On a manifold with topology $R^d \times B^b$ where B is compact, consider quantum fields where the fluctuation operator A in (5.4) has the general form of a direct sum of operators D and B

$$A^{ab}(x, y) = D^{ab}(x) + B^{ab}(y) \tag{5.10}$$

with coordinates x on R^d and y on B^b. Assume that the eigenvalues ω_n associated with the eigenfunctions $\psi_n(y)$ of B^{ab} are discrete:

$$B^{ab}\psi_n(y) = \omega_n^{ab}\psi_n(y) \tag{5.11}$$

Decomposing the field $\varphi^a(x, y)$ in terms of $\psi_n(y)$

$$\varphi^a(x, y) = \sum_n \varphi_n^a(x)\psi_n(y) \tag{5.12}$$

one obtains for the quadratic part of the action

$$\frac{1}{2}\int d^dx d^by \varphi^a A^{ab} \varphi^b = \frac{1}{2}\int d^dx[\varphi_n^a f_{nm} D^{ab} \varphi_m^b + \omega_n^{ab} f_{nm} \varphi_n^a \varphi_m^b] \tag{5.13}$$

where $f_{nm} = \int d^by\psi_n(y)\psi_m(y)$. When φ_n are properly normalized $f_{nm} = \delta_{nm}$ (we will make such a choice here) the resulting theory in terms of the new fields φ_n^a will involve massive fields

Journal of Physics: Conference Series **330** (2011) 012003 doi:10.1088/1742-6596/330/1/012003

with masses determined by the eigenvalue matrix λ_{in}^{ab} of the operators $\nabla^2 + M_i$ [see (5.7)], even if the fields in terms of the old variables appeared massless. We will take the smallest eigenvalue to be given by $n = 0$ and assume that its only degeneracy is labeled by the indices a and b. Assume also that the operator D^{ab} is simply minus the Laplacian ∇_d^2 on R^d times δ^{ab}, the $n = 0$ mode is then governed by the action whose quadratic term is

$$\frac{1}{2}\int d^dx[\varphi_0^a(-\nabla_d^2)\delta^{ab}\varphi_0^b + \omega_0^{ab} f_{00}\varphi_0^a\varphi_0^b] \tag{5.14}$$

Thus this appears like a d-dimensional field with an apparent mass matrix ω_0^{ab}. For the case of an N component $\lambda\phi^4$ theory the action after this decomposition takes the form

$$S[\bar{\phi} + \varphi] = \int d^dx[\frac{1}{2}\varphi_n^a(-\nabla_d^2\delta^{ab} + \omega_n^{ab})\varphi_n^b$$

$$+\frac{\lambda}{6} g_{n\ell m}^a \varphi_\ell^b\varphi_m^b + \frac{\lambda}{4!} f_{kn\ell m}\varphi_k^a\varphi_n^a\varphi_\ell^b\varphi_m^b], \tag{5.15}$$

where $g_{n\ell m}^a = \int d^by\bar{\varphi}^a\psi_n\psi_\ell\psi_m$ and $f_{kn\ell m} = \int d^by\psi_k\psi_n\psi_\ell\psi_m$. The effective action is now given by the functional integral

$$e^{-\Gamma[\bar{\phi}]} = \int [d\varphi_n^a]e^{-S[\bar{\phi}+\varphi]}. \tag{5.16}$$

The interesting case occurs when the lowest eigenvalue approaches zero. At low energy the Appelquist-Carazzone decoupling theorem assures us that with higher modes decoupled from the dynamics, the infrared behavior is governed by the lowest band. We are then left with a purely lower dimensional theory. The higher modes do play a role in the ultraviolet divergences present in the theory (e.g. renormalization problem in Kaluza-Klein theories) and therefore determine the high energy running of coupling constants, but the infrared region of the theory is governed by the lower d-dimensional theory.

5.2. Correlation length and effective infrared dimension

The above result of dimensional reduction from a formal derivation of mode decoupling can be understood in a more physical way by using the concept of effective infrared dimensions (EIRD). By effective IR dimension we mean the dimension of space or spacetime wherein the system at low energy effectively behaves. One well-known example is the Kaluza-Klein theory of unification and cosmology. There, starting with an 11-dimensional spacetime with full diffeomorphism symmetry, after spontaneous compactification it reduces at energy below the Planck scale to the physical four dimensional space with GL(4,R) covariance and a seven-dimensional internal space with symmetry group containing the standard $SU_3 \times SU_2 \times U_1$ subgroups of strong and electroweak interactions. For observers today of very low energy the effective IR dimension of spacetime is four, even though the complete theory is eleven dimensional. By the same token, Einstein's theory of general relativity can presumably be regarded as the effective IR limit of an otherwise more complete and fundamental theory of quantum gravity, such as the superstring theory. For curved-space symmetry breaking considerations, the EIRD which the system "feels" is governed by a parameter η which is the ratio of the correlation length Ξ and the scale length L of the background space $\eta \equiv \Xi/L$. For compact spaces like S^4, L is simply 2π times the "radius" of S^4, the only geometric scale present. For product spaces $R^d \times B^b$ with some compact space B, there are two scale lengths: L_b is finite in the compact dimensions and $L_d = \infty$ in the non-compact dimensions. Examples are Kaluza-Klein theories d=4, b=1, 6, 7, finite temperature field theory (imaginary time formalism, $L = \beta$ inverse temperature) d=3, b=1, and Einstein or the spatially-closed Robertson-Walker universe d=1, b=3.

The symmetry behavior of the system (described here by a $\lambda\phi^4$ scalar field as example) is determined by the *correlation length* Ξ defined as the inverse of the *effective mass* M_{eff} related to the effective potential V_{eff} by (we use subscript eff to denote quantities including higher loop corrections)

$$\Xi^{-2} = \frac{\partial^2 V_{eff}}{\partial \bar{\phi}^2} |_{\bar{\phi}_{min}} \equiv M^2_{eff} = \left({curvature-induced\ mass\ M^2_{1,2} \atop + \ radiative\ corrections} \right) \qquad (5.17)$$

It measures the curvature of the effective potential at a minimum energy state ($\bar{\phi} = 0$ for the symmetric state, or the false vacuum, $\bar{\phi} = \bar{\phi}_{min}$ for the broken-symmetry state or the true vacuum.) The effective mass is defined to include radiative corrections to the same order corresponding to the effective potential. (This quantity is called the generalized susceptibility function in condensed matter physics.) The critical point of a system is reached when $\Xi \to \infty$ or $M_{eff} \to 0$. In flat or open spaces or for bulk systems, the critical point can be reached without restriction from the geometry (note that in dynamical situations, exponential expansion can effectively introduce a finite size effect equivalent to event horizons, see [48]). However, in spaces with compact dimensions, the correlation length of fluctuations can only extend to infinity in the remaining non-compact dimensions, and thus the critical behavior becomes effectively equivalent to a lower d-dimensional system. One can also think of Ξ as the Compton wavelength $\Lambda = 2\pi/M_{eff}$ of a system of quasi-particles with effective mass M_{eff}. Any fine structure of the background spacetime with scale L is relevant only if $\Lambda \leq L$. Thus when Λ is small or $\eta \ll 1$, (far away from critical point, at higher energy, with higher mode contributions) it sees the details of a spacetime of full dimensionality. At this wavelength, the apparent size of the universe is large in both compact and non-compact dimensions. When $\Lambda \to \infty$ or $\eta \gg 1$ (near critical point, IR limit, lowest mode dominant) finer features in the compact dimensions will not be important. The apparent 'size' of the universe will be measured by the non-compact directions and the EIRD is measured by the number of non-compact dimensions. The value of η getting very large is an indication of when dimensional reduction can take place.

Notice that in flat space (R^d) critical phenomena the effective potential V_{eff} (free energy density) depends on the coupling constants of fields which run with energy and temperature. In curved-space coupling parameters run also with curvature or the scale length of the space. This makes the concept of EIRD even more interesting, as there is now an interplay between Ξ and L, and η can either decrease or increase with curvature. For example, for $\lambda\phi^4$ fields in the Einstein universe near the symmetric state $\bar{\phi} = 0$, the EIRD is equal to 1 but near the global minimum of broken symmetry state $\bar{\phi} = \bar{\phi}_{min}$ is equal to 4. Near the symmetric state, $\eta \gg 1$ signifies reduction of EIRD to one. This is consistent with the theorem of Hohenberg, Mermin and Wagner (for statistical mechanics on a lattice) and Coleman (for continuum field theory) which states that in dimensions less than or equal to two, the infrared divergence of the scalar field is so severe that there could be no possibility of spontaneous symmetry breaking: the only vacuum expectation value for $\bar{\phi}$ allowed is zero. Away from the region $\bar{\phi} \simeq 0$ the one-dimensional behavior no longer prevails. Indeed a global minimum of the effective potential exists at $\bar{\phi}_{min}$. Near $\bar{\phi}_{min}$, $\eta \ll 1$ and decreases with curvature. Thus the apparent size of the universe near the global minimum actually increases with increasing curvature. There is therefore no dimensional reduction and the system has a full 4-dimensional IR behavior. A transition to the asymmetric ground state is not precluded as symmetry breaking via tunneling is in principle possible. The complete picture extending from $\bar{\phi} = 0$ to $\bar{\phi} = \bar{\phi}_{min}$ is a combination of one dimensional and four dimensional infrared behavior. Similar arguments can be applied to other spacetimes or field theories. Using this notion one can understand, for example, why it is often said that at high temperatures (small radius limit of S^1) the finite temperature theory becomes an effective three-dimensional theory.

Journal of Physics: Conference Series **330** (2011) 012003 doi:10.1088/1742-6596/330/1/012003

5.3. Dimensional reduction: an eigenvalue analysis

In the above we have introduced the notion of effective infrared -dimension and suggested the ratio of the correlation length Ξ to the geometric scale of spacetime L as a measure of the conditions for the system to behave effectively as in a reduced dimension in the infrared regime. We suggested that for product spaces $R^d \times B^b$ with some noncompact dimension d, the effective IR dimension is usually just d. We will now verify this assertion by analyzing the spectrum of the fluctuation operators in these spacetimes directly.

Consider for simplicity *direct product spaces.* Examples are a) physical cosmological spacetimes with topology $R^1(time) \times S^3(or\ R^3, H^3, T^3)$, b) Kaluza-Klein cosmology with M^4 (Minkowski) $\times S^7$ (or other internal space), c) finite-temperature (imaginary-time) theory has $R^3 \times S^1$. Take a simple example $S^2 \times S^1$ for illustration. (This could be the spatial geometry of a "handled" Gowdy universe.) Similar reasoning can be extended to a wide range of product spaces.

From (5.7) the wave operator A governing the fluctuation fields in the large N limit is $A \equiv \nabla^2 + M_2^2$, where ∇^2 is the Laplace-Beltrami operator on a general curved spacetime. For $S^2 \times S^1$ with radii a_2 and a_1 respectively, ∇^2 is a sum of the total angular momentum operator L on S^2, and L_z on S^1

$$\nabla^2 = \frac{L^2}{a_2^2} + \frac{L_z^2}{a_1^2} \tag{5.18}$$

The eigenfunction is a product of $Y_{\ell m}(\theta, \phi)e^{in\chi}$ belonging to the eigenvalues

$$\kappa_{\mathbf{n}}^2 = \frac{\ell(\ell+1)}{a_2^2} + \frac{n^2}{a_1^2}\,, \tag{5.19}$$

where $\mathbf{n} = (\ell, m, n), \ell = 0 \cdot \cdot \infty, m = -\ell$ to $\ell, n = 0 \pm 1, \cdots$ The eigenvalues of A are then $\lambda_{\mathbf{n}} = \kappa_{\mathbf{n}}^2 + M_2^2$. In the infrared region, we are interested in the contribution of the lowest eigenvalue (zero mode) to the effective potential. We will consider the two limiting cases of 1) $S^2 \times R^1$ and 2) $R^2 \times S^1$ obtained when a_1 and $a_2 \to \infty$ respectively, and show that the EIRD is equal to 1 and 2 respectively.

The effective potential V_{eff} can be constructed from the ζ function $\zeta(0)$ and its derivative $\zeta'(0)$ by (5.8).

$$\zeta(\nu) = \sum_{\mathbf{n}}(\mu^{-2}\lambda_{\mathbf{n}})^{-\nu} = \mu^{2\nu}\sum_{\ell,n}(2\ell+1)[\frac{\ell(\ell+1)}{a_2^2} + \frac{n^2}{a_1^2} + M_2^2]^{-\nu} \tag{5.20}$$

where $2\ell+1$ is the degeneracy of m. In the limiting cases the summation over discrete quantum numbers will be replaced by integrals of the form $\int k^{D-1}dk$. We will derive the dimensionality of the reduced system by finding D. Thus in case a) $a_1 \to \infty$, the lowest eigenvalues belong to the band $\ell = 0$ as $a_1 \to \infty$, $k = n/a_1$ assumes continuous value and $\zeta(\nu)$ becomes

$$\zeta(\nu) \sim a_1 \int_{-\infty}^{\infty} dk(k^2 + M_2^2)^{-\nu},\ a_1 \to \infty, \quad M_2 a_2 << 1 \tag{5.21}$$

This is a one-dimensional integral. In case (b) $a_2 \to \infty$, the lowest eigenvalues come from the lowest band n=0 given by the first term in (5.19), where ℓ assumes continuous values as $a_2 \to \infty$. Now call $k = \ell/a_2, \zeta(\nu)$ becomes $(n=0)$

$$\zeta(\nu) \sim 2a_2 \int_o^{\infty} dkk(k^2 + M_2^2)^{-\nu},\ a_2 \to \infty\,,\ M_2 a_1 << 1 \tag{5.22}$$

The extra factor of k comes from the degeneracy of m for nonzero ℓ. The integral is thus 2-dimensional, as expected.

Physically, the two limiting cases may represent two different symmetry states of the system: one with cylindrical symmetry where the quantum states with good quantum number m are eigenstates of the L_z operator, the other with spherical symmetry with good quantum numbers ℓ associated with the L^2 operator (on S^2). As a result of symmetry breaking the original system can end up in one of these states with a different symmetry. External perturbations with a particular symmetry and sufficient magnitude will influence the selection of the end state of the system. A simple example in elementary quantum mechanics is the Zeeman versus the Paschen-Bach effects associated with many electron atoms subjected to an external magnetic field. The total magnetic quantum number will become a good quantum number at very strong fields. More sophisticated examples can be found in the consideration of classes of internal spaces admitting larger symmetry groups with the strong and electroweak gauge groups as subgroup in the Kaluza-Klein theory. By the same analysis, it is not difficult to see that the EIRD of any homogeneous cosmology with compact spatial section is one, the spatial metrics of the Einstein-Rosen waves $S^2 \times R^1$, the Kantowski-Sachs universe with spatial metric $d\ell^2 = dz^2 + a^2 d\Omega_2^2$, black hole spacetimes and the whole class of stationary axisymmetric metrics in relativistic astrophysics have EIRD=2. Similarly, the (imaginary time) finite temperature theory $R^3 \times S^1$ has EIRD three and the Kaluza-Klein theory on $M^4 \times B^b$ with compact internal space B has EIRD four.

It is important to recognize that this infrared dimensional reduction feature is not due directly to the topology of the underlying space, but rather the 'finite size' of the system. Spaces which are not direct product spaces but which can approach product spaces in the limit of extreme deformations or large perturbations have similar IR behavior. In the example of direct product spaces shown above, $S^2 \times S^1$ can be shown to be the extreme deformation limit of S^3: The eigenfunctions of S^3 which are the hyperspherical harmonics $Y_{n\ell m}(\chi, \theta, \phi)$ can be expressed as a product of the spherical harmonics $Y_{n\ell m}(\theta, \phi)$ and the Gegenbauer polynomial $G_{n\ell}(\chi)$. The eigenfunctions $e^{in\chi}$ on S^1 we saw above are obtained as limits of $G_n(\chi)$ as the angular momentum of S^2 is decoupled from the"radial" equations governing $G_{n\ell}$. Another example given in [47] is the (static) Taub universe which is a deformed Einstein Universe. An eigenvalue analysis there provides a vivid illustration of this infrared dimensional reduction effect. To the extent that the large scale features of a quantum system shows up near criticality this reveals the surprising effect of its underlying spacetime structure.

6. Summary

In our first paper we explored macroscopic quantum phenomena (MQP) from a large N perspective, where N is the number of particles or components in a many-body quantum system. In this paper we explore MQP from the perspective of correlations and interactions which exist between components of the quantum system, how the coupling strength varies (runs) with energy or scale, especially how its infrared behavior at the critical point determines or affects the macroscopic features of a quantum system. We began with a review of the way how N, the number of particles in a quantum system, enters in the thermodynamic functions in quantum statistical mechanics. The (grand) partition function of a system at finite temperature (and chemical potential) in the (grand) canonical ensemble contains an explicit dependence of N, with proper factors accounting for whether the particles are distinguishable or identical. We note that the meaning of quantum in QSM refers specifically to spin-statistics (boson or fermions) and distinguishability, whereby classical and quantum are clearly divided. This restricted sense of quantumness in QSM is different from the other meanings or criteria of 'quantumness' which define MQP, i.e., quantum coherence, fluctuations, correlation and entanglement associated with the full many-body wave function of a quantum system.

We then explore the issue of how MQP can be measured or understood from the behavior of quantum correlations which exist in a quantum system and how the interaction strengths change with energy or scale, under ordinary situations and when the system is near its critical point. Noting that quantum correlation is further related to quantum fluctuations and quantum coherence, we used three examples from known results to illustrate these aspects: a) the existence of an H-theorem at the next-to-leading order large N expansion for an $O(N)$ quantum mechanical model, with entropy generation signifying the emergence of macroscopic (thermodynamic) properties. b) the nonequilibrium dynamics of N atoms in an optical lattice under the large $\mathcal{N}$ (field components), $2PI$ and second order perturbative expansions, illustrating how N and $\mathcal{N}$ enter in these three aspects of quantum correlations, fluctuations and coupling strength. c) the infrared behavior of an interacting quantum system, we discuss the conditions where dimensional reduction shows up. The effective IR dimension is determined by the spectrum of the Laplacian of an interacting quantum system, in particular, whether a lowest eigenmode exists. On the same theme, we also discuss how the effective field theory concept bears on MQP: the running of the coupling parameters with energy or scale imparts a dynamical-dependent and an interaction-sensitive definition of 'macroscopia'.

Quantum entanglement will be the theme of our third essay where we intend to focus on how this uniquely quantum feature shows up in quantum many-body systems and inquire to what extent it quantifies a quantum system as micro, meso versus macro. In this enquiry we will necessarily also touch on related subjects such as 'small systems' and emergent concepts in the new field known as quantum thermodynamics (ostensibly this is not the thermodynamics of quantum systems, treated in QSM textbooks, where quantum pertains to spin-statistics of particles). We can see why this is relevant to MQP in a simplified way: thermodynamics is a macroscopic theory valid when the volume V and number of constituents N both go to infinity while their ratio N/V remains constant. Quantum dynamics is used for the description of micro-objects or small systems. As one increases the 'size' (number and volume, say) of a quantum system, in what parameter ranges will a thermodynamic description of such a many body quantum system become useful, at least begin to make sense? These inquires into the meaning of MQP will unavoidably lead us to confront some foundational issues of quantum and statistical mechanics.

Acknowledgments

We thank Esteban Calzetta for discussions at the initial stage of this work. BLH enjoyed the warm hospitality of the National Center for Theoretical Sciences (South) and the Department of Physics of National Cheng-Kung University, Tainan, Taiwan while part of this work was done. This work is supported by the National Science Council of Taiwan under grant NSC 97-2112-M-006-004-MY3, the US NSF grants PHY-0801368 to the University of Maryland. Two of the authors spent some time at the KITPC program on Quantum Open Systems supported by the Project of Knowledge Innovation Program (PKIP) of Chinese Academy of Sciences, Grant No. KJCX2.YW.W10.

[1] Chou C H, Hu B L and SubaşıY 2011 *J. Physics* (Conf. Series) for DICE 2010 meeting, ed. Elze T. *Preprint* quant-ph/1106.0556v1
[2] Huang K 1987 *Statistical Mechanics* 2nd ed. (Wiley)
[3] Zurek W H 1991 *Phys. Today* **44** 36
[4] Guilini D et al 2003 *Decoherence and the Appearance of a Classical World in Quantum Theory* 2nd ed. (Berlin: Springer-Verlag)
[5] Pathria R K and Beale P D 1996 *Statistical Mechanics* 2nd ed. (Butterworth-Heinemann)
[6] Calzetta E, Ho Kwan-yuet and Hu B L 2010 *J. Phys.* **B**: At. Mol. Opt. Phys. **43**, 095004.
[7] Hu B L 2009 *J. Phys. Conf. Ser.* **174** 012015 *Preprint* gr-qc/09030878
[8] Calzetta E, Hu B L and Mazzitelli F D 2001 *Physics Report* **352** 459-520 *Preprint* hep-th/0102199
[9] Hu B L 1991 Coarse-Graining and Backreaction in Inflationary and Minisuperspace Cosmology *Invited*

Lectures at the Seventh International Latin-American Symposium on General Relativity (SILARG VII) Proceeding appeared as *Relativity and Gravitation: Classical and Quantum* J. D'Olivio et al (eds) (Singapore: World Scientific)
[10] Hu B L 2005 The Universe as an Ultimate Macroscopic Quantum Phenomenon, *Invited talk at the Quantum Physics of Nature (QUPON) Conference* Vienna, Austria
[11] Jacobson T A 1995 Introduction to Black Hole Microscopy *Utrecht* THU-95/23, *Preprint* hep-th/9510026
[12] Calzetta E and Hu B L 1993 Decoherence of correlation histories *Directions in General Relativity* vol. 2 Hu B L and Jacobson T A (eds) (Cambridge: Cambridge Univ. Press) *Preprint* hep-th/9501040
[13] Verstraete F, Popp M and Cirac J I 2004 *Phys. Rev. Lett.* **92** 027901
[14] Calzetta E and Hu B L 1999 *Phys. Rev.* D **61** 025012
[15] Calzetta E and Hu B L 1995 Correlations, decoherence, dissipation and noise in quantum field theory *Heat Kernel Techniques and Quantum Gravity* vol. 4 of *Discourses in Mathematics and Its Applications* ed Fulling S A (A&M Univ. Press, College Station, TX) *Preprint* hep-th/9501040
[16] Calzetta E and Hu B L 2000 *Phys. Rev.* D **61** 025012
[17] Calzetta E A and Hu B L 2003 *Phys. Rev.* D **68** 065027
[18] Mihaila B, Athan T, Cooper F, Dawson J and Habib S 2000 *Phys. Rev.* D **62** 125015
[19] Mihaila B, Dawson J and Cooper F 2001 *Phys. Rev.* D **63** 096003
[20] Balescu R 1975 *Equilibrium and Nonequilibrium Statistical Mechanics* (New York: John Wiley & Sons) Balescu R 1997 *Statistical Dynamics, Matter out of Equilibrium* (London: Imperial College Press)
[21] Baym G and Kadanoff L 1962 *Quantum Statistical Mechanics* (New York: Benjamin)
[22] Calzetta E and Hu B L 1988 *Phys. Rev.* D **37** 2878
[23] Schwinger J 1960 *Field Theory Methods in non-field-theory contexts* Brandeis Summer Institute in Theoretical Physics. Keldysh L V 1964 *Zh. Eksp. Teor. Fiz.* **47** 1515 (1965 *Sov. Phys. JETP* **20** 1018)
[24] Cornwall J M, Jackiw R and Tomboulis E 1974 *Phys. Rev.* D **10** 2428
[25] Hu B L 1989 *Physica* A **158** 399
[26] Hu B L 1994 Quantum statistical fields in gravitation and cosmology *Proc. Third International Workshop on Thermal Field Theory and Applications* Kobes R and Kunstatter (eds) (Singapore: World Scientific) *Preprint* gr-qc/9403061
[27] Hu B L and Sinha S 1995 *Phys. Rev.* D **51** 1587
[28] Kac M and Logan J 1979 Fluctuations *Fluctuation Phenomena* eds Montroll E and Lebowitz J (New York: Elsevier)
[29] Spohn H 1991 *Large Scale Dynamics of Interacting Particles* (Berlin: Springer-Verlag)
[30] Habib S 1990 *Phys. Rev.* D **42** 2566
[31] Aarts G, Ahrensmeier D, Baier R, Berges J and Serreau J 2002 *Phys. Rev.* D **66** 045008
[32] Hu B L and Pavon D 1986 *Phys. Lett.* B **180** 329
[33] Habib S, Kluger Y, Mottola E and Paz J 1996 *Phys. Rev. Lett.* **76** 4660
[34] Habib S 2004 Gaussian dynamics is classical dynamics *Preprint* quant-ph/0406011
[35] Calzetta E and Hu B L 2008 *Nonequilibrium Quantum Field Theory* (Cambridge: Cambridge University Press)
[36] Fisher M P A, Weichman P B, Grinstein G and Fisher D S 1989 *Phys. Rev.* B **40** 546 Oosten D van, Straten P van der and Stoof H T C 2001 *Phys. Rev.* A **63** 053601
[37] Rey A M, Hu B L, Calzetta E, Roura A and Clark C W 2004 *Phys. Rev.* A **69** 033610
[38] Morgan S A 2000 *J. Phys* B **57** 3847
[39] Hutchinson D A W *et al.* 2000 *J. Phys* B **57** 3825
[40] Hohenberg P and Martin P 1965 *Ann. Phys. (N.Y.)* **34** 291
[41] Shi H and Griffin A 1998 *Phys. Rep* **304** 1 Griffin A 1995 *Phys. Rev.* B **53** 9341
[42] Giorgini S 1997 *Phys. Rev.* A **57** 2949 Giorgini S 2000 *Phys. Rev.* A **61** 063615
[43] Proukakis N P and Burnett K 1995 *J. Res. Natl. Inst. Stand. Technol.* **101** 457 Proukakis N P and Burnett K 1998 *Phys. Rev.* A **57** 1230 Kohler T and Burnett K 2002 *Phys. Rev.* A **65** 033601
[44] Calzetta E, Hu B L and Ramsey S A 2000 *Phys. Rev.* D **61** 125013
[45] Wright E M, Wong T, Collett M J, Tan S M and Walls D F 1997 *Phys. Rev.* A **56** 591
[46] Smerzi A and Raghavan S 2000 *Phys. Rev.* A **61** 063601
[47] Hu B L and O'Conner D J 1986 *Phys. Rev. Lett.* **56** 1613
[48] Hu B L 1987 Dynamical Finite Size Effect, Inflationary Cosmology and Thermal Particle Production, invited talk given at the CAP-NSERC Summer Institute in Theoretical Physics, Edmonton, Canada. Proceedings edited by K. Khanna, G. Kunstatter and H. Umezawa (Singapore: World Scientific Publishing Co. 1988).

Journal of Physics: Conference Series **330** (2011) 012004 doi:10.1088/1742-6596/330/1/012004

Quantum states and local projective measurement in a relativistic field

Shih-Yuin Lin

Department of Physics, National Changhua University of Education, Changhua 50007, Taiwan

E-mail: sylin@cc.ncue.edu.tw

Abstract. In an explicitly covariant system with point-like detectors coupled to a relativistic quantum field, the Gaussian states of the combined system and the reduced states of the detectors derived from them are consistent, but not have to be covariant, in different frames of different observers. Even in the presence of spatially local projective measurements, which result instantaneous wave functional collapses on different time-slices in different reference frames, the consistency still holds.

1. Introduction

Recently the new field on the relativistic quantum information (RQI) is emerging between relativity and quantum information science (QIS) [1]. People are applying the knowledge newly developed in the QIS to relativistic systems, to broaden the range of applications of the QIS on the one hand, and to deepen understanding on both relativity and quantum physics on the other. The central issue of the RQI is the consistency of relativistic locality and quantum nonlocality. The latter in quantum mechanics (QM) manifests when combining quantum entanglement of two or more parties with quantum measurement on one of these parties. This is why most of the efforts in RQI have been put on quantum entanglement in different reference frames: How to transform the measures or degrees of entanglement covariantly [1, 2]? How to entangle and what to be entangled for quantum fields living in a non-trivial spacetime such as the Rindler space or the black hole geometry [3, 4]? What are the entanglement dynamics of quantum fields in dynamical background spacetime such as the expanding universe [5, 6, 7], and the entanglement dynamics of atoms or detectors as an open quantum system [8, 9, 10, 11, 12, 13]? In the extreme cases such as black holes or even worm holes, can the RQI offer new insights into them [14, 15, 16, 17]?

As much attention has been drawn to quantum entanglement, quantum measurement in relativistic systems is also of great interest. Aharonov and Albert have shown that indirect measurement on quantum objects localized in space and time is consistent with relativistic QM [18] and quantum states make sense only within a given frame. In Ref. [19] Aharonov and Albert further concluded that there is no covariant description of measurement in terms of time evolution of quantum states. Even in the simplest case they illustrated, with no wave function collapse between initial and final time-slices, they found that quantum states have to be parametrized by many local times if the system consists of many local quantum objects. Diósi pointed out that in this case it would be simpler to describe such processes using Heisenberg operators, which are naturally covariant in a relativistic model [20].

One kind of models quite popular for studying entanglement dynamics in RQI is the detector theory [21, 22], where the detectors (point-like quantum objects with internal degrees of freedom) are always coupled with a relativistic quantum field locally at their positions. One may wonder whether instantaneous projective measurement is still consistent in the presence of relativistic quantum fields [23]. This question concerns those direct projective measurements in the interaction region rather than in the asymptotic region, while such kind of measuring process cannot be described by a Hamiltonian. To our knowledge, a general argument for the consistency in such situations is still lacking. Thus we are looking at the consistency in a specific detector-field model below by analyzing the evolution of the Gaussian states with or without projective measurement in a direct extension of standard QM.

A brief introduction to our toy model and the mode function expansion we are using are given in Sections 2 and 3 respectively. Then we discuss the time-slicing independence of the reduced state of a point-like object without being measured in Section 4. The consistency of quantum states of the combined system and the reduced states of the detectors when we extend the instantaneous projective measurement in QM to relativistic quantum field theory are illustrated in Sections 5 and 6. Finally in Section 7 we summarize our findings and apply them to the case of quantum teleportation using two entangled detectors with one at rest, the other uniformly accelerated in Minkowski space.

2. RSG detector theory in Minkowski space

To make our discussions more concrete, let us consider a model with one or more Raine-Sciama-Grove (RSG) detectors [24] coupled to a massless scalar field in (1+1)D Minkowski space, described by the action

$$
\begin{aligned}
S &= -\frac{1}{2}\int d^2x\sqrt{-g}\partial_\alpha\Phi\partial^\alpha\Phi + \sum_{\mathbf{d}}\left\{\frac{1}{2}\int d\tau_{\mathbf{d}}\left[(\partial_{\mathbf{d}}Q_{\mathbf{d}})^2 - \omega_{\mathbf{d}}^2 Q_{\mathbf{d}}^2\right] + \right.\\
&\left. \lambda\int d\tau_{\mathbf{d}}\partial_{\mathbf{d}}Q_{\mathbf{d}}\int d^2x\Phi(x^0,x^1)\delta^2(x^\alpha - z_{\mathbf{d}}^\alpha(\tau_{\mathbf{d}}))\right\}, \qquad (1)
\end{aligned}
$$

where $\alpha = 0, 1$, g is the determinant of the metric tensor $g_{\alpha\beta}$ of the background spacetime, Φ is a massless scalar field, $Q_{\mathbf{d}}$ is the harmonic oscillator (HO) inside the detector $\mathbf{d}$ with $\mathbf{d} = A$ for one-detector case, $\mathbf{d} = A, B$ for two-detector case, also $\tau_{\mathbf{d}}$ and $z_{\mathbf{d}}^\alpha(\tau_{\mathbf{d}})$ are the proper time and the prescribed trajectory of the detector $\mathbf{d}$, respectively, and $\partial_{\mathbf{d}} \equiv d/d\tau_{\mathbf{d}}$. The combined system in (1) is a linear system. It is so simple that in some cases analytic results have been obtained in the whole parameter range without any approximation, but still it is complicated enough to give nontrivial results and insights.

The conjugate momenta of Φ_{x^1} and $Q_{\mathbf{d}}$ can be obtained from the action (1), then one can write down the Hamiltonian defined on the whole time-slice $x^1 \in \mathbf{R}$ and evolving in time x^0. Quantizing the Hamiltonian gives the Schrödinger equation, whose solutions are the wave functionals of the detectors and the field $\psi[Q_{\mathbf{d}}, \Phi_{x^1}; x^0]$. However, different observers may have different reference frames, where time-slices are differently foliated. Since the field is defined on the whole time-slices, the Hamiltonian as well as the quantum state or the wave functional of the field in the Schrödinger equation are also defined on the whole time-slice. So the quantum states and their histories for the same system could look very different in different frames. Moreover, if the postulate of projective measurement in QM can be extended to relativistic quantum field theory, then the wave functional of a field on the whole time-slice would be collapsed by a measurement localized in space if the relevant degrees of freedom are entangled. This implies that the wave functionals in different frames could collapse on different time-slices passing through the same measurement event, if the event is local in spacetime.

Journal of Physics: Conference Series **330** (2011) 012004 doi:10.1088/1742-6596/330/1/012004

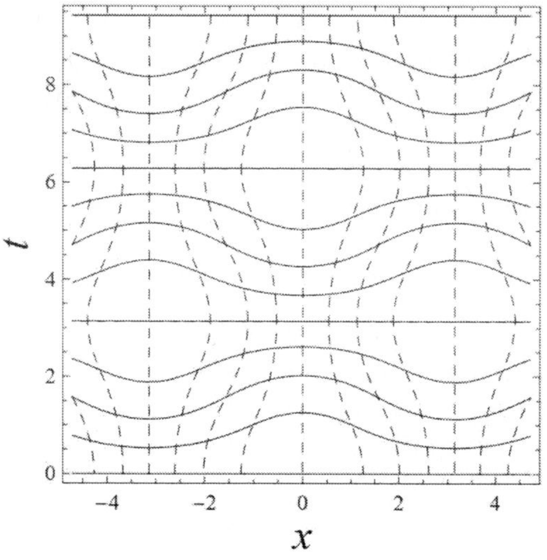

Figure 1. The alternative coordinate (η, ξ) given by (2) with $A = 1/2$ in t-x diagram of (1+1)D Minkowski space. The solid and dashed curves are constant η and ξ hypersurfaces, respectively.

It has been shown by Aharonov and Albert [18] that quantum states in relativistic quantum mechanics make sense only in a given frame. Quantum states defined on two different time slices but intersecting at some spacetime points could be very different even for the sector of the dynamical variables defined right on the intersecting region [19]. Thus one can compare two wave functionals of a field defined in two different frames only when the time-slices they are living on are totally the same. For example, in an alternative coordinate in (1+1) dimensional Minkowski space,

$$\eta = t - A \sin t \cos x, \qquad \xi = x - A \sin x \cos t, \tag{2}$$

with constant $A < 1$ (illustrated in Figure 1,) η-slices and conventional t-slices will overlap at $t = \eta = n\pi$ with integer n. Off those moments, η-slices are the wavy ones in Figure 1, where t-slices would be the horizontal straight lines. So if the initial state is assigned at $t = 0$ time-slice, both the observer in conventional t-x frame and the observer in the η-ξ frame will agree with the same initial state. Then at $t = \eta = \pi, 2\pi, 3\pi, \ldots$, two observers can make a clear comparison on the quantum states defined on that time-slice.

3. Mode functions

From the Hamiltonian the Heisenberg equations of motion for operators read

$$\partial_{\mathbf{d}} \hat{Q}_{\mathbf{d}}(\tau_{\mathbf{d}}) + \omega_{\mathbf{d}}^2 \hat{Q}_{\mathbf{d}}(\tau_{\mathbf{d}}) = -\lambda \partial_{\mathbf{d}} \hat{\Phi}(z_{\mathbf{d}}^{\alpha}(\tau_{\mathbf{d}})), \tag{3}$$

$$\sqrt{-g}\Box\hat{\Phi}(x^{\alpha}) = \lambda \sum_{\mathbf{d}} \int d\tau_{\mathbf{d}} \partial_{\mathbf{d}} \hat{Q}_{\mathbf{d}} \delta^2\left(x^{\alpha} - z_{\mathbf{d}}^{\alpha}(\tau_{\mathbf{d}})\right) \tag{4}$$

where $\Box \equiv \sqrt{-g}^{-1}\partial_{\alpha}\sqrt{-g}g^{\alpha\beta}\partial_{\beta}$. By virtue of the linearity of the system, operators at x^0 are linear combinations of the operators defined at the initial moment [25]:

$$\hat{Q}_{\mathbf{d}}(\tau_{\mathbf{d}}(x^0)) = \sum_{\mathbf{d}'} \left[\phi_{\mathbf{d}}^{\mathbf{d}'}(\tau_{\mathbf{d}})\hat{Q}_{\mathbf{d}'}^{[0]} + f_{\mathbf{d}}^{\mathbf{d}'}(\tau_{\mathbf{d}})\hat{P}_{\mathbf{d}'}^{[0]}\right] + \int dy \left[\phi_{\mathbf{d}}^{y}(\tau_{\mathbf{d}})\hat{\Phi}_{y}^{[0]} + f_{\mathbf{d}}^{y}(\tau_{\mathbf{d}})\hat{\Pi}_{y}^{[0]}\right], \tag{5}$$

$$\hat{\Phi}_{x^1}(x^0) = \sum_{\mathbf{d}'} \left[\phi_{x^1}^{\mathbf{d}'}(x^0)\hat{Q}_{\mathbf{d}'}^{[0]} + f_{x^1}^{\mathbf{d}'}(x^0)\hat{P}_{\mathbf{d}'}^{[0]}\right] + \int dy \left[\phi_{x^1}^{y}(x^0)\hat{\Phi}_{y}^{[0]} + f_{x^1}^{y}(x^0)\hat{\Pi}_{y}^{[0]}\right], \tag{6}$$

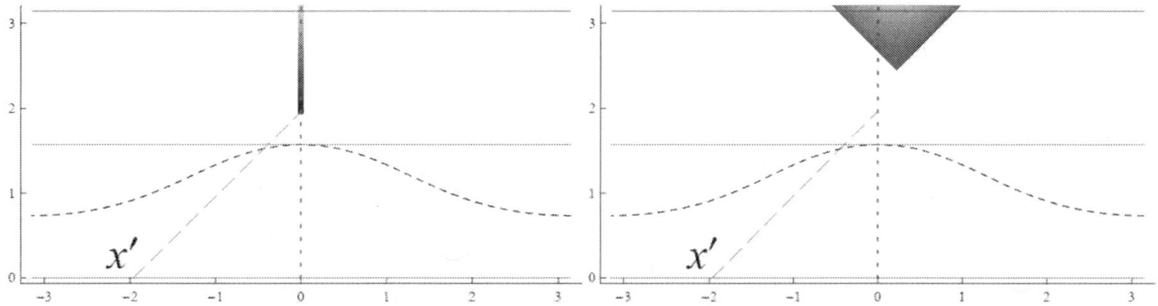

Figure 2. The diagrams represent the mode functions of the oscillators inside the detector $\phi_A^{x'[20]}$ (left), which generate the retarded field (the inhomogeneous solution) $\phi_x^{x'(1)[20]}$ (right). Here the gray horizontal lines denote the t-slices at $t = \eta = 0$, $t = \pi/2$, and $t = \eta = \pi$, the dashed wavy lines are η-slices with $\eta = (\pi - 1)/2$ (here we choose $A = 1/2$ in (2)), and the dotted vertical lines are worldlines of the detector A at $x = \xi = 0$. These mode functions are independent of the data on the t_1- or η_1-slices except those right at the position of the point-like detector $(t, x) = (\pi/2, 0)$, where the local measurement was done. The long-dashed lines in the two plots denote the vacuum fluctuations started from $x = x'$ at $t = 0$.

from which $\hat{P}_{\mathbf{d}}(x^0)$ and $\hat{\Pi}_{x^1}(x^0)$ can be derived. Here $\hat{O}_\mu^{[n]} \equiv \hat{O}_\mu(t_n)$ and all the "mode functions" $\phi_\nu^\mu(x^0)$ and $f_\nu^\mu(x^0)$ are real functions of time, which can be related to those in k-space in Ref. [25] (with different initial conditions.) Inserting the above expansions into (4), one obtains the field equations for mode functions similar to (4), which gives $\phi^\mu_{x^1}(x^0) = \phi^{\mu(0)}_{x^1}(x^0) + \phi^{\mu(1)}_{x^1}(x^0)$ where, for proper initial conditions, the homogeneous solutions are $\phi^{\mathbf{d}(0)}_{x^1} = 0$ and

$$\phi^{y^1(0)}_{x^1}(T) = \int \frac{dk}{2\pi} e^{ik(x^1-y^1)} \cos\omega_k T = \frac{1}{2}\left[\delta(x^1 - y^1 + T) + \delta(x^1 - y^1 - T)\right], \tag{7}$$

with $\omega_k = |k|$. Then one would see that the homogeneous solutions $\phi^{y^1(0)}_{x^1}$ can be interpreted as vacuum fluctuations of the field, and $\phi^{\mu(1)}_{x^1}(x^0)$ behave like retarded relativistic fields sourced by the point-like detectors. Inserting the solutions of $\phi^\mu_{x^1}$ into (3), one finds that $\phi^{\mathbf{d}'}_{\mathbf{d}}$ behaves like those HO damped by the back reaction to the field, and $\phi^y_{\mathbf{d}}$ behave like damped HO driven by vacuum fluctuations of the field $\phi^{y(0)}_{z^1_{\mathbf{d}}(\tau_{\mathbf{d}})}$ at the positions of the detectors $z^1_{\mathbf{d}}(\tau_{\mathbf{d}})$, both HO are living in the point-like detectors and not extended in space.

Comparing the expansions (5) and (6) of two equivalent continuous evolutions, one from t_0 to t_1 then from t_1 to t_2, the other from t_0 all the way to t_2, one can see that the mode functions have the following identities,

$$\phi_\nu^{\mu[20]} = \sum_{\mathbf{d}'}\left[\phi_\nu^{\mathbf{d}'[21]}\phi_{\mathbf{d}'}^{\mu[10]} + f_\nu^{\mathbf{d}'[21]}\pi_{\mathbf{d}'}^{\mu[10]}\right] + \int dx'\left[\phi_\nu^{x'[21]}\phi_{x'}^{\mu[10]} + f_\nu^{x'[21]}\pi_{x'}^{\mu[10]}\right], \tag{8}$$

$$f_\nu^{\mu[20]} = \sum_{\mathbf{d}'}\left[\phi_\nu^{\mathbf{d}'[21]}f_{\mathbf{d}'}^{\mu[10]} + f_\nu^{\mathbf{d}'[21]}p_{\mathbf{d}'}^{\mu[10]}\right] + \int dx'\left[\phi_\nu^{x'[21]}f_{x'}^{\mu[10]} + f_\nu^{x'[21]}p_{x'}^{\mu[10]}\right], \tag{9}$$

where $F^{[mn]} \equiv F(t_m - t_n)$, $\pi^\mu_{\mathbf{d}}(\tau_{\mathbf{d}}(t)) \equiv \partial_{\mathbf{d}}\phi^\mu_{\mathbf{d}}(\tau_{\mathbf{d}}(t)) + \lambda\phi^\mu_{x^1_{\mathbf{d}}(t)}(t)$, $\pi^\mu_{x^1}(t) \equiv \partial_t\phi^\mu_{x^1}(t)$, $p^\mu_{\mathbf{d}}(\tau_{\mathbf{d}}(t)) \equiv \partial_{\mathbf{d}} f^\mu_{\mathbf{d}}(\tau_{\mathbf{d}}(t)) + \lambda f^\mu_{x^1_{\mathbf{d}}(t)}(t)$ and $p^\mu_{x^1}(t) \equiv \partial_t f^\mu_{x^1}(t)$ with $\mu, \nu = \{\mathbf{d}\} \cup \{x^1\}$. Similar identities for π^μ_ν and p^μ_ν can be derived straightforwardly from (8) and (9). Such identities can be interpreted as the Huygens' principle of the mode functions, and can be verified by inserting particular solutions of the mode functions into the identities.

Journal of Physics: Conference Series **330** (2011) 012004 doi:10.1088/1742-6596/330/1/012004

4. Time-slicing scheme and quantum states

Suppose at $t_0 = \eta_0 = 0$ (when $\tau_{\mathbf{d}} \equiv 0$ for all detectors) the combined system of the detectors and the field is initially in a Gaussian state, which could be pure or mixed. Then the quantum state will always be Gaussian because of the linearity of the interaction. It will be the most convenient to work with the quantum state in the (K, Δ)-representation of the Wigner functional [26], which is the double Fourier transform of the conventional Wigner functional,

$$\begin{aligned}\rho[\mathbf{K}, \mathbf{\Delta}] &= \int \mathcal{D}\mathbf{\Sigma} e^{\frac{i}{\hbar}\mathbf{K}\cdot\mathbf{\Sigma}} \bar{\rho}\left[\mathbf{\Sigma} - (\mathbf{\Delta}/2), \mathbf{\Sigma} + (\mathbf{\Delta}/2); x^0\right] \\ &= \exp\frac{-1}{2\hbar}\left[K^\mu \mathcal{Q}_{\mu\nu} K^\nu - 2\Delta^\mu \mathcal{R}_{\mu\nu} K^\nu + \Delta^\mu \mathcal{P}_{\mu\nu}\Delta^\nu\right], \end{aligned} \tag{10}$$

where $\bar{\rho}[(Q_{\mathbf{d}}, \Phi_{x^1}), (Q'_{\mathbf{d}}, \Phi'_{x^1}); x^0]$ is the density matrix in the conventional representation, $\mu, \nu = \{\mathbf{d}\} \cup \{x^1\}$, and the time-dependent factors $\mathcal{Q}_{\mu\nu}$, $\mathcal{P}_{\mu\nu}$, and $\mathcal{R}_{\mu\nu}$ are actually the symmetric two-point correlators $\langle\, A, B\,\rangle \equiv \langle\, AB + BA\,\rangle/2$, for they are obtained by $\langle\, \hat{\Phi}_\mu, \hat{\Phi}_\nu\,\rangle = \frac{\hbar\delta}{i\delta K^\mu}\frac{\hbar\delta}{i\delta K^\nu}\rho[K, \Delta]\big|_{\Delta=K=0} = \hbar\mathcal{Q}_{\mu\nu}$, $\langle\, \hat{\Pi}_\mu, \hat{\Pi}_\nu\,\rangle = \frac{i\hbar\delta}{\delta\Delta^\mu}\frac{i\hbar\delta}{\delta\Delta^\nu}\rho[K, \Delta]\big|_{\Delta=K=0} = \hbar\mathcal{P}_{\mu\nu}$, and $\langle\, \hat{\Pi}_\mu, \hat{\Phi}_\nu\,\rangle = \frac{i\hbar\delta}{\delta\Delta^\mu}\frac{\hbar\delta}{i\delta K^\nu}\rho[K, \Delta]\big|_{\Delta=K=0} = \hbar\mathcal{R}_{\mu\nu}$, with $\hat{\Phi}_{\mathbf{d}} \equiv \hat{Q}_{\mathbf{d}}$ and $\hat{\Pi}_{\mathbf{d}} \equiv \hat{P}_{\mathbf{d}}$. Without projective measurement, the form of the time-dependent factors $\mathcal{Q}_{\mu\nu}$, $\mathcal{P}_{\mu\nu}$, and $\mathcal{R}_{\mu\nu}$ in terms of two-point correlators will keep unchanged from initial to final time-slices.

The reduced state of a detector (say, the detector A) at some moment x^0 is obtained by tracing out all other degrees of freedom in the density matrix,

$$\bar{\rho}_R(Q_A, Q'_A; x^0) = \mathrm{Tr}_\Phi \bar{\rho}[(Q_A, \Phi_{x^1}), (Q'_A, \Phi'_{x^1}); x^0] \equiv \int \mathcal{D}\Phi_{x^1}\bar{\rho}[(Q_A, \Phi_{x^1}), (Q'_A, \Phi_{x^1}); x^0], \tag{11}$$

or in (K, Δ)-representation,

$$\begin{aligned}\rho_R[K^A, \Delta^A; x^0] &= \rho[(K^A, K^{x^1}), (\Delta^A, \Delta^{x^1}); x^0]\big|_{K^{x^1}=\Delta^{x^1}=0} \\ &= \exp\frac{-1}{2\hbar}\left[K^A \mathcal{Q}_{AA}(x^0) K^A - 2\Delta^A \mathcal{R}_{AA}(x^0) K^A + \Delta^A \mathcal{P}_{AA}(x^0)\Delta^A\right]. \end{aligned} \tag{12}$$

The reduced state ρ_R of the detector can be completely determined by the two-point correlators of the detector A, namely, $\mathcal{Q}_{AA}(x^0) = \langle\, \hat{Q}_A^2(\tau_A(x^0))\,\rangle$, $\mathcal{P}_{AA}(x^0) = \langle\, \hat{P}_A^2(\tau_A(x^0))\,\rangle$, and $\mathcal{R}_{AA}(x^0) = \langle\, \hat{Q}_A(\tau_A(x^0)), \hat{P}_A(\tau_A(x_0))\,\rangle$, which are actually parametrized by the proper time τ_A of the detector A. So when the fields and possibly other detectors defined on different time-slices but passing through the worldline of the detector A at the same spacetime point associated with its proper time τ_A are traced out, the same reduced state of A will be obtained. In other words, the reduced state of a detector is independent of the time-slicing outside the detector.

Looking more closely, one can see that the two-point correlators of the detector A are combinations of the mode functions and the initial data, the latter is in the form of correlators evaluated at the initial moment $t = t_0$. E.g.,

$$\langle\, \hat{Q}_A^2(\tau_A)\,\rangle = \phi_A^A(\tau_A)\phi_A^A(\tau_A)\langle\, (\hat{Q}_A^{[0]})^2\,\rangle_0 + \int dx dy\, \phi_A^x(\tau_A)\phi_A^y(\tau_A)\langle\, \hat{\Phi}_x^{[0]}, \hat{\Phi}_y^{[0]}\,\rangle_0 + \ldots. \tag{13}$$

Here $\langle\ \rangle_n$ denotes the expectation values taken from the quantum state right after $t = t_n$. One can see that all the relevant degrees of freedom are those in the past lightcone of the spacetime point $z^\alpha(\tau_A)$ where the reduced state of the detector is defined (cf. Figure 2). So we can say that the reduced state is causal and independent of different degrees of freedom living on different time-slices in different frames because those degrees of freedom are outside the past lightcone and so irrelevant.

Journal of Physics: Conference Series **330** (2011) 012004 doi:10.1088/1742-6596/330/1/012004

The reduced states of two or more detectors and the degree of entanglement derived from them are also parametrized by $\tau_{\mathbf{d}}$ only and independent of time-slicing outside the detectors, provided that there is no projective measurement on any detector or field during the evolution. As we have observed in [10], the evolution of the two-point correlators of the detectors in one coordinate (e.g. $\langle \hat{Q}_A(\tau_A(t)), \hat{Q}_B(\tau_B(t)) \rangle$ in t) could be quite different from the ones in a different coordinate (e.g. $\langle \hat{Q}_A(\tau_A(\eta)), \hat{Q}_B(\tau_B(\eta)) \rangle$ in η), since τ_A and τ_B evolve in different ways in different frames.

5. Projective measurement on detectors

Suppose the detector A is located at $x = 0$ in Figure 1 and started to evolve at its proper time $\tau_A = t = 0$. If a local measurement is done on the detector at some moment $0 < t_1 < \pi$, then which time-slice, t_1- or η_1-slice ($\eta_1 \equiv \eta(t_1)$), will the wave functional of the combined system collapse on? If both collapses occur for different observers, will the two post-measurement states (PMS) be "identical" [23]?

Suppose a Gaussian measurement is done on the detector A at $t = t_1 \in (0, \pi)$ when the quantum state on the t_1-slice collapses to $\tilde{\rho} = \rho_A \otimes \rho_{\bar{A}}$ for the observer in the t-x frame, where we assume

$$\rho_A = \exp -\frac{1}{4\hbar}\left[g_A \left(\Delta^A\right)^2 + \frac{1}{g_A}\left(K^A\right)^2\right] \tag{14}$$

is a squeezed coherent state for the detector A with constant g_A. From (14) one obtains

$$\rho_{\bar{A}} = \exp \frac{-1}{2\hbar}\left[K^{\bar{\mu}}\tilde{\mathcal{Q}}_{\bar{\mu}\bar{\nu}}K^{\bar{\nu}} - 2\Delta^{\bar{\mu}}\tilde{\mathcal{R}}_{\bar{\mu}\bar{\nu}}K^{\bar{\nu}} + \Delta^{\bar{\mu}}\tilde{\mathcal{P}}_{\bar{\mu}\bar{\nu}}\Delta^{\bar{\nu}}\right], \tag{15}$$

with $\bar{\mu}, \bar{\nu} = \{x\} \cup \{\mathbf{d}\} - \{A\}$, $\hbar\mathcal{Q}_{\bar{\mu}\bar{\nu}} = \langle \hat{\Phi}_{\bar{\mu}}^{[1]}, \hat{\Phi}_{\bar{\nu}}^{[1]} \rangle_1$, $\hbar\mathcal{R}_{\bar{\mu}\bar{\nu}} = \langle \hat{\Phi}_{\bar{\mu}}^{[1]}, \hat{\Pi}_{\bar{\nu}}^{[1]} \rangle_1$, and $\hbar\mathcal{P}_{\bar{\mu}\bar{\nu}} = \langle \hat{\Pi}_{\bar{\mu}}^{[1]}, \hat{\Pi}_{\bar{\nu}}^{[1]} \rangle_1$, where

$$\langle \hat{\Theta}_{\bar{\mu}}^{[1]}, \hat{\Theta}_{\bar{\nu}}^{[1]} \rangle_1 = \langle \hat{\Theta}_{\bar{\mu}}^{[10]}, \hat{\Theta}_{\bar{\nu}}^{[10]} \rangle_0 + \frac{I_A^{[1,0]}(\hat{\Theta}_{\bar{\mu}}^{[10]}, \hat{\Theta}_{\bar{\nu}}^{[10]})}{J_A^{[1,0]}}, \tag{16}$$

with $\Theta = \Phi$ or Π, $\mathcal{O}_\mu^{[ns]} \equiv \mathcal{O}_\mu(t_n - t_s)$, and $I_{\mathbf{d}}^{[n,s]}$ and $J_{\mathbf{d}}^{[n,s]}$ being combinations of two-point correlators [23]. For the observer in the η-ξ frame, similar projection occurs but we assume the wave functional collapses on the η_1-slice instead. Then, started at t_1 and η_1, both PMS evolve to $t_2 = \pi = \eta_2$, when t and η-slices overlap so two observers can make a comparison between these two quantum states.

In the conventional (t, x) coordinates of Minkowski space, the two-point correlators at t_2 determining the wave functional can be expressed as combinations of the mode-functions evolving from t_1 to t_2, together with the initial data on the t_1-slice in the form of the correlators of the field at spatial points on the slice, e.g.,

$$\hbar\tilde{\mathcal{Q}}_{xy}(t_2) = \langle \hat{\Phi}_x^{[2]}, \hat{\Phi}_y^{[2]} \rangle_2 = \mathrm{Tr}\hat{\Phi}_x^{[21]}\hat{\Phi}_y^{[21]}\tilde{\rho} \sim \int dx'dy'\phi_x^{x'[21]}\phi_y^{y'[21]}\langle \hat{\Phi}_{x'}^{[1]}, \hat{\Phi}_{y'}^{[1]} \rangle_1 + \cdots \tag{17}$$

where x' and y' are points on the t_1-slice. It turns out that the dependence on the data on the t_1-slice in (17) is removable by expressing the $\langle \, \rangle_1$ correlators as (16) with (5) and (6) inserted, then using (8), (9) and similar identities for π_μ^ν and p_μ^ν to replace the $\int dx'$ and $\int dy'$ terms. The two-point correlators, and therefore the (K, Δ) representation of the quantum state on the t_2-slice, end up with functionals of $F_x^{\mu[20]}$, $F_x^{A[21]}$, $F_A^{\mu[20]}$, $F_A^{\mu[10]}$, $F_A^{A[21]}$ with $F = \phi, f, \pi, p$, and the initial data in the two-point correlators evaluated at t_0.

The quantum state at η_2 in the η-ξ frame has exactly the same functional form, except that the mode functions here are those in the η-ξ frame. Now $F_A^{\mu[20]}$, $F_A^{\mu[10]}$, and $F_A^{A[21]}$ are HO in the point-like detector, $F_x^{A[21]}$ and $F_x^{A[20]}$ are retarded fields sourced from the detector, while $F_x^{x'[20]}$ are superpositions of vacuum fluctuations and retarded fields. All of them are explicitly covariant and independent of the data on the t_1-slice outside the detector. Thus the quantum states collapsed in two different frames are identical up to a coordinate transformation when compared on the same time-slice at $t_2 = \eta_2 = \pi$.

The result in the case with two successive measurements on a single detector at $t = t_1$ and t_2, $t_1 < t_2 < t_3 = \pi$ is similar. The two-point correlators at t_3 can be witten in functionals of the covariant functions and the initial data at t_0. They are independent of the data on t_1-, η_1-, t_2-, or η_2-slice outside the detector, so the quantum state at t_3 is still independent of the time-slices on which the wave functional collapsed.

What happens if two measurement events are done separately on two detectors? Consider the case with two spatially separated detectors, Q_A is put at $x = -\pi$ and Q_B at $x = 0$, so both are at rest in the t-x and the η-ξ frames. Suppose a local measurement is done on Q_A at t_1 and $\eta_1(t_1, -\pi) = t_1 + A\sin t_1$, $0 < A < 1$, another local measurement on Q_B at $t_2 > t_1$ but $\eta_2(t_2, 0) = t_2 - A\sin t_2 < \eta_1$: the two measurement events are spacelike separated so that the time order of these two events can be altered in coordinate transformation. In this case, the two-point functions at $t_3 = \eta_3 = \pi$ can still be written in functionals independent of the data on the t_1-, η_1-, t_2-, or η_2-slice outside the detectors, e.g. in the t-x frame,

$$\begin{aligned}\langle\, \hat\Phi_x, \hat\Phi_y \,\rangle_3 \;=\;& \frac{\hbar}{2} g_B \phi_x^{B[32]}\phi_y^{B[32]} + \frac{\hbar}{2g_B} f_x^{B[32]} f_y^{B[32]} + \frac{\hbar}{2} g_A \phi_x^{A[31]}\phi_y^{A[31]} + \frac{\hbar}{2g_A} f_x^{A[31]} f_y^{A[31]} + \\ & \langle\, \hat\Upsilon_x^{[0]}, \hat\Upsilon_y^{[0]} \,\rangle_0 + \frac{1}{J_A^{[1,0]}} I_A^{[1,0]} \left(\hat\Upsilon_x^{[0]}, \hat\Upsilon_y^{[0]}\right) + \frac{1}{J_B^{[2,1]}} I_B^{[2,1]} \left(\hat\Upsilon_x^{[1]}, \hat\Upsilon_y^{[1]}\right) \end{aligned} \tag{18}$$

where

$$\begin{aligned}\hat\Upsilon_x^{[0]} \;\equiv\;& \hat\Phi_\mu^{[0]}(\phi_x^{\mu[30]} - \phi_x^{A[31]}\phi_A^{\mu[10]} - f_x^{A[31]}\pi_A^{\mu[10]} - \phi_x^{B[32]}\phi_B^{\mu[20]} - f_x^{B[32]}\pi_B^{\mu[20]}) + \\ & \hat\Pi_\mu^{[0]}(f_x^{\mu[30]} - \phi_x^{A[31]} f_A^{\mu[10]} - f_x^{A[31]} p_A^{\mu[10]} - \phi_x^{B[32]} f_B^{\mu[20]} - f_x^{B[32]} p_B^{\mu[20]}). \end{aligned} \tag{19}$$

By expressing all correlators $\langle\,\rangle_1$ in $J_B^{[2,1]}$ and $I_B^{[2,1]}$ in terms of $\langle\,\rangle_0$ by (16), both $J_B^{[2,1]}$ and $I_B^{[2,1]}$ will become functionals of $\hat\Upsilon_x^{[0]}$, too. So the final expression of the wave functional at t_3 depends only on the covariant mode functions in (18) and (19) and vacuum fluctuations from the t_0-slice. Further, using computer algebra it is straightforward to verify that when the retarded mutual influences $F_B^{A[21]}$ vanish, $I_A^{[1,2]}(\hat\Upsilon_\xi^{[2]}, \hat\Upsilon_{\xi'}^{[2]})/J_A^{[1,2]} + I_B^{[2,0]}(\hat\Upsilon_\xi^{[0]}, \hat\Upsilon_{\xi'}^{[0]})/J_B^{[2,0]}$ obtained in the η-ξ frame has the same functional form as $I_A^{[1,0]}(\hat\Upsilon_x^{[0]}, \hat\Upsilon_y^{[0]})/J_A^{[1,0]} + I_B^{[2,1]}(\hat\Upsilon_x^{[1]}, \hat\Upsilon_y^{[1]})/J_B^{[2,1]}$ in the t-x frame. Thus the time order of the spacelike separated measurement events does not matter.

If the two measurement events are timelike separated, then different observers in different frames will recognize the same time order of the two events. In this case the retarded mutual influences will come into play so the resulted PMS will be different from those in the reversed time order, while the consistency of quantum states at the final time-slice is obvious after knowing the results in the previous cases.

6. Reduced state of a detector with its entangled partner being measured

Let us consider a similar setup as the previous case but now only a local measurement is performed on the detector A at some $t_1 \in (0, \pi)$. Then at t_1 or η_1 the quantum state collapses to $\rho_A \otimes \rho_{\bar A}$ on t- or η-slice, depending on the observer's frame. The reduced PMS of the detector B still has the same form as (12) with A there replaced by B. Before the detector B enters the future lightcone of the measurement event on A, namely, when $t = t_M < t_1 + \pi$ in t-x

frame, the two-point correlators of the detector B is either in the original, uncollapsed form, e.g. $\langle \hat{Q}_B^2(t_M - t_0) \rangle_0$, or in the collapsed form evolved from the PMS with the correlators evaluated at t_1 or η_1,

$$\langle \hat{Q}_B^2(t_M) \rangle = \left\langle \left[\sum_{\mathbf{d}} \left(\phi_B^{\mathbf{d}[M1]} \hat{Q}_{\mathbf{d}}^{[1]} + f_B^{\mathbf{d}[M1]} \hat{P}_{\mathbf{d}}^{[1]} \right) + \int dx \left(\phi_B^{x[M1]} \hat{\Phi}_x^{[1]} + f_B^{x[M1]} \hat{\Pi}_x^{[1]} \right) \right]^2 \right\rangle_1$$
$$= \frac{\hbar}{2} g_A \left(\phi_B^{A[M1]} \right)^2 + \frac{\hbar}{2 g_A} \left(f_B^{A[M1]} \right)^2 + \langle (\tilde{\Upsilon}_B^{[M0]})^2 \rangle_0 + \frac{1}{J_A^{[1,0]}} I_A^{[1,0]} \left(\tilde{\Upsilon}_B^{[M0]}, \tilde{\Upsilon}_B^{[M0]} \right), \tag{20}$$

depending on the observer's frame. Here we have used the Huygens' principles (8) and (9), and defined

$$\tilde{\Upsilon}_B^{[M0]} \equiv \hat{\Phi}_\mu^{[0]} \left[\phi_B^{\mu[M0]} - \phi_B^{A[M1]} \phi_A^{\mu[10]} - f_B^{A[M1]} \pi_A^{\mu[10]} \right] + \hat{\Pi}_\mu^{[0]} \left[f_B^{\mu[M0]} - \phi_B^{A[M1]} f_A^{\mu[10]} - f_B^{A[M1]} p_A^{\mu[10]} \right]. \tag{21}$$

Note that before the detector B enters the lightcone, $\phi_B^{A[M1]} = f_B^{A[M1]} = 0$, such that $\tilde{\phi}_B^{\mu[M0]} = \phi_B^{\mu[M0]}$, $\tilde{f}_B^{\mu[M0]} = f_B^{\mu[M0]}$, and $\tilde{\Upsilon}_B^{[M0]} = \hat{Q}_B^{[M0]}$.

No matter in which frame the system is observed, the correlators in the reduced state of the detector B must have become the collapsed ones like (20) at the moment the detector B is entering the future lightcone of the measurement event on the detector A, namely, when $t_M = t_1 + \pi$ in the t-x frame. After that moment the retarded mutual influences will reach B so those $F_B^{A[M1]}$ become nonzero and get involved in the correlators of B, though some information of measurement has entered the correlators of B via I_A and J_A much earlier.

7. Summary and Application

In summary, we found that the Gaussian states of the detector-field system in different frames if started with the same initial state defined on the same fiducial time-slice will evolve to the same state on the same final time-slice (up to a spatial coordinate transformation). Neither the choice of coordinate between the initial and final time-slices nor the presence of local projective measurements during the evolution will violate this consistency. We believe such a consistency could be more generic in relativistic quantum field theory. Furthermore, the reduced states of the detectors are parametrized by the proper times of the detectors and independent of the time-slicing outside the detectors. If one detector of an entangled pair is measured, then the reduced state of the other detector will have different histories for different observers before the moment it enters the future lightcone of the measurement event. After that moment the reduced state of the other detector becomes identical.

An application of the above analysis is quantum teleportation in a relativistic system. Following the setup in Ref. [27] by Shiokawa for quantum teleportation, suppose Alice is at rest in Minkowski space and Rob is uniformly accelerated, each holds one Unruh-DeWitt (UD) detector of an entangled pair (A and B), and Alice holds the third UD detector (C) whose quantum state is unknown. Suppose Alice is teleporting the quantum state of C to Rob. At the moment t_1 before Alice goes beyond the event horizon of Rob (Figure 3 (left)) Alice performs a joint measurement on A and C. Then the wave functional of the combined system (including the quantum fields that these UD detectors are coupled with) will collapse at that moment either on the Minkowski time-slice (gray horizontal line in Figure 3 (left)), or on the quasi-Rindler time-slice (the gray long-dashed line in the same plot: it almost overlaps the Rindler time-slice in the R-wedge but the part in the L-wedge has been bent to the region with positive t to make the whole time-slice located after the initial time-slice), or whatever time-slice may be passing through the same measurement event in some observer's frame. All the PMS of the combined

Journal of Physics: Conference Series **330** (2011) 012004 doi:10.1088/1742-6596/330/1/012004

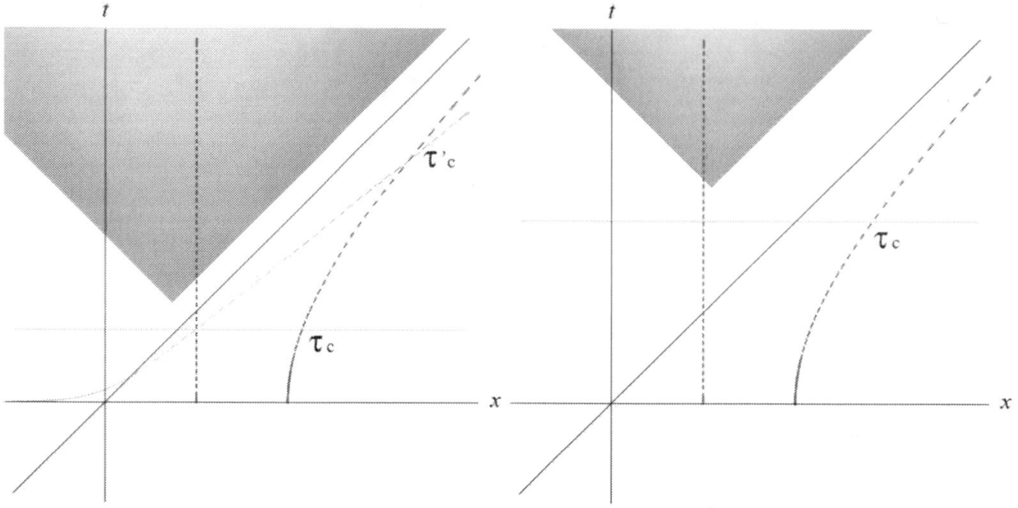

Figure 3. Setup for quantum teleportation from Alice (dotted dark line) to Rob (dashed dark curve). The straight line $t = x$ is the event horizon of Rob, and the shaded region represents the retarded field from the detector A after the joint measurement on A and C by Alice, or the future lightcone of the joint measurement event. The joint measurement is done before the moment that Alice goes beyond the event horizon of Rob (left), or after that moment (right).

system in different frames will evolve to the same state when compared on the same time-slice after the measurement, according to the analysis in Section 5.

The reduced state of the detector B consists of the two-point correlators of Q_B and P_B. It has a sudden change from the uncollapsed to the collapsed one at $\tau_B = \tau_c$ in Figure 3 (left), as observed in the conventional Minkowski frame, or at τ_c', as observed in the quasi-Rindler frame. Such a sudden change occurs at different spacetime points for different observers. In other words, when observed at some moment $\tau_B = \tau_M$ before Rob enters the future lightcone of the joint measurement event, those correlators of B may either in the uncollapsed form, or in a collapsed form as shown in (20), depending on the observer. In the latter case, the correlators of B consist of the mode functions $F_B^{B[M0]}$, $F_B^{x[M0]}$, and $F_x^{B[M0]}$ $(F = \phi, f, \pi, p)$, the information about the outcome of the joint measurement on A (in I_A and J_A), as well as the two-point correlators evaluated on the initial time-slice at t_0 only. Either collapsed or uncollapsed all the correlators of B are independent of the data on the time-slice that the wave functional collapsed on except those right at the position of the detector A.

If Rob never performs any further measurement on B before entering the future lightcone of the joint measurement event done by Alice, certainly he will have no idea that B is in the uncollapsed or collapsed state. But at the moment Rob is entering the future lightcone of the joint measurement event, all these reduced state of B must have become identical, namely, the collapsed one, with the same combination of the mode functions. So quantum teleportation after Rob receives the classical information sent by Alice will give a definite result consistent in all frames.

Suppose Rob performs a measurement on B before entering the future lightcone of the joint measurement event to see which state B is in. In some observer's frames B's reduced state at the moment of Rob's measurement is in the collapsed state, then the outcome will have some dependence of the outcome of A since A and B were entangled initially. In other observer's frames, B is still in the uncollapsed state right before the measurement, then this measurement will result a wave functional collapse of the combined system, later in these frames the outcome of Alice's joint measurement will have the same correlation with the outcome of B. Both kinds of histories interpreted by different observers will be consistent *a posteriori*, but both could not help Rob to conclude B was in the uncollapsed or collapsed state right before the measurement. Rob cannot obtain the full knowledge about the quantum state of B by only one single measurement. If Rob and Alice share an ensemble of many copies of the entangled pairs, in the reference frames that Rob performs the measurement first, using the outcomes Rob can recognize the reduced state of B as uncollapsed by quantum state tomography. In the reference frames that Rob performs the measurement after Alice, the exact collapsed state of one copy of the entangled pair is different from another after the measurements by Alice and not predictable either by Alice or by Rob, while the distribution of the collapsed state is determined by the uncollapsed state of A and B right before Alice's measurement. So the reduced state of detector B recognized by Rob from his measurement on this ensemble of the collapsed states will have no difference from the uncollapsed one. Rob still cannot tell whether the quantum state of each copy of B before his measurement is collapsed or not.

If the joint measurement on A and C is performed after Alice went beyond the event horizon of Rob (Figure 3 (right)), the mutual influences F_A^B will never reach Rob, though it appears that some information of measurement could enter the collapsed reduced state of B via I_A and J_A in (20). Similar to the previous case with Rob still outside the future lightcone of the joint measurement, the functional form of the correlators in the reduced state of B will be suddenly changed at the moment of the projective measurement, and different observers will recognize different spacetime points where this change occurs. But again Rob will never know whether the state of B is in the uncollapsed state or collapsed. When τ_B goes to infinity or after Rob makes a measurement on B, these reduced states in different frames will become identical.

Above discussion indicates that in ultraweak coupling limit with mutual influences neglected, the approximated reduced state of B after Rob enters the future lightcone of the measurement event will be almost independent of the time-slice on which the wave functional collapsed. Indeed, even the wave functional is collapsed by hand on the future lightcone, numerical calculation shows that the approximated reduced state of B after Rob enters the lightcone is almost identical to the one collapsed on the t-slice, within the error of the ultraweak coupling approximation. This will largely simplify the numerical work for quantum teleportation between Alice and Rob.

Acknowledgments

The author would like to thank Bei-Lok Hu for illuminating discussions and useful comments. This work is supported by the National Science Council of Taiwan under the grant NSC 99-2112-M-018-001-MY3 and in part by the National Center for Theoretical Sciences, Taiwan.

References

[1] Peres A and Terno D R 2004 *Rev. Mod. Phys.* **76** 93
[2] Soo C and Lin C-Y 2004 *Int. J. Quantum Info.* **2** 183
[3] Alsing P M and Milburn G J *Phys. Rev. Lett.* **91** 180404
[4] Fuentes I and Mann R B 2005 *Phys. Rev. Lett.* **95** 120404
[5] Ball J L, Fuentes I and Schuller F P 2006 *Phys. Lett.* **A 359** 2550
[6] Campo D and Parentani R 2008 *Phys. Rev.* D **78** 065044
[7] Lin S-Y, Chou C-H and Hu B L 2010 *Phys. Rev.* D **81** 084018
[8] Reznik B 2003 *Found. Phys.* **33** 167
[9] Massar S and Spindel P 2006 *Phys. Rev.* D **74** 085031
[10] Lin S-Y, Chou C-H Chou and B. L. Hu 2008 *Phys. Rev.* D **78** 125025
[11] Lin S-Y and Hu B L 2009 *Phys. Rev.* D **79** 085020
[12] Lin S-Y and Hu B L 2010 *Phys. Rev.* D **81** 045019
[13] Ostapchuk D C M, Lin S-Y, Mann R B and Hu B L 2011 *Entanglement Dynamics with Non-Uniformly Accelerated Detectors* in preparation.

[14] Hayden P and Preskill J 2007 *JHEP* **0709** 120
[15] Lin S-Y and Hu B L 2008 *Class. Quantum Grav.* **25** 154004
[16] Deutsch D 1991 *Phys. Rev.* D **44** 3197
[17] Ralph T C and Meyers 2010 C R *Phys. Rev.* A **82** 062330
[18] Aharonov Y and Albert D Z 1981 *Phys. Rev.* D **24** 359
[19] Aharonov Y and Albert D Z 1984 *Phys. Rev.* D **29** 228
[20] Diósi L 1991 *Phys. Rev.* A **43** 17
[21] Unruh W G 1976 *Phys. Rev.* D **14** 870
[22] DeWitt B S 1979 *General Relativity: an Einstein Centenary Survey* ed Hawking S W and Israel W (Cambridge: Cambridge University Press)
[23] Lin S-Y 2010 *Is Projective Measurement Consistent in a Relativistic Quantum Field?* arXiv:1104.0772.
[24] Raine D J, Sciama D W and Grove P G *Proc. R. Soc. Lond.* A **435** 205
[25] Lin S-Y and Hu B L 2006 *Phys. Rev.* D **73** 124018
[26] Unruh W G and Zurek W H 1989 *Phys. Rev.* D **40** 1071
[27] Shiokawa K 2009 Preprint arXiv:0910.1715

Journal of Physics: Conference Series **330** (2011) 012005 doi:10.1088/1742-6596/330/1/012005

Poincaré gauge theory with even and odd parity dynamic connection modes: isotropic Bianchi cosmological models

Fei-Hung Ho[1] **and James M. Nester**[1,2,3]

[1]Department of Physics, National Central University, Chungli 320, Taiwan
[2]Institute of Astronomy, National Central University, Chungli 320, Taiwan
[3]Center for Mathematics and Theoretical Physics, National Central University, Chungli 320, Taiwan

E-mail: `93242010@cc.ncu.edu.tw`, `nester@phy.ncu.edu.tw`

Abstract. The Poincaré gauge theory of gravity has a metric compatible connection with independent dynamics that is reflected in the torsion and curvature. The theory allows two good propagating spin-0 modes. Dynamical investigations using a simple expanding cosmological model found that the oscillation of the 0^+ mode could account for an accelerating expansion similar to that presently observed. The model has been extended to include a 0^- mode and more recently cross parity couplings. We investigate the dynamics of this model in a situation which is simple, non-trivial, and yet may give physically interesting results that might be observable. We consider homogeneous cosmologies, more specifically, isotropic Bianchi class A models. We find an effective Lagrangian for our dynamical system, a system of first order equations, and present some typical dynamical evolution.

1. Introduction

All the known fundamental physical interactions can be formulated in a common framework: as *local gauge theories*. However the standard theory of gravity, Einstein's general relativity (GR), based on the spacetime metric, is a rather unnatural gauge theory. Physically (and geometrically) it is reasonable to consider gravity as a gauge theory of the local Poincaré symmetry of Minkowski spacetime. A theory of gravity based on local spacetime geometry gauge symmetry, the quadratic Poincaré gauge theory of gravity (PG, aka PGT) was worked out some time ago.

We briefly sketch this theory, and how the search for good dynamical propagating modes led to focusing on the two scalar modes, and then how cosmological models were used to reveal the dynamics of the resulting model.

There is no known fundamental reason why the gravitational coupling should respect parity. With this in mind, the general quadratic PG theory has recently had a major extension to include all possible odd parity couplings. The appropriate odd parity couplings have been incorporated, in particular, into the special case of the dynamically favored two scalar mode model. Here we show how a simple effective Lagrangian can reveal in a cosmological context the dynamics of this extended model.

2. The Poincaré gauge theory

In the Poincaré gauge theory of gravity [1–7], the two sets of local gauge potentials are, for "translations", the orthonormal co-frame $\vartheta^\alpha = e^\alpha{}_i dx^i$, where the metric is $g = -\vartheta^0 \otimes \vartheta^0 + \delta_{ab}\vartheta^a \otimes \vartheta^b$, and, for "rotations", the metric-compatible (Lorentz Lie-algebra valued) connection 1-forms $\Gamma^{\alpha\beta} = \Gamma^{[\alpha\beta]}{}_i dx^i$.

The associated field strengths are the torsion and curvature 2-forms

$$T^\alpha := d\vartheta^\alpha + \Gamma^\alpha{}_\beta \wedge \vartheta^\beta = \frac{1}{2} T^\alpha{}_{\mu\nu} \vartheta^\mu \wedge \vartheta^\nu, \tag{1}$$

$$R^{\alpha\beta} := d\Gamma^{\alpha\beta} + \Gamma^\alpha{}_\gamma \wedge \Gamma^{\gamma\beta} = \frac{1}{2} R^{\alpha\beta}{}_{\mu\nu} \vartheta^\mu \wedge \vartheta^\nu, \tag{2}$$

which satisfy the respective Bianchi identities:

$$DT^\alpha \equiv R^\alpha{}_\beta \wedge \vartheta^\beta, \qquad DR^\alpha{}_\beta \equiv 0. \tag{3}$$

Turning from kinematics to dynamics, the PG Lagrangian density is generally taken to have the standard quadratic Yang-Mills form, which leads to quasi-linear second order equations for the gauge potentials. Qualitatively,

$$\mathscr{L}[\vartheta, \Gamma] \sim \kappa^{-1}[\Lambda + \text{curvature} + \text{torsion}^2] + \varrho^{-1}\text{curvature}^2, \tag{4}$$

where Λ is the cosmological constant, $\kappa = 8\pi G/c^4$, and ϱ^{-1} has the dimensions of action. The field equations obtained by variation w.r.t. the two gauge potentials have the respective general forms

$$\Lambda + \text{curvature} + D\text{torsion} + \text{torsion}^2 + \text{curvature}^2 \sim \text{source energy-momentum density}, \tag{5}$$

$$\text{torsion} + D\text{curvature} \sim \text{source spin density}. \tag{6}$$

From these two equations, with the aid of the Bianchi identities (3), one can obtain, respectively, the conservation of source energy-momentum and angular momentum statements.

Earlier investigations considered only models with even parity terms; the models had in general 10 dimensionless scalar coupling constants. The recent BHN investigation [8] systematically considered all the possible odd parity Lagrangian terms, introducing 7 new pseudoscalar coupling constants. Not all of these coupling constants are physically independent, since there are 3 topological invariants: the (odd parity) Nieh-Yan [9] identity

$$d(\vartheta^\alpha \wedge T_\alpha) \equiv T^\alpha \wedge T_\alpha + R_{\alpha\beta} \wedge \vartheta^{\alpha\beta}, \tag{7}$$

the (even parity) Euler 4-form $R^{\alpha\beta} \wedge R^{\gamma\delta}\eta_{\alpha\beta\gamma\delta}$, and the (odd parity) Pontryagin 4-form $R^\alpha{}_\beta \wedge R^\beta{}_\alpha$.

Early PG Investigations (especially [3, 10]) of the linearized theory identified six possible dynamic connection modes; they carry spin-$2^\pm$, spin-$1^\pm$, spin-$0^\pm$. A good dynamic mode should transport positive energy and should not propagate outside the forward null cone. The linearized investigations found that at most three modes can be simultaneously dynamic; all the acceptable cases were tabulated; many combinations of three modes are satisfactory to linear order. Complementing this, the Hamiltonian analysis revealed the related constraints [11]. Then detailed investigations of the Hamiltonian and propagation [12–15] concluded that effects due to nonlinearities in the constraints could be expected to render all of these cases physically unacceptable except for the two "scalar modes", carrying spin-0^+ and spin-0^-.

In order to further investigate the dynamical possibilities of these PG scalar modes, Friedmann-Lemaître-Robinson-Walker (FLRW) cosmological models were considered. Using a $k = 0$ model it was found that the 0^+ mode naturally couples to the acceleration of the universe and could account for the present day observations [16, 17]; this model was then extended to include the 0^- mode [18].

After developing the general odd parity PG theory, in BHN [8] the two scalar torsion mode PG Lagrangian was extended to include the appropriate pseudoscalar coupling constants.

The BHN Lagrangian [8] has the specific form

$$\mathscr{L}_{\text{BHN}}[\vartheta, \Gamma] = \frac{1}{2\kappa}\left[-2\Lambda + a_0 R + b_0 X - \frac{1}{2}\sum_{n=1}^{3} a_n {}^{(n)}T^2 + 3\sigma_2 T_\mu P^\mu\right] - \frac{1}{2\varrho}\left[\frac{w_6}{12}R^2 - \frac{w_3}{12}X^2 + \frac{\mu_3}{12}RX\right], \tag{8}$$

where R is the scalar curvature and X is the pseudoscalar curvature (specifically $X/6 = R_{[0123]}$ is the magnitude of the one component of the totally antisymmetric curvature), and $T_\mu \equiv T^\alpha{}_{\alpha\mu} = {}^{(2)}T^\alpha{}_{\alpha\mu}$, $P_\mu \equiv \frac{1}{2}\epsilon_{\mu\nu}{}^{\alpha\beta}T^\nu{}_{\alpha\beta} = \frac{1}{2}\epsilon_{\mu\nu}{}^{\alpha\beta(3)}T^\nu{}_{\alpha\beta}$ are the torsion trace and axial vectors.

In the aforementioned work the general field equations were worked out, and then specialized to find the most general 2-scalar mode PG FLRW cosmological model. Here we will consider manifestly isotropic Bianchi models.

3. The PG scalar mode isotropic Bianchi I and IX cosmological models

For earlier PG cosmological investigations see Minkevich and coworkers, e.g., [19–23] and Goenner & Müller-Hoissen [24]; for recent work see [8, 16–18, 25–28].

For the usual FLRW models, although they are actually homogeneous and isotropic, the representation is not manifestly so. Indeed they are merely manifestly isotropic-about-a-chosen-point. In contrast, the representation used here for the isotropic (class A) Bianchi I and IX models have the virtue of being *manifestly homogeneous* and *manifestly isotropic*.

For homogeneous, isotropic Bianchi type I and IX (respectively equivalent to FLRW $k = 0$ and $k = +1$) cosmological models the isotropic orthonormal coframe has the form

$$\vartheta^0 := dt, \qquad \vartheta^a := a\sigma^a, \tag{9}$$

where $a = a(t)$ is the scale factor and σ^j depends on the (not needed here) spatial coordinates in such a way that

$$d\sigma^i = \zeta\epsilon^i{}_{jk}\sigma^j \wedge \sigma^k, \tag{10}$$

where $\zeta = 0$ for Bianchi I and $\zeta = 1$ for Bianchi IX, thus $\zeta^2 = k$, the sign of the FLRW Riemannian spatial curvature.

Because of isotropy, the only non-vanishing connection one-form coefficients are of the form

$$\Gamma^a{}_0 = \psi(t)\,\sigma^a, \qquad \Gamma^a{}_b = \chi(t)\epsilon^a{}_{bc}\,\sigma^c. \tag{11}$$

Here $\epsilon_{abc} := \epsilon_{[abc]}$ is the usual 3 dimensional Levi-Civita anti-symmetric symbol.

From the definition of the curvature (2), one can now find all the nonvanishing curvature 2-form components:

$$R^a{}_b = \dot\chi dt \wedge \epsilon^a{}_{bc}\sigma^c + [\psi^2 - \chi^2]\sigma^a \wedge \sigma_b + \chi\zeta\epsilon^a{}_{bc}\epsilon^c{}_{ij}\sigma^i \wedge \sigma^j, \tag{12}$$

$$R^a{}_0 = \dot\psi dt \wedge \sigma^a - \chi\psi\sigma^b \wedge \epsilon^a{}_{bc}\sigma^c + \psi\zeta\epsilon^a{}_{bc}\sigma^b \wedge \sigma^c. \tag{13}$$

Consequently, the scalar and pseudoscalar curvatures are, respectively,

$$R = 6[a^{-1}\dot\psi + a^{-2}(\psi^2 - [\chi-\zeta]^2 + \zeta^2)], \tag{14}$$

$$X = 6[a^{-1}\dot\chi + 2a^{-2}\psi(\chi-\zeta)]. \tag{15}$$

Because of isotropy, the only nonvanishing torsion tensor components are of the form

$$T^a{}_{b0} = u(t)\delta^a_b, \qquad T^a{}_{bc} = -2x(t)\epsilon^a{}_{bc}, \tag{16}$$

where u and x are the scalar and pseudoscalar torsion, respectively. From the definition of the torsion (1), one can find the relation between the torsion components and the gauge variables:

$$u = a^{-1}(\dot a - \psi), \qquad x = a^{-1}(\chi - \zeta). \tag{17}$$

Note that $a^{-1}\psi = H - u$, where $H = a^{-1}\dot a$ is the *Hubble function*.

From the isotropic assumption, the material energy momentum tensor must have the perfect fluid form. In our present work we focus on the late time behavior. Accordingly we assume that the fluid pressure can be neglected, so that the gravitating material behaves like dust with a density satisfying $\rho a^3 =$ constant. Also, although we expect the spin density to play an important role in the early universe, it is reasonable to assume that the material spin density is negligible at late times.

Journal of Physics: Conference Series **330** (2011) 012005 doi:10.1088/1742-6596/330/1/012005

4. Effective Lagrangian

The dynamical equations for these homogeneous cosmologies could be obtained by imposing the Bianchi symmetry on the general field equations found in [8] from their *Lagrangian density*. On the other hand dynamical equations can (following the same procedure used in [18]) be obtained directly and independently from a classical mechanics type effective *Lagrangian*, which in this case can be simply obtained by restricting the BHN Lagrangian density to the Bianchi symmetry (this step is where the manifestly homogeneous representation plays an essential role). This procedure is known to successful give the correct dynamical equations for all Bianchi class A models (which includes our cases) in GR (see [29]), and it is conjectured to also work equally as well for the PG theory. Our calculations explicity verify this property for the isotropic Bianchi I and IX models. Indeed the equations we obtained in this way are equivalent to those found (at a later date) by BHN for their FLRW models (which are equivalent to our isotropic Bianchi models) by restricting to FLRW symmetry their general dynamical PG equations. This has proved to be a useful cross check.

The *effective Lagrangian* is

$$\begin{aligned} L_{\text{eff}} = (2\kappa)^{-1}(a_0 R + b_0 X - 2\Lambda)a^3 &+ \frac{3}{2\kappa}(-a_2 u^2 + 4a_3 x^2 + 4\sigma_2 u x)a^3 \\ &- \frac{1}{24\varrho}(w_6 R^2 - w_3 X^2 + \mu_3 R X)a^3. \end{aligned} \tag{18}$$

It should be noted that the parameter restrictions $a_2 < 0$, $w_6 < 0$, $w_3 > 0$, and $\mu_3^2 + 4w_3 w_6 < 0$ are *necessary* for the *least action principle*, which requires *positive* quadratic-kinetic-terms.

In the following we often take for simplicity units such that $\kappa = 1 = \varrho$. These factors can be easily restored in the final results by noting that in the Lagrangian they occur in conjunction with certain PG parameters. Hence in the final results one need merely make the replacements

$$\{a_0, a_2, a_3, b_0, \Lambda, \sigma_2\} \to \kappa^{-1}\{a_0, a_2, a_3, b_0, \Lambda, \sigma_2\}, \quad \{w_3, w_6, \mu_3\} \to \varrho^{-1}\{w_3, w_6, \mu_3\}. \tag{19}$$

The effective gravitational Lagrangian has the usual form of a sum of terms homogeneous in "velocities": $L_{\text{eff}} = L_0 + L_1 + L_2$; the associated *energy function* is thus

$$\begin{aligned} \mathcal{E}_{\text{eff}} &:= \frac{\partial L_{\text{eff}}}{\partial \dot\psi}\dot\psi + \frac{\partial L_{\text{eff}}}{\partial \dot\chi}\dot\chi + \frac{\partial L_{\text{eff}}}{\partial \dot a}\dot a - L_{\text{eff}} = L_2 - L_0 \\ &= a^3 \Bigg\{ -3(a_0 - \frac{1}{2}a_2)u^2 - 3a_0 H^2 + 3x^2(a_0 - 2a_3) \\ &\quad + 6uH(a_0 - \frac{1}{2}a_2) + 6(b_0 + \sigma_2)x(H - u) - 3a_0\frac{\zeta^2}{a^2} + \Lambda \\ &\quad - \frac{w_6}{24}\left[R^2 - 12R\left\{(H-u)^2 - x^2 + \frac{\zeta^2}{a^2}\right\}\right] \\ &\quad + \frac{w_3}{24}\left[X^2 + 24Xx(H-u)\right] \\ &\quad - \frac{\mu_3}{24}\left[RX - 6X\left\{(H-u)^2 - x^2 + \frac{\zeta^2}{a^2}\right\} + 12Rx(H-u)\right]\Bigg\}. \end{aligned} \tag{20}$$

Since L_{eff} is time independent, the energy function is conserved. This constant *energy* has the form $-a^3\rho$ where ρ can be identified as the material *energy density*. This relation is consistent with the vanishing divergence of the source fluid energy-momentum tensor, which has been assumed to have the dust form with vanishing pressure.

Making use of the formulas for the torsion and curvature components in terms of the gauge variables (14,15,17), we now obtain the Euler-Lagrange equations. The dynamical equations are the ψ equation:

$$\frac{d}{dt}\frac{\partial L_{\rm eff}}{\partial\dot\psi} = \frac{\partial L_{\rm eff}}{\partial\psi}, \qquad \Longrightarrow$$

$$-\frac{w_6}{2}\dot R - \frac{\mu_3}{4}\dot X = -\left[3(2a_0 - a_2) - w_6 R - \frac{\mu_3}{2}X\right]u - \left[6(b_0+\sigma_2) - \frac{\mu_3}{2}R + w_3 X\right]x. \tag{21}$$

the χ equation:

$$\frac{d}{dt}\frac{\partial L_{\rm eff}}{\partial\dot\chi} = \frac{\partial L_{\rm eff}}{\partial\chi}, \qquad \Longrightarrow$$

$$-\frac{\mu_3}{4}\dot R + \frac{w_3}{2}\dot X = -\left[6(b_0+\sigma_2) - \frac{\mu_3}{2}R + w_3 X\right]u + \left[6(a_0 - 2a_3) - w_6 R - \frac{\mu_3}{2}X\right]x, \tag{22}$$

and the a equation:

$$\frac{d}{dt}\frac{\partial L_{\rm eff}}{\partial\dot a} = \frac{\partial L_{\rm eff}}{\partial a}, \qquad \Longrightarrow$$

$$\begin{aligned}
-3(a_2\dot u - 2\sigma_2\dot x) &= (a_0R + b_0X - 3\Lambda) + 6H(a_2u - 2\sigma_2x)\\
&\quad+\frac{3}{2}(-a_2u^2 + 4a_3x^2 + 4\sigma_2ux) - \frac{1}{24}(w_6R^2 - w_3X^2 + \mu_3RX)\\
&\quad+\frac{a_0}{2}\left[-6(H-u)^2 + 6x^2 - 6\frac{\zeta^2}{a^2}\right] + 6b_0x(H-u)\\
&\quad-\frac{1}{24}(2w_6R + \mu_3X)\left[-6(H-u)^2 + 6x^2 - 6\frac{\zeta^2}{a^2}\right]\\
&\quad+\frac{1}{24}(2w_3X - \mu_3R)\left[12x(H-u)\right].
\end{aligned} \tag{23}$$

The above equations (21,22,23) are 3 *second order* equations for the gauge potentials a, ψ, χ. However they can in an alternative way be used as part of a set of 6 *first order* equations along with the Hubble relation $\dot a = aH$ and the following two relations, obtained by taking the time derivatives of the torsion (17) and using the curvature definitions (14,15):

$$\dot x = -Hx - \frac{X}{6} - 2x(H-u), \tag{24}$$

$$\dot H - \dot u = \frac{R}{6} - H(H-u) - (H-u)^2 + x^2 - \frac{\zeta^2}{a^2}. \tag{25}$$

One advantage of such a reformulation is that the variables are now all observables.

Our 6 first order dynamical equations and the energy constraint equation can now be put in the form

$$\dot a = aH, \tag{26}$$

$$\dot H = \frac{1}{6a_2}(\tilde a_2R - 2\tilde\sigma_2X) - 2H^2 + \frac{\tilde a_2 - 4\tilde a_3}{a_2}x^2 - \frac{\zeta^2}{a^2} + \frac{\rho}{3a_2} + \frac{4\Lambda}{3a_2}, \tag{27}$$

$$\dot u = -\frac{1}{3a_2}(a_0R + \tilde\sigma_2X) - 3Hu + u^2 - \frac{4a_3}{a_2}x^2 + \frac{\rho}{3a_2} + \frac{4\Lambda}{3a_2}, \tag{28}$$

$$\dot x = -\frac{X}{6} - (3H - 2u)x, \tag{29}$$

$$-\frac{w_6}{2}\dot R - \frac{\mu_3}{4}\dot X = \left[3\tilde a_2 + w_6R + \frac{\mu_3}{2}X\right]u + \left[-6\tilde\sigma_2 + \frac{\mu_3}{2}R - w_3X\right]x, \tag{30}$$

$$-\frac{\mu_3}{4}\dot R + \frac{w_3}{2}\dot X = \left[-6\tilde\sigma_2 + \frac{\mu_3}{2}R - w_3X\right]u - \left[12\tilde a_3 + w_6R + \frac{\mu_3}{2}X\right]x, \tag{31}$$

$$
\begin{aligned}
\rho \;=\;& 3(-\frac{1}{2}\tilde{a}_2 + 2\tilde{a}_3)x^2 + \frac{3a_2}{2}\left[H^2 + \frac{\zeta^2}{a^2}\right] - \Lambda \\
& - \left(6\tilde{\sigma}_2 + w_3 X - \frac{\mu_3 R}{2}\right) x(H-u) + \frac{1}{24}(w_6 R^2 - w_3 X^2 + \mu_3 RX) \\
& - \left(\frac{3}{2}\tilde{a}_2 + \frac{w_6 R}{2} + \frac{\mu_3 X}{4}\right)\left[(H-u)^2 - x^2 + \frac{\zeta^2}{a^2}\right],
\end{aligned} \tag{32}
$$

where we have introduced certain *modified* parameters $\tilde{a}_2$, $\tilde{a}_3$ and $\tilde{\sigma}_2$ with the definitions

$$
\tilde{a}_2 := a_2 - 2a_0, \qquad \tilde{a}_3 := a_3 - \frac{a_0}{2}, \qquad \tilde{\sigma}_2 := \sigma_2 + b_0. \tag{33}
$$

The last two dynamical equations (30,31) could of course easily be resolved for $\dot{R}$ and $\dot{X}$, but for some puposes the form given above is more convenient.

It is remarkable that here in these effective Lagrange equations and the associated conserved energy we get (for this case with dust as the source) exactly the correct equations for our model—*without including any explicit source coupling.*

From the above first order equations (following the method used in [18]) we have found the associated linear equations and the late time normal modes. The details of these results will be reported in another work. In the following we will show some typical evolution.

4.1. Numerical demonstration

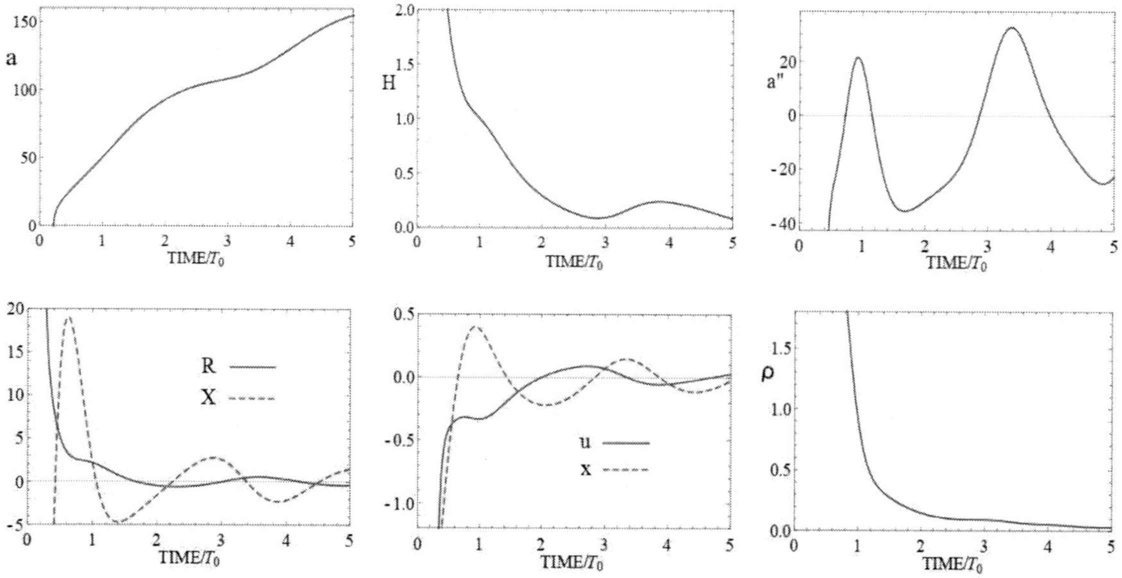

Figure 1. Evolution of the expansion factor a, the Hubble function, H, the second time derivative of the expansion factor, $\ddot{a}$, the scalar curvature and the pseudoscalar curvature, R and X, the torsion and the axial torsion function, u and x and the energy density, ρ.

We present the results of a numerical evolution of our cosmological model. For all these calculations we take $\Lambda = 0$. We need to look into the scaling features of this model before we can obtain the sort of evolution results we seek on a cosmological scale. In terms of fundamental units we can scale the variables and the parameters as $\kappa = 8\pi G = 1$. So the variables and the scaled parameters w_6 and

w_3 become dimensionless (from the Newtonian limit $a_0 = 1$). However, as we are interested in the cosmological scale to see changes of the age of our Universe, let us introduce a dimensionless constant T_0, which represents the magnitude of the Hubble time ($T_0 = H_0^{-1} \doteq 4.41504 \times 10^{17}$ seconds). With this scaling, all the field equations are kept unchanged while the period $T \to T_0 T$.

We take the parameters as

$$a_0 = 1, \quad a_2 = 0.83, \quad a_3 = -0.35, \quad w_6 = 1.1, \quad w_3 = 0.091, \quad \sigma_2 = 0.12 \quad \text{and} \quad \mu_3 = -0.18. \tag{34}$$

The behavior of the 6 equations has been observed with several sets of initial values. We plot a typical case in (Fig. 1): first, evolution of the expansion factor a, second, the Hubble function, H, third, the second time derivative of the expansion factor, $\ddot{a}$, fourth, the scalar curvature and the pseudoscalar curvature, R and X, fifth, the torsion and the axial torsion function, u and x, sixth, the energy density, ρ.

5. Concluding discussion

We have been investigating the dynamics of the Poincaré gauge theory of gravity. Recently, the model with two good propagating modes carrying spin 0^+, 0^- (referred to as the scalar and pseudoscalar modes) has been extended to include pseudoscalar constants that couple the two different parity modes. It should be mentioned that here are many interesting physical and mathematical details related to torsion and parity and the type of models considered that we could not discuss here which are covered in references [8, 17, 18]. Here we have considered the dynamics of this BHN model in the context of manifestly homogeneous and isotropic cosmological models. We found an effective Lagrangian and a system of first order equations. In these models, at late times the acceleration oscillates. It can be positive at the present time. It should be noted that the 0^+ torsion does not directly couple to any known form of matter, but it does couple directly to the Hubble expansion, and thus can directly influence the acceleration of the universe. On the other hand, the 0^- couples directly to fundamental fermions; with the newly introduced pseudoscalar coupling constants it too can directly influence the cosmic acceleration.

Acknowledgments

This work was supported by the National Science Council of the R.O.C. under the grants NSC-98-2112-M-008-008 and NSC-99-2112-M-008-004 and in part by the National Center of Theoretical Sciences (NCTS).

References

[1] Hehl F W, von der Heyde P, Kerlick G D and Nester J M 1976 *Rev. Mod. Phys.* **48** 393
[2] Hehl F W 1980 in *Proc. of 6th Course of the International School of Cosmology and Gravitation on Spin, Torsion and Supergravity*, eds P G Bergmann and V de Sabatta (New York: Plenum) p 5
[3] Hayashi K and Shirafuji T 1980 *Prog. Theor. Phys.* **64** 866, 883, 1435, 2222
[4] Mielke E W 1987 *Geometrodynamics of Gauge Fields* (Berlin: Akademie-Verlag)
[5] Hehl F W, McCrea J D, Mielke E W and Neeman Y 1995 *Phys. Rep.* **258** 1
[6] Gronwald F and Hehl F W 1996 "On the Gauge Aspects of Gravity" in Proc. 14th Course of the School of Gravitation and Cosmology (Erice), eds P G Bergman, V de Sabbata and H J Treder (Singapore: World Scientific) pp 148–98
[7] Blagojević M 2002 *Gravitation and Gauge Symmetries* (Bristol: Institute of Physics)
[8] Baekler P, Hehl F W and Nester J M 2011 *Phys. Rev.* D **83** 024001
[9] Nieh H T and Yan M L 1982 *J. Math. Phys.* **23** 373
[10] Sezgin E and van Nieuwenhuizen P 1980 *Phys. Rev.* D **21** 3269
[11] Blagojević M and Nicolić I A 1983 *Phys. Rev.* D **28** 2455
[12] Chen H, Nester J M and Yo H J 1998 *Acta Phys. Pol.* B **29** 961
[13] Hecht R, Nester J M and Zhytnikov V V 1996 *Phys. Lett.* A **222** 37
[14] Yo H J and Nester J M 1999 *Int. J. Mod. Phys.* D **8** 459
[15] Yo H J and Nester J M 2002 *Int. J. Mod. Phys.* D **11** 747
[16] Yo H J and Nester J M 2007 *Mod. Phys. Lett.* A **22** 2057
[17] Shie K F, Nester J M and Yo H J 2008 *Phys. Rev.* D **78** 023522
[18] Chen H, Ho F H, Nester J M, Wang C H and Yo H J 2009 *J. Cosmol. Astropart. Phys.* JCAP10(2009)027

[19] Minkevich A V 1980 *Phys. Lett.* A **80** 232
[20] Minkevich A V 1983 *Phys. Lett.* A **95** 422
[21] Minkevich A V and Nemenman I M 1995 *Class. Quantum Grav.* **12** 1259
[22] Minkevich A V and Garkun A S 2006 *Class. Quantum Grav.* **23** 4237
[23] Minkevich A V, Garkun A S and Kudin V I 2007 *Class. Quantum Grav.* **24** 5835
[24] Goenner H and Müller-Hoissen F 1984 *Class. Quantum Grav.* **1** 651
[25] Li X-z, Sun C-b and Xi P 2009 *Phys. Rev.* D **79** 027301
[26] Li X-z, Sun C-b and Xi P 2009 *J. Cosmol. Astropart. Phys.* JCAP04(2009)015
[27] Ao X-c, Li X-z and Xi P 2010 *Phys. Lett.* B **694** 186
[28] Wang C H and Wu Y H 2009 *Class. Quantum Grav.* **26** 045016
[29] Ashtekar A and Samuel J 1991 *Class. Quantum Grav.* **8** 2191

7th Biennial Conf. on Classical and Quantum Relativistic Dynamics of Particles and Fields IOP Publishing
Journal of Physics: Conference Series **330** (2011) 012006 doi:10.1088/1742-6596/330/1/012006

Quantum stress tensor fluctuation effects in inflationary cosmology

Jen-Tsung Hsiang[1], **Chun-Hsien Wu**[2], **L. H. Ford**[3], **Kin-Wang Ng**[4]

[1] Department of Physics, National Dong Hwa University, Hualien 94701, Taiwan
[2] Department of Physics, Soochow University, 70 Linhsi Road, Shihlin, Taipei 111, Taiwan
[3] Institute of Cosmology, Department of Physics and Astronomy, Tufts University, Medford, MA 02155, USA
[4] Institute of Physics, Academia Sinica, Nankang, Taipei 11529, Taiwan

E-mail: [1]cosmology@gmail.com, [2]chunwu@scu.edu.tw, [3]ford@cosmos.phy.tufts.edu, [4]nkw@phys.sinica.edu.tw

Abstract. We review several related investigations of the effects of the quantum stress tensor of a conformal field in inflationary cosmology. Particular attention will be paid to the effects of quantum stress tensor fluctuations as a source of density and tensor perturbations in inflationary models. These effects can possibly depend upon the total expansion factor during inflation, and hence be much larger than one might otherwise expect. They have the potential to contribute a non-scale invariant and non-Gaussian component to the primordial spectrum of perturbations, and might be observable.

1. Introduction

The gravitational effects of quantum matter fields are often described by a semiclassical theory whereby a suitable renormalized expectation value of the quantum stress tensor operator becomes the source of gravity. The semiclassical Einstein equation may be written as

$$G_{\mu\nu} = 8\pi\ell_p^2 \langle T_{\mu\nu} \rangle \,, \tag{1}$$

where ℓ_p is the Planck length and $\langle T_{\mu\nu} \rangle$ is the renormalized expectation value. This theory clearly has a wide domain of applicability, as it includes classical general relativity theory as a special case. Furthermore, it is expected to be useful in describing non-classical effects, such as quantum violation of the classical energy conditions.

However, the semiclassical theory is limited in that it does not describe quantum fluctuations of gravity. These fluctuations can arise directly from the dynamical degrees of freedom of the gravitational field itself, the "active" fluctuations. They can also be driven by quantum fluctuations of the quantum stress tensor, the "passive" fluctuations. The latter will be the primary focus of this paper.

The fluctuations of quantum stress tensors and their physical effects have been discussed by several authors in recent years [1, 2, 3, 4, 5, 6, 7]. For a recent review with further references, see Ref. [8]. Quantum stress tensor fluctuations necessarily have a skewed, highly non-Gaussian, probability distribution, although the explicit form of this distribution has only been found in

Journal of Physics: Conference Series **330** (2011) 012006 doi:10.1088/1742-6596/330/1/012006

two-dimensional spacetime models [9]. The basic features of the probability distribution are a lower bound and an infinite, positive tail. The lower bound is at the quantum inequality bound for expectation values in an arbitrary state. For the present purposes, the most important result is the non-Gaussian character, which implies that cosmological perturbations driven by stress tensor fluctuations will be non-Gaussian. Here we will summarize several related pieces of work and try to illustrate their interconnections, but for more details, the reader is referred to the original articles.

2. The spacetime geometry of inflationary cosmology

Here we briefly sketch the assumptions about the spacetime in the version of inflation which we consider. We take the metric to be that of a spatially flat Friedmann-Robertson-Walker model

$$ds^2 = -dt^2 + a^2(t)\,(dx^2 + dy^2 + dz^2) = a^2(\eta)\,(-d\eta^2 + dx^2 + dy^2 + dz^2)\,. \tag{2}$$

Here t is the comoving time, and η the conformal time. The inflationary period will be taken to be de Sitter spacetime with approximately constant curvature, so that

$$a = e^{Ht} = -\frac{1}{H\eta}\,. \tag{3}$$

We may set $a = 1$ at the end of inflation, which occurs at $t = 0$, or $\eta = -1/H$. Most inflationary models are insensitive to initial conditions, but that will not be the case for the effects which we discuss. It will be necessary to impose an initial condition that the effects of quantum fluctuations vanish at some initial time, which will be taken to be $\eta = \eta_0$. The total expansion factor of the universe from this time to the end of inflation is

$$S = \frac{1}{H|\eta_0|}\,. \tag{4}$$

This factor will play a key role in our subsequent discussion.

We assume that inflation ends quickly, with efficient reheating to a subsequent radiation dominated epoch. This assumption is not crucial, but simplifies the discussion. In this case, the radiation immediately after inflation is described by a reheating energy of E_R, where

$$H^2 = \frac{8\pi}{3}\,\ell_p^2\,E_R^4\,. \tag{5}$$

3. Semiclassical effects on gravity waves

In this section, we will discuss an effect in the semiclassical theory, dealing with the expectation value of a stress tensor rather than its fluctuations. However the result has features in common with the stress tensor fluctuation effects to be discussed later. The effects of $\langle T_{\mu\nu}\rangle$ for a conformal field on the propagation of gravity waves in de Sitter spacetime was recently treated in Ref. [10]. The calculations were based on a formalism developed by Horowitz and Wald [11]. It is assumed that at early times, $\eta < \eta_0$, there is a linearly polarized plane gravitational wave of the form

$$h_\mu^\nu = c_0\, e_\mu^\nu\,(1 + ik\eta)\,e^{i(\mathbf{k}\cdot\mathbf{x} - k\eta)}\,, \tag{6}$$

where c_0 is a constant, e_μ^ν is the polarization tensor, and k is the wavenumber. Next the coupling to the expectation value of the quantized matter field is assumed to be switched on at $\eta = \eta_0$. Its effect is to add a correction term $h'^\nu_\mu(x)$, which has the same functional form as $h_\mu^\nu(x)$, but differs

Journal of Physics: Conference Series **330** (2011) 012006 doi:10.1088/1742-6596/330/1/012006

in amplitude and phase. Most importantly, its amplitude at the end of inflation depends upon the expansion factor S and upon k. In the case that the conformal field is the electromagnetic field, the fractional correction is

$$\Gamma = \left| \frac{h'_{\mu}{}^{\nu}}{h_{\mu}{}^{\nu}} \right| = \frac{1}{5\pi} \ell_p^2 \, H \, S \, k \, . \tag{7}$$

This effect grows with increasing S and k, but its total magnitude is limited by the requirement that $\Gamma \lesssim 1$ for the one-loop approximation to hold.

Nonetheless, this effect could have observational consequences in the form of a modification of the tensor perturbations predicted by inflation. Inflationary models predict a nearly scale invariant spectrum of both scalar and tensor perturbations, both of which arise from the quantum fluctuations of nearly free fields and are Gaussian in character. The tensor perturbations result from the fluctuations of quantized linear perturbations of de Sitter spacetime [12, 13, 14]. They have not yet been observed, but are expected to leave signatures in the cosmic microwave background that could be seen in the future. The effect described by Eq. (7) modifies the normalization of the vacuum graviton modes in a way which breaks the scale invariance, and increases the power on shorter wavelengths.

4. Negative power spectra

One of the remarkable properties of quantum stress tensor fluctuations is that they can introduce negative power spectra of fluctuations. In most statistical processes, this is not possible. The well-known Wiener-Khinchin [15, 16] theorem states that the Fourier transform of a correlation function is a power spectrum. A corollary of this theorem is that the power spectrum can normally be written as the expectation value of a squared quantity, and hence must be positive. However, the latter result can fail in quantum field theory, and negative power spectra are possible [17]. The basic reason is that quantum correlation functions can be highly singular at coincident points, and the expectation value in the Wiener-Khinchin theorem may not exist.

A simple example is the spectrum of vacuum energy density fluctuations of the quantized electromagnetic field in flat spacetime. The energy density correlation function is

$$C_0(\tau, r) = \langle \rho(t, \mathbf{x}) \, \rho(t', \mathbf{x}') \rangle = \frac{(\tau^2 + 3r^2)(r^2 + 3\tau^2)}{\pi^4 (r^2 - \tau^2)^6} \, , \tag{8}$$

where $\tau = t - t'$ and $r = |\mathbf{x} - \mathbf{x}'|$. Its spatial Fourier transform is

$$\hat{C}_0(\tau, k) = -\frac{k^4 \sin(k\,\tau)}{960 \pi^5 \, \tau} \, , \tag{9}$$

and the power spectrum is

$$P_0(k) = \hat{C}_0(0, k) = -\frac{k^5}{960 \pi^5} \, . \tag{10}$$

The negative sign in the power spectrum essentially interchanges correlations and anticorrelations, compared to what one would have with a positive spectrum of the same functional form.

5. Density perturbations in inflation

Just as fluctuations of the free graviton field in de Sitter spacetime can lead to observable tensor perturbations, fluctuations of the inflaton field can produce density perturbations [18, 19, 20, 21,

Journal of Physics: Conference Series **330** (2011) 012006 doi:10.1088/1742-6596/330/1/012006

22]. The resulting spectrum of nearly scale invariant, Gaussian fluctuations has apparently been observed in the temperature fluctuations of the cosmic microwave background [23]. However, quantum stress tensor fluctuations can produce an additional, non-scale invariant, non-Gaussian contribution, which was studied in Refs. [24, 25]. This effect arises from the quantum fluctuations of the comoving energy density of a conformal field in its vacuum state. In a dynamic model treated in Ref. [25], these fluctuations couple to the dynamics of the inflaton field and produce a spectrum of density perturbations proportional to S^2,

$$\mathcal{P}(k) = \frac{8\ell_p^4\, k^2\, S^2}{75}\,. \tag{11}$$

Note that

$$\mathcal{P}(k) = 4\pi k^3\, P(k)\,, \tag{12}$$

is the quantity usually called the power spectrum in cosmology, and which is independent of k for a scale invariant spectrum. Thus the effect of the quantum stress tensor fluctuations is a blue-tilted, non-Gaussian contribution.

The fact that this contribution has not yet been observed can be interpreted as placing an upper bound on S, which is found to be

$$S \lesssim 10^{42} \left(\frac{10^{12}\ \mathrm{GeV}}{E_R}\right)^3 . \tag{13}$$

This bound is consistent with adequate inflation to solve the horizon and flatness problems, which requires $S > 10^{23}$.

6. Gravity waves from stress tensor fluctuations in inflation

In addition to the scalar (density) perturbations, stress tensor fluctuations can also create tensor perturbations, which are gravity waves [26]. Here the fluctuations of spatial components of the stress tensor can couple to the transverse, tracefree components of the gravitational field. The result at the end of inflation is a spectrum of gravity wave fluctuations with a power spectrum given by

$$\mathcal{P}(k) = -\frac{4\ell_p^4\, k^2\, H^2\, S^2}{3\pi}(1 + k^2\, H^{-2})\,. \tag{14}$$

This is an example of a negative power spectrum, as well as one which is strongly blue-tilted. This result assumes that the coupling between the quantum stress tensor and the gravitational field is switched on suddenly at $\eta = \eta_0$. However, more gradual switching leads to qualitatively similar results. If S were sufficiently large, then the resulting tensor perturbations in the CMB spectrum should have already been observed. This leads to the constraint

$$S \lesssim 10^{46} \left(\frac{10^{12} GeV}{E_R}\right)^3 , \tag{15}$$

which is somewhat weaker than the bound, Eq. (13), which comes from density perturbations.

However, gravity wave fluctuations are potentially detectable at much smaller scales than are the density perturbations. The shorter wavelengths of the primordial density perturbation spectrum have presumably been erased by nonlinear classical effects by now, but gravity waves interact very weakly and should still exist. This raises the possibility of detecting the primordial gravity waves from stress tensor fluctuations in gravity wave detectors, in the form of background noise corresponding to the spectrum given in Eq. (14). LIGO has placed limits [27] of $h \lesssim 10^{-24}$ on scales of the order of $10^2\, km$, leading to the constraint

$$S < 10^{23} \left(\frac{10^{10} GeV}{E_R} \right)^3 . \tag{16}$$

This result is compatible with adequate inflation to solve the horizon and flatness problems only if

$$E_R \lesssim 10^{10}\, GeV . \tag{17}$$

This would put a non-trivial constraint on inflationary models. An even more exciting possibility is that Earth or space-based gravity wave detectors might have a chance of detecting the effects of quantum stress tensor fluctuations during inflation.

7. The transplanckian issue

There is an important qualification to all of the results which depend upon S, the expansion factor during inflation. This is that quantum field modes above the Planck scale in energy are required. Consider, for example, the example in the previous section of a gravity wave with a present proper wavelength of $100km$. If we assume $E_R \approx 10^{10}\, GeV$, then there has been an expansion by a factor of about 10^{23} since reheating to redshift to the current CMB temperature. This, combined with an additional expansion of at least 10^{23} during inflation, implies that this mode would have had a proper wavelength of about $10^{-3}\ell_p$ at the beginning of inflation. The quantum field would have to have fluctuations on this scale to create such a gravity wave mode.

This raises questions as to whether the framework of quantum field theory on a fixed or weakly fluctuating background can be trusted at this scale. However, an even more extreme version of the transplanckian issue arises in black hole thermodynamics. Hawking's [28] derivation of of black hole radiance relies upon modes which begin far above the Planck energy. The fact that the Hawking effect gives a beautiful unification of gravity, thermodynamics, and quantum theory can be considered to be a powerful argument to take transplanckian modes seriously. It is true that it is possible to derive the Hawking effect without transplanckian modes [29, 30], but only at the price of introducing modified dispersion relations which break local Lorentz symmetry and hence postulate new physics. An analogous prescription in cosmology is to consider modes only after they redshift below the Planck energy in the comoving frame [25]. There has been an extensive discussion of the possible role of transplanckian modes in inflationary cosmology. (See Ref. [25] for a lengthy list of references.)

The dependence of the gravity wave spectrum upon a positive power of S might seems to contradict a theorem due to Weinberg [31], which was generalized by Chaicherdsakul [32]. This theorem states that radiative corrections during inflation should not grow faster than a logarithm of the scale factor. However, as was discussed in more detail in Ref. [25], the effect of density perturbations growing as a power of S is really due to high frequency modes at the initial time, and is hence always large rather than growing. This applies to all the effects discussed in the present paper.

8. Discussion

Here we have reviewed three distinct, but related, quantum effects in inflation which depend upon the expansion factor S during inflation. The first is the effect of the expectation value of a conformal field upon the amplitude of gravity waves in de Sitter spacetime [10]. The second is the effect of energy density fluctuations of such a field upon the spectrum of primordial density perturbations [24, 25]. The third effect is the role of quantum stress tensor fluctuations in generating a spectrum of primordial gravity waves [26]. The latter two effects are associated with non-Gaussian fluctuations, and all three effects grow as S increases and produce a blue-tilted spectrum, in which there is more power on shorter wavelengths. The gravity waves produced by stress tensor fluctuations are an example of a negative power spectrum, which is usually forbidden, but is possible in quantum field theory [17].

All three effects have the potential to produce observable effects, either in the cosmic microwave background radiation, or in gravity wave detectors. However, all three effects also rely upon the existence of transplanckian modes. As such, they have the possibility to open an observational window on transplanckian physics.

Acknowledgments

We would like to thank Da-Shin Lee and Hoi-Lai Yu for valuable discussions. This work is partially supported by the National Science Council, Taiwan, ROC under the Grants NSC98-2112-M-001-009-MY3 and NSC99-2112-M-031-002-MY3, and by the U.S. National Science Foundation under Grant PHY-0855360.

References

[1] Wu C-H and Ford L H 2001 *Phys. Rev. D* **64** 045010 quant-ph/0012144.
[2] Borgman J and Ford L H 2004 *Phys. Rev. D* **70** 064032 gr-qc/0307043.
[3] Hu B L and Verdaguer E 2004 *Living Rev. Rel.* **7** 3 gr-qc/0307032.
[4] Ford L H and Woodard R P 2005 *Class. Quant. Grav.* **22** 1637 gr-qc/0411003.
[5] Ford L H and Roman T A 2005 *Phys. Rev. D* **72** 105010 gr-qc/0506026
[6] Thompson R T and Ford L H 2006 *Phys. Rev. D* **74** 024012 gr-qc/0601137.
[7] Perez-Nadal G, Roura A and Verdaguer E 2010 *JCAP* 1005:036 arXiv:0911.4870.
[8] Ford L H and Wu C H 2008 *AIP Conf. Proc.* **977** 145 arXiv:0710.3787.
[9] Fewster C J, Ford L H, and Roman T A 2010 *Phys. Rev. D* **81** 121901(R) arXiv:1004.0179.
[10] Hsiang J-T, L H Ford, Lee D-S, and Yu H-L 2011 *Phys. Rev. D* **83** 084027 arXiv:1012.1582.
[11] Horowitz G T and Wald R M 1980 *Phys. Rev. D* **21** 1462.
Horowitz G T and Wald R M 1982 *Phys. Rev. D* **25** 3408 .
[12] Starobinsky A A 1979 *JETP Lett.* **30** 682.
[13] Abbott L F and Wise M B 1984 *Nucl. Phys. B* **244** 541.
[14] Allen B 1988 *Phys. Rev. D* **37** 2078.
[15] Wiener N 1930 *Acta. Math Stockholm* **55** 117.
[16] Khinchin A 1934 *Math. Ann.* **109** 604.
[17] Hsiang J-T, Wu C-H, and Ford L H 2011 *Phys. Lett. A* **375** 2296 arXiv:1012.3226.
[18] Mukhanov V and Chibisov G 1981 *JETP Lett.* **33** 532.
[19] Guth A H and Pi S-Y 1982 *Phys. Rev. Lett.* **49** 1110.
[20] Hawking S W 1982 *Phys. Lett. B* **115** 295.
[21] Starobinsky A A 1982 *Phys. Lett. B* **117** 175.
[22] Bardeen J M, Steinhardt P J, and Turner M S 1983 *Phys. Rev. D* **28** 679.
[23] Komatsu E 2011 *et al., Astrophys. J. Suppl* **192** 18 arXiv:1001.4538.
[24] Wu C-H, Ng K-W, and Ford L H 2007 *Phys. Rev. D* **75** 103502 gr-qc/0608002.
[25] Ford L H, Miao S-P, Ng K-W, Woodard R, and Wu C-H 2010 *Phys. Rev. D* **82** 043501 arXiv:1005.4530.
[26] Wu C-H, Hsiang J-T, Ford L H, and Ng K-W, arXiv:1105.1155.
[27] Sutton P J 2008 *J. Phys. Conf. Ser.* **110** 062024.
[28] Hawking S W 1975 *Commun. Math. Phys.* **43** 199.
[29] Unruh W G 1995 *Phys. Rev. D* **51** 2827 gr-qc/9409008.
[30] Corley S and Jacobson T 1996 *Phys. Rev. D* **54** 1568 hep-th/9601073.
[31] Weinberg S 2005 *Phys. Rev. D* **72** 043514.
Weinberg S 2006 *Phys. Rev. D* **72** **74** 023508.
[32] Chaicherdsakul K 2007 *Phys. Rev. D* **75** 063522.

Journal of Physics: Conference Series **330** (2011) 012007 doi:10.1088/1742-6596/330/1/012007

Relativistic quantum mechanics on the light-front consistent with quantum field theory

A. N. Kvinikhidze[1,2] **and B. Blankleider**[2]

[1] A. Razmadze Mathematical Institute, Tbilisi, Georgia

[2] School of Chemical and Physical Sciences, Flinders University, Bedford Park, SA 5042, Australia

Abstract.
A relativistic quantum mechanics (RQM) on the light-front (LF) is derived from quantum field theory (QFT) through the use of a unitary transformation. Cluster separability of the derived RQM is proved by solving the so-called decoupling equation for the disconnected parts of the unitary transformation relating the Poincaré generators of RQM to the ones of QFT. For processes involving baryon-like number conservation ($NN \to NN$, $Nd \to Nd$, etc.), we show that the two theories, RQM and QFT, are equivalent.

1. Introduction

Dirac's work on the three forms of relativistic dynamics [1] has led to the present well established approaches of relativistic quantum mechanics (RQM) whose formulation is based on the realization of Poincaré algebra in the space of a fixed number of particles [2, 3, 4, 5, 6, 7]; in particular, they are derived by constructing 10 Poincaré generators in the space of few-body states. Specific questions addressed in each of the papers [2, 3, 4, 5, 6, 7] are briefly discussed in [8], while an overview of the whole subject is given in [9].

Although this RQM has been applied to a number of relativistic few-body problems, the question remains as to how RQM is related to quantum field theory (QFT) where, as is well known, Lorentz invariance requires an interaction Hamiltonian, H_I, that allows for particle creation and annihilation. Indeed, the basic question is whether RQM is consistent with QFT. An important step in answering this question has been taken in [10, 11] where a unitary transformation (UT) was introduced to block diagonalizes the QFT Hamiltonian, H, in such a way that the subspace, η, of a fixed number of "nucleons" is decoupled from the rest of the space, λ. We will consider the somewhat simpler case where the subspace η consists of particles (we shall refer to them "nucleons") obeying baryon-like number conservation, so that the QFT in question forbids transitions between states of different numbers of nucleons (if there are no anti-nucleons in these states).

Here we consider the RQM based on the front form of dynamics which stands out by having the maximal number of *kinematical* generators (seven versus six in other forms of dynamics). These generators make up the stability subgroup corresponding to the transformations which do not change the hyperplane of constant LF time, $x^+ = t + z = 0$, and consist of the three components of vector $\underline{P} = (P^+, P_\perp) = (P^0 + P^3, P^1, P^2)$ where P^μ is the four-momentum, plus four generators constructed from the components of the angular momentum $\vec{J}$ and boost $\vec{K}$

Journal of Physics: Conference Series **330** (2011) 012007 doi:10.1088/1742-6596/330/1/012007

operators, as follows: J^3, K^3, $E^1 = -K^1 + J^2$ and $E^2 = -K^2 - J^1$. RQM in the front form of dynamics has been formulated and applied to the study of mesons and baryons in [12, 13, 14, 15]. Note that in the instant form of dynamics (i.e. with the usual time) there are only six kinematical generators, $\vec{P}$ and $\vec{J}$.

In the present paper we show that the theory corresponding to the diagonalized Hamiltonian[1], $H' = UHU^\dagger$, is equivalent to the initial one corresponding to H, and moreover, that the Hamiltonian of the new theory (H') in the η subspace, has all the essential properties of a Hamiltonian of a *correct* RQM. By the latter, we mean a number of fundamental properties of Quantum Mechanics (discussed below), including so-called "cluster separability" which guarantees, for example, that a two-body potential in the three-body system is the same as the two-body potential in an isolated two-body system. The achievement of cluster separability is particularly significant as it has been a long-standing problem in RQM theories [2, 3, 4, 5, 6, 7]. Our proof of cluster separability is based on an exact solution of the so-called *decoupling equation* for the disconnected parts of the UT.

Although the issues addressed in this paper are of a fundamental theoretical nature, it is worth noting that they may be of practical interest as well; in particular, the results of the paper should be useful for the modern theory of nuclear forces (see [16] and the references therein) whose construction in most studies is based on the UT method [10, 11].

2. Unitary transformation method

In 1954, Okubo introduced a UT [10, 11] to block diagonalize the Hamiltonian of usual equal-time QFT so that the subspace of a fixed number of nucleons is decoupled from the rest of the space. Since then, it has been shown [8] that the UT actually diagonalizes all ten Poincaré generators of the QFT. As a result, it has become apparent that the unitary transformed generators realize Poincaré algebra in the subspace of the fixed number of nucleons.

In the present paper, we apply the UT method to the case of LF QFT, and in this way, derive LF RQM by block diagonalizing the 10 Poincaré generators given by LF QFT (which of course do not conserve the number of nucleons). However, further steps need to be performed in order to establish that these diagonalized generators indeed constitute what can rightfully be called *relativistic Quantum Mechanics on the LF*; in particular, the diagonalized generators should: (i) be able to be split into free and interaction parts[2] in such a way that the interaction parts are diagonal in the number of nucleons, and the resultant theory generates the same scattering amplitude as the one determined by the QFT we started with, and (ii) produce matrix elements of the transformed Hamiltonian whose disconnected parts have a "quantum mechanical structure". The latter requirement is clarified in the discussions below, but means, for example, that the fully disconnected part would be the sum of the LF energies of the nucleons, the pair interaction potential in the three-body system would be the same as in the isolated two-body system, etc. To emphasise the significance of requirement (ii), note that it is badly violated in the Tamm-Dankoff diagonalization method [17, 18]. It is important to emphasize that this "quantum mechanical structure" does not follow from Lorentz invariance alone; indeed, in order to establish this structure, we needed to calculate the disconnected parts of the Hamiltonian H' *non-perturbatively*, using a previously unknown factorisation property of the UT derived from the *decoupling equation*, Eq. (4).

The stability subgroup Poincaré generators of LF QFT, P^+, $P^\perp$, J^3, K^3, E^1 and E^2, commute

[1] Recall that in LF dynamics the Hamiltonian is the difference of the zero'th and the third components of the four-momentum generator of the Poincaré group: $H = P^- = P^0 - P^3$.

[2] Eigenstates of the free part of the four-momentum generator determine the asymptotic states of the nucleons which are needed apart from the Hamiltonian to construct the scattering amplitude.

Journal of Physics: Conference Series **330** (2011) 012007 doi:10.1088/1742-6596/330/1/012007

with the UT, so they do not even change under this transformation:[3]

$$U\underline{P}U^\dagger = \underline{P}, \quad UJ^3U^\dagger = J^3, \quad UK^3U^\dagger = K^3, \quad UE^1U^\dagger = E^1, \quad UE^2U^\dagger = E^2 \tag{1}$$

All the three remaining dynamical generators, the Hamiltonian P^-, $F^1 = -K^1 - J^2$, and $F^2 = -K^2 + J^1$, can be block diagonalized by the same UT. Thus any of these generators, and in particular the transformed Hamiltonian, has the following block diagonal form

$$H' = UHU^\dagger = \begin{pmatrix} \eta H'\eta & 0 \\ 0 & \lambda H'\lambda \end{pmatrix} \tag{2}$$

We follow the commonly used parametrization of the UT as originally introduced in [10]:

$$U = \begin{pmatrix} \eta U\eta & \eta U\lambda \\ \lambda U\eta & \lambda U\lambda \end{pmatrix} = \begin{pmatrix} (1+A^\dagger A)^{-1/2}\eta & (1+A^\dagger A)^{-1/2}A^\dagger \\ -(1+AA^\dagger)^{-1/2}A & (1+AA^\dagger)^{-1/2}\lambda \end{pmatrix} \tag{3}$$

where $A = \lambda A\eta$ (i.e. $\eta A\lambda = \eta A\eta = \lambda A\lambda = 0$) is determined from the *decoupling equation*,

$$(\lambda - A)H(\eta + A) = 0. \tag{4}$$

The first problem is to show that the theories corresponding to the Hamiltonians H' and H are equivalent to each other. This is a trivial problem in the standard setting where H_0', the free part of the Hamiltonian $H' = UHU^\dagger$, is obtained from the initial free Hamiltonian H_0, using the same transformation, $H_0' = UH_0U^\dagger$; in that case, the whole transformation is equivalent to a transition to new fundamental fields, $\Psi'(x) = U\Psi(x)U^\dagger$, which satisfy the same field-theoretical equations and the same commutation relations as the fields $\Psi(x)$ of the initial theory. Another way of seeing equivalence in this setting (with $H_0' = UH_0U^\dagger$) is to notice that the inhomogeneous term in the equation for an in-state (or out-state), is the UT-ed eigenstate of H_0, which in turn, leads to the equation $|\Psi^{(in)'}\rangle = U|\Psi^{(in)}\rangle$. This relation between scattering states of theories corresponding to the Hamiltonians H' and H, leaves the S-matrix, $S = \langle\Psi^{(out)}|\Psi^{(in)}\rangle$, unchanged.

However, in order to stay in the subspace η of a fixed number of particles, the *nonstandard setting* is needed where the free part of the transformed Hamiltonian H' is identified with the initial free Hamiltonian, $H_0' = H_0$ ($\neq UH_0U^\dagger$). The way the Hamiltonian is split into free and interaction parts *makes a difference* for the scattering problem, but not for the bound state problem because the latter has an unambiguous solution (up to a normalization) for the bound state vector, $|B'\rangle = U|B\rangle$. The bound-state energy spectrum thus does not change under the UT.

One needs to show that the on-shell scattering amplitude corresponding to the interaction $V' = H' - H_0$, coincides with the one produced by the initial Hamiltonian H. This can be done by comparing the scattering states of the two theories. For the NN scattering in-states one has the straightforward relation

$$|\Psi_q'^{(+)}\rangle = \lim_{\epsilon\to 0}\frac{i\epsilon}{P_q^- - H' + i\epsilon}|q_1, q_2\rangle = U\lim_{\epsilon\to 0}\frac{i\epsilon}{P_q^- - H + i\epsilon}U^\dagger|q_1, q_2\rangle \tag{5}$$

where

$$P_q^- = \frac{q_{1\perp}^2 + m^2}{q_1^+} + \frac{q_{2\perp}^2 + m^2}{q_2^+}, \tag{6}$$

[3] Note that the UT introduced by Okubo [10] was for the instant form of dynamics and therefore commutes with only six of the Poincaré generators, the 3-momentum, $\vec{P}$ and the angular momentum $\vec{J}$, so that, $U\vec{P}U^\dagger = \vec{P}$ and $U\vec{J}U^\dagger = \vec{J}$.

m is a dressed nucleon mass. It is obvious from Eq. (5) that the operator $U^\dagger$ on the right hand side may make a non-trivial contribution to the connection between $|\Psi_q'^{(+)}\rangle$ and the in-state of the initial theory

$$|\Psi_q^{(+)}\rangle = \lim_{\epsilon\to 0} \frac{i\epsilon}{P_q^- - H + i\epsilon}|q_1, q_2\rangle. \tag{7}$$

As described below, we have derived the contribution of $U^\dagger$ by solving the decoupling equation, Eq. (4), and found that its contribution is indeed non-trivial, so that $|\Psi_q'^{(+)}\rangle \neq U|\Psi_q^{(+)}\rangle$. This is in sharp contrast to all previous studies where $U^\dagger$ was assumed to be replaceable by the identity operator. We will see that $U^\dagger$ contributes via the disconnected part of the matrix element:

$$\langle q_1', q_2'|\frac{i\epsilon}{P_q^- - H + i\epsilon}U^\dagger|q_1, q_2\rangle^{disc} \neq \langle q_1', q_2'|\frac{i\epsilon}{P_q^- - H + i\epsilon}|q_1, q_2\rangle^{disc} \tag{8}$$

where the disconnected part of any matrix element is defined as the sum of all terms corresponding to the transition between two given subgroups of particles, $(q_1, ...q_s)(q_{s+1}, ...q_m)$ and $(p_1, ...p_l)(p_{l+1}, ...p_n)$, in the initial and final states respectively such that there are no particle exchanges between these two independent transitions. Correspondingly the total momentum is conserved in each of these two transitions:

$$\begin{aligned}&\langle p_1, ...p_l; p_{l+1}, ...p_n|O|q_1, ...q_s; q_{s+1}, ...q_m\rangle^{disc}\\ &= \delta^3(\underline{p}_1 + ... + \underline{p}_l - \underline{q}_1 - ... - \underline{q}_s)\delta^3(\underline{p}_{l+1} + ... + \underline{p}_n - \underline{q}_{s+1} - ... - \underline{q}_m)O(p, q)\end{aligned} \tag{9}$$

where the semicolons separate the quantum numbers of particles from the independent subgroups and . $\delta^3(\underline{p}) = \delta(p^+)\delta^2(p^\perp)$.

The paper is based on the conjecture about factorization of the disconnected parts of the matrix elements of the η projected part, ηU, of the unitary transformation U,

$$\begin{aligned}&\langle p_1, ...p_l; p_{l+1}, ...p_n|\eta U|q_1, ...q_s; q_{s+1}, ...q_m\rangle^{disc}\\ &= \langle p_{l+1}, ...p_n|\eta U|q_{s+1}, ...q_m\rangle\langle p_1, ...p_l|\eta U|q_1, ...q_s\rangle.\end{aligned} \tag{10}$$

It results from the rigorous analysis of the decoupling Eq. (4) (see appendix A) and holds for only matrix elements of U corresponding to the transitions from arbitrary initial state to *only* η final subspace. As we will see below matrix elements of $\lambda U\eta$ factorise in a different way,

$$\begin{aligned}\langle\lambda, \eta|U|\eta_1, \eta_2\rangle^{disc} &= +\langle\lambda|U|\eta_1\rangle\langle\eta|U|\eta_2\rangle\\ \langle\lambda_1, \lambda_2|U|\eta_1, \eta_2\rangle^{disc} &= -\langle\lambda_1|U|\eta_1\rangle\langle\lambda_2|U|\eta_2\rangle\end{aligned} \tag{11}$$

where we considered two different type of λ states, $\langle\lambda, \eta|$ and $\langle\lambda_1, \lambda_2|$. The difference from Eq. (10) is in the minus sign in the second of the equations (11).

The factorization property (10) leads to the following connection between the two-body scattering states, $|\Psi_q'^{(+)}\rangle$ and $|\Psi_q^{(+)}\rangle$ (see appendix B):

$$\frac{1}{Z'}\lim_{\epsilon\to 0}\frac{i\epsilon}{P_q^- - H' + i\epsilon}|q_1, q_2\rangle = \frac{1}{Z}U\lim_{\epsilon\to 0}\frac{i\epsilon}{P_q^- - H + i\epsilon}|q_1, q_2\rangle, \tag{12}$$

where the renormalization constants Z' and Z are defined as the residues of the single particle propagators (at the pole energy $q^- = \dfrac{q_\perp^2 + m^2}{q^+}$) corresponding to the single particle matrix elements of the resolvent operators

$$\langle\underline{p}|\frac{1}{q^- - H' + i\epsilon}|\underline{q}\rangle \sim \delta^3(\underline{p} - \underline{q})\frac{Z'}{q^- - \dfrac{q_\perp^2 + m^2}{q^+} + i\epsilon} \tag{13}$$

$$\langle \underline{p}|\frac{1}{q^- - H + i\epsilon}|\underline{q}\rangle \sim \delta^3(\underline{p}-\underline{q})\frac{Z}{q^- - \dfrac{q_\perp^2+m^2}{q^+} + i\epsilon}. \tag{14}$$

Note that since H' is diagonal (with respect to the number of particles) in the η subspace, the corresponding self-energy, $\Sigma'(q^-)$, does not depend on the LF "energy", q^-, and therefore the single particle wave function does not get renormalized, $Z'_q = 1$ (see Eq. (B.8) in the Appendix). In Appendix B it is proved that the RQM corresponding to the Hamiltonian H' produces the same on-shell NN scattering amplitude as the one produced by the initial QFT.

The factorization property (10) is crucial for the analysis of the structure of the few-body interaction, H'. To this end note the following property of the disconnected parts of the matrix elements of H,

$$\begin{aligned}&\langle p_1,...p_l;p_{l+1},...p_n|H|q_1,...q_s;q_{s+1},...q_m\rangle^{disc}\\ &= \langle p_{l+1},...p_n|H|q_{s+1},...q_m\rangle\langle p_1,...p_l|q_1,...q_s\rangle\\ &+\langle p_{l+1},...p_n|q_{s+1},...q_m\rangle\langle p_1,...p_l|H|q_1,...q_s\rangle,\end{aligned} \tag{15}$$

which follows from the general representation for the Hamiltonian in the QFT on the LF

$$H = \int \mathcal{H}(\psi(x))d^3x, \tag{16}$$

where $d^3x = dx^-d^2x^\perp$. Lorentz invariance does not allow an interaction Hamiltonian in the form of the product of "usual" interaction Hamiltonians

$$\left[\int \mathcal{H}_1(\psi(x))d^3x\right]\left[\int \mathcal{H}_2(\psi(x))d^3x\right] \tag{17}$$

as in the latter the two operators are at the same LF time x^+, yet at different 3-coordinate $\underline{x}$. The property (15) is at the heart of the convolution approach [17, 18] where it enables one to sum all possible terms contributing to the disconnected parts of the Green functions (including ones corresponding to so-called same-time dressing). Just the full sum solves the longstanding renormalization problem in time ordered perturbation theory (TOPT), see [17, 18] and references therein. Using the property (15) in combination with Eq. (10) one gets the same property for H'. In the subspace η we have

$$\begin{aligned}&\langle p_1,...p_s;p_{s+1},...p_n|\eta H'|q_1,...q_s;q_{s+1},...q_n\rangle^{disc}\\ &= \langle p_{s+1},...p_n|H'|q_{s+1},...q_n\rangle\langle p_1,...p_s|q_1,...q_s\rangle\\ &+\langle p_{s+1},...p_n|q_{s+1},...q_n\rangle\langle p_1,...p_s|H'|q_1,...q_s\rangle.\end{aligned} \tag{18}$$

Eq. (18) is all one needs to analyse the structure of the equations in the η-space. Such analysis shows that the "free" part of the equations is determined by the sum of the LF energies of the particles with physical (not bare) masses. It is determined by the fully disconnected part of the corresponding matrix element of the Hamiltonian H' and in the two nucleon case it is

$$\begin{aligned}\langle p_1;p_2|\eta H'|q_1;q_2\rangle^{disc} &= \langle p_1|H'|q_1\rangle\delta^3(\underline{p}_2-\underline{q}_2) + \langle p_2|H'|q_2\rangle\delta^3(\underline{p}_1-\underline{q}_1)\\ &= \delta^3(\underline{p}_1-\underline{q}_1)\delta^3(\underline{p}_2-\underline{q}_2)\left(\frac{q_{1\perp}^2+m^2}{q_1^+}+\frac{q_{2\perp}^2+m^2}{q_2^+}\right)\end{aligned} \tag{19}$$

Journal of Physics: Conference Series **330** (2011) 012007 doi:10.1088/1742-6596/330/1/012007

The pair interaction (say of particles 2 and 3) in the three-body system is determined by the sum of all disconnected diagrams corresponding to the first particle in flight

$$
\begin{aligned}
&\langle p_1; p_2, p_3|\eta H'|q_1; q_2, q_3\rangle^{disc} \\
&= \langle p_1|H'|q_1\rangle\delta^3(p_2-q_2)\delta^3(p_3-q_3) + \langle p_2,p_3|H'|q_2,q_3\rangle\delta^3(p_1-q_1) \\
&= \delta^3(p_1-q_1)\delta^3(p_2-q_2)\delta^3(p_3-q_3)\frac{q_{1\perp}^2+m^2}{q_1^+} + \left(\langle p_2,p_3|H'|q_2,q_3\rangle^{disc} + V_{23}^{(2)}\right)\delta^3(p_1-q_1) \\
&= \delta^3(p_1-q_1)\delta^3(p_2-q_2)\delta^3(p_3-q_3)\left(\sum_{i=1}^{3}\frac{q_{i\perp}^2+m^2}{q_i^+}\right) + V_{23}^{(2)}\delta^3(p_1-q_1) \qquad (20)
\end{aligned}
$$

where here and below $V^{(n)}$ is the n-body force potential defined as the connected part of the n-body matrix element of H'. It is remarkable that the pair interaction in the three-body system is just the same as in the case of the isolated system of two particles. Recall that in the corresponding equations of TOPT, the pair interaction kernel (the sum of irreducible diagrams where the third particle is in flight) cannot even be *expressed* in terms of the isolated two-body kernel.

A similar situation is present in the more complicated four-body sector; using Eq. (18) one can show that there is no two-pair simultaneous interaction in the four-body system - one has only the sum of pair interactions:

$$
\begin{aligned}
&\langle p_1, p_2; p_3, p_4|\eta H'|q_1, q_2; q_3, q_4\rangle^{disc} \\
&= \langle p_1,p_2|H'|q_1,q_2\rangle\delta^3(p_3-q_3)\delta^3(p_4-q_4) + \langle p_3,p_4|H'|q_3,q_4\rangle\delta^3(p_1-q_1)\delta^3(p_2-q_2) \\
&= \delta^3(p_1-q_1)\delta^3(p_2-q_2)\delta^3(p_3-q_3)\left(\sum_{i=1}^{4}\frac{q_{i\perp}^2+m^2}{q_i^+}\right) \\
&+V_{12}^{(2)}\delta^3(p_3-q_3)\delta^3(p_4-q_4) + V_{34}^{(2)}\delta^3(p_1-q_1)\delta^3(p_2-q_2). \qquad (21)
\end{aligned}
$$

The properties observed in Eqs. (19), (20), and (21) hold for the general case of arbitrary number of particles and lead to the equations of quantum mechanical form. They can be summarized in the compact form of a second quantized QM:

$$
\begin{aligned}
H' &= \int d^3p\, a_p^\dagger\frac{p_\perp^2+m^2}{p^+}a_p + \int d^3p_1 d^3p_2 d^3q_1 d^3q_2\; a_{p_1}^\dagger a_{p_2}^\dagger V^{(2)}(\underline{p}_1,\underline{p}_2,\underline{q}_1,\underline{q}_2)a_{q_1}a_{q_2} \\
&+ \int d^3p_1 d^3p_2 d^3q_1 d^3q_2\; a_{p_1}^\dagger a_{p_2}^\dagger a_{p_3}^\dagger V^{(3)}(\underline{p}_1,\underline{p}_2,\underline{p}_3,\underline{q}_1,\underline{q}_2,\underline{q}_3)a_{q_1}a_{q_2}a_{q_3} \\
&+ \int d^3p_1 d^3p_2 d^3q_1 d^3q_2\; a_{p_1}^\dagger a_{p_2}^\dagger a_{p_3}^\dagger a_{p_4}^\dagger V^{(4)}(\underline{p}_1,\underline{p}_2,\underline{p}_3,\underline{p}_4,\underline{q}_1,\underline{q}_2,\underline{q}_3,\underline{q}_4)a_{q_1}a_{q_2}a_{q_3}a_{q_4} + \ldots \qquad (22)
\end{aligned}
$$

where $d^3p = dp^+d^2p^\perp$ (which should not be confused with a similar notation for the coordinate phase space, $d^3x = dx^-d^2x^\perp$), $a_p^\dagger$, a_p are the nucleon creation and annihilation operators, $V^{(n)}$ are n-body force potentials. Note that all these potentials depend only on $\underline{p}_i, \underline{q}_j$, the three-dimensional part of the momenta, which are integration variables in Eq. (22); in particular, they do not depend on the total off-shell LF energy, in contrast to the Tamm-Dankoff or quasipotential approaches. This is also the property of usual QM.

3. Comments

As mentioned in Sect. 2, all ten Poincaré generators of QFT are block-diagonalized by a unique UT. Seven of the unitary transformed generators, P'^+, $P'^\perp$, J'^3, K'^3, E'^1 and E'^2, are kinematic

Journal of Physics: Conference Series **330** (2011) 012007 doi:10.1088/1742-6596/330/1/012007

(as are the initial ones, P^+, $P^\perp$, J^3, K^3, E^1 and E^2), because they are diagonal in the number of particles and commute with the UT, see Eq. (1). The three remaining generators, the Hamiltonian, P'^- and the generators F'^1, F'^2, are dynamical. Note that the factorisation property of the UT allows one to show that the few-body structure of F'^1 and F'^2 is similar to the structure of P'^- studied in the paper and represented in a compact form in Eq. (22).

Note that the property of cluster separability is violated in [2] because, by construction, pair interaction parts of a three-body potential in the momentum representation are not diagonal with respect to the three-momentum of a particle in flight. They are accompanied with δ-functions which makes them diagonal with respect to the space components of the *boosted* momenta of particles in flight. These boosts are different in the initial and final states because they are determined by total momentum of the three *on-mass-shell* particles in the initial and final states respectively, which is why the momentum of a particle in flight is not conserved. This violation is important in problems where the three-body system cannot be fixed at rest, like, for example, in the case of electron scattering off a three-body bound state. It would be interesting to discuss our results in the context of "packing operators" [6, 7] which were devised to handle the clustering problem present in [2].

Appendix A. Disconnected parts of the matrix elements of A

To prove Eq. (10), we first need to use the *decoupling equation*, Eq. (4), to derive expressions for the disconnected parts of the matrix elements of A. We begin by proving that the disconnected parts of the matrix elements of $A^\dagger$ are given by

$$\begin{aligned} \langle \eta_1, \eta_2' | A^\dagger | \lambda_1, \eta_2 \rangle^{disc} &= \langle \eta_1 | A^\dagger | \lambda_1 \rangle \langle \eta_2' | \eta_2 \rangle \\ \langle \eta_1, \eta_2 | A^\dagger | \lambda_1, \lambda_2 \rangle^{disc} &= \langle \eta_1 | A^\dagger | \lambda_1 \rangle \langle \eta_2 | A^\dagger | \lambda_2 \rangle. \end{aligned} \tag{A.1}$$

There is a direct way to carefully check whether the anzatz (A.1) satisfies Eq. (4), but this is a tedious and error-prone task. A more efficient way is to identify our disconnectedness problem with the one generated by the sum of two independent hamiltonians. This trick proved very efficient in the construction of disconnected amplitudes in TOPT [17, 18]. In the current context, a simple example of this approach is to express Eq. (15) for the disconnected matrix elements of the Hamiltonian H as

$$\langle s_1', s_2' | H | s_1, s_2 \rangle^{disc} = \langle s_1', s_2' | (H_1 + H_2) | s_1, s_2 \rangle \tag{A.2}$$

where H_1 and H_2 are independent hamiltonians that depend on their own nucleon and meson fields, $\psi_1 \equiv \{\Psi_1, \Phi_1\}$ and $\psi_2 \equiv \{\Psi_2, \Phi_2\}$, respectively, in exactly the same way that H depends on its fields $\psi \equiv \{\Psi, \Phi\}$. Because particles of "type 1" do not interact with particles of "type 2", the corresponding Hamiltonian is $H_1 + H_2$. The state, $|s_1, s_2\rangle$ on the right hand side of Eq. (A.2) is given by the direct product $|s_1, s_2\rangle = |s_1\rangle \times |s_2\rangle$ where $|s_1\rangle$ and $|s_2\rangle$ are particle Fock states associated with "type 1" and "type 2" particles, respectively. Similarly $\langle s_1', s_2'| = \langle s_1'| \times \langle s_2'|$. It follows that

$$\begin{aligned} \langle s_1', s_2' | (H_1 + H_2) | s_1, s_2 \rangle &= \langle s_1' | H_1 | s_1 \rangle \langle s_2' | s_2 \rangle + \langle s_2' | H_2 | s_2 \rangle \langle s_1' | s_1 \rangle \\ &= \langle s_1' | H | s_1 \rangle \langle s_2' | s_2 \rangle + \langle s_2' | H | s_2 \rangle \langle s_1' | s_1 \rangle \end{aligned} \tag{A.3}$$

which is just Eq. (15) in compact notation. In a similar way we express Eq. (A.1) for the disconnected matrix elements of $A^\dagger$ as

$$\langle \eta_1, \eta_2 | A^\dagger | s_1, s_2 \rangle^{disc} = \langle \eta_1, \eta_2 | (A_1^\dagger + A_2^\dagger + A_1^\dagger A_2^\dagger) | s_1, s_2 \rangle \tag{A.4}$$

where like H_i ($i = 1$ or 2), A_i is determined in the theory of "type i" nucleons and mesons in the same way as A is determined by Eq. (4) in the initial theory with the fields Ψ and Φ.

There is an important subtlety regarding Eq. (A.4) - it is valid only if the final state is from the η-subspace; e.g., it is not valid for the final state $\langle\lambda_1,\eta_2|$. Indeed,

$$\begin{aligned}&\langle\lambda_1,\eta_2|A^\dagger|s_1,s_2\rangle^{disc}=0\\&\neq\langle\lambda_1,\eta_2|(A_1^\dagger+A_2^\dagger+A_1^\dagger A_2^\dagger)|s_1,s_2\rangle=\langle\lambda_1,\eta_2|A_2^\dagger|s_1,s_2\rangle\\&=\langle\eta_2|A_2^\dagger|s_2\rangle\langle\lambda_1|s_1\rangle.\end{aligned}\tag{A.5}$$

Eq. (A.4) can be written for an arbitrary final state in the following form

$$\langle s_1',s_2'|A^\dagger|s_1,s_2\rangle^{disc}=\langle s_1',s_2'|(\eta A_1^\dagger+\eta A_2^\dagger+A_1^\dagger A_2^\dagger)|s_1,s_2\rangle\tag{A.6}$$

where we have taken ito account that $\langle s_1',s_2'|\eta A_1^\dagger A_2^\dagger=\langle s_1',s_2'|A_1^\dagger A_2^\dagger$. It is straightforward to rewrite Eq. (A.6) for the oparator A

$$\langle s_1',s_2'|A|s_1,s_2\rangle^{disc}=\langle s_1',s_2'|(A_1\eta+A_2\eta+A_1A_2)|s_1,s_2\rangle.\tag{A.7}$$

To prove Eq. (A.7), we show that it leads to the decoupling equation for disconnected matrix elements; indeed, recalling the left hand side of Eq. (4), we have that

$$\begin{aligned}&[\langle\lambda|(\lambda-A)H(\eta+A)|\eta\rangle]^{disc}=[\langle\lambda|(\lambda-A\eta)H(\eta+A)|\eta\rangle]^{disc}=\\&\langle\lambda|(1-A_1\eta-A_2\eta-A_1A_2\eta)(H_1+H_2)(1+A_1+A_2+A_1A_2)|\eta\rangle.\end{aligned}\tag{A.8}$$

Please note that we inserted the η-projector next to A to validate the use of Eq. (A.4). There are two types of λ-states, $|\lambda_1,\eta_2\rangle$ and $|\lambda_1,\lambda_2\rangle$; we should proceed separately for each of them in Eq. (A.8). For the final $\langle\lambda_1,\lambda_2|$ state we have

$$\begin{aligned}&[\langle\lambda_1,\lambda_2|(\lambda-A)H(\eta+A)|\eta\rangle]^{disc}\\&=\langle\lambda_1,\lambda_2|(1-A_1\eta-A_2\eta-A_1A_2\eta)(H_1+H_2)(1+A_1+A_2+A_1A_2)|\eta\rangle\\&=\langle\lambda_1,\lambda_2|(1-A_1)H_1(1+A_1)A_2+(1\leftrightarrow2)]|\eta\rangle\\&=\langle\lambda_1|(1-A)H(1+A)|\eta_1\rangle\langle\lambda_2|A|\eta_2\rangle+(1\leftrightarrow2)=0\end{aligned}\tag{A.9}$$

where we have used Eq. (4) for the full amplitude, and the fact that

$$\begin{aligned}\langle\lambda_1,\lambda_2|A_i\eta&=0,\qquad A_1A_1=\lambda A_1\eta\lambda A_1\eta=0,\\\langle\lambda_1,\lambda_2|X_1N_1B_1|\eta\rangle&=\langle\lambda_1,\lambda_2|X_2N_2B_2|\eta\rangle=0.\end{aligned}\tag{A.10}$$

Note that Eq. (A.9) is just the decoupling equation for the case of disconnected matrix elements. In the same way for the final $\langle\lambda_1,\eta_2'|$ state we have

$$\begin{aligned}&[\langle\lambda_1,\eta_2'|(\lambda-A)H(\eta+A)|\eta\rangle]^{disc}=\\&\langle\lambda_1,\eta_2'|(1-A_1-A_2-A_1A_2)(H_1+H_2)(1+A_1+A_2+A_1A_2)|\eta\rangle\\&=\langle\lambda_1,\eta_2'|(1-A_1)H_1(1+A_1)|\eta_1,\eta_2\rangle\\&=\langle\lambda_1|(1-A)H(1+A)|\eta_1\rangle\langle\eta_2'|\eta_2\rangle=0.\end{aligned}\tag{A.11}$$

Having established Eq. (A.4) for operator A, we now derive a similar factorization for $A^\dagger A$:

$$\begin{aligned}&\langle p_1,...p_l;p_{l+1},...p_n|A^\dagger A|q_1,...q_l;q_{l+1},...q_n\rangle^{disc}\\&=\langle p_1,...p_l;p_{l+1},...p_n|[A_1^\dagger A_1+A_2^\dagger A_2+(A_1^\dagger A_1)(A_2^\dagger A_2)]|q_1,...q_l;q_{l+1},...q_n\rangle\end{aligned}\tag{A.12}$$

where in the second line the nucleons $p_1, ...p_l; q_1, ...q_l$ in the final and initial state are of "type 1", the nucleons $p_{l+1}, ...p_n; q_{l+1}, ...q_n$ in the final and initial state are of "type 2", and the opertors A_i are defined via the fields ψ_i in the same way as A is defined via the fields ψ. Indeed, writing (A.12) as

$$\begin{aligned}\langle\eta_1', \eta_2'|A^\dagger A|\eta_1, \eta_2\rangle^{disc} = \langle\eta_1'|A^\dagger A|\eta_1\rangle\langle\eta_2'|\eta_2\rangle + \langle\eta_2'|A^\dagger A|\eta_2\rangle\langle\eta_1'|\eta_1\rangle \\ + \langle\eta_1'|A^\dagger A|\eta_1\rangle\langle\eta_2'|A^\dagger A|\eta_2\rangle,\end{aligned} \tag{A.13}$$

we have that

$$\begin{aligned}\langle\eta_1', \eta_2'|A^\dagger A|\eta_1, \eta_2\rangle^{disc} &= \sum_\lambda \langle\eta_1', \eta_2'|A^\dagger|\lambda\rangle^{disc}\langle\lambda|A|\eta_1, \eta_2\rangle^{disc} \\
&= \sum_{\lambda,\eta''} \langle\eta_1', \eta_2'|A^\dagger|\lambda_1, \eta_2''\rangle^{disc}\langle\lambda_1, \eta_2''|A|\eta_1, \eta_2\rangle^{disc} \\
&+ \sum_{\lambda,\eta''} \langle\eta_1', \eta_2'|A^\dagger|\eta_1'', \lambda_2\rangle^{disc}\langle\eta_1'', \lambda_2|A|\eta_1, \eta_2\rangle^{disc} \\
&+ \sum_{\lambda} \langle\eta_1', \eta_2'|A^\dagger|\lambda_1, \lambda_2\rangle^{disc}\langle\lambda_1, \lambda_2|A|\eta_1, \eta_2\rangle^{disc} \\
&= \sum_\lambda \langle\eta_1'|A^\dagger|\lambda_1\rangle\langle\lambda_1|A|\eta_1\rangle\langle\eta_2'|\eta_2\rangle \\
&+ \sum_\lambda \langle\eta_2'|A^\dagger|\lambda_2\rangle\langle\lambda_2|A|\eta_2\rangle\langle\eta_1'|\eta_1\rangle \\
&+ \sum_\lambda \langle\eta_1'|A^\dagger|\lambda_1\rangle\langle\eta_2'|A^\dagger|\lambda_2\rangle\langle\lambda_1|A|\eta_1\rangle\langle\lambda_2|A|\eta_2\rangle \\
&= \langle\eta_1'|A^\dagger A|\eta_1\rangle\langle\eta_2'|\eta_2\rangle + \langle\eta_2'|A^\dagger A|\eta_2\rangle\langle\eta_1'|\eta_1\rangle \\
&+ \langle\eta_1'|A^\dagger A|\eta_1\rangle\langle\eta_2'|A^\dagger A|\eta_2\rangle.\end{aligned} \tag{A.14}$$

Next, one can prove the factorization (10) of the matrix elements of ηU. The most nontrivial part is to show the factorisation of $\eta U \eta$, using Eq. (A.12) in

$$\langle\eta_1, \eta_2|U|\eta_1, \eta_2\rangle^{disc} = \langle\eta_1, \eta_2|(1 + A^\dagger A)^{-1/2}|\eta_1, \eta_2\rangle^{disc}. \tag{A.15}$$

Let us check Eq. (10) in the η subspace, i.e., for the case of n nucleon states $\langle p_1, ...p_l; p_{l+1}, ...p_n|U|q_1, ...q_l; q_{l+1}, ...q_n\rangle$,

$$\begin{aligned}&\langle p_1, ...p_l; p_{l+1}, ...p_n|U|q_1, ...q_l; q_{l+1}, ...q_n\rangle^{disc} \\
&= \langle p_1, ...p_l; p_{l+1}, ...p_n|(1 + A^\dagger A)^{-1/2}|q_1, ...q_l; q_{l+1}, ...q_n\rangle^{disc} \\
&= \langle p_1, ...p_l; p_{l+1}, ...p_n|(1 + [A^\dagger A]^{disc})^{-1/2}|q_1, ...q_l; q_{l+1}, ...q_n\rangle^{disc}\end{aligned} \tag{A.16}$$

where we have used the notation $[A^\dagger A]^{disc}$

$$\begin{aligned}&\langle p_1, ...p_l; p_{l+1}, ...p_n|[A^\dagger A]^{disc}|q_1, ...q_l; q_{l+1}, ...q_n\rangle \\
&= \langle p_1, ...p_l; p_{l+1}, ...p_n|A^\dagger A|q_1, ...q_l; q_{l+1}, ...q_n\rangle^{disc}.\end{aligned} \tag{A.17}$$

Eq. (A.16) follows from the fact that the disconnected part of the matrix element of any iteration of operator $A^\dagger A$ is equal to the iteration of the disconnected parts of the matrix elements of $A^\dagger A$. This statement for the second iteration mathematically looks as follows:

$$\langle f|(A^\dagger A)(A^\dagger A)|i\rangle^{disc} = \sum_j \langle f|(A^\dagger A)|j\rangle^{disc}\langle j|(A^\dagger A)|i\rangle^{disc}. \tag{A.18}$$

Using Eq. (A.12) in Eq. (A.16) one gets Eq. (10)

$$\begin{aligned}
&\langle p_1, ...p_l; p_{l+1}, ...p_n|U|q_1, ...q_l; q_{l+1}, ...q_n\rangle^{disc} \\
&= \langle p_1, ...p_l; p_{l+1}, ...p_n|(1 + A^\dagger A)^{-1/2}|q_1, ...q_l; q_{l+1}, ...q_n\rangle^{disc} \\
&= \langle p_1, ...p_l; p_{l+1}, ...p_n|(1 + [A^\dagger A]^{disc})^{-1/2}|q_1, ...q_l; q_{l+1}, ...q_n\rangle^{disc} \\
&= \langle p_1, ...p_l; p_{l+1}, ...p_n|(1 + A_1^\dagger A_1 + A_2^\dagger A_2 + A_1^\dagger A_1 A_2^\dagger A_2)^{-1/2}|q_1, ...q_l; q_{l+1}, ...q_n\rangle^{disc} \\
&= \langle p_1, ...p_l; p_{l+1}, ...p_n|(1 + A_1^\dagger A_1)^{-1/2}(1 + A_2^\dagger A_2)^{-1/2}|q_1, ...q_l; q_{l+1}, ...q_n\rangle \\
&= \langle p_1, ...p_l|(1 + A_1^\dagger A_1)^{-1/2}|q_1, ...q_l\rangle \times \langle p_{l+1}, ...p_n|(1 + A_2^\dagger A_2)^{-1/2}|q_{l+1}, ...q_n\rangle \\
&= \langle p_1, ...p_l|U_1|q_1, ...q_l\rangle \times \langle p_{l+1}, ...p_n|U_2|q_{l+1}, ...q_n\rangle \\
&= \langle p_1, ...p_l|U|q_1, ...q_l\rangle \times \langle p_{l+1}, ...p_n|U|q_{l+1}, ...q_n\rangle
\end{aligned} \tag{A.19}$$

There are two types of disconnected parts corresponding to $\eta U \lambda$,

$$\begin{aligned}
\langle \eta_1, \eta_2|U|\lambda_1, \eta_2\rangle^{disc} &= \langle \eta_2|U|\eta_2\rangle\langle \eta_1|U|\lambda_1\rangle \\
\langle \eta_1, \eta_2|U|\lambda_1, \lambda_2\rangle^{disc} &= \langle \eta_2|U|\lambda_2\rangle\langle \eta_1|U|\lambda_1\rangle,
\end{aligned} \tag{A.20}$$

whose factorisation is not hard to show:

$$\begin{aligned}
&\langle \eta_1, \eta_2|U|\lambda_1, \eta_2\rangle^{disc} = \langle \eta_1, \eta_2|(1 + A^\dagger A)^{-1/2}A^\dagger|\lambda_1, \eta_2\rangle^{disc} \\
&= \sum_{\eta'} \langle \eta_1, \eta_2|(1 + A^\dagger A)^{-1/2}|\eta_1', \eta_2'\rangle^{disc}\langle \eta_1', \eta_2'|A^\dagger|\lambda_1, \eta_2\rangle^{disc} \\
&= \sum_{\eta'} \langle \eta_2|(1 + A^\dagger A)^{-1/2}|\eta_2\rangle\langle \eta_1|(1 + A^\dagger A)^{-1/2}|\eta_1'\rangle\langle \eta_1'|A^\dagger|\lambda_1\rangle \\
&= \langle \eta_2|(1 + A^\dagger A)^{-1/2}|\eta_2\rangle\langle \eta_1|(1 + A^\dagger A)^{-1/2}A^\dagger|\lambda_1\rangle \\
&= \langle \eta_2|U|\eta_2\rangle\langle \eta_1|U|\lambda_1\rangle, \\
&\langle \eta_1, \eta_2|U|\lambda_1, \lambda_2\rangle^{disc} = \langle \eta_1, \eta_2|(1 + A^\dagger A)^{-1/2}A^\dagger|\lambda_1, \lambda_2\rangle^{disc} \\
&= \sum_{\eta'} \langle \eta_1, \eta_2|(1 + A^\dagger A)^{-1/2}|\eta_1', \eta_2'\rangle^{disc}\langle \eta_1', \eta_2'|A^\dagger|\lambda_1, \lambda_2\rangle^{disc} \\
&= \sum_{\eta'} \langle \eta_2|(1 + A^\dagger A)^{-1/2}|\eta_2'\rangle\langle \eta_1|(1 + A^\dagger A)^{-1/2}|\eta_1'\rangle\langle \eta_1'|A^\dagger|\lambda_1\rangle\langle \eta_2'|A^\dagger|\lambda_2\rangle \\
&= \langle \eta_2|(1 + A^\dagger A)^{-1/2}A^\dagger|\eta_2\rangle\langle \eta_1|(1 + A^\dagger A)^{-1/2}A^\dagger|\lambda_1\rangle \\
&= \langle \eta_2|U|\lambda_2\rangle\langle \eta_1|U|\lambda_1\rangle.
\end{aligned} \tag{A.21}$$

Appendix B. Equivalence of few-body RQM to QFT

Here we prove that the RQM corresponding to Hamiltonian H' produces the same on-shell scattering amplitudes as the ones produced by the QFT corresponding to Hamiltonian H.

The two-body states, $|N_q\rangle = \alpha|\Psi_q^{(+)}\rangle$ and $|N_q'\rangle = |\Psi_q'^{(+)}\rangle$, are normalized as

$$\langle N_p|N_q\rangle = \langle N_p'|N_q'\rangle = \delta^3(\underline{p}_1 - \underline{q}_1)\delta^3(\underline{p}_2 - \underline{q}_2) \tag{B.1}$$

$$|N_q\rangle = Z^{-1} \lim_{\epsilon \to 0} \frac{i\epsilon}{P_q^- - H + i\epsilon}|q_1, q_2\rangle \tag{B.2}$$

$$|N_q'\rangle = \lim_{\epsilon \to 0} \frac{i\epsilon}{P_q^- - H' + i\epsilon}|q_1, q_2\rangle \tag{B.3}$$

where the renormalization constant Z_q is defined as the residue at the single nucleon pole of the nucleon resolvent:

$$\langle \underline{p}|\frac{1}{p^- - H + i\epsilon}|\underline{q}\rangle \sim \delta^3(\underline{p}-\underline{q})\frac{Z_q}{p^- - \dfrac{q_\perp^2 + m^2}{q^+} + i\epsilon} \tag{B.4}$$

The renormalization constant Z'_q in Eq. (B.3) is equal to unity. Indeed it is defined as the residue of the single particle propagator at the pole, $p^- = \dfrac{q_\perp^2 + m^2}{q^+}$, corresponding to the single particle intermediate state contribution

$$\langle p_1|\frac{1}{p^- - H' - i\epsilon}|q_1\rangle \sim \delta^3(\underline{p}+\underline{q})\frac{Z'_q}{p^- - \dfrac{q_\perp^2 + m^2}{q^+} + i\epsilon} \tag{B.5}$$

Since H' is diagonal in the η space, the corresponding self-energy, $\Sigma'(p^-, \underline{p})$,

$$\Sigma'(p^-, \underline{p})\delta^3(\underline{p}-\underline{q}) = \langle \underline{p}|\left(H' - \frac{q_\perp^2 + m^2}{q^+}\right)|\underline{q}\rangle \tag{B.6}$$

does not depend on the "energy", p^-, and therefore the single particle wave function does not get renormalized, $Z'_q = 1$. Another way to see this is to realize that the one nucleon eigenstate of H' is the bare nucleon

$$|\Psi_p'^{(1)}\rangle = U|\Psi_p^{(1)}\rangle = |p\rangle, \tag{B.7}$$

then from the definition (B.5) of Z' one gets

$$\begin{aligned} \delta^3(p_1 - q_1)Z'_{q_1} &= \int d^3p\langle p_1|\Psi_p'^{(1)}\rangle\langle\Psi_p'^{(1)}|q_1\rangle = \int d^3p\langle p_1|p\rangle\langle p|q_1\rangle \\ &= \delta^3(p_1 - q_1) \end{aligned} \tag{B.8}$$

The relation between the normalized two-body states $|N_q\rangle$ and $|N'_q\rangle$ is given as

$$\begin{aligned} |N'_q\rangle &= \lim_{\epsilon\to 0}\frac{i\epsilon}{E_q - H' + i\epsilon}|q_1, q_2\rangle = U\lim_{\epsilon\to 0}\frac{i\epsilon}{E_q - H + i\epsilon}U^\dagger|q_1, q_2\rangle \\ &= \int (dk) U\lim_{\epsilon\to 0}\frac{i\epsilon}{E_q - E_k + i\epsilon}|N_k\rangle\langle N_k|U^\dagger|q_1, q_2\rangle \\ &= \int (dk) U\lim_{\epsilon\to 0}\frac{i\epsilon}{E_q - E_k + i\epsilon}|N_k\rangle\frac{1}{Z}\langle k_2, k_1|\frac{-i\epsilon}{E_k - H - i\epsilon}U^\dagger|q_1, q_2\rangle \\ &= \int (dk)\frac{1}{Z}U\lim_{\epsilon\to 0}\frac{i\epsilon}{E_q - E_k + i\epsilon}|N_k\rangle\langle k_2, k_1|\frac{-i\epsilon}{E_k - H - i\epsilon}U^\dagger|q_1, q_2\rangle^{disc} \\ &= \frac{1}{Z}F_U(q)U|N_q\rangle \end{aligned} \tag{B.9}$$

where $(dp) = d^3p_1d^3p_2$, in the second last line of Eq. (B.9) we have thrown away the connected part of $\langle k_2, k_1|\dfrac{-i\epsilon}{E_k - H - i\epsilon}U^\dagger|q_1, q_2\rangle$ because it does not contribute in the limit of $\epsilon \to 0$, it is avoided by keeping the energy of the state $|q_1, q_2\rangle$ below the pion production threshold [8].

$$\begin{aligned} \delta^3(k_1 - q_1)\delta^3(k_2 - q_2)F_U(q) &= \lim_{\epsilon\to 0}\langle k_2, k_1|\frac{-i\epsilon}{E_k - H - i\epsilon}U^\dagger|q_1, q_2\rangle^{disc} \\ &= \lim_{\epsilon\to 0}\langle k_2, k_1|U^\dagger\frac{-i\epsilon}{E_k - H' - i\epsilon}|q_1, q_2\rangle^{disc} \end{aligned} \tag{B.10}$$

Journal of Physics: Conference Series **330** (2011) 012007 doi:10.1088/1742-6596/330/1/012007

Note that in contrast to [8] we have a non-trivial contribution from $U^\dagger - 1$ to $F_U(q)$. Using our technique of calculating disconnected parts of matrix elements we find that $F_U(q) = Z$:

$$\delta^3(k_1 - q_1)\delta^3(k_2 - q_2)F_U(q) = \lim_{\epsilon\to 0}\langle k_2, k_1|U^\dagger \frac{-i\epsilon}{E_k - H' - i\epsilon}|q_1, q_2\rangle^{disc}$$
$$= \lim_{\epsilon\to 0}\langle k_2, k_1|U_1^\dagger U_2^\dagger \frac{-i\epsilon}{E_k - H_1' - H_2' - i\epsilon}|q_1, q_2\rangle$$
$$= \lim_{\epsilon\to 0}\langle k_2, k_1|U_1^\dagger U_2^\dagger|q_1, q_2\rangle \frac{-i\epsilon}{E_k - E_q - i\epsilon}$$
$$= \langle k_2|U^\dagger|q_2\rangle\langle k_1|U^\dagger|q_1\rangle = \langle k_2|\Psi_{q_2}\rangle\langle k_1|\Psi_{q_1}\rangle = Z\delta^3(k_1 - q_1)\delta^3(k_2 - q_2) \tag{B.11}$$

where we have used, $|\Psi'_{q_i}\rangle = |q_i\rangle$, $H'|q_i\rangle = \frac{q_{i\perp}^2 + m^2}{q_i^+}|q_i\rangle$, in the parametrization (3)

$$\langle k_i|U^\dagger|q_i\rangle = \langle k_i|U^\dagger|q_i\rangle^* = \langle k_i|\Psi_{q_i}\rangle = \sqrt{Z}\delta^3(k_i - q_i) \tag{B.12}$$

Note the trick used in the second line of Eq. (B.11): the disconnected part of a matrix element of an operator in the theory corresponding to the Hamiltonian $H(\psi)$ is replaced by a matrix element in the theory corresponding to the sum of the commuting with each other Hamiltonians $H(\psi_1) + H(\psi_2)$ which leads to the replacements, $U^\dagger\eta \to U_1^\dagger U_2^\dagger\eta$ and $H'\eta \to (H_1' + H_2')\eta$.

The missing terms in [8] imply the wrong expression for $\langle k|U^\dagger|q\rangle$,

$$\langle k|U^\dagger|q\rangle = \delta(k - q) \neq \delta(k - q)\sqrt{Z} \tag{B.13}$$

Then finally we have

$$|N_q'\rangle = U|N_q\rangle \tag{B.14}$$

To compare with [8] using Eq. (B.2) and Eq. (B.3) we get

$$\lim_{\epsilon\to 0}\frac{i\epsilon}{E_q - H' + i\epsilon}|q_1, q_2\rangle = \frac{1}{Z}U\lim_{\epsilon\to 0}\frac{i\epsilon}{E_q - H + i\epsilon}|q_1, q_2\rangle, \tag{B.15}$$

which *differs by factor Z* from the analogous relation in [8]. Just this factor resulting from $\sqrt{Z}$ associated with each nucleon leg should be factored out to derive a finite unitary scattering amplitude.

In the case of the NN scattering problem only two-nucleon scattering states $|\Psi_q^{(+)}\rangle$ and $|\Psi_q'^{(+)}\rangle$ (of the initial and UT-ed theories respectively) were to be related to each other. In Eq. (B.14) it has been shown that they are related to each other by UT, $|\Psi_q'^{(+)}\rangle = U|\Psi_q^{(+)}\rangle$ (if they are normalized in the same way, say, to the unity), thereby it has been shown that the two theories produce the same scattering amplitude if it is *on shell*. Thus the S-matrices corresponding to the initial and UT-ed Hamiltonians coincide with each other:

$$\langle p|S'|q\rangle = \langle|\Psi_p'^{(-)}|\Psi_q'^{(+)}\rangle = \langle|\Psi_p^{(-)}|U^\dagger U|\Psi_q^{(+)}\rangle = \langle|\Psi_p^{(-)}|\Psi_q^{(+)}\rangle = \langle p|S|q\rangle \tag{B.16}$$

In the NNN scattering problem one should also prove, analogous to Eq. (B.14), the relation between the scattering states where two nucleons are bound in the asymptotics. This will be shown in a more detailed paper.

References

[1] Dirac P A M 1949 *Rev. Mod. Phys.* **21** 392.
[2] Bakamjian B and Thomas L H 1953 *Phys. Rev.* **92** 1300.
[3] Foldy L L 1961 *Phys. Rev.* **122** 275.
[4] Foldy L L and Krajcik R A 1974 *Phys. Rev. Lett.* **32** 1025
Foldy L L and Krajcik R A 1975 *Phys. Rev. D* **12** 1700.
[5] Carlson J, Pandharipande V R and Shiavilla R S 1993 *Phys. Rev. C* **47** 484
[6] Sokolov S N 1977 *Dokl. Akad. Nauk* **233** 575
Sokolov S N 1978 *Teor. Mat. Fiz.* **36** 193
Sokolov S N and Shatnii A N 1978 *ibid.* **37** 291.
[7] Coester F and Polyzou W 1982 *Phys. Rev. D* **26** 1348.
[8] Krüger A and Glöckle W 1999 *Phys. Rev. C* **59** 1919.
[9] Keister B D and Polyzou W N 1991 *Adv. Nucl. Phys.* **20** 225.
[10] Okubo S 1954 *Prog. Theor. Phys. (Japan)* **12** 603.
[11] Fukuda N, Sawada K and Taketani M 1954 *Prog. Theor. Phys. (Japan)* **12** 156.
[12] Terent'ev M V 1976 *Sov. J. Nucl. Phys.* **24** 106.
[13] Berestetsky V B and Terentev M V 1976 *Sov. J. Nucl. Phys.* **24** 547.
[14] Berestetsky V B and Terentev M V 1977 *Sov. J. Nucl. Phys.* **25** 347.
[15] Kondratyuk L A and Terentev M V 1980 *Sov. J. Nucl. Phys.* **31** 561.
Epelbaum E, Ulf-Meissner G, Gloeckle W 2003 *Nucl. Phys.* **A714** 535.
[16] Epelbaum E 2006 Prog. Part. Nucl. Phys. **57**, 654.
[17] Kvinikhidze A N and Blankleider B 1993 *Phys. Rev. C* **48** 25.
[18] Kvinikhidze A N and Blankleider B 1993 *Phys. Lett.* **B307** 7.

Journal of Physics: Conference Series **330** (2011) 012008 doi:10.1088/1742-6596/330/1/012008

Diverse applications of the gauging of equations method

B. Blankleider[1], **A. N. Kvinikhidze**[1,2], **and T. Skawronski**[1]

[1]School of Chemical and Physical Sciences, Flinders University, Bedford Park, SA 5042, Australia,

[2]The Mathematical Institute of Georgian Academy of Sciences, Tbilisi, Georgia

Abstract.

The gauging of equations method was developed some years ago in order to extract gauge invariant electromagnetic currents from strongly interacting few-body systems described by integral equations. Although the method is by now well-established for this purpose, it is not so well known that the same approach can be used to solve a wide variety of problems. The purpose of this paper is to demonstrate the true power of this method by using it to solve key problems in three very different areas of study; in particular, we show (i) how generalized parton distributions (GPDs) can be determined in the case where hadrons are described in terms of their partonic degrees of freedom through solutions of dynamical equations, (ii) how to construct gauge invariant two-nucleon currents in cutoff effective field theory (EFT), even though the cutoff usually violates the gauge invariance of the underlying Lagrangian, and (iii) how to construct a potential model of πN scattering that is crossing symmetric and approximately unitary.

It is noteworthy that in each case, the gauging of equations method produces results that are consistent with the relevant requirements of quantum field theory. For example, in extracting GPDs from the model of strong interactions defined by dynamical equations, all possible mechanisms contributing to the GPDs are taken into account, so that all GPD sum rules are satisfied automatically. Similarly, in the construction of electromagnetic currents in cutoff EFT, gauge invariance is a consequence of the fact that an external photon is effectively coupled to every part of every strong interaction diagram in the model.

1. Introduction

Experimentally, photons have long been used as probes to study the structure of hadronic systems. Theoretically, the corresponding reactions are described by matrix elements of the electromagnetic (EM) current operator, J^μ, and it must ultimately be the goal of the theorist to construct models where such matrix elements describe the experimental data as accurately as possible. In this respect, an essential constraint on J^μ is that it should be introduced in a gauge invariant way. Note that gauge invariance is a stronger constraint than just current conservation, $\partial_\mu J^\mu(x) = 0$: beyond current conservation, it requires commutation relations of the form $[J^0, \Psi] = \hat{e}\Psi$, where the matrix $\hat{e}$ extracts the charge of the field Ψ. This difference is particularly important for an effective field theory (EFT), where an additional constraint increases its predictive power.

Gauge invariance is a fundamental property of any theory at the Lagrangian level, and it is essential to preserve this gauge invariance in the formalism leading to matrix elements that

Journal of Physics: Conference Series **330** (2011) 012008 doi:10.1088/1742-6596/330/1/012008

describe physical strong interaction processes and their coupling to an external electromagnetic field. Here we are concerned with descriptions where the strong interaction processes are described by dynamical integral equations. These may be, for example, four-dimensional Bethe-Salpeter type equations that sum up, non-perturbatively, an infinite number of Feynman diagrams, or they may be their properly reduced three-dimensional versions. Gauge invariance in such equations is manifested in the form of Ward-Takahashi identities between certain Green functions. Because one uses models to approximate the full theory, gauge invariance is not guaranteed *a priori*; for this reason, much effort has been devoted to the question of how to impose this invariance within such strong interaction models [1, 2, 3, 4, 5, 6, 7, 8, 9].

Here we present a method for constructing gauge invariant currents that is applicable to any type of strongly interacting systems which are described nonperturbatively by integral equations. This method involves a direct gauging of the equations themselves in such a way that one gets a linear equation for a Green function of the strongly interacting system with the external photon effectively attached everywhere, so that gauge invariance and thereby currrent conservation is guaranteed.

A notorious problem plaguing many field-theoretical approaches based on the nonperturbative summing of Feynman diagrams is that of overcounting [10, 11, 6] . An important feature of the gauging of equations method is that it not only guarantees gauge invariance, but it also attaches photons everywhere without introducing overcounting. In this sense the actual effect of the method exceeds expectations, and perhaps it is for this reason that this additional feature of the method has been somewhat under-appreciated. For example, a group intensively engaged in calculations of nucleon form-factors in the quark-diquark model fell victim to the overcounting problem [12, 13], despite the fact that such overcounting was discussed in the context of the gauging of equations method a few years earlier [11, 5]. Even a specific Comment [14] on the overcounting in Refs. [12, 13] turned out not to be convincing, and it took another 5 years until the problem was finally recognised and the corresponding expressions corrected in further publications since 2005 [15, 16, 17].

Given the power of the gauging of equations method just outlined, it is gratifying that it can be used on a wide variety of problems. Its use in generating electromagnetic currents of standard three-dimensional (3D) and four-dimensional (4D) few-body problems is by now well documented [2, 3, 4, 5, 18, 6, 7, 8, 19, 20, 9]. The purpose of the present paper is to demonstrate the versatility of this method by applying it to three very different areas of intense current interest - generalized parton distributions (GPDs), nuclear effective field theory (EFT), and potential models of πN scattering. The paper therefore has three main parts. The first part, Section II, is devoted to the construction of the GPDs along with the gauge invariant currents within the model of strong interactions defined by dynamical equations. Here the gauging of equations method provides a powerful approach where gauge invariance and GPD sum rules are implemented in a theoretically correct fashion, and with all possible overcounting avoided. In the second part, Section III, for the first time a gauge invariant approach to cutoff-regularized two-nucleon (NN) EFT is constructed. It is noteworthy that our approach is able to construct gauge invariant currents and preserve power counting, in spite of the fact that such regularization destroys the gauge invariance built into the Lagrangian. Finally in Section IV, the latest application of the gauging of equations method is presented: we provide an approximate solution to the long-standing problem of constructing a non-perturbative potential model of πN scattering that, simultaneously, respects both crossing symmetry and unitarity. In particular, the derived πN amplitude has exact crossing symmetry while the accuracy of its unitarity depends on the numerical ability of the model to fit data. The key to this achievement is the idea of "double-gauging" a dressed nucleon propagator with pions.

Journal of Physics: Conference Series **330** (2011) 012008 doi:10.1088/1742-6596/330/1/012008

2. Generalized parton distributions.

The structure of hadrons is comprehensively described through their GPDs [21, 22, 23, 24, 25, 26, 27, 28] (for reviews see Refs. [29, 30]). GPDs are usually defined in terms of light front correlation functions $\langle P'|\bar{q}_\beta(0)q_\alpha(y)|P\rangle$, where the bilocal operator $\bar{q}_\beta(0)q_\alpha(y)$ is a product of quark, gluon, or other parton fields, and $|P\rangle$ is a hadronic state with momentum P. However, perhaps the most generally applicable way to specify GPDs is through the density matrix defined as [22, 23]

$$\rho_{\alpha\beta}(P', P, k) = \int d^4y\, e^{ik\cdot y} \langle P'|T\bar{q}_\beta(0)q_\alpha(y)|P\rangle_C \tag{1}$$

where 'T' stands for usual time (y^0) ordering, and subscript 'C' indicates that only connected contributions are retained. Eq. (1) describes the virtual (off-shell) scattering of partons off the hadronic state, as illustrated in Fig. 1. Then, for the study of deeply virtual Compton scattering (DVCS) where one is restricted to the light front ($y^+ = 0$), Eq. (1) can be used to define the light front distribution function

$$\rho_{\alpha\beta}(P', P, \underline{\mathbf{k}}) \equiv \int \frac{dk^-}{2\pi}\, \rho_{\alpha\beta}(P', P, k) = \int d^4y\, e^{ik\cdot y} \langle P'|T\bar{q}_\beta(0)q_\alpha(y)|P\rangle_C\, \delta(y^+). \tag{2}$$

where $\underline{\mathbf{k}} = (k^+, \mathbf{k}^\perp)$. A further integration over $d^2k^\perp$ then leads to the usual definition of GPDs.[1]

On the other hand, for studies where rotational invariance is important, like investigations of the shape of a nucleon [33, 34, 35], Eq. (1) can be used to define a rotationally invariant distribution function on the $y^0 = 0$ surface

$$\rho_{\alpha\beta}(P', P, \mathbf{k}) \equiv \int \frac{dk^0}{2\pi}\, \rho_{\alpha\beta}(P', P, k) = \int d^4y\, e^{ik\cdot y} \langle P'|T\bar{q}_\beta(0)q_\alpha(y)|P\rangle_C\, \delta(y^0). \tag{3}$$

Also, the same expression for $\rho_{\alpha\beta}$ as Eq. (1), but where $\bar{q}_\beta$ and q_α are nucleon fields and $|P\rangle$ is the state vector corresponding to an atomic nucleus, can be similarly used to describe the nucleonic structure of nuclei.

Using Eq. (1) as the basic starting point for the description of GPDs and other quantities, this paper addresses the question of how to extract the density matrix $\rho_{\alpha\beta}$ in the case where the hadron's structure is modeled by a dynamical equation describing the mutual scattering of its constituents. As a concrete example, we consider the case of a meson or diquark modeled by the Bethe-Salpeter equation, as well as the case of a baryon, modeled as the bound state of three quarks, whose bound state wave function is found by solving a Faddeev-like three-body scattering equation. For this purpose we consider both a four-dimensional (4D) formulation of the three-body scattering equations, and a covariant three-dimensional (3D) one using the "spectator approach" of Gross [36, 37, 38].

[1] Here one also needs to replace the time-ordered product of fields by an ordinary product, a step justified for the diagonal case $P = P'$ in Ref. [31], and for the general case in Ref. [32].

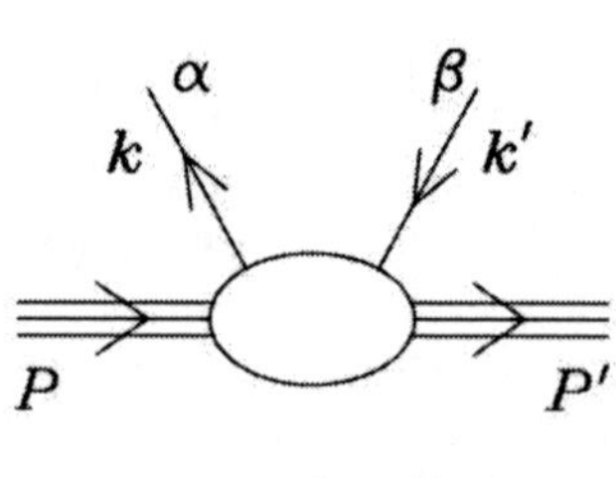

Figure 1. Virtual quark-hadron scattering as described by the density matrix $\rho_{\alpha\beta}$ of Eq. (1).

Journal of Physics: Conference Series **330** (2011) 012008 doi:10.1088/1742-6596/330/1/012008

An important requirement in any extraction of $\rho_{\alpha\beta}$ is the preservation of sum rules which relate GPDs to electromagnetic form factors. In this regard, the density matrix defined by Eq. (1) must satisfy the sum rule of Eq. (25) which relates $\rho_{\alpha\beta}$ to the electromagnetic vertex function Γ^μ. Charge and current conservation then lead to two further sum rules: Eq. (28), and Eq. (29) - we shall refer to these three sum rules collectively as 'GPD sum rules'. In any given model, of course, these sum rules are not guaranteed. Indeed, the nonperturbative nature of the model considered here (nonperturbative solutions of scattering equations), makes the task of extracting $\rho_{\alpha\beta}$, while preserving the GPD sum rules, particularly challenging.

Some years ago a similar problem presented itself: how to determine the electromagnetic currents of hadronic systems described by nonperturbative solutions of scattering equations. Here, of course, the currents had to obey charge and current conservation. The solution to this problem came with the development of the 'gauging of equations method' [2, 3, 4, 5, 6, 7]. This method not only solves the problem of extracting charge and current conserving electromagnetic currents, but it does so in accordance with strict adherence to theory: the external electromagnetic field is attached to all possible places in the nonperturbative strong interaction processes defined by the dynamic equation model. A further feature of the method is that it automatically takes care of all the overcounting problems that plague 4D approaches [10, 8, 4]. Indeed, the gauging of equations method has since been instrumental in the construction of current conserving electromagnetic interactions [39, 40, 41, 9], and in enabling careful analyses of the overcounting problem [14, 42, 9].

In the present paper we exploit the fact that the gauging of equations method can be applied not only to matrix elements of the electromagnetic current operator, but to all other field operators as well. In particular, by applying this method to the case of matrix elements of bilocal operators, we are able to construct the density matrix $\rho_{\alpha\beta}$, and thus GPDs, of hadronic systems described by dynamical equations. Moreover, just like 'gauging equations with a local electromagnetic operator' results in currents where the electromagnetic field is attached to all possible places in the strong interaction model, 'gauging equations with a bilocal operator' results in GPDs where the two external legs $\bar{q}_\beta$ and q_α (bilocal field) originate from cutting all possible bare quark propagators in the strong interaction model. It is this completeness of 'cutting bare propagators' that makes the resulting $\rho_{\alpha\beta}$ satisfy the GPD sum rules.

More specifically, by applying the gauging of equations method for bilocal fields to a dynamical equation describing a hadronic state, we obtain an expression for the density matrix $\rho_{\alpha\beta}$ that obeys the GPD sum rules whenever the input quantities - the distribution functions that result from cutting all bare propagators in (i) the dressed quark propagator, and (ii) the quark-quark potential - themselves satisfy corresponding sum rules. Because the gauging of equations method for bilocal fields cuts *all* bare propagators in the model, the resulting density matrix not only satisfies the GPD sum rules, but it does so in the theoretically correct way.

For the concrete model of a baryon described by Faddeev-like equations, the gauging of equations method leads to Eq. (42) for the density matrix $\rho_{\alpha\beta}$. This expression satisfies the GPD sum rules whenever the inputs satisfy corresponding sum rules, and can be used directly for practical calculations.

2.1. *Gauging*

By 'gauging' we shall mean the transformation $G \to G^U$ where G is the n-point Green function

$$G(x_1, \ldots, x_n) = \langle 0|Tq(x_1)\ldots\bar{q}(x_n)|0\rangle, \tag{4}$$

and G^U is the 'gauged' Green function defined as the corresponding $(n+2)$-point function

$$G^U(x_1, \ldots, x_n; x, y) = \langle 0|Tq(x_1)\ldots\bar{q}(x)q(y)\ldots\bar{q}(x_n)|0\rangle_c \tag{5}$$

Journal of Physics: Conference Series **330** (2011) 012008 doi:10.1088/1742-6596/330/1/012008

where subscript c indicates that all contributions having a disconnected piece $\langle 0|T\bar{q}(x)q(y)|0\rangle$ are excluded. To be definite, we shall refer to quark parton distributions in hadrons and thus take all the $q(x_1), \ldots, q(x_n)$, $q(x)$ and $q(y)$ to represent quark fields; however, it should be understood that our discussion applies equally well, for example, to gluon distributions in which case $q(x)$ and $q(y)$ would represent gluon fields, and to nucleon distributions within a nucleus, in which case $q(x_1), \ldots, q(x_n)$, $q(x)$ and $q(y)$ would all represent spinor nucleon fields. To make the connection with GPDs more clear, it is also useful to write Eq. (5) with two of the spinor components made explicit:

$$G^U_{\alpha\beta}(x_1, \ldots, x_n; x, y) = \langle 0|Tq(x_1) \ldots \bar{q}_\beta(x)q_\alpha(y) \ldots \bar{q}(x_n)|0\rangle_c. \tag{6}$$

The definition of the gauged Green function given in Eq. (5) can be considered as an extension to bilocal fields of the definition used in the case of coupling to a local field, for example the electromagnetic field, which involves the $(n+1)$-point function

$$G^\mu(x_1, \ldots, x_n; z) = \langle 0|Tq(x_1) \ldots \bar{q}(z)\hat{\Gamma}^\mu q(z) \ldots \bar{q}(x_n)|0\rangle \tag{7}$$

where $J^\mu(z) = \bar{q}(z)\hat{\Gamma}^\mu q(z)$ is the electromagnetic current operator. Indeed the gauging method used in this paper is closely based on the one developed for the $(n+1)$-point function of Eq. (7) [4, 5, 6, 7].

We are interested in the case where the Green function G is modelled nonperturbatively as the solution to an integral equation of the form

$$G = G_0^P + G_0VG \tag{8}$$

where G_0 is a product of dressed single particle propagators, G_0^P is the antisymmetrized version of G_0, and V is the interaction kernel (in the $3 \to 3$ processes of main interest here, V consists of all possible 3-particle irreducible diagrams). In the 'gauging of equations method', the $(n+1)$-point Green function G^μ is obtained by 'gauging' Eq. (8) with a local (vector) field as [4, 5, 6, 7]

$$G^\mu = G_0^{\mu P} + G_0^\mu VG + G_0V^\mu G + G_0VG^\mu. \tag{9}$$

Similarly, we obtain the $(n+2)$-point Green function G^U by 'gauging' Eq. (8) with a bilocal (spinor) field as

$$G^U = G_0^{UP} + G_0^U VG + G_0V^U G + G_0VG^U. \tag{10}$$

The one-to-one correspondence between these two types of gauging is self-evident, with each of the above two equations leading to solutions (for G^μ and G^U) that are of identical form. When necessary to distinguish between these two types of gauging, we shall refer to the transformation $G \to G^U$ according to Eq. (10) as 'U-gauging', and the transformation $G \to G^\mu$ according to Eq. (9) as 'μ-gauging'.

Although this method of gauging is designed specifically for nonperturbative approaches, its idea is rooted in perturbation theory: for example, to any diagram $\mathcal{D}$ contributing to

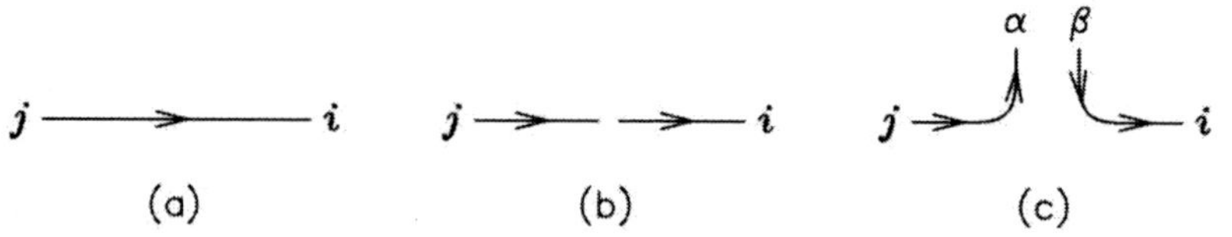

Figure 2. Pictorial representation of 'U-gauging'. (a) Bare particle propagator. (b) Bare particle propagator cut through the middle. (c) Inner cut legs form a bilocal external field.

Journal of Physics: Conference Series **330** (2011) 012008 doi:10.1088/1742-6596/330/1/012008

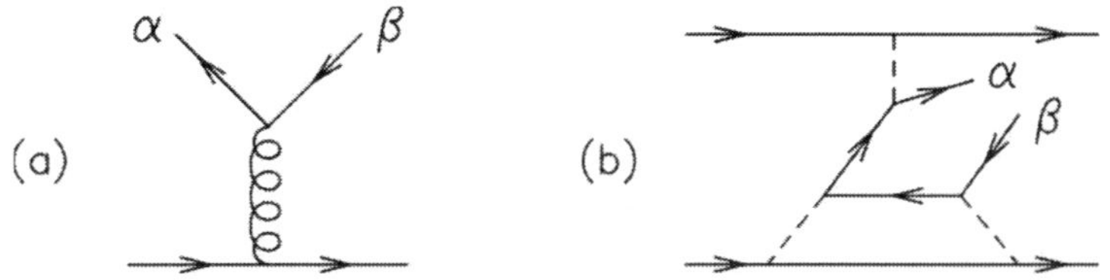

Figure 3. Example of diagrams which cannot be obtained by gauging. (a) One gluon exchange contribution. Joining the end points α and β gives a tadpole diagram which is zero. (b) A three-pion (dashed lines) contribution which, upon joining together the end-points α and β, gives a $2 \to 2$ diagram which is forbidden by G-parity.

$G(x_1, \ldots, x_n)$, there corresponds a sum of diagrams $\mathcal{D}^U_{\alpha\beta}$ belonging to $G^U_{\alpha\beta}(x_1, \ldots, x_n; x, y)$, each of which can be obtained by replacing a bare propagator $d_0(u-v) = \langle 0_f|Tq_f(u)\bar{q}_f(v)|0_f\rangle$ in $\mathcal{D}$ (subscript f indicates a free field or state) by $d^U_{0,\alpha\beta}(u, v; x, y) = \langle 0_f|Tq_f(u)\bar{q}_\beta(x)q_\alpha(y)\bar{q}_f(v)|0_f\rangle_c$ where subscript c ensures that only the contribution corresponding to Fig. 2(c) is retained. Thus Eq. (10) is just the statement that G^U is obtained from G by inserting operator $\bar{q}_\beta(x)q_\alpha(y)$ within every bare propagator of every Feynman diagram of the theory. This insertion corresponds pictorially to cutting the bare propagator into two, as illustrated in Fig. 2. Unfortunately not all contributions to the G^U defined by Eq. (5) can be obtained by gauging G in this way. However, in many problems, including those discussed in the present paper, either only diagrams obtained by gauging are of interest, or the missing ones are easily taken into account. Two examples of diagrams that contribute to $G^U_{\alpha\beta}$ but that cannot be obtained by U-gauging G are given in Fig. 3

For those used to the functional integral approach it might be more transparent to define U-gauging by a functional derivative with respect to an external bilocal field $\mathcal{B}(x, y)$ introduced into the action $S[q, A]$ (where A represents gluons or other fields) as $S[q, A] \to S[q, A, \mathcal{B}] = S[q, A] + \int d^4x\, d^4y\, \bar{q}(x)\mathcal{B}(x, y)q(y)$ [43]. This modification amounts to the replacement of the inverse bare quark propagator as

$$d_0^{-1}(x-y) \;\to\; d_{0\mathcal{B}}^{-1}(x, y) = d_0^{-1}(x-y) - i\mathcal{B}(x, y) \tag{11}$$

in the perturbation theory generated by $S[q, A]$, in order to obtain the perturbation theory generated by $S[q, A, \mathcal{B}]$. The usual Green function, the vacuum expectation of the time ordered product of field operators (excluding vacuum loops), can be written in terms of a functional integral as

$$\begin{aligned} G(\{x_i\},\{y_j\},\{z_k\}) &= \langle 0|T \ldots q(x_i) \ldots \bar{q}(y_j) \ldots A(z_k) \ldots |0\rangle \\ &= \frac{\int Dq\, D\bar{q}\, DA\, e^{iS[q,A]} \ldots q(x_i) \ldots \bar{q}(y_j) \ldots A(z_k) \ldots}{\int Dq\, D\bar{q}\, DA\, e^{iS[q,A]}}. \end{aligned} \tag{12}$$

When modified by the presence of the external bilocal field $\mathcal{B}$, this Green function becomes

$$G_{\mathcal{B}}(\{x_i\}, \{y_j\}, \{z_k\}) = \frac{\int Dq\, D\bar{q}\, DA\, e^{iS[q,A,\mathcal{B}]} \ldots q(x_i) \ldots \bar{q}(y_j) \ldots A(z_k) \ldots}{\int Dq\, D\bar{q}\, DA e^{iS[q,A,\mathcal{B}]}}. \tag{13}$$

Introducing the generating functionals

$$Z[\mathcal{B}] = \int Dq\, D\bar{q}\, DA\, e^{iS[q,A,\mathcal{B}]}, \qquad Z = Z[0] = \int Dq\, D\bar{q}\, DA\, e^{iS[q,A]}, \tag{14}$$

the gauged Green function can then be defined as

$$G^U(\{x_i\},\{y_j\},\{z_k\};x,y) = -i\, Z^{-1}\frac{\delta Z[\mathcal{B}]G_\mathcal{B}(\{x_i\},\{y_j\},\{z_k\})}{\delta\mathcal{B}(x,y)}\bigg|_{c,\mathcal{B}=0}$$

$$= -\,i\, Z^{-1}Z[\mathcal{B}]\,\frac{\delta G_\mathcal{B}(\{x_i\},\{y_j\},\{z_k\})}{\delta\mathcal{B}(x,y)}\bigg|_{c,\mathcal{B}=0}$$

$$-\,i\, Z^{-1}\frac{\delta Z[\mathcal{B}]}{\delta\mathcal{B}(x,y)}G_\mathcal{B}(\{x_i\},\{y_j\},\{z_k\})\bigg|_{c,\mathcal{B}=0}$$

$$= -\,i\,\frac{\delta G_\mathcal{B}(\{x_i\},\{y_j\},\{z_k\})}{\delta\mathcal{B}(x,y)}\bigg|_{\mathcal{B}=0} + d(x,y)G(\{x_i\},\{y_j\},\{z_k\})|_c\,. \tag{15}$$

The last term in the Eq. (15) should be discarded as it contributes only to those disconnected terms which are forbiden by the meaning of subscript c. Note that there are still other disconnected contributions to Eq. (15) which should be kept. So the proper functional definition of gauging is just the functional derivative:

$$G^U(\{x_i\},\{y_j\},\{z_k\};x,y) = -i\,\frac{\delta G_\mathcal{B}(\{x_i\},\{y_j\},\{z_k\})}{\delta\mathcal{B}(x,y)}\bigg|_{\mathcal{B}=0}. \tag{16}$$

In this sense Eq. (10) is just a statement of the product rule for derivatives. Note that Eq. (11) for the modified bare propagator is obvious, but can be derived from Eq. (13) as well. The gauged bare propagator can be derived using

$$\frac{\delta d_{0\mathcal{B}}^{-1}(x_1,x_2)}{\delta\mathcal{B}(x,y)} = -i\frac{\delta\mathcal{B}(x_1,x_2)}{\delta\mathcal{B}(x,y)} = -i\delta(x_1-x)\delta(y-x_2) \tag{17}$$

and

$$\frac{\delta(d_{0\mathcal{B}}\,d_{0\mathcal{B}}^{-1})}{\delta\mathcal{B}} = \frac{\delta d_{0\mathcal{B}}}{\delta\mathcal{B}}d_{0\mathcal{B}}^{-1} + d_{0\mathcal{B}}\frac{\delta d_{0\mathcal{B}}^{-1}}{\delta\mathcal{B}} = 0. \tag{18}$$

One finds that

$$\begin{aligned} d_{0\mathcal{B}}^U(x_1,x_2;x,y) &\equiv -i\frac{\delta d_{0\mathcal{B}}(x_1,x_2)}{\delta\mathcal{B}(x,y)} = d_{0\mathcal{B}}(x_1-x)d_{0\mathcal{B}}(y-x_2) \\ d_0^U(x_1,x_2;x,y) &= -i\,\frac{\delta d_{0\mathcal{B}}(x_1,x_2)}{\delta\mathcal{B}(x,y)}\bigg|_{\mathcal{B}=0} = d_0(x_1-x)d_0(y-x_2). \end{aligned} \tag{19}$$

As expected from the discussion above, the antisymmetrizing contribution, $d_0(x_1-x_2)d_0(y-x)$, does not appear in the expression for the gauged free propagator $d_0^U(x_1,x_2;x,y)$. It is also clear from Eq. (19) how the gauging of the bare quark propagator corresponds, diagrammatically, to cutting the bare propagator into two pieces, as illustrated in Fig. 2. Finally, Eq. (16), together with the product rule for derivatives, enables one to see that gauging any complicated diagram corresponds to cutting bare propagators entering this diagram in all possible ways.

2.2. *Extracting GPDs from three-body scattering equations*

2.2.1. GPD sum rules As discussed above, all GPDs can be obtained from the density matrix $\rho_{\alpha\beta}$ defined in Eq. (1). In turn, $\rho_{\alpha\beta}$ can be found from the $(n+2)$-point Green function $G^U_{\alpha\beta}$, as given by Eq. (6), by inserting a complete set of states on either side of the operator $\bar{q}_\beta(x)q_\alpha(y)$ and then taking residues at the bound state poles corresponding to the physical states $|P\rangle$ and

$|P'\rangle$. By writing $G^U_{\alpha\beta} = G\Gamma^U_{\alpha\beta}G$ and recognizing that $G \sim i\Psi_P\bar{\Psi}_P/(P^2 - M^2)$ in the vicinity of the $P^2 = M^2$ pole, one obtains

$$\rho_{\alpha\beta}(P', P, k) = \int d^4y\, e^{ik\cdot y}\, \langle P'|T\bar{q}_\beta(0)q_\alpha(y)|P\rangle_c = \bar{\Psi}_{P'}\Gamma^U_{\alpha\beta}\Psi_P \tag{20}$$

where Ψ_P is the bound state wave function corresponding to state $|P\rangle$:

$$\Psi_P(x_1, x_2, x_3) = \langle 0|Tq(x_1)q(x_2)q(x_3)|P\rangle, \tag{21}$$

and

$$\Gamma^U_{\alpha\beta} = G^{-1}G^U_{\alpha\beta}G^{-1} \tag{22}$$

is the corresponding bound state vertex function. To find $G^U_{\alpha\beta}$ we shall use the U-gauging method described in the previous section. This is in direct analogy to what was done in Refs. [4, 5, 6, 7] to find the electromagnetic bound state current which is given as

$$j^\mu(P', P) = \langle P'|\bar{q}(0)\hat{\Gamma}^\mu q(0)|P\rangle = \bar{\Psi}_{P'}\Gamma^\mu\Psi_P \tag{23}$$

where

$$\Gamma^\mu = G^{-1}G^\mu G^{-1}. \tag{24}$$

The close similarity between the definitions of G^U in Eq. (5) and G^μ in Eq. (7) is embodied in the sum rule

$$\sum_{\alpha,\beta}\int \frac{d^4k}{(2\pi)^4}\,\rho_{\alpha\beta}(P', P, k)\hat{\Gamma}^\mu_{\beta\alpha} = j^\mu(P', P), \tag{25}$$

which follows from Eq. (20) and Eq. (23). Furthermore, since the bound state current satisfies current conservation,

$$(P' - P)^\mu j^\mu(P', P) = 0, \tag{26}$$

and charge conservation,

$$j^\mu(P, P) = 2QP^\mu, \tag{27}$$

where Q is the total charge of the three-body bound state, the density matrix $\rho_{\alpha\beta}$ satisfies two further sum rules,

$$(P' - P)^\mu \sum_{\alpha,\beta}\int \frac{d^4k}{(2\pi)^4}\,\rho_{\alpha\beta}(P', P, k)\hat{\Gamma}^\mu_{\beta\alpha} = 0, \tag{28}$$

and

$$\sum_{\alpha,\beta}\int \frac{d^4k}{(2\pi)^4}\,\rho_{\alpha\beta}(P, P, k)\hat{\Gamma}^\mu_{\beta\alpha} = 2QP^\mu. \tag{29}$$

Note that $|P\rangle$ is an eigenstate of the conserved charge operator $\hat{Q} = \int d^3x J^0(x)$ with corresponding eigenvalue Q; that is, Q is a physical quantity corresponding to the conserved Noether current $J^\mu(x) = \bar{q}(x)\hat{\Gamma}^\mu q(x)$. Although we take J^μ to be the conserved electromagnetic current operator, it is clear that essentially the same expressions will hold for conserved isotopic vector currents (CVC), conserved axial currents (CAC), and partially conserved axial currents (PCAC) for which the right-hand side (RHS) of Eq. (26) and Eq. (28) would be non-zero [44].

2.2.2. GPDs of two-body bound states Before discussing the GPDs of three-body bound states described by Faddeev-like equations, it is useful to first demonstrate the main ideas of this approach on the simpler case of two-body bound states described by the Bethe-Salpeter (BS) equation. In this respect we note that GPDs in Bethe-Salpeter approaches have already received some attention in the literature [45, 46, 47].

To describe the scattering of two distinguishable particles (e.g., a quark and an anti-quark) one can use the integral equation for the two-body Green function G, Eq. (8) (but with the antisymmetrization superscript "P" dropped from this and subsequent equations). Then μ-gauging Eq. (8) gives Eq. (9) which can be solved to give [6]

$$G^\mu = G\Gamma^\mu G, \tag{30}$$

$$\Gamma^\mu = \Gamma_1^\mu d_2^{-1} + d_1^{-1}\Gamma_2^\mu + V^\mu, \tag{31}$$

where d_i is the propagator of particle i, Γ_i^μ is its electromagnetic vertex function, and V^μ is the gauged two-body potential. The two-body bound state current can then be found by taking left and right residues of G^μ at the bound state poles:

$$j^\mu = \bar{\phi}(d_1^\mu d_2 + d_1 d_2^\mu + d_1 d_2 V^\mu d_1 d_2)\phi \tag{32}$$

where $\psi = d_1 d_2 \phi$ defines the two-body bound state vertex function ϕ in terms of the two-body bound state wave function ψ, and $d_i^\mu = d_i \Gamma_i^\mu d_i$ is the μ-gauged propagator of particle i. This agrees with the result first derived by Gross and Riska [1]. An alternative but equivalent approach is to start with the Bethe-Salpeter equation for ϕ:

$$\phi = V d_1 d_2 \phi. \tag{33}$$

This equation can then be gauged to obtain [6]

$$\phi^\mu = T(d_1 d_2)^\mu \phi + (1 + T d_1 d_2) V^\mu d_1 d_2 \phi \tag{34}$$

where T is two-body t matrix. Taking the residue at the bound state pole of T one again obtains Eq. (32). As shown in Ref. [1], the j^μ of Eq. (32) obeys charge and current conservation if the inputs d_i^μ and V^μ obey both Ward and Ward-Takahashi identities.

Replacing μ-gauging by U-gauging in the above derivations, one obtains the equation corresponding Eq. (32):

$$\rho = \bar{\phi}(d_1^U d_2 + d_1 d_2^U + d_1 d_2 V^U d_1 d_2)\phi \tag{35}$$

where ρ is the matrix whose components are $\rho_{\alpha\beta}$. It is clear that the ρ of Eq. (35) and the j^μ of Eq. (32) will obey the GPD sum rule of Eq. (25) if the input pairs (d_i^μ, d_i^U), and (V^μ, V^U) each obey corresponding sum rules. This aspect is an essential ingredient of our approach and will be discussed in more detail below. The other two GPD sum rules, Eq. (28) and Eq. (29), are then automatically satisfied because of the charge and current conservation properties of j^μ. Eq. (35) is illustrated in Fig. 4.

2.2.3. GPDs of three-body bound states The sum rules for GPDs, derived in subection A above, constitute important constraints satisfied by the exact theory of strong interactions. As such, it is desirable that the same sum rules be also satisfied by models that seek to approximate the exact theory. In Ref. [6, 7] we showed that for nonperturbative strong interaction models described by integral equations, the gauging of equations method, when applied to obtain G^μ and therefore the bound state electromagnetic current j^μ, results in both charge and current conservation being satisfied. We shall now show that the same gauging method, when applied

Journal of Physics: Conference Series **330** (2011) 012008 doi:10.1088/1742-6596/330/1/012008

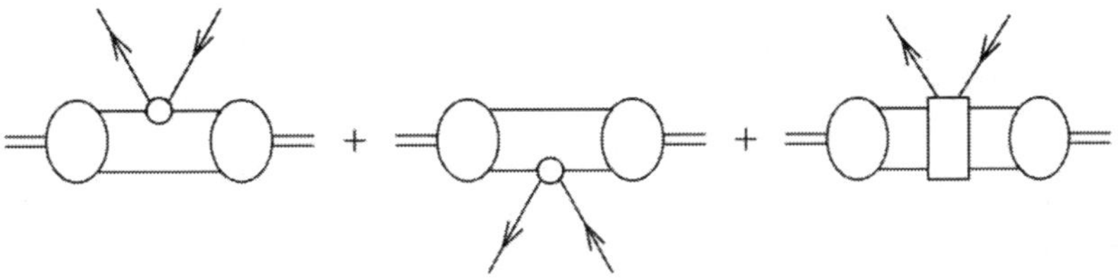

Figure 4. The two-body density matrix of Eq. (35) specifying the GPDs of two-body bound states.

to obtain G^U and therefore $\rho_{\alpha\beta}$, results additionally in the GPD sum rules of Eq. (25), Eq. (28), and Eq. (29) being satisfied.

Here we apply the U-gauging procedure to the case of three identical particles, so that our expression for $\rho_{\alpha\beta}$ can be used directly for calculations of GPDs in the nucleon or nucleon distributions in ^{3}He. We assume that the interaction kernel V defining G [Eq. (8)] is given as a sum of only two-particle interactions, so that

$$V = \sum_{i=1}^{3} \frac{1}{2} v_i d_i^{-1} \tag{36}$$

where v_i is the (fully antisymmetric) interaction potential between particles j and k (where ijk is a cyclic permutation of 123), d_i is the dressed propagator of particle i, and the 1/2 is a factor arising from antisymmetry [7]. As shown in Ref. [7], the μ-gauging of Eq. (8) then gives the following expression for G^μ:

$$G^\mu = G\Gamma^\mu G, \tag{37}$$

$$\Gamma^\mu = \frac{1}{6}\sum_{i=1}^{3}\left(\Gamma_i^\mu d_j^{-1} d_k^{-1} + \frac{1}{2} v_i^\mu d_i^{-1} - \frac{1}{2} v_i \Gamma_i^\mu\right). \tag{38}$$

Taking left and right residues of G^μ then gives the bound state electromagnetic current of three identical particles:

$$j^\mu(P', P) = \frac{1}{6}\sum_{i=1}^{3} \bar{\Phi}_{P'}\, d_j d_k \left(d_i^\mu d_j^{-1} d_k^{-1} + \frac{1}{2} v_i^\mu d_i - \frac{1}{2} v_i d_i^\mu\right) d_j d_k\, \Phi_P \tag{39}$$

where $\Psi_P = d_1 d_2 d_3 \Phi_P$ defines the three-body bound state vertex function Φ_P in terms of the three-body bound state wave function Ψ_P. Replacing μ-gauging by U-gauging in the above gives the three-body density matrix

$$\rho(P', P, k) = \frac{1}{6}\sum_{i=1}^{3} \bar{\Phi}_{P'}\, d_j d_k \left(d_i^U d_j^{-1} d_k^{-1} + \frac{1}{2} v_i^U d_i - \frac{1}{2} v_i d_i^U\right) d_j d_k\, \Phi_P. \tag{40}$$

Again, it is clear that the ρ of Eq. (40) and the j^μ of Eq. (39) will obey the sum rule of Eq. (25) if the input pairs (d_i^μ, d_i^U), and (v_i^μ, v_i^U) each obey corresponding sum rules, and that the sum rules of Eq. (28) and Eq. (29) are also satisfied because the j^μ of Eq. (39) satisfies both charge and current conservation. The contributions to ρ, as given by Eq. (40), are illustrated in Fig. 5. It is noteworthy that the last term of Eq. (40) (last term of Fig. 5), comes with a negative sign; as discussed in Ref. [7], this subtraction term is necessary to remove overcounted contributions that are present in the first term.

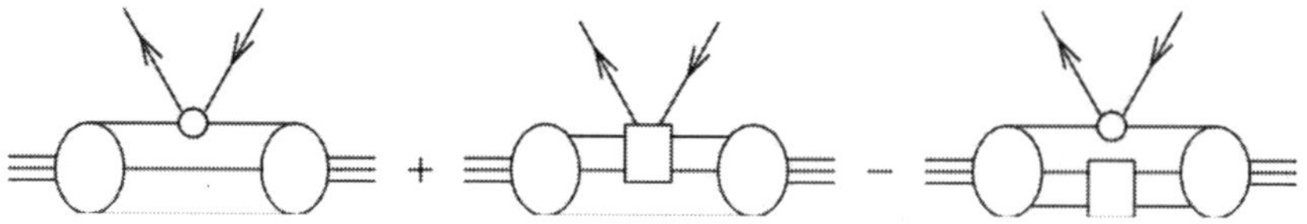

Figure 5. Contributions to the three-body density matrix of Eq. (40) specifying the GPD's of three-body bound states.

As in the two-body case discussed above, one can give an alternative derivation of j^μ and ρ by gauging the bound state equation for the vertex function. This has the advantage that one gauges directly the equation that one actually solves numerically; this way, for example, any approximations used in the solution of the equation will appropriately be taken into account in the gauged result. Indeed, because of the numerical difficulty of solving 4D integral equations, one is often interested to perform a 3D reduction of the original 4D approach. If one invokes such a reduction, it turns out that it is better to reduce the dimension of 4D integral equations *first*, and *then* gauge the resulting 3D bound state equations in order to deduce the expressions for j^μ and ρ, rather than gauge at the 4D level first, and then try and reduce to three-dimensions the 4D expressions obtained for j^μ and ρ. This point was examined in detail in Ref. [48] in the context of Gross's 3D spectator approach [36, 37, 38]. Here we shall present the results of gauging only the 4D bound state equation, and refer the reader to Ref. [48] for the analogous 3D case.

As is well known from three-body theory, the bound state vertex function Φ can be specified by writing it as a sum of Faddeev components, $\Phi = \Phi_1 + \Phi_2 + \Phi_3$, where

$$\Phi_1 = -t_1 d_2 d_3 P_{12} \Phi_1. \tag{41}$$

In Eq. (41) t_1 is the two-body t matrix between particles 2 and 3 and P_{12} is the operator that interchanges particles 1 and 2. The two-body potential v_1 and corresponding t matrix t_1 are defined to be antisymmetric ($P_{23}v_1 = -v_1$ and $P_{23}t_1 = -t_1$) so that full antisymmetry of the three-body wave function $\Psi = d_1 d_2 d_3 \Phi$ is ensured.

In order to find the expression for $\rho_{\alpha\beta}$, as defined above, we can follow Ref. [5] and directly U-gauge Eq. (41). After some algebra we obtain

$$\begin{aligned}\rho(P', P, k) =& \bar{\Phi}_1^{P'} d_1 \left(d_2^U d_3 + d_2 d_3^U\right) \left(\frac{1}{2} - P_{12}\right) \Phi_1^P \\ &+ \bar{\Phi}_1^{P'} \left(P_{12} - \frac{1}{2}\right) d_2 d_3 d_1 v_1^U d_2 d_3 (P_{12} - \frac{1}{2}) \Phi_1^P,\end{aligned} \tag{42}$$

with a corresponding bound state electromagnetic current

$$\begin{aligned}j^\mu(P', P) =& \bar{\Phi}_1^{P'} d_1 \left(d_2^\mu d_3 + d_2 d_3^\mu\right) \left(\frac{1}{2} - P_{12}\right) \Phi_1^P \\ &+ \bar{\Phi}_1^{P'} \left(P_{12} - \frac{1}{2}\right) d_2 d_3 d_1 v_1^\mu d_2 d_3 (P_{12} - \frac{1}{2}) \Phi_1^P.\end{aligned} \tag{43}$$

The exact expression for ρ, Eq. (42), corresponds to cutting (as in Fig. 2) every one of the infinite number of bare particle propagators that exist in the nonperturbative expression for the strong interaction Green function G, Eq. (8). In the same way the corresponding expressions for the electromagnetic current j^μ, Eq. (43), corresponds to attaching a photon to all the bare propagators in the nonperturbative theory. It is this completeness of the photon attachment that guarantees the charge and current conservation properties of j^μ. It is evident that the ρ of Eq. (42) and the j^μ of Eq. (43) will fulfill the GPD sum rules whenever the gauged input quantities, d^U and v^U, satisfy corresponding sum rules.

2.2.4. Gauged propagator The gauged one-particle propagators d^μ and d^U form one of the basic inputs to the equations describing the three-body bound state current j^μ, Eq. (39), and the density matrix ρ, Eq. (40), respectively. In momentum space these gauged propagators are defined as

$$d^\mu(p',p) = \int d^4x\, d^4y\, e^{i(p'\cdot x - p\cdot y)}\langle 0|Tq(x)\bar{q}(0)\hat{\Gamma}^\mu q(0)\bar{q}(y)|0\rangle \tag{44}$$

and

$$d^U_{\alpha\beta}(p',p,k) = \int d^4x\, d^4y\, d^4z\, e^{i(p'\cdot x + k\cdot z - p\cdot y)}\langle 0|Tq(x)\bar{q}_\beta(0)q_\alpha(z)\bar{q}(y)|0\rangle_c. \tag{45}$$

Writing the spinor indices of $q(x)$ and $\bar{q}(y)$ explicitly, the subscript c in Eq. (45) means that the disconnected part $\langle 0|Tq_i(x)\bar{q}_j(y)|0\rangle\, \langle 0|Tq_\alpha(z)\bar{q}_\beta(0)|0\rangle$ is excluded. On the other hand, the remaining disconnected part, $\langle 0|Tq_i(x)\bar{q}_\beta(0)|0\rangle\, \langle 0|Tq_\alpha(z)\, \bar{q}_j(y)|0\rangle$, does contribute to the U-gauged propagator as

$$d^U_{i\alpha,j\beta}(p',p,k)_{\rm disc} = (2\pi)^4\delta^4(k-p)d_{i\beta}(p')\, d_{\alpha j}(p). \tag{46}$$

Similarly to Eq. (25), the two types of gauged propagator are related by a sum rule:

$$\sum_{\alpha,\beta}\int \frac{d^4k}{(2\pi)^4}\, d^U_{\alpha\beta}(p',p,k)\hat{\Gamma}^\mu_{\beta\alpha} = d^\mu(p',p). \tag{47}$$

In addition, d^μ satisfies the Ward-Takahashi (WT) identity

$$(p'-p)^\mu d^\mu(p',p) = ie[d(p) - d(p')], \tag{48}$$

and the Ward identity

$$d^\mu(p,p) = -ie\frac{\partial d(p)}{\partial p_\mu}, \tag{49}$$

where e is the charge of the particle (for isodoublet particle fields e is a 2×2 matrix). We thus can specify two further sum rules for $d^U_{\alpha\beta}$ as

$$(p'-p)^\mu \sum_{\alpha,\beta}\int \frac{d^4k}{(2\pi)^4}\, d^U_{\alpha\beta}(p',p,k)\hat{\Gamma}^\mu_{\beta\alpha} = ie[d(p) - d(p')] \tag{50}$$

and

$$\sum_{\alpha,\beta}\int \frac{d^4k}{(2\pi)^4}\, d^U_{\alpha\beta}(p,p,k)\hat{\Gamma}^\mu_{\beta\alpha} = -ie\frac{\partial d(p)}{\partial p_\mu}. \tag{51}$$

Also, writing

$$d^\mu(p',p) = d(p')\Gamma^\mu d(p), \tag{52a}$$
$$d^U_{\alpha\beta}(p',p,k) = d(p')\Gamma^U_{\alpha\beta} d(p), \tag{52b}$$

where Γ^μ and $\Gamma^U_{\alpha\beta}$ are one-body vertex functions, the WT and Ward identities give

$$(p'-p)^\mu\Gamma^\mu(p',p) = ie[d^{-1}(p') - d^{-1}(p)] \tag{53}$$

and

$$\Gamma^\mu(p,p) = ie\frac{\partial d^{-1}(p)}{\partial p_\mu} \tag{54}$$

respectively, while Eq. (47) implies that

$$\sum_{\alpha,\beta}\int\frac{d^4k}{(2\pi)^4}\Gamma^U_{\alpha\beta}(p',p,k)\hat{\Gamma}^\mu_{\beta\alpha} = \Gamma^\mu(p',p). \tag{55}$$

Note that, to save on notation, we use the same symbols to denote the one-body vertex functions of Eq. (52) and the three-body bound state vertex functions of Eq. (22) and Eq. (24) - the type of vertex function meant should be clear from the context; moreover, it is easy to see that Eq. (55) in fact holds true for both cases. Combining Eq. (53) and Eq. (54) with Eq. (55) gives the sum rules

$$(p'-p)^\mu\sum_{\alpha,\beta}\int\frac{d^4k}{(2\pi)^4}\Gamma^U_{\alpha\beta}(p',p,k)\hat{\Gamma}^\mu_{\beta\alpha} = ie[d^{-1}(p') - d^{-1}(p)], \tag{56}$$

$$\sum_{\alpha,\beta}\int\frac{d^4k}{(2\pi)^4}\Gamma^U_{\alpha\beta}(p,p,k)\hat{\Gamma}^\mu_{\beta\alpha} = ie\frac{\partial d^{-1}(p)}{\partial p_\mu}, \tag{57}$$

which relate the vertex function Γ^U to the propagator d.

In the absence of dressing, in which case we write Eqs. (52) as

$$d^\mu_0(p',p) = d_0(p')\Gamma^\mu_0\, d_0(p), \tag{58a}$$

$$d^U_{0,\alpha\beta}(p',p,k) = d_0(p')\Gamma^U_{0,\alpha\beta}\, d_0(p), \tag{58b}$$

where $d_0 = i/(\not{p} - m)$ is the bare particle propagator, and Eq. (58b) with all spinor indices revealed reads

$$d^U_{0,i\alpha,j\beta}(p',p,k) = \sum_{k,l} d_{0,ik}(p')\Gamma^U_{0,k\alpha,l\beta}\, d_{0,lj}(p), \tag{59}$$

one finds that

$$\Gamma^\mu_0 = \hat{\Gamma}^\mu, \tag{60a}$$

$$\Gamma^U_{0,k\alpha,l\beta} = (2\pi)^4\delta(k-p)\delta_{k\beta}\delta_{l\alpha}. \tag{60b}$$

As mentioned previously, the density matrix ρ, as specified by Eq. (42), will satisfy the GPD sum rules only if the gauged inputs d^U and v^U, will themselves satisfy corresponding sum rules. For the case of d^U, the above discussion shows that, *in the exact theory*, d^U does indeed satisfy such sum rules, namely, Eq. (47), Eq. (50), and Eq. (51). However, what is of practical interest in our approach is to construct strong interaction *models* of d^U such that Eq. (47), Eq. (50), and Eq. (51) are still satisfied. In this respect, one might think of modeling d^U by just its disconnected part, as given by Eq. (46):

$$d^U_{i\alpha,j\beta}(p',p,k) \equiv (2\pi)^4\delta^4(k-p)d_{i\beta}(p')\, d_{\alpha j}(p). \tag{61}$$

In this case the left hand side of the sum rule of Eq. (47) would give

$$\sum_{\alpha,\beta}\int\frac{d^4k}{(2\pi)^4}d^U_{\alpha\beta}(p',p,k)\hat{\Gamma}^\mu_{\beta\alpha} = d(p')\hat{\Gamma}^\mu d(p). \tag{62}$$

Since $\hat{\Gamma}^\mu \neq \Gamma^\mu$, the RHS of Eq. (62) is not equal to $d^\mu(p',p)$, so the sum rules of Eq. (47), Eq. (50) and Eq. (51) will not be satisfied; indeed, only in the case of no dressing, for which Eq. (60a) holds, does such a model for d^U satisfy the required sum rules.

This shows that the connected part of d^U needs to be taken into account if we want the GPD sum rules for ρ to be satisfied in a model with dressed propagators. One way that this can be achieved is to construct a purely phenomenological d^U which satisfies the sum rules of Eq. (47), Eq. (50), and Eq. (51). However, a theoretically more rigorous way would be to first construct a model for the dressed propagator d using the Dyson-Schwinger (DS) equation, and then U-gauging the DS equation. The details of constructing d^U in this way will be discussed elsewhere (although Fig. 6, discussed later, provides a graphical example of such an approach).

2.2.5. Gauged potential The last inputs needed to be considered are the gauged two-body potentials v^μ and v^U. As for the gauged propagators, their construction will depend on the nature of the model chosen for potential v. However, independently of the model chosen for v, it is a requirement of our approach that v^μ is constructed so that it satisfy the WT identity

$$q_\mu v^\mu(p_1'p_2',p_1p_2) = i[e_1 v(p_1'-q,p_2';p_1p_2) - v(p_1'p_2';p_1+q,p_2)e_1 + e_2 v(p_1',p_2'-q;p_1p_2) - v(p_1'p_2';p_1,p_2+q)e_2] \quad (63)$$

and the Ward identity

$$v^\mu(p',P-p';p,P-p) = -i\left[e_1\frac{\partial v(p',P-p';p,P-p)}{\partial p'_\mu} + \frac{\partial v(p',P-p';p,P-p)}{\partial p_\mu}e_1 + (e_1+e_2)\frac{\partial v(p',P-p';p,P-p)}{\partial P_\mu}\right] \quad (64)$$

where e_i is the charge of particle i. In the case where v is defined through a finite sum of Feynman diagrams or through a dynamical equation, applying the μ-gauging procedure to the equation defining v will give a v^μ that automatically satisfies Eq. (63) and Eq. (64).

With an appropriately constructed v^μ, all that is left is to construct the U-gauged potential v^U so that it satisfy the sum rule

$$\sum_{\alpha,\beta}\int\frac{d^4k}{(2\pi)^4}\,v^U_{\alpha\beta}(p_1'p_2',p_1p_2,k)\hat{\Gamma}^\mu_{\beta\alpha} = v^\mu(p_1'p_2',p_1p_2). \quad (65)$$

Again, if v is defined through a finite sum of Feynman diagrams or through a dynamical equation, applying the U-gauging procedure to the equation defining v will give a v^U that automatically satisfies Eq. (65).

In the case that v is not given in terms of Feynman diagrams or a dynamical equation, one can easily construct v^μ and v^U *ad hoc*, in order to satisfy Eq. (63), Eq. (64), and Eq. (65).

2.3. Discussion

In this section we have demonstrated our U-gauging method by obtaining the density matrix $\rho_{\alpha\beta}$ of Eq. (1), and therefore GPDs, for the specific cases of a two-quark bound state described by the Bethe-Salpeter equation, and a three-quark bound state described by 4D Faddeev-like equations. The method corresponds to cutting all possible bare quark propagators in the strong interaction model, and ensures that GPD sum rules are satisfied automatically. Also, because our formulation is Lorentz covariant, the derived GPDs obey polynomiality [30]. However, it is important to emphasize that our procedure is general, applying to any type of dynamical equation, as well as to various types of distributions. For example, the U-gauging

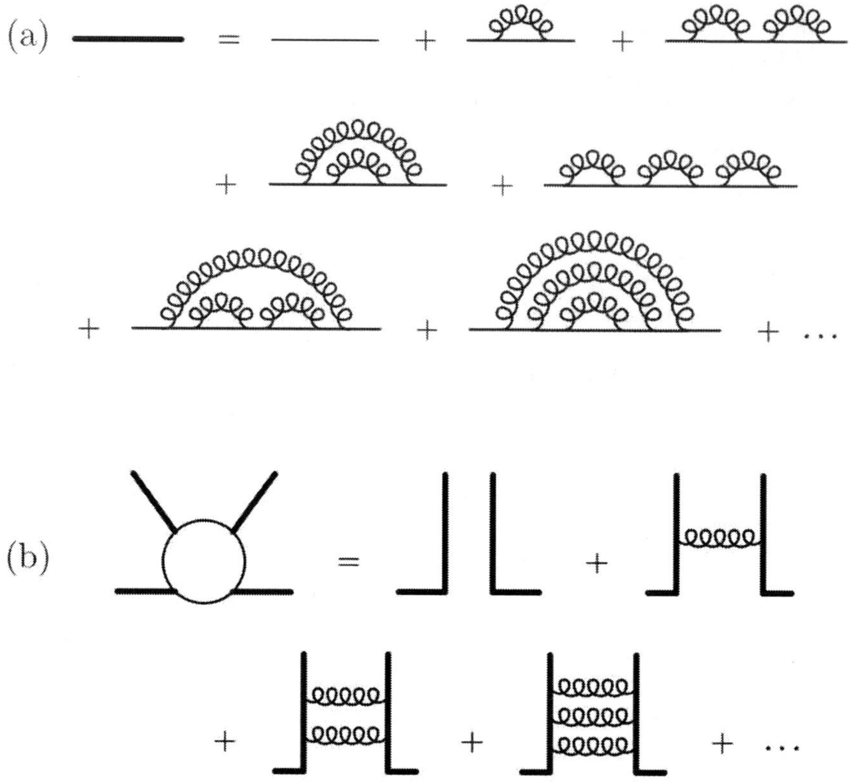

Figure 6. (a) The dressed propagator d in rainbow approximation. (b) The distribution function $d^U = d\Gamma^U d$ obtained by cutting all possible bare propagators in the rainbow perturbation series for d shown in (a). The propagator d defined by (a) and the vertex function Γ^U defined by (b) will satisfy the GPD sum rules of Eq. (55), Eq. (56) and Eq. (57).

procedure can be applied to effectively cut *all* the bare propagators existing within a given dynamical equation model, not just those of quarks. In this way one could derive the model's generalized gluon distribution [30], pion distribution, etc. With sum rules corresponding to current conservation being ensured, one can, for example, selectively turn on and off U-gauging of different constituents in order to study the fraction of a hadron's quantum number (e.g. spin) carried by a quark, gluon, meson, etc.

The whole approach holds also for the 3D spectator formalism [48], and furthermore, can be applied to nucleon distributions in nuclei, i.e., with nucleons replacing quarks in the above formulation. Similarly, one could consider matrix elements as in Eq. (1), but taken between three-particle states other than $\langle P'|$ and $|P\rangle$. Apart from these three-body bound states, one can have states of three free particles and those where one particle is free while the other two form a bound sub-system. All these transitions are considered in Ref. [4, 5, 6, 7] for the case of electromagnetic currents, and can be used directly to find GPDs simply through the replacements

$$d^\mu \to d^U, \qquad v^\mu \to v^U. \tag{66}$$

Off-diagonal transitions $N \to \pi N$ in context of GPDs have also been suggested and their importance emphasized in Ref. [29]. In this regard we should mention that applying techniques of the present paper to the πNN equations derived in Ref. [49, 50, 10], and their electromagnetic currents [8], would lead to a comprehensive description of the GPDs for the transitions $NN \to \pi NN, d \to \pi d$, etc.

With the same techniques, mesonic correction to nucleon GPDs can be calculated. For example, within the NJL model of the nucleon, one can use the static approximation where the mass of the exchanged quark is taken to infinity. Then the quark-diquark Green function, which enters the expression for the mesonic corrections [51], is calculated algebraically, making the calculation of corrections very practical. In Ref. [51], we have already shown how to calculate mesonic corrections to electromagnetic currents of the NJL model, and using the U-gauging method proposed here one could calculate the mesonic corrections to GPDs so that all sum rules are satisfied. The important point is that our approach enables these mesonic corrections to enter calculations of GPDs and currents consistently.

3. Gauge invariant currents in NN EFT with cutoff regularization.

In this section we shall apply the gauging of equations method to two-nucleon (NN) EFT, a subject at the heart of modern approaches to nuclear physics [52], and show that it is instrumental in constructing gauge invariant currents when a momentum cutoff is introduced into the theory [53]. To this end we first briefly summarise the main ideas behind NN EFT as first proposed by Weinberg [54, 55, 56].

Nuclear forces cannot be as yet derived directly from the fundamental theory of strong interactions, quantum chromodynamics (QCD), due to that part of low energy physics which should be treated non-perturbatively. Presently, nuclear forces are most rigorously studied within the framework of chiral EFT whose fundamental degrees of freedom are pionic and nucleonic fields. EFT is based on the most general Lagrangian which approximately satisfies chiral and other symmetries of QCD, and provides a systematic and model independent way of analyzing the properties of hadronic systems at low energies. The pion and single-baryon sector are treated purely perturbatively in EFT, whereas for the two (or more) nucleon sector a non-perturbative problem is involved. The reason for this was pointed out by Weinberg [54, 55, 56]: the scattering amplitude experiences an enhancement stemming from the propagators in the purely nucleonic intermediate states. As a result, perturbation theory is applicable only to the kernel of the corresponding equation for the scattering amplitude, while the equation itself needs to be solved non-perturbatively.

The non-perturbative nature of the scattering problem makes it impossible to implement a symmetry preserving regularization such as dimensional regularization. The latter is the method of dealing with infinities which appear in perturbative quantum field theory - it eliminates power-law divergences and isolates all logarithmic divergences so that gauge invariance and chiral symmetry are preserved.

Thus, for calculations of NN scattering (in which case the perturbative series for the T-matrix is useless due to its divergence at the energies of interest), the use of cut-off regularization cannot be avoided, despite the fact that it destroys the gauge invariance of the theory at the level of the Lagrangian. Here we will show how cutoff regularization can nevertheless be carried out in the framework of the chiral effective theory without destroying gauge invariance of the matrix elements corresponding to the processes involving electromagnetic interaction.

An important aspect of our approach is that the used cut-off regularized EFT is explicitly constructed to be equivalent to EFT in the renormalization scheme: both the NN scattering amplitude and the NN EM currents in the cutoff scheme are identical to the ones in the renormalization scheme. This equivalence ensures that the potentials and current vertices have the same decomposition into long and short range terms (corresponding to the Goldstone meson exchange, and contact interaction parts, respectively) as in the renormalization scheme, and that power counting applies to the new NN potentials and current vertices. The currents

Journal of Physics: Conference Series **330** (2011) 012008 doi:10.1088/1742-6596/330/1/012008

in the new approach are manifestly gauge invariant, i.e., the current vertices obey usual WT identities at each stage of the calculation. The gauging of equations method is instrumental in this achievement.

3.1. Derivation of currents using the gauging of equations method

Here we show how to derive the *exact* currents of cutoff-regularized EFT; that is, the currents which we derive using cutoff EFT are identical to those which would be produced in EFT within the standard subtractive renormalization scheme. We achieve our formulation utilising the gauging of equations method [6, 7].

The exact current is defined via the quantity T^μ which results from attaching an external photon to all possible places within the scattering amplitude T. To be concrete, we take T to be the NN scattering amplitude based on an underlying chiral Lagrangian with explicit nucleon and pion degrees of freedom; moreover, we assume the scattering amplitude satisfies the Lippmann-Schwinger (LS) equation in the subtractive renormalization scheme:

$$T = K + KG_0T \tag{67}$$

where G_0 is the non-relativistic propagator of two non-interacting nucleons, and K is the NN potential expressed in terms of bare coupling constants, i.e., those which determine the Lagrangian. To be more clear, the scattering amplitude cannot be written as a solution of the LS equation whose driving term is the renormalized potential because the renormalization procedure distorts the form of the LS equation. For this reason, we start with the LS equation where the potential K is extracted directly from the Lagrangian, and includes an infinite number of counter terms which cancel infinities of the loop integrals. In our approach, the kernel K used to generate the finite scattering amplitude T is meant to be the total sum of the NN irreducible diagrams (not a part of them). For the purposes of introducing a cutoff, we then construct an exact effective potential V for which it is apparent that all the infinities are cancelled.

Although, as mentioned above, the kernel K extracted from the chiral Lagrangian contains pion exchanges and contact interaction terms (the latter encoding the heavier meson exchanges and other integrated out higher energy physics), our derivation of the cutoff scheme is applicable also for the phenomenological approach where the contact term in K (short range part of the interaction) is mimicked by usual finite ρ meson exchange, i.e. $K = K_\pi + K_\rho$.

We emphasize that in this section of the paper a symbolic equation like that of Eq. (67) represents a non-relativistic three-dimensional integral equation; in particular, we take all quantities T, K, G_0, etc., as operators in relative three-momentum space with an overall total four-momentum delta function removed. Thus Eq. (67) is shorthand for

$$T(\mathbf{p}', \mathbf{p}, P) = K(\mathbf{p}', \mathbf{p}, P) + \int d\mathbf{k}'\, d\mathbf{k}\, K(\mathbf{p}', \mathbf{k}', P)\, G_0(\mathbf{k}', \mathbf{k}; P)\, T(\mathbf{k}, \mathbf{p}, P) \tag{68}$$

where $P = (E, \mathbf{P})$ is the total four-momentum (made up of the off-shell total energy E and total momentum $\mathbf{P} = \mathbf{p}_1 + \mathbf{p}_2 = \mathbf{p}'_1 + \mathbf{p}'_2$, $\mathbf{p} = \frac{1}{2}(\mathbf{p}_1 - \mathbf{p}_2)$ is the relative momentum of the two nucleons in initial state (similarly $\mathbf{p}'$, $\mathbf{k}$, and $\mathbf{k}'$), and

$$T(\mathbf{p}', \mathbf{p}, P) \equiv \langle \mathbf{p}'|T(P)|\mathbf{p}\rangle, \qquad K(\mathbf{p}', \mathbf{p}, P) \equiv \langle \mathbf{p}'|K(P)|\mathbf{p}\rangle, \tag{69a}$$

$$G_0(\mathbf{k}', \mathbf{k}; P) \equiv \langle \mathbf{k}'|G_0(P)|\mathbf{k}\rangle = \delta(\mathbf{k}'_1 - \mathbf{k}_1)\, \frac{1}{E - \mathbf{k}_1^2/2M - \mathbf{k}_2^2/2M + i\eta}. \tag{69b}$$

The chiral effective Lagrangian also involves the electromagnetic vector-potential A_μ in a gauge invariant way (in a way more general than that due to minimal coupling). In general, this means that the strong interaction vertices in the Lagrangian, and the corresponding ones with an attached external photon, are related to each other via WT identities. Given such a Lagrangian, attaching an external photon everywhere to the solution of Eq. (67) is carried out by gauging the LS equation, Eq. (67):

$$T^\mu = K^\mu + K^\mu G_0 T + K G_0^\mu T + K G_0 T^\mu. \tag{70}$$

This equation is likewise shorthand for a three-dimensional equation, similar to Eq. (68), but involves the gauged quantities

$$T^\mu(\mathbf{p}', P'; \mathbf{p}, P) \equiv \langle \mathbf{p}'|T^\mu(P', P)|\mathbf{p}\rangle, \quad K^\mu(\mathbf{p}', P'; \mathbf{p}, P) \equiv \langle \mathbf{p}'|K^\mu(P', P)|\mathbf{p}\rangle, \tag{71a}$$

$$G_0^\mu(\mathbf{k}', P'; \mathbf{k}, P) \equiv \langle \mathbf{k}'|G_0^\mu(P', P)|\mathbf{k}\rangle = \langle \mathbf{k}'|G_0(P')\Gamma_0^\mu(P', P)G_0(P)|\mathbf{k}\rangle \tag{71b}$$

where $P' = P + q$ for an incoming photon of momentum q, and $\Gamma_0^\mu = G_0^{-1} G_0^\mu G_0^{-1}$ is the single nucleon current as extracted from the initial Lagrangian:

$$\langle \mathbf{p}'|\Gamma_0^\mu(P', P)|\mathbf{p}\rangle = \delta(\mathbf{p}_2' - \mathbf{p}_2)\Gamma_1^\mu(\mathbf{p}_1', \mathbf{p}_1) + \delta(\mathbf{p}_1' - \mathbf{p}_1)\Gamma_2^\mu(\mathbf{p}_2', \mathbf{p}_2) \tag{72}$$

where Γ_i^μ is the electromagnetic vertex function of nucleon i:

$$\Gamma_i^\mu(\mathbf{p}', \mathbf{p}) = ie_i \left(1, \frac{\mathbf{p}' + \mathbf{p} + i\boldsymbol{\sigma}_i \times \mathbf{q}}{2M}\right) \tag{73}$$

where $\mathbf{q} = \mathbf{p}' - \mathbf{p}$. Solving Eq. (70), one derives the gauged scattering amplitude T^μ:

$$T^\mu = (1 + TG_0)K^\mu(1 + G_0 T) + TG_0^\mu T. \tag{74}$$

As in the previous sections, the quantity obtained through the use of the gauging of equations method, T^μ in this case, is guaranteed to satisfy the WT identity. This identity has the form of Eq. (63) and can be conveniently written in symbolic form as

$$q_\mu T^\mu = [\Gamma_0, T] \tag{75}$$

where

$$\begin{aligned}\langle \mathbf{p}'|\Gamma_0(P', P)|\mathbf{p}\rangle &\equiv ie_1\delta(\mathbf{p}_2' - \mathbf{p}_2) + ie_2\delta(\mathbf{p}_1' - \mathbf{p}_1) \\ &= ie_1\delta(\mathbf{p}' - \mathbf{p} - \mathbf{q}/2) + ie_2\delta(\mathbf{p}' - \mathbf{p} + \mathbf{q}/2).\end{aligned} \tag{76}$$

It is obvious that a naive introduction of a cutoff into Eq. (67) and Eq. (74) will destroy the gauge invariance [Eq. (75) will no longer hold]. It is the goal of this section to introduce a cutoff in a way that preserves the gauge invariance.

We begin by introducing an effective potential V in the cutoff scheme, by requiring that it generate, via the LS equation, the same *off shell* NN scattering amplitude, T, as the one determined in the underlying theory by the solution of Eq. (67). This requirement is important to enable the cutoff theory to describe not only elastic NN scattering, but NN currents as well, because the latter are determined in the underlying theory (at least partly) by the half-off-shell solution of Eq. (67). We achieve this requirement by expressing the LS equation, Eq. (67), as a set of two equations

$$T = V + VG_0\Theta T \tag{77a}$$

$$V = K + KG_0\bar{\Theta}V \tag{77b}$$

where $\langle \mathbf{p}'|\Theta(P)|\mathbf{p}\rangle = \delta(\mathbf{p}' - \mathbf{p})\theta(\Lambda - |\mathbf{p}|)$ is the sharp cutoff function with cutoff momentum Λ. Note that $\Theta + \bar{\Theta} = 1$ and $[G_0, \Theta] = 0$.

Although the couple of equations, Eqs. (77), result only from simple algebra, a non-trivial aspect of this decomposition is that it preserves the power counting properties of the potentials. That is, according to Eqs. (77b), for sufficiently small relative momenta, $|\mathbf{p}'|, |\mathbf{p}| < \Lambda$, potentials $V(E, \mathbf{p}', \mathbf{p})$ and $K(E, \mathbf{p}', \mathbf{p})$ have the same long range parts (only their short range parts are different - see Ref. [53]). This fact is at the heart of our derived EFT theory with cutoff, namely, in that this theory has the same predictive power as the underlying theory given by the initial Lagrangian.

Although Eq. (77b) may be of no use to derive the exact effective potential V for the EFT with cutoff [one cannot use perturbation theory, because high momenta are involved in Eq. (77b)], if the potential of the underlying theory, K, is given, then for sufficiently small momenta $|\mathbf{p}|$ and $|\mathbf{p}'|$, the unknown part of V is restricted to only zero-range potentials (consisting of the delta function and its derivatives in coordinate space) with the same number of coupling constants as in the underlying theory (these are not known a priori and are supposed to be extracted from experimental data.)

As a next step, we need to gauge both Eq. (77a) and Eq. (77b) to derive the five-point function T^μ of Eq. (74). Gauging Eqs. (77) we obtain

$$\tilde{T}^\mu = V^\mu + V^\mu G_0\Theta T + V(G_0\Theta)^\mu T + VG_0\Theta\tilde{T}^\mu \qquad (78a)$$

$$V^\mu = K^\mu + K^\mu G_0\bar{\Theta}V + K(G_0\bar{\Theta})^\mu V + KG_0\bar{\Theta}V^\mu \qquad (78b)$$

where, for the moment, we use a tilde notation to distinguish $\tilde{T}^\mu$ from the T^μ of Eq. (74), until it is explicitly shown, below, that the two are in fact equal to each other. The solution of Eqs. (78) is

$$\tilde{T}^\mu = (1 + TG_0\Theta)V^\mu(1 + G_0\Theta T) + T(G_0\Theta)^\mu T \qquad (79)$$

where

$$V^\mu = (1 + VG_0\bar{\Theta})K^\mu(1 + G_0\bar{\Theta}V) + V(G_0\bar{\Theta})^\mu V. \qquad (80)$$

In general, even if the input to a dynamical equation is specified, once the equation is gauged, one will still need to specify how exactly to attach a "photon" to the input. Thus, although the input to Eqs. (77) is known, consisting of the kernel K and the cutoff modified free Green functions $G_0\Theta$ and $G_0\bar{\Theta}$, the corresponding gauged inputs K^μ, $(G_0\Theta)^\mu$ and $(G_0\bar{\Theta})^\mu$, appearing in the gauged result of Eq. (79) and Eq. (80), still need to be defined. In the case of K^μ this is easy, as it is specified by the underlying Lagrangian and will automatically satisfy local gauge invariance, i.e., it will fulfill the WT identity $q_\mu K^\mu = [\Gamma_0, K]$. It is less obvious what to take for the gauged input $(G_0\bar{\Theta})^\mu$; however, we shall require it to likewise obey the WT identity, as only in this way can we ensure the necessary WT identity for V^μ [via Eq. (80)]. We shall also require that

$$(G_0\Theta)^\mu + (G_0\bar{\Theta})^\mu = G_0^\mu \qquad (81)$$

if only to guarantee that $\tilde{T}^\mu = T^\mu$. Indeed, let us now derive $\tilde{T}^\mu = T^\mu$ using Eq. (81) (even though from the above described gauging procedure, this may be obvious). Using Eq. (79) and Eq. (80) we have that

$$\begin{aligned}\tilde{T}^\mu &= (1 + TG_0\Theta)V^\mu(1 + G_0\Theta T) + T(G_0\Theta)^\mu T \\ &= (1 + TG_0\Theta)\left[(1 + VG_0\bar{\Theta})K^\mu(1 + G_0\bar{\Theta}V) + V(G_0\bar{\Theta})^\mu V\right] \\ &\quad \times (1 + G_0\Theta T) + T(G_0\Theta)^\mu T \\ &= (1 + TG_0\Theta + TG_0\bar{\Theta})K^\mu(1 + G_0\Theta T + G_0\bar{\Theta}T) + T(G_0\bar{\Theta})^\mu T + T(G_0\Theta)^\mu T \\ &= (1 + TG_0)K^\mu(1 + G_0T) + TG_0^\mu T = T^\mu. \end{aligned} \qquad (82)$$

We can derive the WT identity for $\tilde{T}^\mu$ if we assume that $(G_0\Theta)^\mu$ satisfies the WT identity

$$q_\mu(G_0\Theta)^\mu = [\Gamma_0, G_0\Theta], \qquad (83)$$

and if we can prove the WT identity for V: $q_\mu V^\mu = [\Gamma_0, V]$. Of course, we already know that $\tilde{T}^\mu$ must satisfy the WT identity because, as just shown, $\tilde{T}^\mu = T^\mu$ and T^μ satisfies the WT identity of Eq. (75). Nevertheless, we shall still derive the WT identity for $\tilde{T}^\mu$ just to see it as a result of our gauging, i.e., without using the above derivation of $\tilde{T}^\mu = T^\mu$. Indeed, combining the fact that $q_\mu G_0^\mu = [\Gamma_0, G_0]$ with Eq. (81) and Eq. (83), we obtain the WT for $(G_0\bar{\Theta})^\mu$,

$$q_\mu (G_0\bar{\Theta})^\mu = [\Gamma_0, G_0\bar{\Theta}], \tag{84}$$

and hence the WT identity for V^μ:

$$\begin{aligned} q_\mu V^\mu &= (1 + VG_0\bar{\Theta})q_\mu K^\mu(1 + G_0\bar{\Theta}V) + Vq_\mu(G_0\bar{\Theta})^\mu V \\ &= (1 + VG_0\bar{\Theta})[\Gamma_0, K](1 + G_0\bar{\Theta}V) + V[\Gamma_0, G_0\bar{\Theta}]V \\ &= (1 + VG_0\bar{\Theta})\Gamma_0 V - V\Gamma_0(1 + G_0\bar{\Theta}V) + V[\Gamma_0, G_0\bar{\Theta}]V \\ &= [\Gamma_0, V]. \end{aligned} \tag{85}$$

The WT identity for $\tilde{T}^\mu$ then follows from Eq. (79).

All that is now left, is the explicit construction of the gauged input $(G_0\Theta)^\mu$. In this respect, we note that Eq. (79) and the WT relation between V^μ and V, Eq. (85), are important to the construction of effective currents with maximal predictive power. It is then crucial that the single nucleon current of the theory with cutoff, $(G_0\Theta)^\mu$, not only satisfy the WT identity of Eq. (83), but that it also preserve the predictive power of the underlying theory. It is found that the following form for $(G_0\Theta)^\mu$ satisfies both these requirements and provides for the effective interaction current, V^μ, to have the same long-short range decomposition as K^μ [53]:

$$\langle \mathbf{p}'|G_0^{-1}(G_0\Theta)^\mu G_0^{-1}|\mathbf{p}\rangle = \frac{1}{2}\delta(\mathbf{p}_2' - \mathbf{p}_2)\,\mathcal{A}_1^\mu + (1 \leftrightarrow 2) \tag{86}$$

where

$$\begin{aligned} \mathcal{A}_1^\mu &= \Gamma_1^\mu(\mathbf{p}_1', \mathbf{p}_1)\left[\theta(\Lambda - p) + \theta(\Lambda - p')\right] \\ &\quad + \left[0, \mathbf{\Gamma}_1(\mathbf{p}', \mathbf{p})\right] \frac{\theta(\Lambda - p') - \theta(\Lambda - p)}{p'^2 - p^2}(E_{cm}'M - \mathbf{p}'^2 + E_{cm}M - \mathbf{p}^2) \end{aligned} \tag{87}$$

In the above expression, $p = |\mathbf{p}|$, $p' = |\mathbf{p}'|$, and E_{cm} (E_{cm}') is the energy of the two initial (final) nucleons in their own centre of mass system. The detailed derivation of Eq. (87) can be found in Ref. [53], and is based on a number of constraints, one of which is the WT identity, Eq. (83).

3.2. How to calculate different, photon involving, reactions

The full dynamics of NN reactions involving an external photon is contained in the gauged NN scattering amplitude, T^μ, discussed above. The electromagnetic bound state form factors, the NN bound state photodisintegration amplitude, and the NN bremsstrahlung amplitude, can all be derived from Eq. (79) by using the bound-state pole structure of the scattering amplitude,

$$T(\mathbf{p}', \mathbf{p}, P) \sim \frac{\phi_P(\mathbf{p}')\bar{\phi}_P(\mathbf{p})}{E - E_b} \qquad \text{as} \qquad E \to E_b, \tag{88}$$

where E_b is the bound-state energy, and the equation for the bound state vertex function ϕ,

$$\phi = VG_0\Theta\phi. \tag{89}$$

Journal of Physics: Conference Series **330** (2011) 012008 doi:10.1088/1742-6596/330/1/012008

3.2.1. The NN bound state current: $B + \gamma \to B'$ Although the physical NN system has only one bound state, the deuteron, we shall keep the presentation more general by formally allowing different two-body bound states, B and B', in the initial and final channels, respectively. Taking residues of Eq. (79) at the initial and final bound state poles, we obtain the bound state current

$$\langle P'|J^\mu(0)|P\rangle = \bar{\phi}_{P'}[G_0\Theta V^\mu G_0\Theta + (G_0\Theta)^\mu]\phi_P. \tag{90}$$

Using the explicit form for the single nucleon EM vertex, Eq. (87), one can easily write the expanded form of Eq. (90); for example, for the zeroth component of the current

$$\begin{aligned}\langle P'|J^0(0)|P\rangle &= \int d\mathbf{p}'\, d\mathbf{p}\, \bar{\phi}_{P'}(\mathbf{p}') \frac{\theta(\Lambda - p')\, V^0(\mathbf{p}', P'; \mathbf{p}, P)\, \theta(\Lambda - p)}{(E_b' - \mathbf{p}'^2/M)(E_b - \mathbf{p}^2/M)} \phi_P(\mathbf{p}) \\ &+ \frac{ie_1}{2} \int d\mathbf{p}\, \bar{\phi}_{P'}(\mathbf{p}') \frac{\theta(\Lambda - p') + \theta(\Lambda - p)}{(E_b' - \mathbf{p}'^2/M)(E_b - \mathbf{p}^2/M)} \phi_P(\mathbf{p}) + (1 \leftrightarrow 2)\end{aligned} \tag{91}$$

where E_b' and E_b are binding energies in the final and initial state, respectively. We note that the bound state vertex functions, $\phi_P(\mathbf{p})$, in fact do not depend on the total momentum P due to Galillei invariance (i.e., they depend only on the relative momentum of the nucleons, $\mathbf{p}$); also, the first (6-dimensional) integral of Eq. (91) corresponds to exchange currents with $\mathbf{p}', \mathbf{p}$ being independent integration variables, while the second (3-dimensional) integral corresponds to the impulse approximation with $\mathbf{p}' = \mathbf{p} + \mathbf{q}/2$.

3.2.2. The NN bound state photodisintegration current: $B + \gamma \to N + N$ To find the bound state photodisintegration current, j_0^μ, in addition to the photon attachments contained in T^μ, one also needs photon attachments to the two free final state nucleons. To include the latter, it is sufficient to gauge the quantity G_0T:

$$(G_0T)^\mu = G_0^\mu T + G_0 T^\mu, \tag{92}$$

take the residue at the initial bound state pole, and then to "chop off" the final legs:

$$\begin{aligned} j_0^\mu &= G_0^{-1}[G_0^\mu + G_0(1 + TG_0\Theta)V^\mu G_0\Theta + G_0 T(G_0\Theta)^\mu]\phi \\ &= [G_0^{-1}G_0^\mu + (1 + TG_0\Theta)V^\mu G_0\Theta + T(G_0\Theta)^\mu]\phi. \end{aligned} \tag{93}$$

3.2.3. The NN Bremsstrahlung current: $N + N \to N + N + \gamma$ To find the Bremsstrahlung current, j_{00}^μ, in addition to the photon attachments contained in T^μ, one also needs photon attachments to the free initial- and final-state nucleons. For this, it is sufficient to gauge the quantity G_0TG_0 and the to "chop off" both the inital and final legs:

$$\begin{aligned} j_{00}^\mu &= G_0^{-1}(G_0TG_0)^\mu G_0^{-1} = T^\mu + G_0^{-1}G_0^\mu T + TG_0^\mu G_0^{-1} \\ &= (1 + TG_0\Theta)V^\mu(1 + G_0\Theta T) + T(G_0\Theta)^\mu T + \Gamma_0^\mu G_0 T + TG_0\Gamma_0^\mu. \end{aligned} \tag{94}$$

3.3. Discussion

In the context of two-nucleon EFT, we have developed a cutoff scheme for strong interaction amplitudes and their EM currents that is equivalent to, and has the same predictive power as, two-nucleon EFT in the renormalization scheme.

In the renormalization scheme, we have an NN potential K with long and short range parts, whose short range couplings are determined from a comparison of the generated amplitude with the effective range expansion. We also have the interaction current K^μ which is given by the initial Lagrangian, and is restricted only by local gauge invariance, i.e. by the WT identity $q_\mu K^\mu = [\Gamma_0, K]$. The physical NN scattering amplitude and NN currents then are derived via Eq. (67) and Eq. (74) respectively.

In the cutoff scheme we also have an NN potential V with long and short range parts, whose short range couplings are determined from a comparison with the effective range expansion. We also have an interaction current V^μ which is restricted only by local gauge invariance, i.e., by the WT identity $q_\mu V^\mu = [\Gamma_0, V]$. The physical NN scattering amplitude and NN currents are derived via the cutoff versions of Eq. (67) and Eq. (74), namely, Eq. (77a) and Eq. (79), respectively. The currents in our cutoff scheme are conserved.

The essential difference between the approaches with interactions K and V, is in that K is given directly by the initial Lagrangian, whereas V is related to K in a complicated way. This difference is of no consequence in the philosophy of EFT because K and V have the same long range parts. The short range parts of K and V are determined by comparing the NN scattering amplitude with predictions of experiment, while the only actual way of determining both K^μ and V^μ is by relating them to K and V via their respective WT identities.

The new technical element of our EFT with cutoff is the single nucleon current, $(G_0\Theta)^\mu$, which satisfies the WT identity, Eq. (83), and whose specific form is given by Eq. (87). It prescribes a specific gauge invariant way for implementing the cutoff in the impulse approximation.

Eq. (87) and its derivation in Ref. [53] can be used as a guide for the gauge invariant treatment of other theories where a cutoff is a necessary attribute for actual calculations (see for example the NJL model of Ref. [57]). Gauge invariance can be maintained by special regularization [analogous to the one of Eq. (87)] of the integrals corresponding to loops with attached photons, as is the case of the present paper. This regularization will depend on the way the loops without photons are regularized.

4. Crossing symmetric model of pion-nucleon scattering

Our final example demonstrates how the gauging of equations method can be used to attach external particles other than photons to non-perturbative processes. In particular, we use it to obtain a complete attachment of *two* external pions to a dressed nucleon propagator of an underlying nonperturbative πN potential model, and in this way, generate a crossing symmetric πN amplitude. The formulation automatically provides expressions also for the crossing symmetric and gauge invariant pion photoproduction and Compton scattering amplitudes. We show that our amplitudes are also unitary if they coincide on-shell with the amplitudes obtained by attaching one pion to the dressed πNN vertex of the same potential model. It is left to future numerical calculations to determine the accuracy with which this coincidence can be achieved by adjusting parameters of the underlying πN model. To this extent, our gauging method provides an approximate solution to the long-standing problem of constructing a potential model that is at the same time unitary and crossing symmetric [58].

4.1. Underlying strong interaction πN model

The πN scattering amplitude t is often described using the set of equations [59, 60, 61]

$$t = fg\bar{f} + t_b, \qquad t_b = v + t_b G_0 v, \tag{95a}$$

$$\bar{f} = \bar{f}_0 + \bar{f}_0 G_0 t_b, \qquad f = f_0 + t_b G_0 f_0, \tag{95b}$$

$$g = g_0 + g_0 \Sigma g, \qquad \Sigma = \bar{f}_0 G_0 f, \tag{95c}$$

where f (f_0) is the dressed (bare) $N \to \pi N$ vertex, $\bar{f}$ ($\bar{f}_0$) is the dressed (bare) $\pi N \to N$ vertex, g (g_0) is the dressed (bare) nucleon propagator, G_0 is the disconnected πN propagator, and t_b (v) is the "non-pole" πN t-matrix (potential) with the pole term $fg\bar{f}$ ($f_0 g_0 \bar{f}_0$) removed. Although these equations provide an exact description in full field theory, their main feature is that they allow one to preserve unitarity when making models for the potential v and bare vertex f_0. However, like all potential models, these equations suffer from a lack of crossing symmetry, a property whose importance has been emphasized for more than 50 years [62, 63, 64].

Similarly, the pion photoproduction amplitude t^γ is often descsribed by a set of equations that essentially result from Eq. (95a) and Eq. (95b) by replacing the initial pion with a photon [65, 66, 67]:

$$t^\gamma = fg\bar{f}^\gamma + t_b^\gamma, \qquad t_b^\gamma = v^\gamma + t_b G_0 v^\gamma, \tag{96a}$$

$$\bar{f}^\gamma = \bar{f}_0^\gamma + \bar{f}_0 G_0 t_b^\gamma, \tag{96b}$$

where $\bar{f}^\gamma$ ($\bar{f}_0^\gamma$) is the dressed (bare) $\gamma N \to N$ vertex, and t_b^γ (v^γ) is the pion photoproduction amplitude (Born term) with the pole term $fg\bar{f}^\gamma$ ($f_0 g_0 \bar{f}_0^\gamma$) removed. Once again the feature of these equations is that they respect unitarity. This time, however, these equations suffer not only from a lack of crossing symmetry, but also from the breaking of manifest gauge invariance (because the photon is not coupled to all places in the underlying field theory). We shall refer to Eqs. (95) and Eqs. (96) as the *standard* description.

Below we shall present new equations for πN scattering, pion photoproduction, and Compton scattering, that are based on the potential model of Eqs. (95), but that preserve crossing symmetry and manifest gauge invariance. Our approach is based on the idea of coupling external pions and photons to all possible places in the dressed propagator g of Eq. (95c), and is achieved using the gauging of equations method [6, 68]. Like the standard description, our approach is exact in full field theory;[2] however, just opposite to the standard description, when models are made for the potential v and bare vertex f_0, our approach preserves crossing symmetry and gauge invariance at the expense of unitarity. The lack of built-in unitarity is not surprising since our approach effectively sums the full perturbation series in a way that is different from the usual method of iterating a kernel. Nevertheless, we show that our amplitudes will satisfy unitarity whenever the crossing symmetric πN amplitude coincides, on-shell, with the one obtained by attaching one pion to the dressed πNN vertex f of Eq. (95b).

4.2. Single-gauged amplitude

In Refs. [6, 68] we introduced a technique for attaching an external photon to all possible places (vertices, propagators, potentials, etc.) within a strongly interacting system described by dynamical equations. The completeness of the attachment led to the gauge invariance of the resulting electromagnetic currents. Here we use the same technique to also attach first one external pion, and then in the next section, a second external pion, in order to achieve our goal of deriving a crossing symmetric πN scattering amplitude.

We begin by applying our gauging technique to the πN Green function G generated by the non-pole potential v:

$$G = G_0 + G_0 v G. \tag{97}$$

Denoting by G^μ the 5-point function resulting from a complete attachment of an external pion to G, Eq. (97) is "gauged"[3] to obtain

$$G^\mu = G_0^\mu + G_0^\mu v G + G_0 v^\mu G + G_0 v G^\mu \tag{98}$$

which is easily solved to get

$$G^\mu = G\Lambda^\mu G, \qquad \Lambda^\mu = \Lambda_0^\mu + v^\mu \tag{99}$$

[2] For simplicity of presentation, we ignore any terms that cannot be obtained by the attachment of external pions or photons. Such contributions, if present, are gauge invariant and crossing symmetric on their own, and can therefore be separately added to our derived amplitudes.

[3] We shall use "gauging" to mean the process of attaching any external particle, not just a gauge boson.

Journal of Physics: Conference Series **330** (2011) 012008 doi:10.1088/1742-6596/330/1/012008

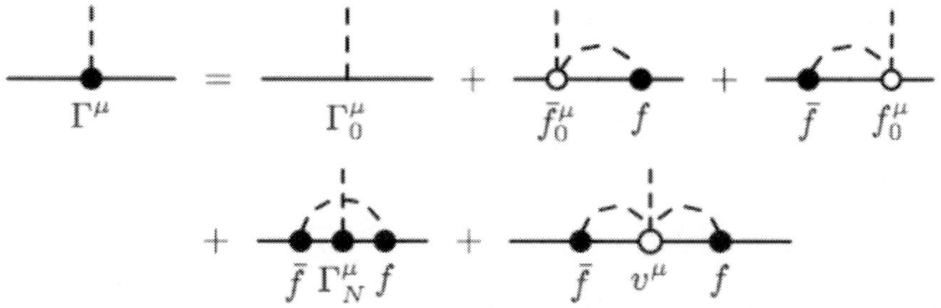

Figure 7. The dressed πNN vertex function Γ^μ resulting from a complete pion attachment to the dressed nucleon propagator g defined by Eq. (95c).

where $\Lambda_0^\mu \equiv G_0^{-1}G_0^\mu G_0^{-1}$ is a vertex function derived by attaching an external pion to the disconnected πN propagator G_0. Note that $G_0 = g_\pi g_N$ where g_π is the pion propagator and g_N is the effective nucleon propagator (in exact field theory $g_N = g$, but in practical calculations where there are no explicit two-pion states, one usually takes g_N to be just the pole part of g). As G-parity conservation forbids a three-pion vertex, $\Lambda_0^\mu = \Gamma_N^\mu g_\pi^{-1}$ where $\Gamma_N^\mu \equiv g_N^{-1}g_N^\mu g_N^{-1}$. Thus

$$\Lambda^\mu = \Gamma_N^\mu g_\pi^{-1} + v^\mu. \tag{100}$$

The gauged potential v^μ can be constructed phenomenologically, or derived by gauging a specific model for v. Similarly, the gauging of the dressed nucleon propagator g of Eq. (95c), gives the 3-point function g^μ:

$$g^\mu = g\Gamma^\mu g, \qquad\qquad \Gamma^\mu = \Gamma_0^\mu + \Sigma^\mu \tag{101}$$

where Γ^μ is the dressed πNN vertex function, $\Gamma_0^\mu \equiv g_0^{-1}g_0^\mu g_0^{-1}$ is the bare vertex function, and Σ^μ is the gauged dressing. Gauging $\Sigma = \bar f_0 G f_0$ then leads to a simple intuitive expression for the πNN dressed vertex function:

$$\Gamma^\mu = \Gamma_0^\mu + \bar f_0^\mu G_0 f + \bar f G_0 f_0^\mu + \bar f G_0^\mu f + \bar f G_0 v^\mu G_0 f, \tag{102}$$

which is illustrated in Fig. (1).

The main results of this section come from the gauging of the dressed πNN vertrex f:

$$\begin{aligned}(G_0 f g)^\nu = (G f_0 g)^\nu &= G^\nu f_0 g + G f_0^\nu g + G f_0 g^\nu \\ &= G\Lambda^\nu G_0 f g + G f_0^\nu g + G_0 f g \Gamma^\nu g.\end{aligned} \tag{103}$$

Cutting off the external legs immediately gives the amplitude $T^\nu \equiv G_0^{-1}(G_0 f g)^\nu g^{-1}$:

$$T^\nu = (1 + t_b G_0)(\Lambda_0^\nu G_0 f + v^\mu G_0 f + f_0^\nu) + f g \Gamma^\nu. \tag{104}$$

If superscript ν corresponds to an external photon, then $T^\nu Z$ is the properly normalized manifestly gauge invariant pion photoproduction amplitude, originally derived in Ref. [69] using a more involved approach. If superscript ν corresponds to an external pion, then $T^\nu Z$ is a "hybrid" πN scattering amplitude where the *initial* state pion is due to gauging and the final state pion is due to the original standard description. In a similar way, vertex $\bar f$ can be gauged to obtain the hybrid amplitude $\bar T^\mu$ where the *final* state pion is due to gauging. One can express these amplitudes in a "pole plus non-pole" form analogous to Eq. (95a):

$$T^\nu = f g \Gamma^\nu + T_b^\nu, \qquad T_b^\nu = (1 + t_b G_0)V^\nu, \qquad V^\nu = f_0^\nu + \Lambda^\nu G_0 f, \tag{105a}$$

$$\bar T^\mu = \Gamma^\mu g \bar f + \bar T_b^\mu, \qquad \bar T_b^\mu = \bar V^\mu (1 + G_0 t_b), \qquad \bar V^\mu = \bar f_0^\mu + \bar f G_0 \Lambda^\mu. \tag{105b}$$

Journal of Physics: Conference Series **330** (2011) 012008 doi:10.1088/1742-6596/330/1/012008

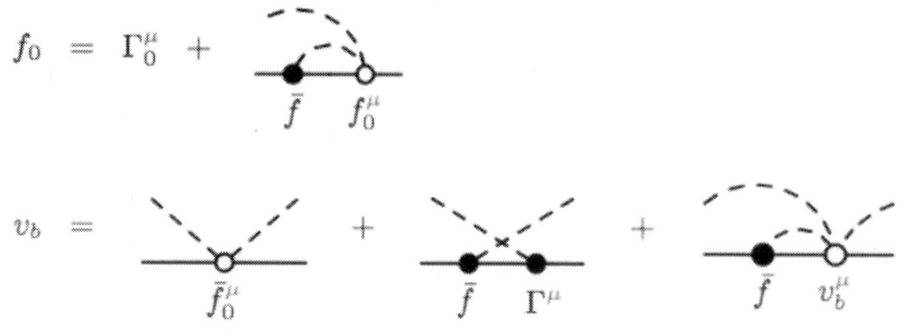

Figure 8. In exact field theory, relations connecting the bare πNN vertex f_0 and "non-pole" πN potential v of the standard description, Eqs. (95), to the corresponding gauged quantities f_0^μ and v^μ of the single-gauged description - see the first two of Eqs. (108).

It follows that amplitudes T_b^ν and $\bar{T}_b^\mu$ satisfy non-standard Bethe-Salpeter equations

$$T_b^\nu = V^\nu + vG_0T_b^\nu, \qquad \bar{T}_b^\mu = \bar{V}^\mu + \bar{T}_b^\mu G_0 v \tag{106}$$

where the kernel and inhomogeneous terms involve πN potentials of different origin. The amplitudes are clearly not crossing symmetric.

We can now express Eq. (102) in the following two ways analogous to Eq. (95b):

$$\Gamma^\nu = \bar{F}_0^\nu + \bar{f}G_0V^\nu, \qquad \bar{F}_0^\nu = \Gamma_0^\nu + \bar{f}_0^\nu G_0 f, \tag{107a}$$
$$\Gamma^\mu = F_0^\mu + \bar{V}^\mu G_0 f, \qquad F_0^\mu = \Gamma_0^\mu + \bar{f}G_0 f_0^\mu, \tag{107b}$$

where Eq. (107a) and Eq. (107b) are to be used to describe pion (or photon) absorption and creation vertices, respectively. In exact field theory, Γ^μ can be identified with function f of the standard description. Eqs. (107) then imply the following identities (in exact field theory) relating the standard description quantities f_0 and v (typically the model inputs) to their gauged counterparts:

$$f_0 = F_0^\mu, \qquad v = \bar{V}^\mu; \qquad \bar{f}_0 = \bar{F}_0^\nu, \qquad v = V^\nu. \tag{108}$$

The first two of Eqs. (108) are illustrated in Fig. 8.

4.2.1. Unitarity For ease of presentation, we discuss unitarity within the framework of time ordered perturbation theory for which

$$G_0(E^+) - G_0(E^-) = -2\pi i\delta(E - H_0) \tag{109}$$

where H_0 is the free Hamiltonian and $E^\pm = E \pm i\epsilon$. We consider only 2-body unitarity and thus restrict the discussion to energies E below the two-pion threshold. In this energy region the potential v and bare vertex f_0 are real, and the standard description of Eqs. (95) will therefore satisfy the following unitarity relations:

$$t - t^\dagger = t^\dagger \delta\, t, \qquad t_b - t_b^\dagger = t_b^\dagger \delta\, t_b, \tag{110a}$$
$$f - f^\dagger = t_b^\dagger \delta f, \qquad \bar{f} - \bar{f}^\dagger = \bar{f}^\dagger \delta t_b, \tag{110b}$$
$$g - g^\dagger = g^\dagger (\Sigma - \Sigma^\dagger)\, g, \qquad \Sigma - \Sigma^\dagger = \bar{f}^\dagger \delta\, f, \tag{110c}$$
$$G - G^\dagger = (1 + G_0^\dagger t_b^\dagger)\, \delta\, (1 + t_b G_0), \tag{110d}$$

Journal of Physics: Conference Series **330** (2011) 012008 doi:10.1088/1742-6596/330/1/012008

where δ is shorthand for $-2\pi i\delta(E-H_0)$ and where identical quantities with and without a dagger represent the same functions of E^- and E^+, respectively (i.e., a dagger does not mean Hermitean conjugate, but rather, $T \equiv T(E^+)$ and $T^\dagger \equiv T(E^-)$). Applying these relations to Eqs. (105) and Eqs. (107) one obtains the analogous unitarity relations for the hybrid amplitudes:

$$T^\nu - T^{\nu\dagger} = t^\dagger\delta\, T^\nu, \qquad T_b^\nu - T_b^{\nu\dagger} = t_b^\dagger\delta\, T_b^\nu, \tag{111a}$$

$$\bar{T}^\mu - \bar{T}^{\mu\dagger} = \bar{T}^{\mu\dagger}\delta\, t, \qquad \bar{T}_b^\mu - \bar{T}_b^{\mu\dagger} = \bar{T}_b^{\mu\dagger}\delta\, t_b, \tag{111b}$$

$$\Gamma^\mu - \Gamma^{\mu\dagger} = \bar{T}_b^{\mu\dagger}\delta f, \qquad \Gamma^\nu - \Gamma^{\nu\dagger} = \bar{f}^\dagger\delta T_b^\nu. \tag{111c}$$

In the case of gauging with photons, Eq. (111a) provides just the usual statement of Watson's theorem for the gauge invariant pion photoproduction amplitude T^ν. However, in the case of gauging with pions, Eq. (111a) differs from the usual statement of unitarity for the hybrid amplitude in that a $T^{\nu\dagger}$ has been replaced with a $t^\dagger$ from the standard description. Thus the only way for a hybrid πN amplitude T^ν to be unitary is for it to coincide, on-shell, with the standard amplitude t.

4.3. Double-gauged amplitude

To obtain a crossing symmetric πN scattering amplitude, we attach two external pions to the dressed nucleon propagator g; that is, we first gauge g to obtain g^μ as in Eq. (101), and then gauge Eq. (101) to obtain $g^{\mu\nu}$ with indices ν and μ denoting the initial- and final-state pion, respectively. Thus

$$g^{\mu\nu} = gT^{\mu\nu}g, \qquad T^{\mu\nu} = \Gamma^\mu g\Gamma^\nu + \Gamma^\nu g\Gamma^\mu + \Gamma_0^{\mu\nu} + \Sigma^{\mu\nu} \tag{112}$$

where $ZT^{\mu\nu}$ is the properly normalized crossing symmetric πN amplitude. In Eq. (112)

$$\Gamma_0^{\mu\nu} = \left(g_0^{-1}g_0^\mu g_0^{-1}\right)^\nu = g_0^{-1}g_0^{\mu\nu}g_0^{-1} - \Gamma_0^\mu g_0\Gamma_0^\nu - \Gamma_0^\nu g_0\Gamma_0^\mu, \tag{113a}$$

$$\Sigma^{\mu\nu} = \bar{f}_0^{\mu\nu}G_0 f + \bar{f}G_0 f_0^{\mu\nu} + \bar{f}G_0\Lambda^{\mu\nu}G_0 f + A^{\mu\nu} + A^{\nu\mu} \tag{113b}$$

where

$$\Lambda^{\mu\nu} = \Gamma_N^{\mu\nu}g_\pi^{-1} + v^{\mu\nu}, \tag{114a}$$

$$A^{\mu\nu} = \bar{V}^\mu G V^\nu = \bar{V}^\mu G_0 T_b^\nu = \bar{T}_b^\mu G_0 V^\nu. \tag{114b}$$

One can thus write the crossing symmetric $T^{\mu\nu}$ in terms of pole and non-pole parts as

$$T^{\mu\nu} = \Gamma^\mu g\Gamma^\nu + T_b^{\mu\nu}, \qquad T_b^{\mu\nu} = V^{\mu\nu} + \bar{T}_b^\mu G_0 V^\nu, \tag{115a}$$

$$\begin{aligned} V^{\mu\nu} &= \Gamma^\nu g\Gamma^\mu + \Gamma_0^{\mu\nu} + \bar{f}_0^{\mu\nu}G_0 f + \bar{f}G_0 f_0^{\mu\nu} + \bar{f}G_0\Lambda^{\mu\nu}G_0 f + A^{\nu\mu} \\ &= \left(\bar{F}_0^\nu\right)^\mu + \Gamma^\nu g\Gamma^\mu + \bar{f}G_0\left(V^\nu\right)^\mu. \end{aligned} \tag{115b}$$

We note that the second of Eqs. (115a) still has the basic structure of a Bethe-Salpeter equation although neither the two non-pole πN potentials $V^{\mu\nu}$ and V^ν, nor the two non-pole t matrices $T_b^{\mu\nu}$ and $\bar{T}_b^\mu$ are of the same origin. It is also important to note that Eqs. (115) apply also to the cases where either one or both of the superscripts μ and ν refer to the gauging by photons. That is, these equations provide a unified, crossing symmetric description of pion-nucleon elastic scattering, pion photoproduction, and Compton scattering. Morever, the electromagnetic amplitudes are due to a complete attachment of photons and are therefore manifestly gauge invariant.

4.3.1. Unitarity In the crossing symmetric formulation, the πN potential $V^{\mu\nu}$ of Eq. (115b) is real below two-pion threshold, as is the hybrid potential V^{ν}. It is thus straightforward to obtain the 2-body unitarity relations from Eq. (115a) and the unitarity relations for the hybrid amplitudes, Eqs. (111). One obtains

$$T^{\mu\nu} - T^{\mu\nu\dagger} = \bar{T}^{\mu\dagger}\delta\, T^{\nu}, \qquad T_b^{\mu\nu} - T_b^{\mu\nu\dagger} = \bar{T}_b^{\mu\dagger}\delta\, T_b^{\nu}. \tag{116}$$

As for the hybrid case, these have the same form as usual unitarity relations, and differ from them only in that they contain πN t matrices of different origin. The task of achieving exact 2-body unitarity for the crossing symmetric amplitudes is therefore a numerical one - the standard potential model of Eqs. (95), and its parameters, need to be adjusted so as to ensure that the single- and double-gauged amplitudes coincide on-shell. Finally, it is worth pointing out that 3-body unitarity could be obtained in a similar fashion by gauging a standard Faddeev-like description of the $\pi\pi N$ system [70].

References

[1] F. Gross and D. O. Riska, Phys. Rev. C **36**, 1928 (1987).
[2] A. N. Kvinikhidze and B. Blankleider, in *invited talk given at the Joint Japan Australia Workshop: Quarks, Hadrons and Nuclei* (unpublished, Adelaide, November 15-24, 1995).
[3] H. Haberzettl, Phys. Rev. C **56**, 2041 (1997), *Preprint* nucl-th/9704057.
[4] A. N. Kvinikhidze and B. Blankleider, Phys. Rev. C **56**, 2963 (1997), *Preprint* nucl-th/9706051.
[5] A. N. Kvinikhidze and B. Blankleider, Phys. Rev. C **56**, 2973 (1997), *Preprint* nucl-th/9706052.
[6] A. N. Kvinikhidze and B. Blankleider, Phys. Rev. C **60**, 044003 (1999), *Preprint* nucl-th/9901001.
[7] A. N. Kvinikhidze and B. Blankleider, Phys. Rev. C **60**, 044004 (1999), *Preprint* nucl-th/9901002.
[8] A. N. Kvinikhidze and B. Blankleider, Phys. Rev. C **59**, 1263 (1999), *Preprint* nucl-th/9810025.
[9] F. Gross, A. Stadler, and M. T. Pena, Phys. Rev. C **69**, 034007 (2004), *Preprint* nucl-th/0311095.
[10] A. N. Kvinikhidze and B. Blankleider, Nucl. Phys. **A574**, 788 (1994), *Preprint* nucl-th/9402010.
[11] A. N. Kvinikhidze and B. Blankleider, Nucl. Phys. **A631**, 559c (1998), *Preprint* nucl-th/9710039.
[12] J. C. R. Bloch, C. D. Roberts, S. M. Schmidt, A. Bender, and M. R. Frank, Phys. Rev. C **60**, 062201 (1999), *Preprint* nucl-th/9907120.
[13] J. C. R. Bloch, C. D. Roberts, and S. M. Schmidt, Phys. Rev. C **61**, 065207 (2000), *Preprint* nucl-th/9911068.
[14] B. Blankleider and A. N. Kvinikhidze, Phys. Rev. C **62**, 039801 (2000), *Preprint* nucl-th/9912003.
[15] R. Alkofer, A. Holl, M. Kloker, A. Krassnigg, and C. D. Roberts, Few Body Syst. **37**, 1 (2005), *Preprint* nucl-th/0412046.
[16] C. D. Roberts, M. S. Bhagwat, A. Holl, and S. V. Wright, Eur. Phys. J. ST **140**, 53 (2007), *Preprint* 0802.0217.
[17] I. Cloet, G. Eichmann, B. El-Bennich, T. Klahn, and C. Roberts, Few Body Syst. **46**, 1 (2009), *Preprint* 0812.0416.
[18] D. R. Phillips and S. J. Wallace, Few Body Syst. **24**, 175 (1998), *Preprint* nucl-th/9708027.
[19] R. Alkofer and L. von Smekal, Phys. Rept. **353**, 281 (2001), *Preprint* hep-ph/0007355.
[20] A. N. Kvinikhidze and B. Blankleider, Phys. Rev. D **68**, 025021 (2003), *Preprint* hep-th/0303038.
[21] D. Mueller, D. Robaschik, B. Geyer, F. M. Dittes, and J. Horejsi, Fortschr. Phys. **42**, 101 (1994), *Preprint* hep-ph/9812448.
[22] X.-D. Ji, Phys. Rev. Lett. **78**, 610 (1997), *Preprint* hep-ph/9603249.
[23] X.-D. Ji, Phys. Rev. D **55**, 7114 (1997), *Preprint* hep-ph/9609381.
[24] A. V. Radyushkin, Phys. Lett. **B380**, 417 (1996), *Preprint* hep-ph/9604317.
[25] A. V. Radyushkin, Phys. Lett. **B385**, 333 (1996), *Preprint* hep-ph/9605431.
[26] A. V. Radyushkin, Phys. Rev. D **56**, 5524 (1997), *Preprint* hep-ph/9704207.
[27] M. Burkardt, Phys. Rev. D **62**, 071503 (2000), *Preprint* hep-ph/0005108.
[28] M. Burkardt, Phys. Rev. D **66**, 119903 (2002).
[29] K. Goeke, M. V. Polyakov, and M. Vanderhaeghen, Prog. Part. Nucl. Phys. **47**, 401 (2001), *Preprint* hep-ph/0106012.
[30] M. Diehl, Phys. Rept. **388**, 41 (2003), *Preprint* hep-ph/0307382.
[31] R. L. Jaffe, Nucl. Phys. **B229**, 205 (1983).
[32] M. Diehl and T. Gousset, Phys. Lett. **B428**, 359 (1998), *Preprint* hep-ph/9801233.
[33] X.-d. Ji, Phys. Rev. Lett. **91**, 062001 (2003), *Preprint* hep-ph/0304037.
[34] G. A. Miller, Phys. Rev. C **68**, 022201 (2003), *Preprint* nucl-th/0304076.

[35] F. Gross and P. Agbakpe, Phys. Rev. C **73**, 015203 (2006), *Preprint* nucl-th/0411090.
[36] F. Gross, Phys. Rev. **186**, 1448 (1969).
[37] F. Gross, Phys. Rev. C **26**, 2203 (1982).
[38] F. Gross, Phys. Rev. C **26**, 2226 (1982).
[39] M. Oettel, Ph.D. thesis, Tübingen University (2000), *Preprint* nucl-th/0012067.
[40] N. Ishii, Nucl. Phys. **A689**, 793 (2001), *Preprint* nucl-th/0004063.
[41] D. R. Phillips, S. J. Wallace, and N. K. Devine, Phys. Rev. C **58**, 2261 (1998), *Preprint* nucl-th/9802067.
[42] M. Oettel, M. Pichowsky, and L. von Smekal, Eur. Phys. J. A **8**, 251 (2000), *Preprint* nucl-th/9909082.
[43] R. T. Cahill and S. M. Gunner, Fizika **B7**, 171 (1998), *Preprint* hep-ph/9812491.
[44] B. Blankleider and A. N. Kvinikhidze, Few Body Syst. Suppl. **12**, 223 (2000), *Preprint* nucl-th/9912075.
[45] B. C. Tiburzi and G. A. Miller, Phys. Rev. D **67**, 054014 (2003), *Preprint* hep-ph/0210304.
[46] B. C. Tiburzi and G. A. Miller, Phys. Rev. D **67**, 054015 (2003), *Preprint* hep-ph/0210305.
[47] S. Noguera, L. Theussl, and V. Vento, Eur. Phys. J. A **20**, 483 (2004), *Preprint* nucl-th/0211036.
[48] A. N. Kvinikhidze and B. Blankleider, Nucl. Phys. **A784**, 259 (2007), *Preprint* hep-ph/0412409.
[49] A. N. Kvinikhidze and B. Blankleider, Phys. Lett. **B307**, 7 (1993).
[50] B. Blankleider and A. N. Kvinikhidze, in *Proceedings of the Sixth International Symposium on Meson-Nucleon Physics and the Structure of the Nucleon*, edited by D. Drechsel, G. Höler, W. Kluge, and B. M. K. Nefkens (Blaubeuren/Tübingen, Germany, 1995), vol. 11 of *PiN Newslett.*, p. 96, *Preprint* nucl-th/9508027.
[51] A. N. Kvinikhidze, M. C. Birse, and B. Blankleider, Phys. Rev. C **66**, 045203 (2002), *Preprint* hep-ph/0110060.
[52] E. Epelbaum, Prog. Part. Nucl. Phys. **57**, 654 (2006), *Preprint* nucl-th/0509032.
[53] A. N. Kvinikhidze, B. Blankleider, E. Epelbaum, C. Hanhart, and M. P. Valderrama, Phys. Rev. C **80**, 044004 (2009), *Preprint* 0904.4128.
[54] S. Weinberg, Phys. Lett. **B251**, 288 (1990).
[55] S. Weinberg, Nucl. Phys. **B363**, 3 (1991).
[56] S. Weinberg, Phys. Lett. **B295**, 114 (1992), *Preprint* hep-ph/9209257.
[57] H. Asami, N. Ishii, W. Bentz, and K. Yazaki, Phys. Rev. C **51**, 3388 (1995).
[58] A. Martin, Phys. Rev. **161**, 1528 (1967).
[59] S. Morioka and I. R. Afnan, Phys. Rev. C **26**, 1148 (1982).
[60] F. Gross and Y. Surya, Phys. Rev. C **47**, 703 (1993).
[61] V. Pascalutsa and J. A. Tjon, Phys. Rev. C **61**, 054003 (2000), *Preprint* nucl-th/0003050.
[62] G. F. Chew and F. E. Low, Phys. Rev. **101**, 1570 (1956).
[63] R. J. McLeod, Phys. Rev. C **29**, 1098 (1984).
[64] C. Fernandez-Ramirez, E. Moya de Guerra, and J. M. Udias, Phys. Lett. **B660**, 188 (2008).
[65] S. Nozawa, B. Blankleider, and T. S. H. Lee, Nucl. Phys. **A513**, 459 (1990).
[66] Y. Surya and F. Gross, Phys. Rev. C **53**, 2422 (1996).
[67] V. Pascalutsa and J. A. Tjon, Phys. Rev. **C70**, 035209 (2004), *Preprint* nucl-th/0407068.
[68] A. N. Kvinikhidze and B. Blankleider, Phys. Rev. C **60**, 044004 (1999), *Preprint* nucl-th/9901002.
[69] C. H. M. van Antwerpen and I. R. Afnan, Phys. Rev. **C52**, 554 (1995), *Preprint* nucl-th/9407038.
[70] I. R. Afnan and B. C. Pearce, Phys. Rev. C **35**, 737 (1987).

Journal of Physics: Conference Series **330** (2011) 012009 doi:10.1088/1742-6596/330/1/012009

Holographic Dual of the Reissner-Nordström Black Hole

Chiang-Mei Chen[1] **and Jia-Rui Sun**[2]

[1]Department of Physics and Center for Mathematics and Theoretical Physics, National Central University, Chungli 320, Taiwan
[2]Department of Physics, National Central University, Chungli 320, Taiwan

E-mail: [1]`cmchen@phy.ncu.edu.tw`, [2]`jrsun@phy.ncu.edu.tw`

Abstract. In this article we give a brief, nevertheless, comprehensive review on recent studies in the quantum gravity description of the Reissner-Nordström (RN) black hole from the perspective of the AdS/CFT correspondence. We survey the known evidence supporting a two-dimensional conformal field theory (CFT) description holographically dual to the RN black hole.

1. Introduction

The holographic principle [1, 2] provides a novel visual angle to deal with problems of quantum gravity. It allows us to study the quantum gravity without quantizing the gravity itself directly, but instead, to find a lower dimensional equivalently dual field theory which contains the same information as that of the gravity. However, until now there are only few explicit realizations of the holographic principle which have been constructed, for examples the $\mathrm{AdS}_5/\mathrm{CFT}_4$ [3, 4, 5] and the $\mathrm{AdS}_3/\mathrm{CFT}_2$ [6] correspondences. Recent progress made along this direction was initiated from the study of the holographic description of the extremal Kerr black hole—the Kerr/CFT correspondence [7], which showed that the near horizon extremal Kerr (NHEK) black hole is dual to the chiral part of a 2D CFT with central charge $c_L = 12J$ and temperature $T_L = 1/(2\pi)$, where J is the angular momentum of the Kerr black hole. This work was later generalized to the near-extremal and nonextremal Kerr black holes by analyzing the scattering amplitudes of probe fields both from the gravity and the dual CFT sides [8, 9, 10], for earlier calculation on greybody factor see [11, 12]. The NHEK geometry containing an AdS_2 factor together with an S^1 bundle belongs to the warped AdS_3 geometry, thus the Kerr/CFT correspondence is a kind of the warped $\mathrm{AdS}_3/\mathrm{CFT}_2$ duality. The extensions to other spacetimes which contain a warped AdS_3 structure include, for examples, [13, 14, 15, 16, 17, 18, 19, 20, 21, 22, 23] (see the recent review [24] and the references therein).

The method in the Kerr/CFT correspondence, however, cannot be directly applied to non-rotating charged black holes, e.g., the 4D Reissner-Nordström (RN) black hole, which only contains an apparent AdS_2 geometry in the extremal limit, while the S^1 bundle is nontrivially encoded in the electromagnetic potential. To study its corresponding CFT dual, so far there are two possible approaches. One is to embed the RN black hole into a higher dimensional spacetime, where part of the background $U(1)$ gauge field forms an extra S^1 circle which acts like a kind of rotation in higher dimensional spacetime, and the related analysis can be classified

Journal of Physics: Conference Series **330** (2011) 012009 doi:10.1088/1742-6596/330/1/012009

as the Kerr/CFT type, i.e. the warped AdS_3/CFT_2 correspondence [25, 26, 27]. Another way, following the Kerr case studied in [28], is to reduce the 4D extremal RN black hole into 2D and then to analyze its 2D effective action based on the boundary counterterm method in the AdS_2/CFT_1 correspondence [29]. These two approaches reveal different sections of the dual CFT: the former one for the left-moving sector and the latter one for the right-moving sector. This interpretation was further confirmed by the studies of CFT dual for the nonextremal RN black hole based on the fact that there exists a hidden 2D conformal symmetry in the background geometry which can be probed by a scalar field at low frequencies [30, 31].

To what extent shall we believe in this CFT description for black holes? As we have mentioned in the beginning, the holographic principle tells us the dual field theory should contain the same information as that of the gravity. That means the generating functionals of the two theories are equivalent. In other words, besides the matching of entropies of the black hole and the CFT, the n-point correlation functions need also match with each other. However, until now only 2-point (absorption cross section of scattering) and 3-point [32, 33] correlation functions have been checked. The matching of higher point correlators are expected but it requires further study.

This paper is organized as follows. In section II, we give a brief introduction to the 4D RN black hole and its 5D uplifted counterpart, and then in section III, we describe the CFT dual of the (near-)extremal RN black hole in the warped AdS_3/CFT_2 and in the AdS_2/CFT_1 correspondences, respectively. In section IV, we move to the CFT dual for nonextremal RN black hole, by probing the hidden conformal symmetry. Finally, we come to the summary and discussion.

2. Review of the RN black hole

2.1. 4D RN black hole

The 4D Reissner-Nordström solution for a charged black hole is

$$\begin{aligned} ds^2 &= -\left(1-\frac{2M}{r}+\frac{Q^2}{r^2}\right)dt^2+\frac{dr^2}{1-\frac{2M}{r}+\frac{Q^2}{r^2}}+r^2d\Omega_2^2, \\ A_{[1]} &= \frac{Q}{r}\,dt, \qquad F_{[2]}=\frac{Q}{r^2}\,dt\wedge dr, \end{aligned} \tag{1}$$

where M and Q are the mass and electric charge of the black hole.[1] It is the uniquely spherically symmetric electro-vacuum solution to the equations of motion (Einstein and Maxwell) of the 4D Einstein-Maxwell theory

$$I_4=\frac{1}{16\pi G_4}\int d^4x\sqrt{-g_4}\left(R_4-F_{[2]}^2\right). \tag{2}$$

The RN black hole is a thermodynamical system with the corresponding Hawking temperature and Bekenstein-Hawking entropy

$$T_H=\frac{r_+-r_-}{4\pi r_+^2}, \qquad S_{BH}=\frac{\pi r_+^2}{G_4}, \tag{3}$$

where $r_\pm = M\pm\sqrt{M^2-Q^2}$ are radii of outer and inner horizons.

The near horizon geometry of the near-extremal RN black hole is [34, 35, 36]

$$\begin{aligned} ds^2 &= -\frac{\rho^2-\rho_0^2}{Q^2}d\tau^2+\frac{Q^2}{\rho^2-\rho_0^2}d\rho^2+Q^2d\Omega_2^2, \\ A_{[1]} &= -\frac{\rho}{Q}\,d\tau, \qquad F_{[2]}=\frac{1}{Q}\,d\tau\wedge d\rho, \end{aligned} \tag{4}$$

[1] Note that the parameters M and Q have dimension of length, the corresponding ADM mass and the electric charge (dimensionless) are $\mathcal{M}=M/G_4$, $\mathcal{Q}=Q/\sqrt{G_4}$.

Journal of Physics: Conference Series **330** (2011) 012009 doi:10.1088/1742-6596/330/1/012009

where ρ_0 is a small parameter which labels the derivation from extremality and $\rho_0 = 0$ is the extremal limit. The above geometry is $\mathrm{AdS}_2 \times S^2$, and $\rho = \rho_0$ is the horizon, while $\rho \to \infty$ is the AdS_2 boundary. In the extremal limit, $T_H = 0$ and $S_{BH} = \pi Q^2/G_4$.

2.2. 5D uplifted RN black hole

The 4D RN black hole can be uplifted into a 5D solution. Let's consider the 5D Einstein-Maxwell theory

$$I_5 = \frac{1}{16\pi G_5} \int d^5x \sqrt{-\hat{g}} \left(\hat{R} - \frac{1}{12} \hat{F}_{[3]}^2 \right), \tag{5}$$

and then make the Kaluza-Klein (KK) reduction of the following particular assumption (with a consistent truncation of a constant KK scalar)

$$ds_5^2 = (dy + \mathcal{A}_\mu dx^\mu)^2 + ds_4^2, \qquad \hat{A}_{[2]} = \hat{A}_{[1]} \wedge dy, \tag{6}$$

where the extra dimension is a circle of radius ℓ, i.e. $y \sim y + 2\pi\ell$ and $G_5 = 2\pi\ell G_4$. Consequently the reduced 4D effective action recovers (2) with the identifications

$$4F_{[2]}^2 = \mathcal{F}_{[2]}^2 + \hat{F}_{[2]}^2 = 4\mathcal{F}_{[2]}^2, \qquad \Rightarrow \qquad \mathcal{A}_{[1]} = A_{[1]}, \quad \hat{A}_{[1]} = \sqrt{3} A_{[1]}. \tag{7}$$

Therefore, the 5D uplifted counterpart of the RN solution (1) is (it is more convenient to use the "normalized" coordinate $\chi = y/\ell$ which has the usual period $\chi \sim \chi + 2\pi$)

$$\begin{aligned} ds^2 &= -f(r)dt^2 + \frac{dr^2}{f(r)} + r^2 d\Omega_2^2 + \left(\ell d\chi - \frac{Q}{r} dt \right)^2, \\ \hat{A}_{[2]} &= -\sqrt{3} \frac{Q}{r} \ell \, dt \wedge d\chi. \end{aligned} \tag{8}$$

The near horizon near extremal geometry is [27]

$$\begin{aligned} ds^2 &= Q^2 \left[-(\rho^2 - \rho_0^2) d\tau^2 + \frac{d\rho^2}{\rho^2 - \rho_0^2} + d\Omega_2^2 + \frac{\ell^2}{Q^2} \left(d\hat{\chi} + \frac{Q}{\ell} \rho d\tau \right)^2 \right], \\ \hat{A}_{[2]} &= \sqrt{3}\, Q \ell \rho \, d\tau \wedge d\hat{\chi}, \end{aligned} \tag{9}$$

where $\hat{\chi} = \chi - t/\ell$. Note that the size of the extra dimensional circle ℓ is in principle a free parameter, thus solution (9) contains a warped $\mathrm{AdS}_3 \times S^2$ geometry in general (when $\ell = Q$, it becomes $\mathrm{AdS}_3 \times S^2$). In the near extremal limit,

$$T_H = \epsilon \frac{\rho_0}{2\pi Q^2}, \qquad S_{BH} = \frac{A_5}{4G_5} = \frac{A_4}{4G_4} = \frac{\pi}{G_4}(Q^2 + 2\epsilon Q \rho_0). \tag{10}$$

The symbol ϵ is the limiting parameter for taking the extremal limit, namely $\epsilon \to 0$ corresponds to the extremal case.

3. (Near-)Extremal RN/CFT

In this section, we will review the CFT description for the (near-)extremal 4D RN and 5D uplifted RN black holes. Here we will only introduce the basic idea and list the main results, for detailed calculation, please refer to [29, 25, 26, 27].

3.1. Warped AdS_3/CFT_2 correspondence

After embedding the 4D RN black hole into 5D, the near horizon extremal geometry (9) contains the warped AdS_3 structure, thus the studying of its dual CFT_2 description is rather straightforward. One can use the techniques in the Kerr/CFT correspondence to calculate the left hand central charge[2] and the left hand temperature

$$c_L = \frac{6Q^3}{G_4 \ell}, \qquad T_L = \frac{\ell}{2\pi Q}. \tag{11}$$

For generality, it is convenient to keep the parameter ℓ, while the authors in [25, 26] simply chose a particular value of $\ell = 1$ and the other typical choice $\ell = Q$ has been adopted in [27]. Using the Cardy formula, one can check that the CFT entropy reproduces the correct black hole entropy

$$S_{CFT} = \frac{\pi^2}{3} c_L T_L = \frac{\pi Q^2}{G_4}. \tag{12}$$

We can further check this warped AdS_3/CFT_2 correspondence of the 4D RN black hole at the level of the 2-point correlation function in the near-extremal limit [27]. From the gravity side this is simply the absorption cross section of probe fields. For a massless and neutral scalar field of the form

$$\Phi(t, r, \theta, \phi, \chi) = \mathrm{e}^{-i\omega t + in\phi + ik\chi} S(\theta) R(r), \tag{13}$$

the solution of the angular part is just the spherical harmonics $Y_{ln}(\theta, \phi)$, and the absorption cross section can be computed for a particular limit $\omega - k/\ell \simeq 0$ as

$$\sigma_{\rm abs} = \left| \frac{\mathcal{F}^{(\rm in)}(r_H)}{\mathcal{F}^{(\rm in)}(\infty)} \right| \sim z_0^{2\beta} \sinh(2\pi b) \left| \Gamma\left(\frac{1}{2} + \beta - ia\right)\right|^2 \left| \Gamma\left(\frac{1}{2} + \beta - i(2b - a)\right)\right|^2, \tag{14}$$

where the parameters are defined as

$$\begin{aligned} z_0 &= \frac{r_+ - r_-}{r_+} = \epsilon \frac{2\rho_0}{Q} = 2\epsilon T_R / T_L, \\ a &= Q(2\omega - k/\ell) \simeq Qk/\ell, \qquad b = \frac{1}{2}\omega Q + \frac{Q(\omega - k/\ell)}{z_0}, \\ \beta &= \sqrt{(l + 1/2)^2 - 6Q^2 \omega(\omega - k/\ell)} \simeq l + 1/2, \end{aligned} \tag{15}$$

and $\mathcal{F}^{(\rm in)}(r_H)$ is the ingoing flux at the horizon, while $\mathcal{F}^{(\rm in)}(\infty)$ is the ingoing flux at spatial infinity. Together with the left and right hand temperatures of the dual 2D CFT $T_L = \frac{\ell}{2\pi Q}$ and $T_R = \frac{\ell \rho_0}{2\pi Q^2}$.

From the CFT side, the absorption cross section corresponds to the 2-point function of the operator dual to the bulk probed scalar field

$$P_{\rm abs} \sim T_L^{2h_L - 1} T_R^{2h_R - 1} \left(\mathrm{e}^{\pi(\tilde{\omega}_L + \tilde{\omega}_R)} \pm \mathrm{e}^{-\pi(\tilde{\omega}_L + \tilde{\omega}_R)}\right) |\Gamma(h_L + i\tilde{\omega}_L)|^2 \, |\Gamma(h_R + i\tilde{\omega}_R)|^2, \tag{16}$$

where the conformal weight of the dual operator is $h_L = h_R = \frac{1}{2} + \beta$ and

$$\tilde{\omega}_L = \frac{\omega_L - q_L \mu_L}{2\pi T_L}, \qquad \tilde{\omega}_R = \frac{\omega_R - q_R \mu_R}{2\pi T_R}. \tag{17}$$

[2] For the NHEK geometry, there is another boundary condition discussed in [22] to compute the right hand central charge directly from the 4D solution. Recently, another attempt was proposed in [37] to calculate both the left and right hand central charges of NHEK at the same time.

The absorption cross sections (14) and (16) match with each other by identifying the parameters $q_L = q_R = 0 = \mu_L = \mu_R$ and

$$\tilde{\omega}_L = a \quad \Rightarrow \quad \omega_L \simeq k, \tag{18}$$
$$\tilde{\omega}_R = 2b - a \quad \Rightarrow \quad \omega_R \simeq 2\ell(\omega - k/\ell) \to 0. \tag{19}$$

3.2. AdS_2/CFT_1 correspondence

The action (2) of the 4D near horizon extremal RN black hole can be reduced to the 2D effective one

$$I_2 = \frac{Q^2}{4G_4}\int d^2x\sqrt{-g_2}\left(e^{-2\psi}R_2 + \frac{2}{Q^2} + 2\nabla_\mu e^{-\psi}\nabla^\mu e^{-\psi} - e^{-2\psi}F^2_{[2]}\right), \tag{20}$$

by the following ansatz

$$ds^2 = g_{\mu\nu}dx^\mu dx^\nu + Q^2 e^{-2\psi}d\Omega_2^2, \qquad A = A_\mu dx^\mu, \tag{21}$$

where R_2 is the 2D Ricci scalar associated with $g_{\mu\nu}$, ψ is the dilaton field and $F_{\mu\nu} = \partial_\mu A_\nu - \partial_\nu A_\mu$ is the $U(1)$ gauge field strength. The 2D effective action (20) is a classical gravitational theory which needs a regularization by adding suitable boundary counterterms. This process is equivalent to making a renormalization in the CFT side to ensure normalized stress tensor and current. For the precise implementation, it is convenient to write the 2D asymptotically AdS_2 solution in terms of the Gauss normal coordinate

$$ds^2 = g_{\mu\nu}dx^\mu dx^\nu = e^{-2\psi}dr^2 + \gamma_{tt}(t,r)dt^2, \qquad A = A_t(t,r)dt, \tag{22}$$

where $r \to \infty$ is the AdS_2 boundary. In this coordinates, the general solution includes a free function of time $f(t)$ (for extremal RN $f = 0$)

$$\gamma_{tt} = -\frac{Q^2}{16}\left(e^{r/Q} - f(t)e^{-r/Q}\right)^2, \qquad A_t = -\frac{Q}{4}e^{r/Q}\left(1 - \sqrt{f(t)}e^{-r/Q}\right)^2. \tag{23}$$

The normalized 2D effective action is

$$I_{\text{normalized}} = I_2 + I_{\text{GH}} + I_{\text{counter}}, \tag{24}$$

where I_{GH} is the Gibbons-Hawking term and I_{counter} is the boundary counterterm whose explicit form can be determined by a well-defined variational principle. The normalized boundary stress tensor and currents are defined by

$$T_{ab} = -\frac{2}{\sqrt{-\gamma}}\frac{\delta I}{\delta\gamma^{ab}}, \qquad J^a = -\frac{1}{\sqrt{-\gamma}}\frac{\delta I}{\delta A_a}. \tag{25}$$

The right hand central charge of the dual 1D CFT can be obtained from the variation of T_{ab} with respect to a suitable combination of the diffeomorphism and the $U(1)$ gauge transformations [29], and the result is

$$\begin{aligned}\delta_{\epsilon+\Lambda}T_{tt} &= -\frac{Q}{2G_4}e^{-\psi}(\delta_\epsilon\gamma_{tt} + 2A_t\delta_{\epsilon+\Lambda}A_t)\\ &= 2T_{tt}\partial_t\xi + \xi\partial_t T_{tt}\\ &\quad + \frac{Q^3}{G_4}\left[e^{-3\psi}\partial_t^3\xi - \frac{1}{8}e^{-\psi}\sqrt{f}\,e^{r/Q}\left(1+\sqrt{f}\,e^{-r/Q}\right)\partial_t\xi\right]\left(1 - \sqrt{f}\,e^{-r/Q}\right).\end{aligned} \tag{26}$$

Journal of Physics: Conference Series **330** (2011) 012009 doi:10.1088/1742-6596/330/1/012009

Comparing with the expected transformation for 1D CFT [38]

$$\delta_{\epsilon+\Lambda} T_{tt} = 2T_{tt}\partial_t \xi + \xi \partial_t T_{tt} - \frac{c}{12} L \partial_t^3 \xi, \tag{27}$$

where a dimension of length parameter L is needed to ensure the central charge to be dimensionless. Actually, this parameter L cannot be fixed only from the 2D effective theory. It is associated with the radius ℓ of the "internal" circle to accommodate the background $U(1)$ gauge field; a reasonable choice is $L = -2\ell$. Then the central charge is read out to be

$$c_R = \frac{6Q^3}{G_4 \ell}. \tag{28}$$

Since there is no gravitational anomaly, similar to the near extremal Kerr black hole case [28], c_L should equal to c_R. In addition, the left hand temperature of the dual CFT can be calculated from probing the 2D conformal symmetry of the near horizon extremal RN black hole background by a charged scalar field, which gives $T_L = \ell/(2\pi Q)$. Then using the Cardy formula,

$$S_{\rm CFT} = \frac{\pi^2}{3} c_L T_L = \frac{\pi Q^2}{G_4}, \tag{29}$$

the correct black hole entropy is reproduced.

4. Nonextremal RN/CFT

Having discussed the dual CFT description of the (near-)extremal RN black hole, it is natural to ask what is the holographic dual description for the general nonextremal RN black hole. However, this is not an easy problem since there are no apparent AdS structures in the near horizon geometry of nonextremal RN black holes. The background conformal symmetry, if it still exists, has been hidden. Inspired from the experience of studying the scattering process of probe fields in the black hole background, it was found that a 2D hidden conformal symmetry actually can be probed by a massless neutral scalar field propagating in the nonextremal Kerr black hole background in the low frequency limit [10]. In the following, we will show how this hidden conformal symmetry can be extracted from the generic nonextremal RN black hole.

4.1. CFT for 5D uplifted RN black hole

Like the extremal case, probing of the hidden conformal symmetry in the 5D uplifted RN black hole background (8) is similar to that of the 4D Kerr black hole background. Consider a bulk massless scalar field Φ propagating in the background of (8), after making the field decomposition of (13), the radial part of the Klein-Gordon (KG) equation can be simplified into [30]

$$\partial_r(\Delta \partial_r R) + \left[\frac{[(2Mr_+ - Q^2)\omega - (Qr_+/\ell)k]^2}{(r-r_+)(r_+-r_-)} - \frac{[(2Mr_- - Q^2)\omega - (Qr_-/\ell)k]^2}{(r-r_-)(r_+-r_-)} \right] R = l(l+1)R. \tag{30}$$

where $\Delta = (r-r_+)(r-r_-)$, under the conditions of (a) low frequency and low momentum limit $M\omega \ll 1$, $Mk/\ell \ll 1$ (automatically implies $Q\omega \ll 1$, $Qk/\ell \ll 1$), and (b) near region limit of $r\omega \ll 1$ and $rk/\ell \ll 1$. It turns out that the eq. (30) is indeed of the same form with the Casimir operator of the $SL(2,R)_L \times SL(2,R)_R$ Lie algebra

$$\mathcal{H}^2 = \partial_r(\Delta \partial_r) - \frac{r_+ - r_-}{r - r_+}\left(\frac{T_L + T_R}{4A}\partial_t - \frac{n_L + n_R}{4\pi A}\partial_\chi \right)^2 + \frac{r_+ - r_-}{r - r_-}\left(\frac{T_L - T_R}{4A}\partial_t - \frac{n_L - n_R}{4\pi A}\partial_\chi \right)^2, \tag{31}$$

Journal of Physics: Conference Series **330** (2011) 012009 doi:10.1088/1742-6596/330/1/012009

where $\mathcal{A} = n_L T_R - n_R T_L$. Here the eq. (31) is obtained from the standard Casimir operator in the Poincaré coordinates $(w^\pm, z)$ by the following coordinate transformation

$$\begin{aligned} w^+ &= \sqrt{\frac{r-r_+}{r-r_-}}\exp(2\pi T_R\chi + 2n_+t), \\ w^- &= \sqrt{\frac{r-r_+}{r-r_-}}\exp(2\pi T_L\chi + 2n_-t), \\ z &= \sqrt{\frac{r_+-r_-}{r-r_-}}\exp\left[\pi(T_R+T_L)\chi + (n_+ + n_-)t\right]. \end{aligned} \tag{32}$$

One motivation for making the above coordinate transformation is that it transforms the Poincaré AdS_3 spacetime into the rotating BTZ black hole by simply replacing r with r^2. Comparing eq. (30) with eq. (31), one obtains

$$T_R = \frac{(r_+ - r_-)M\ell}{2\pi Q^3}, \qquad T_L = \frac{(r_+ + r_-)M\ell}{2\pi Q^3} - \frac{\ell}{2\pi Q}, \qquad n_\pm = -\frac{r_+ \mp r_-}{4Q^2}, \tag{33}$$

and the conformal weights of the operator dual to the bulk scalar field $(h_L, h_R) = (l+1, l+1)$.

Even though we can determine the temperatures and conformal weights of the dual CFT, we still don't know how to calculate the central charges in the nonextremal case. However, if we simply assume that the central charges hold the same form as those in the extremal limit, i.e. $c_L = c_R = 6Q^3/(G_4\ell)$, then the Cardy formula indeed gives the correct black hole entropy

$$S_{CFT} = \frac{\pi^2}{3}(c_L T_L + c_R T_R) = \pi(2Mr_+ - Q^2) = \frac{\pi r_+^2}{G_4} = S_{BH}. \tag{34}$$

In addition, the absorption cross sections calculated from the gravity side is [30]

$$P_{\rm abs} \sim T_L^{2h_L-1} T_R^{2h_R-1} \sinh\left(\frac{\omega_L}{2T_L} + \frac{\omega_R}{2T_R}\right)\left|\Gamma\left(h_L + i\frac{\omega_L}{2\pi T_L}\right)\right|^2 \left|\Gamma\left(h_R + i\frac{\omega_R}{2\pi T_R}\right)\right|^2, \tag{35}$$

where

$$\omega_L = \frac{(2M^2 - Q^2)(2M\ell\omega - Qk)}{Q^3}, \qquad \omega_R = \frac{2(2M^2 - Q^2)M\ell\omega - 2M^2Qk}{Q^3}. \tag{36}$$

Finally, the $P_{\rm abs}$ matches with the finite temperature absorption cross section of the operator dual to the scalar field in a 2D CFT.

4.2. CFT for 4D dyonic RN black hole

It would be interesting to find a 2D hidden conformal symmetry for the nonextremal 4D RN black hole, without uplifting it into 5D. This will give further evidence that the 4D RN black hole is dual to a 2D CFT. Indeed, it has been shown that the 2D hidden conformal symmetry does exist in this case; the key point is that the $U(1)$ bundle of the 4D dyonic RN black hole can be probed by a charged scalar field. This can be expected since the $U(1)$ symmetry is actually associated with the gauge symmetry of the background gauge field [31].

The charged scalar field Φ with charge q in 4D dyonic RN black hole background can be decomposed as $\Phi(t,r,\theta,\phi) = e^{-i\omega t + im\phi}S(\theta)R(r)$, then in the near region $r\omega \ll 1$, together with the low frequency and small probe charge limit, the radial part of the KG equation becomes

$$\partial_r(\Delta\partial_r R) + \left[\frac{r_+^4(\omega + A_+ q)^2}{(r-r_+)(r_+-r_-)} - \frac{r_-^4(\omega + A_- q)^2}{(r-r_-)(r_+-r_-)}\right]R = l(l+1)R, \tag{37}$$

where $A_\pm = -Q/r_\pm$ are the gauge potentials at the outer/inner horizons. In the dyonic black hole of electric and magnetic charges Q and P, the radii of the inner and outer horizons are $r_\pm = M \pm \sqrt{M^2 - Q^2 - P^2}$.

To compare the eq. (37) with the Casimir operator (31), we should introduce an operator ∂_χ, which can be naturally considered to act on the "internal" $U(1)$ symmetry space of the probe electrically charged scalar field as $\partial_\chi \Phi = i\ell q \Phi$. Then the temperatures of the dual CFT are

$$T_L = \frac{(r_+^2 + r_-^2)\ell}{4\pi Q r_+ r_-}, \qquad T_R = \frac{(r_+^2 - r_-^2)\ell}{4\pi Q r_+ r_-}. \tag{38}$$

Again, one assume the central charges are identical to those of the extremal 4D RN black hole, namely,

$$c_L = c_R = \frac{6Q(Q^2 + P^2)}{G_4 \ell}, \tag{39}$$

then one again gets

$$S_{CFT} = \frac{\pi^2}{3}(c_L T_L + c_R T_R) = \frac{\pi r_+^2}{G_4} = S_{BH}. \tag{40}$$

The RN/CFT$_2$ correspondence can be further checked by studying the scattering of the probe charged scalar field in the RN black hole background. We will see that the absorption cross section of the scalar field is in agreement with the two point function of its corresponding operators in the dual 2D CFT.

The absorption cross section calculated from the gravity side is of the same form as eq. (35) [31]. We can also check the real-time correlators. The two-point retarded correlator obtained from the gravity side is simply

$$\begin{aligned} G_R \sim\ & \sin\left(\pi h_L + i\frac{\tilde{\omega}_L}{2T_L}\right) \sin\left(\pi h_R + i\frac{\tilde{\omega}_R}{2T_R}\right) \\ & \Gamma\left(h_L - i\frac{\tilde{\omega}_L}{2\pi T_L}\right)\Gamma\left(h_L + i\frac{\tilde{\omega}_L}{2\pi T_L}\right)\Gamma\left(h_R - i\frac{\tilde{\omega}_R}{2\pi T_R}\right)\Gamma\left(h_R + i\frac{\tilde{\omega}_R}{2\pi T_R}\right), \end{aligned} \tag{41}$$

with $h_L = h_R = l + 1$, the explicit form of other parameters can be found in [31].

From the 2D CFT side, the retarded Green function $G_R(\omega_L, \omega_R)$ can be calculated by an analytic continuation from the Euclidean correlator

$$\begin{aligned} G_E(\omega_{EL}, \omega_{ER}) \sim\ & T_L^{2h_L - 1} T_R^{2h_R - 1} e^{i\frac{\tilde{\omega}_{EL}}{2T_L}} e^{i\frac{\tilde{\omega}_{ER}}{2T_R}} \\ & \Gamma\left(h_L - \frac{\tilde{\omega}_{EL}}{2\pi T_L}\right)\Gamma\left(h_L + \frac{\tilde{\omega}_{EL}}{2\pi T_L}\right)\Gamma\left(h_R - \frac{\tilde{\omega}_{ER}}{2\pi T_R}\right)\Gamma\left(h_R + \frac{\tilde{\omega}_{ER}}{2\pi T_R}\right), \end{aligned} \tag{42}$$

(where $\omega_{EL} = i\omega_L$, $\omega_{ER} = i\omega_R$, $\tilde{\omega}_{EL} = \omega_{EL} - iq_L\mu_L$ and $\tilde{\omega}_{ER} = \omega_{ER} - iq_R\mu_R$) by

$$G_E(\omega_{EL}, \omega_{ER}) = G_R(i\omega_L, i\omega_R), \qquad \omega_{EL},\ \omega_{ER} > 0, \tag{43}$$

and the Euclidean frequencies ω_{EL} and ω_{ER} should take discrete values of the Matsubara frequencies at finite temperature

$$\omega_{EL} = 2\pi m_L T_L, \qquad \omega_{ER} = 2\pi m_R T_R, \tag{44}$$

in which m_L, m_R are integers for bosons and half integers for fermions. At these frequencies, we can see that the retarded Green function matches well with the gravity side computation (41) up to a normalization factor depending on the charges q_L and q_R, i.e. the electric charge of the probe scalar field.

5. Summary and discussion

In this short article, we give a brief review of recent progress in the holographic description of the RN black hole, for all the extremal, near extremal and nonextremal cases. We show that, the warped AdS_3/CFT_2 and the AdS_2/CFT_1 descriptions are consistent, which means that the 4D RN black hole is dual to a 2D CFT, and indicates that 1D CFT is actually the chiral part of a 2D CFT. This RN/CFT_2 duality is supported from the matching of entropies, the absorption cross sections and the real-time correlators calculated both from the gravity and the dual 2D CFT side. Nevertheless, there are still some unsolved problems such as why does the method of probing hidden conformal symmetry work? Can we calculate the central charges of the dual CFT in the nonextremal limit? Do the n-point correlators match with each other? Further clarification of these important issues is desirable.

Acknowledgments

This work was supported by the National Science Council of the R.O.C. under the grant NSC 99-2112-M-008-005-MY3 and in part by the National Center of Theoretical Sciences (NCTS).

References

[1] G. 't Hooft, "Dimensional reduction in quantum gravity," arXiv:gr-qc/9310026.
[2] L. Susskind, "The world as a hologram," J. Math. Phys. **36**, 6377 (1995) [arXiv:hep-th/9409089].
[3] J. M. Maldacena, "The large N limit of superconformal field theories and supergravity," Adv. Theor. Math. Phys. **2**, 231 (1998) [Int. J. Theor. Phys. **38**, 1113 (1999)] [arXiv:hep-th/9711200].
[4] S. S. Gubser, I. R. Klebanov and A. M. Polyakov, "Gauge theory correlators from non-critical string theory," Phys. Lett. B **428**, 105 (1998) [arXiv:hep-th/9802109].
[5] E. Witten, "Anti-de Sitter space and holography," Adv. Theor. Math. Phys. **2**, 253 (1998) [arXiv:hep-th/9802150].
[6] J. D. Brown and M. Henneaux, "Central charges in the canonical realization of asymptotic symmetries: an example from three-dimensional gravity," Commun. Math. Phys. **104**, 207 (1986).
[7] M. Guica, T. Hartman, W. Song and A. Strominger, "The Kerr/CFT correspondence," arXiv:0809.4266 [hep-th].
[8] I. Bredberg, T. Hartman, W. Song and A. Strominger, "Black Hole Superradiance From Kerr/CFT," arXiv:0907.3477 [hep-th].
[9] T. Hartman, W. Song and A. Strominger, "Holographic Derivation of Kerr-Newman Scattering Amplitudes for General Charge and Spin," arXiv:0908.3909 [hep-th].
[10] A. Castro, A. Maloney and A. Strominger, "Hidden Conformal Symmetry of the Kerr Black Hole," arXiv:1004.0996 [hep-th].
[11]
[11] M. Cvetic and F. Larsen, "General rotating black holes in string theory: Grey body factors and event horizons," Phys. Rev. D **56**, 4994 (1997) [arXiv:hep-th/9705192].
[12]
[12] M. Cvetic and F. Larsen, "Grey body factors for rotating black holes in four-dimensions," Nucl. Phys. B **506**, 107 (1997) [arXiv:hep-th/9706071].
[13] H. Lu, J. Mei and C. N. Pope, "Kerr/CFT Correspondence in Diverse Dimensions," JHEP **0904**, 054 (2009) [arXiv:0811.2225 [hep-th]].
[14] T. Azeyanagi, N. Ogawa and S. Terashima, "Holographic Duals of Kaluza-Klein Black Holes," JHEP **0904**, 061 (2009) [arXiv:0811.4177 [hep-th]].
[15] D. D. K. Chow, M. Cvetic, H. Lu and C. N. Pope, "Extremal black hole/CFT correspondence in (gauged) supergravities," arXiv:0812.2918 [hep-th].
[16] T. Azeyanagi, N. Ogawa and S. Terashima, "The Kerr/CFT Correspondence and String Theory," Phys. Rev. D **79**, 106009 (2009) [arXiv:0812.4883 [hep-th]].
[17] C.-M. Chen and J. E. Wang, "Holographic duals of black holes in five-dimensional minimal supergravity," arXiv:0901.0538 [hep-th].
[18] A. J. Amsel, G. T. Horowitz, D. Marolf and M. M. Roberts, "No dynamics in the extremal Kerr throat," arXiv:0906.2376 [hep-th].
[19] O. J. C. Dias, H. S. Reall and J. E. Santos, "Kerr-CFT and gravitational perturbations," JHEP **0908**, 101 (2009) [arXiv:0906.2380 [hep-th]].

[20] G. Compere, K. Murata and T. Nishioka, "Central Charges in Extreme Black Hole/CFT Correspondence," JHEP **0905**, 077 (2009) [arXiv:0902.1001 [hep-th]].
[21] X. N. Wu and Y. Tian, "Extremal Isolated Horizon/CFT Correspondence," Phys. Rev. D **80**, 024014 (2009) [arXiv:0904.1554 [hep-th]].
[22] Y. Matsuo, T. Tsukioka and C. M. Yoo, "Another Realization of Kerr/CFT Correspondence," arXiv:0907.0303 [hep-th].
[23] M. Cvetic and F. Larsen, "Greybody Factors and Charges in Kerr/CFT," JHEP **0909**, 088 (2009) [arXiv:0908.1136 [hep-th]].
[24] I. Bredberg, C. Keeler, V. Lysov and A. Strominger, "Cargese Lectures on the Kerr/CFT Correspondence," arXiv:1103.2355 [hep-th].
[25] T. Hartman, K. Murata, T. Nishioka and A. Strominger, "CFT duals for extreme black holes," JHEP **0904**, 019 (2009) [arXiv:0811.4393 [hep-th]].
[26] M. R. Garousi and A. Ghodsi, "The RN/CFT correspondence," arXiv:0902.4387 [hep-th].
[27] C.-M. Chen, Y.-M. Huang and S.-J. Zou, "Holographic Duals of Near-extremal Reissner-Nordstrom Black Holes," JHEP **1003**, 123 (2010) [arXiv:1001.2833 [hep-th]].
[28] A. Castro and F. Larsen, "Near extremal Kerr entropy from AdS_2 quantum gravity," arXiv:0908.1121 [hep-th].
[29] C.-M. Chen, J.-R. Sun, S.-J. Zou, "The RN/CFT Correspondence Revisited," JHEP **1001**, 057 (2010). [arXiv:0910.2076 [hep-th]].
[30] C.-M. Chen, J.-R. Sun, "Hidden Conformal Symmetry of the Reissner-Nordstrom Black Holes," JHEP **1008**, 034 (2010). [arXiv:1004.3963 [hep-th]].
[31] C.-M. Chen, Y.-M. Huang, J.-R. Sun, M.-F. Wu, S.-J. Zou, "On Holographic Dual of the Dyonic Reissner-Nordstrom Black Hole," Phys. Rev. **D82**, 066003 (2010). [arXiv:1006.4092 [hep-th]].
[32] M. Becker, S. Cremonini and W. Schulgin, "Extremal Three-point Correlators in Kerr/CFT," JHEP **1102**, 007 (2011) [arXiv:1004.1174 [hep-th]].
[33] M. Becker, S. Cremonini and W. Schulgin, "Correlation Functions and Hidden Conformal Symmetry of Kerr Black Holes," JHEP **1009**, 022 (2010) [arXiv:1005.3571 [hep-th]].
[34] B. Bertotti, "Uniform electromagnetic field in the theory of general relativity," Phys. Rev. **116**, 1331 (1959).
[35] I. Robinson, "A Solution of the Maxwell-Einstein Equations," Bull. Acad. Pol. Sci. Ser. Sci. Math. Astron. Phys. **7**, 351 (1959).
[36] J. M. Maldacena, J. Michelson and A. Strominger, "Anti-de Sitter fragmentation," JHEP **9902**, 011 (1999) [arXiv:hep-th/9812073].
[37] T. Azeyanagi, N. Ogawa, S. Terashima, "On Non-Chiral Extension of Kerr/CFT," [arXiv:1102.3423 [hep-th]].
[38] A. Castro, D. Grumiller, F. Larsen and R. McNees, "Holographic description of AdS_2 black holes," JHEP **0811**, 052 (2008) [arXiv:0809.4264 [hep-th]].

Journal of Physics: Conference Series **330** (2011) 012010 doi:10.1088/1742-6596/330/1/012010

Maxwell's equations, quantum physics and the quantum graviton

Alexander Gersten and Amnon Moalem
Department of Physics, Ben-Gurion University of the Negev, Beer-Sheva, Israel

Abstract. Quantum wave equations for massless particles and arbitrary spin are derived by factorizing the d'Alembertian operator. The procedure is extensively applied to the spin one photon equation which is related to Maxwell's equations via the proportionality of the photon wavefunction Ψ to the sum $\mathbf{E} + i\mathbf{B}$ of the electric and magnetic fields. Thus Maxwell's equations can be considered as the first quantized one-photon equation. The photon wave equation is written in two forms, one with additional explicit subsidiary conditions and second with the subsidiary conditions implicitly included in the main equation. The second equation was obtained by factorizing the d'Alembertian with 4×4 matrix representation of "relativistic quaternions". Furthermore, scalar Lagrangian formalism, consistent with quantization requirements is developed using derived conserved current of probability and normalization condition for the wavefunction. Lessons learned from the derivation of the photon equation are used in the derivation of the spin two quantum equation, which we call the quantum graviton. Quantum wave equation with implicit subsidiary conditions, which factorizes the d'Alembertian with 8×8 matrix representation of relativistic quaternions, is derived. Scalar Lagrangian is formulated and conserved probability current and wavefunction normalization are found, both consistent with the definitions of quantum operators and their expectation values. We are showing that the derived equations are the first quantized equations of the photon and the graviton.
Key words: Maxwell's equations, one photon quantum equation, quantum graviton

1. Introduction

In the past, relativistic wave equations were successfully obtained by factorizing the Klein-Gordon operator. The Dirac equation was derived from the relativistic condition on the Energy E, mass m, and momentum $\mathbf{p}$:

$$\left(E^2 - c^2\mathbf{p}^2 - m^2c^4\right) I^{(4)}\Psi = 0, \tag{1}$$

where $I^{(4)}$ is the 4×4 unit matrix and Ψ is a four component column (bispinor) wave function. Eq. (1) was factorized into

$$\left[EI^{(4)} + \begin{pmatrix} mc^2I^{(2)} & c\mathbf{p}\cdot\boldsymbol{\sigma} \\ c\mathbf{p}\cdot\boldsymbol{\sigma} & -mc^2I^{(2)} \end{pmatrix}\right]\left[EI^{(4)} - \begin{pmatrix} mc^2I^{(2)} & c\mathbf{p}\cdot\boldsymbol{\sigma} \\ c\mathbf{p}\cdot\boldsymbol{\sigma} & -mc^2I^{(2)} \end{pmatrix}\right]\Psi = 0, \tag{2}$$

where $I^{(2)}$ is the 2×2 unit matrix and $\boldsymbol{\sigma}$ is the Pauli spin one-half vector matrix with the components

$$\sigma_x = \begin{pmatrix} 0 & 1 \\ 1 & 0 \end{pmatrix}, \quad \sigma_y = \begin{pmatrix} 0 & -i \\ i & 0 \end{pmatrix}, \quad \sigma_z = \begin{pmatrix} 1 & 0 \\ 0 & -1 \end{pmatrix}, \quad I^{(2)} = \begin{pmatrix} 1 & 0 \\ 0 & 1 \end{pmatrix}. \tag{3}$$

For massless particles wave equations can be obtained by factorizing the zero mass wave operator. The two component (massless) neutrino equation can be derived from the factorization

$$\left(E^2 - c^2\mathbf{p}^2\right) I^{(2)}\psi = \left[EI^{(2)} - c\mathbf{p}\cdot\boldsymbol{\sigma}\right]\left[EI^{(2)} + c\mathbf{p}\cdot\boldsymbol{\sigma}\right]\psi = 0, \tag{4}$$

where ψ is a two component spinor wavefunction.

In the past there were many attempts to describe Maxwell's equation as the quantum photon equation[1]. In this work our presentation is more complete and contains new elements. We extend them to the derivation of the wave equation of the quantum graviton. Our starting point is the factorization of the d'Alembertian. For spin one the factorization leads to the photon wave function. We present different factorizations which lead to quantum photon equations. In all of them, the somewhat mysterious substitution of the wave function $\boldsymbol{\Psi} = \mathcal{N}\left(\mathbf{E} + i\mathbf{B}\right)$, leads to Maxwell's equations. Above $\mathbf{E}$ and $\mathbf{B}$ are the electric and magnetic fields respectively, and $\mathcal{N}$ is a normalization constant (normalizable only for functions that fall off sufficiently rapidly). Thus Maxwell's equations are the one-photon quantum equations. We construct Lagrangians for the photon equations and evaluate the energy-momentum tensors and probability currents, which appear to be consistent with the definitions of quantum operators and their expectation values. We extend our findings to derivations of other zero mass and arbitrary spin quantum equations. We worked out the details of the zero mass spin two quantum equation which is linked to the quantum graviton.

In Sec. 2 and Sec. 4 the photon wave equation is derived by factorizing the wave operator (the d'Alembertian). From the photon equation Maxwell's equations are derived in Sec. 3. In Sec. 2 and Sec. 4 two procedures will be presented: one with subsidiary conditions, in the second procedure the subsidiary conditions will be integrated into the usual form of Maxwell's equations. In Sec. 4 scalar Lagrangian was formulated and conserved probability current was found, both consistent with the definitions of quantum operators and their expectation values. In Sec. 5 the photon Lagrangian was converted into a scalar Lagrangian for Maxwell's equations. In Sec. 6 relativistic quaternions are introduced in order to work properly with the space-time metric. The factorization of the d'Alembertian is achieved with matrix representations of the relativistic quaternions. In Sections 7-8 a general method for deriving quantum equations for zero mass particles and arbitrary spins is presented. In Sec. 9 we use this method to derive again Maxwell's equations, this time with spherical tensor formulation. In Sec. 10 the quantum graviton (spin 2) wave equation (with explicit and implicit subsidiary conditions) are considered. The quantum wave equation with implicit subsidiary conditions, which factorizes the d'Alembertian with 8×8 matrix representation of relativistic quaternions, is derived. Scalar Lagrangian is formulated and conserved probability current was found, both consistent with the definitions of quantum operators and their expectation values. In Sec. 11 the covariance of the equations is explained. Sec. 12 contains our conclusions. In the following symbols with hat over them (like $\hat{E}$) will denote quantum operators.

2. The photon equation

We shall derive the photon equation from the following decomposition[2],

$$\left(\frac{\hat{E}^2}{c^2} - \hat{\mathbf{p}}^2\right) I^{(3)} = \left(\frac{\hat{E}}{c} I^{(3)} + \hat{\mathbf{p}}\cdot\mathbf{S}\right)\left(\frac{\hat{E}}{c} I^{(3)} - \hat{\mathbf{p}}\cdot\mathbf{S}\right) - \begin{pmatrix} \hat{p}_x^2 & \hat{p}_x\hat{p}_y & \hat{p}_x\hat{p}_z \\ \hat{p}_y\hat{p}_x & \hat{p}_y^2 & \hat{p}_y\hat{p}_z \\ \hat{p}_z\hat{p}_x & \hat{p}_z\hat{p}_y & \hat{p}_z^2 \end{pmatrix}, \tag{5}$$

where $I^{(3)}$ is a 3×3 unit matrix, and $\mathbf{S}$ is a spin one vector matrix with components,

Journal of Physics: Conference Series **330** (2011) 012010 doi:10.1088/1742-6596/330/1/012010

$$S_x = \begin{pmatrix} 0 & 0 & 0 \\ 0 & 0 & -i \\ 0 & i & 0 \end{pmatrix}, \quad S_y = \begin{pmatrix} 0 & 0 & i \\ 0 & 0 & 0 \\ -i & 0 & 0 \end{pmatrix}, \quad S_z = \begin{pmatrix} 0 & -i & 0 \\ i & 0 & 0 \\ 0 & 0 & 0 \end{pmatrix}, \tag{6}$$

and with the properties,

$$[S_x, S_y] = iS_z, \quad [S_z, S_x] = iS_y, \quad [S_y, S_z] = iS_x, \quad \mathbf{S}^2 = 2I^{(3)}. \tag{7}$$

The decomposition (5) can be verified directly by substitution. The matrix on the right hand side of Eq. (5) can be rewritten as:

$$\begin{pmatrix} \hat{p}_x^2 & \hat{p}_x\hat{p}_y & \hat{p}_x\hat{p}_z \\ \hat{p}_y\hat{p}_x & \hat{p}_y^2 & \hat{p}_y\hat{p}_z \\ \hat{p}_z\hat{p}_x & \hat{p}_z\hat{p}_y & \hat{p}_z^2 \end{pmatrix} = \begin{pmatrix} \hat{p}_x \\ \hat{p}_y \\ \hat{p}_z \end{pmatrix} \begin{pmatrix} \hat{p}_x & \hat{p}_y & \hat{p}_z \end{pmatrix}. \tag{8}$$

From Eqs.(5-6, and 8) , the photon equation can be obtained,

$$\left(\frac{\hat{E}^2}{c^2} - \hat{\mathbf{p}}^2\right)\mathbf{\Psi} = \left(\frac{\hat{E}}{c}I^{(3)} + \hat{\mathbf{p}}\cdot\mathbf{S}\right)\left(\frac{\hat{E}}{c}I^{(3)} - \hat{\mathbf{p}}\cdot\mathbf{S}\right)\mathbf{\Psi} - \begin{pmatrix} \hat{p}_x \\ \hat{p}_y \\ \hat{p}_z \end{pmatrix}(\hat{\mathbf{p}}\cdot\mathbf{\Psi}) = 0, \tag{9}$$

where $\mathbf{\Psi}$ is a 3 component (column) wave function. Eq.(9) will be satisfied if the two equations,

$$\left(\frac{\hat{E}}{c}I^{(3)} - \hat{\mathbf{p}}\cdot\mathbf{S}\right)\mathbf{\Psi}^{(+)} = 0, \tag{10}$$

$$\hat{\mathbf{p}}\cdot\mathbf{\Psi}^{(+)} = 0, \tag{11}$$

are simultaneously satisfied. A second possibility is to replace Eq. (9) with,

$$\left(\frac{\hat{E}^2}{c^2} - \hat{\mathbf{p}}^2\right)\mathbf{\Psi} = \left(\frac{\hat{E}}{c}I^{(3)} - \hat{\mathbf{p}}\cdot\mathbf{S}\right)\left(\frac{\hat{E}}{c}I^{(3)} + \hat{\mathbf{p}}\cdot\mathbf{S}\right)\mathbf{\Psi} - \begin{pmatrix} \hat{p}_x \\ \hat{p}_y \\ \hat{p}_z \end{pmatrix}(\hat{\mathbf{p}}\cdot\mathbf{\Psi}) = 0. \tag{12}$$

Eq.(12) will be satisfied if the two equations,

$$\left(\frac{\hat{E}}{c}I^{(3)} + \hat{\mathbf{p}}\cdot\mathbf{S}\right)\mathbf{\Psi}^{(-)} = 0, \tag{13}$$

$$\hat{\mathbf{p}}\cdot\mathbf{\Psi}^{(-)} = 0 \tag{14}$$

are simultaneously satisfied. Above $\mathbf{\Psi}^{(+)}$ and $\mathbf{\Psi}^{(-)}$ refer to forward and backward helicities.

3. Maxwell's equations

Maxwell's equations will be derived from Eqs. (10) and (11). We will show below that if in Eqs. (10) and (11) the quantum operator substitutions,

$$\hat{E} \Longrightarrow i\hbar\frac{\partial}{\partial t}, \quad \hat{\mathbf{p}} \Longrightarrow -i\hbar\mathbf{\nabla}, \tag{15}$$

and the wavefunction substitution

$$\mathbf{\Psi} = \mathcal{N}\left(\mathbf{E} + i\mathbf{B}\right), \tag{16}$$

Journal of Physics: Conference Series **330** (2011) 012010 doi:10.1088/1742-6596/330/1/012010

are made, as a result the Maxwell equations will be obtained. In Eq. (16) **E** and **B** are the electric and magnetic fields respectively, and $\mathcal{N}$ is a constant, with units of $\left[\left(\sqrt{energy}\right)^{-1}\right]$, which will be determined later on (see Eq. (66)). Indeed, one can easily check from Eqs. (6) and (15) that the following identity is satisfied,

$$(\hat{\mathbf{p}}\cdot\mathbf{S})\,\mathbf{\Psi}=\hbar\boldsymbol{\nabla}\times\mathbf{\Psi}. \tag{17}$$

From Eqs. (10) , (11) , (15) and 17) we obtain,

$$i\hbar\frac{\partial}{\partial t}\mathbf{\Psi}=c\hbar\boldsymbol{\nabla}\times\mathbf{\Psi}=c\,(\hat{\mathbf{p}}\cdot\mathbf{S})\,\mathbf{\Psi}=\mathcal{H}\mathbf{\Psi}, \tag{18}$$

$$-i\hbar\boldsymbol{\nabla}\cdot\mathbf{\Psi}=0, \tag{19}$$

where $\mathcal{H}$ can be defined as the Hamiltonian operator. Although the definition of the energy operator $\hat{E}$ in Eq. (15) is only formal, in practice it can be replaced by the Hamiltonian. The definition of the Hamiltonian does not depend on the subsidiary conditions Eq. (19).

The constants $\hbar$ and $\mathcal{N}$ can be cancelled out in Eqs. (18), (19), and after replacing $\mathbf{\Psi}$ by Eq. (16), the following equations are obtained,

$$\boldsymbol{\nabla}\times(\mathbf{E}+i\mathbf{B})=\frac{i}{c}\frac{\partial\,(\mathbf{E}+i\mathbf{B})}{\partial t}, \tag{20}$$

$$\boldsymbol{\nabla}\cdot(\mathbf{E}+i\mathbf{B})=0. \tag{21}$$

If in Eqs. (20) and (21) the electric and magnetic fields are real, the separation into the real and imaginary parts will lead to the Maxwell equations (without sources),

$$\boldsymbol{\nabla}\times\mathbf{E}=-\frac{1}{c}\frac{\partial\mathbf{B}}{\partial t}, \tag{22}$$

$$\boldsymbol{\nabla}\times\mathbf{B}=\frac{1}{c}\frac{\partial\mathbf{E}}{\partial t}, \tag{23}$$

$$\boldsymbol{\nabla}\cdot\mathbf{E}=0, \tag{24}$$

$$\boldsymbol{\nabla}\cdot\mathbf{B}=0. \tag{25}$$

One should note that the Plank constant $\hbar$ was cancelled out earlier, in Eqs. (18) and (19), which explains its absence in the Maxwell equations. Here to be noted that from Eq. (15) and Eq. (6),

$$\hat{E}^*=-\hat{E},\quad \hat{\mathbf{p}}^*=-\hat{\mathbf{p}},\quad \mathbf{S}^*=-\mathbf{S}, \tag{26}$$

and therefore, the complex conjugate of the photon equation Eq. (10) ,

$$\left[\left(\frac{\hat{E}}{c}I^{(3)}-\hat{\mathbf{p}}\cdot\mathbf{S}\right)\mathbf{\Psi}^{(+)}\right]^*=-\left(\frac{\hat{E}}{c}I^{(3)}+\hat{\mathbf{p}}\cdot\mathbf{S}\right)\left(\mathbf{\Psi}^{(+)}\right)^*=0, \tag{27}$$

where $\left(\mathbf{\Psi}^{(+)}\right)^*$ satisfies Eq. (13) for backward helicity.

Journal of Physics: Conference Series **330** (2011) 012010 doi:10.1088/1742-6596/330/1/012010

4. The photon equation with an implicit subsidiary condition

Maxwell's equations without sources are[3],

$$\partial_\mu F^{\mu\nu} = 0, \quad \partial_\mu \tilde{F}^{\mu\nu} = 0, \quad \nu = 0, 1, 2, 3, \tag{28}$$

where the antisymmetric tensor $F^{\mu\nu}$ and its dual $\tilde{F}^{\mu\nu}$ are defined via the electric and magnetic fields $\mathbf{E}$ and $\mathbf{B}$ respectively as,

$$(F^{\mu\nu}) = \begin{pmatrix} 0 & -E_x & -E_y & -E_z \\ E_x & 0 & -B_z & B_y \\ E_y & B_z & 0 & -B_x \\ E_z & -B_y & B_x & 0 \end{pmatrix}, \tag{29}$$

$$\tilde{F}^{\mu\nu} = \frac{1}{2}\epsilon^{\mu\nu\alpha\beta} F_{\alpha\beta} = \begin{pmatrix} 0 & -B_x & -B_y & -B_z \\ B_x & 0 & E_z & -E_y \\ B_y & -E_z & 0 & E_x \\ B_z & E_y & -E_x & 0 \end{pmatrix}, \tag{30}$$

where $\epsilon^{\mu\nu\alpha\beta}$ is the totally antisymmetric tensor ($\epsilon^{0123} = 1, \quad \epsilon^{\mu\nu\alpha\beta} = -\epsilon_{\mu\nu\alpha\beta}$). The sum of the $F^{\mu\nu}$ and $i\tilde{F}^{\mu\nu}$ is the self-dual antisymmetric tensor, which depends only on the combination $\mathbf{\Psi} = \mathcal{N}\left(\mathbf{E}+i\mathbf{B}\right)$,

$$F^{\mu\nu} + i\tilde{F}^{\mu\nu} = \begin{pmatrix} 0 & -E_x - iB_x & -E_y - iB_y & -E_z - iB_z \\ E_x + iB_x & 0 & iE_z - B_z & B_y - iE_y \\ E_y + iB_y & B_z - iE_z & 0 & iE_x - B_x \\ E_z + iB_z & iE_y - B_y & B_x - iE_x & 0 \end{pmatrix}, \tag{31}$$

$$= \frac{1}{\mathcal{N}} \begin{pmatrix} 0 & -\Psi_x & -\Psi_y & -\Psi_z \\ \Psi_x & 0 & i\Psi_z & -i\Psi_y \\ \Psi_y & -i\Psi_z & 0 & i\Psi_x \\ \Psi_z & i\Psi_y & -i\Psi_x & 0 \end{pmatrix} = \frac{1}{\mathcal{N}}\left(R_1\Psi_x + R_2\Psi_y + R_3\Psi_z\right), \tag{32}$$

with,

$$R_1 = \begin{pmatrix} 0 & -1 & 0 & 0 \\ 1 & 0 & 0 & 0 \\ 0 & 0 & 0 & i \\ 0 & 0 & -i & 0 \end{pmatrix}, \quad R_2 = \begin{pmatrix} 0 & 0 & -1 & 0 \\ 0 & 0 & 0 & -i \\ 1 & 0 & 0 & 0 \\ 0 & i & 0 & 0 \end{pmatrix},$$

$$R_3 = \begin{pmatrix} 0 & 0 & 0 & -1 \\ 0 & 0 & i & 0 \\ 0 & -i & 0 & 0 \\ 1 & 0 & 0 & 0 \end{pmatrix}.$$

The self-dual Maxwell equations take the form,

$$\mathcal{N}\partial_\mu\left(F^{\mu\nu} + i\tilde{F}^{\mu\nu}\right) = \partial_\mu\left(R_i^{\mu\nu}\Psi_i\right) = R_i^{\mu\nu}\partial_\mu\Psi_i = -\left(\alpha_\mu\right)^{\nu i}\partial_\mu\Psi_i = 0, \tag{33}$$

where,

$$\alpha_0 = \begin{pmatrix} 1 & 0 & 0 & 0 \\ 0 & 1 & 0 & 0 \\ 0 & 0 & 1 & 0 \\ 0 & 0 & 0 & 1 \end{pmatrix}, \tag{34}$$

$$\alpha_1 = \begin{pmatrix} 0 & -1 & 0 & 0 \\ -1 & 0 & 0 & 0 \\ 0 & 0 & 0 & -i \\ 0 & 0 & i & 0 \end{pmatrix}, \tag{35}$$

$$\alpha_2 = \begin{pmatrix} 0 & 0 & -1 & 0 \\ 0 & 0 & 0 & i \\ -1 & 0 & 0 & 0 \\ 0 & -i & 0 & 0 \end{pmatrix}, \tag{36}$$

$$\alpha_3 = \begin{pmatrix} 0 & 0 & 0 & -1 \\ 0 & 0 & -i & 0 \\ 0 & i & 0 & 0 \\ -1 & 0 & 0 & 0 \end{pmatrix} \tag{37}$$

and the Ψ_i is part of the the four component wave function,

$$\Phi = \begin{pmatrix} 0 \\ \Psi_x \\ \Psi_y \\ \Psi_z \end{pmatrix} \tag{38}$$

The free field photon equation becomes

$$i\hbar \left(\alpha_\mu \partial^\mu \right) \Phi = \left(\frac{\hat{E}}{c} a_0 - \hat{\mathbf{p}} \cdot \boldsymbol{\alpha} \right) \Phi = 0. \tag{39}$$

Compare this expression with our previous result of Eqs. (10) and (11). Note that the α_i ($i = 1, 2, 3$) matrices contain as submatrices the spin one matrices of Eq. (6). The subsidiary condition $\hat{\mathbf{p}} \cdot \boldsymbol{\Psi} = 0$ is added to the first row and first column of the α_i ($i = 1, 2, 3$) matrices, i.e.,

$$\alpha_1 = \begin{pmatrix} 0 & -1 \quad 0 \quad 0 \\ -1 & \\ 0 & (S_x) \\ 0 & \end{pmatrix}, \tag{40}$$

$$\alpha_2 = \begin{pmatrix} 0 & 0 \quad -1 \quad 0 \\ 0 & \\ -1 & (S_y) \\ 0 & \end{pmatrix}, \tag{41}$$

$$\alpha_3 = \begin{pmatrix} 0 & 0 \quad 0 \quad -1 \\ 0 & \\ 0 & (S_z) \\ -1 & \end{pmatrix}. \tag{42}$$

Journal of Physics: Conference Series **330** (2011) 012010 doi:10.1088/1742-6596/330/1/012010

Thus the α_i $(i = 1, 2, 3)$ matrices can be constructed from the photon equation with spin one matrices and the subsidiary condition. In Eq. (39) only differential operators act on the wave function Φ, they eliminate any constant in spacetime coordinates, therefore Φ can be replaced by

$$\Phi = \begin{pmatrix} \lambda \\ \Psi_x \\ \Psi_y \\ \Psi_z \end{pmatrix}, \tag{43}$$

where λ is an arbitrary constant, independent of the space-time coordinates. The Hermitian conjugate of Φ is,

$$\Phi^H = \begin{pmatrix} \lambda^* & \Psi_x^* & \Psi_y^* & \Psi_z^* \end{pmatrix}. \tag{44}$$

With these new definitions one can define the Lagrangian density L,

$$L = \frac{\Phi^H}{2}\left(\hat{E}a_0 - c\hat{\mathbf{p}}\cdot\boldsymbol{\alpha}\right)\Phi - \frac{\Phi^T}{2}\left(\hat{E}a_0 - c\hat{\mathbf{p}}\cdot\boldsymbol{\alpha}^*\right)\Phi^*, \tag{45}$$

where Φ^T is the transpose of Φ,

$$\Phi^T = \begin{pmatrix} \lambda & \Psi_x & \Psi_y & \Psi_z \end{pmatrix}. \tag{46}$$

Using the definitions of Eqs. (40) , (42) and (6) one can reduce Eq. (45) to,

$$\frac{L}{c} = \frac{\mathbf{\Psi}^H}{2}\left(\frac{\hat{E}}{c}I^{(3)} - \hat{\mathbf{p}}\cdot\mathbf{S}\right)\mathbf{\Psi} + \lambda^*\left(\hat{\mathbf{p}}\cdot\mathbf{\Psi}\right) - \frac{\mathbf{\Psi}^T}{2}\left(\frac{\hat{E}}{c}I^{(3)} + \hat{\mathbf{p}}\cdot\mathbf{S}\right)\mathbf{\Psi}^* - \lambda\left(\hat{\mathbf{p}}\cdot\mathbf{\Psi}^*\right), \tag{47}$$

where,

$$\mathbf{\Psi} = \begin{pmatrix} \Psi_x \\ \Psi_y \\ \Psi_z \end{pmatrix}. \tag{48}$$

One can see in Eq. (47) that the constants λ and λ^* play the role of Lagrangian multipliers. By varying the Lagrangian L with respect to λ and λ^* the subsidiary condition Eq. (11) and its complex conjugate are recovered. By varying the Lagrangian L with respect to $\mathbf{\Psi}^*$ Eq. (10) is obtained. By varying L with respect to $\mathbf{\Psi}$, Eq. (27) is recovered. The probability current is evaluated in Appendix A, with the result,

$$\partial_t\left(\mathbf{\Psi}^H\mathbf{\Psi}\right) + c\boldsymbol{\nabla}\cdot\left(\mathbf{\Psi}^H\mathbf{S}\mathbf{\Psi}\right) = 0. \tag{49}$$

Thus $\mathbf{\Psi}^H\mathbf{\Psi}$ is interpreted as the density of probability which should be normalized to unity (and is a constant of motion),

$$\int\!\!\int\!\!\int dxdydz\left(\mathbf{\Psi}^H\mathbf{\Psi}\right) = 1, \tag{50}$$

under the condition that the wavefunction vanishes properly at infinity. This is a realistic condition if the wavefunction is a wave packet. In this case the wavefunctions form a base for an Hilbert space. Please note that the constant in Eq. (43) does not appear in the equations that determine the normalization, namely Eqs. (48),(49) and (50).

Having a Lagrangian, one can compute the corresponding energy-momentum tensor,

$$T_{\mu\nu} = \sum_j \frac{\partial L}{\partial \left(\frac{\partial u_j}{\partial x_\mu}\right)} \frac{\partial u_j}{\partial x_\nu} - L\delta_{\mu\nu}, \tag{51}$$

where u_j stand for the components of $\mathbf{\Psi}$ and $\mathbf{\Psi}^*$. For T^{00} we obtain,

$$\frac{T^{00}}{c} = \frac{T_{00}}{c} = \frac{\mathbf{\Psi}^H}{2}\left(\hat{\mathbf{p}}\cdot\mathbf{S}\right)\mathbf{\Psi} + \frac{\mathbf{\Psi}^T}{2}\left(\hat{\mathbf{p}}\cdot\mathbf{S}\right)\mathbf{\Psi}^* = \frac{\mathbf{\Psi}^H}{2}\frac{\hat{E}}{c}\mathbf{\Psi} - \frac{\mathbf{\Psi}^T}{2}\frac{\hat{E}}{c}\mathbf{\Psi}^*, \tag{52}$$

and

$$\int\int\int dxdydzT^{00} = \frac{1}{2}\int\int\int dxdydz\left[\mathbf{\Psi}^H\hat{E}\mathbf{\Psi} - \mathbf{\Psi}^T\hat{E}\mathbf{\Psi}^*\right]. \tag{53}$$

Applying the energy operator on Eq. (50) one obtains,

$$\hat{E}\int\int\int dxdydz\left(\mathbf{\Psi}^H\mathbf{\Psi}\right) = \int\int\int dxdydz\left[\mathbf{\Psi}^H\hat{E}\mathbf{\Psi} + \mathbf{\Psi}^T\hat{E}\mathbf{\Psi}^*\right] = 0 \tag{54}$$

Hence

$$\int\int\int \mathbf{\Psi}^T\hat{E}\mathbf{\Psi}^* dxdydz = -\int\int\int \mathbf{\Psi}^H\hat{E}\mathbf{\Psi} dxdydz. \tag{55}$$

Substituting Eq. (55) into Eq. (53) we finally obtain,

$$\int\int\int dxdydzT^{00} = \int\int\int dxdydz\mathbf{\Psi}^H\hat{E}\mathbf{\Psi} =< \hat{E} >, \tag{56}$$

which is the expectation value of the energy operator. For the T^{0k} components we have,

$$\frac{T^{0k}}{c} = -\frac{T_{0k}}{c} = \mathbf{\Psi}^H\frac{\hat{p}_k}{2}\mathbf{\Psi} - \mathbf{\Psi}^T\frac{\hat{p}_k}{2}\mathbf{\Psi}^*$$

$$= \mathbf{\Psi}^H\frac{\hat{p}_k}{2}\mathbf{\Psi} + \left(\mathbf{\Psi}^H\frac{\hat{p}_k}{2}\mathbf{\Psi}\right)^* = \mathrm{Re}\left(\mathbf{\Psi}^H\hat{p}_k\mathbf{\Psi}\right), \tag{57}$$

and

$$\int\int\int dxdydz\frac{T^{0k}}{c} = \int\int\int dxdydz\mathbf{\Psi}^H\left(\hat{p}_k\right)\mathbf{\Psi} =< \hat{p}_k >, \quad k = 1,2,3, \tag{58}$$

which are the expectation values of real momenta.

Let us prove that $\hat{E}$ is a self adjoint operator. Indeed, from Eq. (15) $\hat{E}^* = -\hat{E}$, and using Eq. (55) we obtain,

$$\int\int\int dxdydz\left(\hat{E}\mathbf{\Psi}\right)^H\mathbf{\Psi} = -\int\int\int dxdydz\left(\hat{E}\mathbf{\Psi}^H\right)\mathbf{\Psi} \tag{59}$$

$$= -\int\int\int dxdydz\mathbf{\Psi}^T\hat{E}\mathbf{\Psi}^* = \int\int\int dxdydz\mathbf{\Psi}^H\hat{E}\mathbf{\Psi}. \tag{60}$$

In a similar way one can prove that the momentum operator $\hat{p}_k$ (for real momenta) is self adjoint,

$$\int\int\int dxdydz\left(\hat{p}_k\mathbf{\Psi}\right)^H\mathbf{\Psi} = -\int\int\int dxdydz\left(\hat{p}_k\mathbf{\Psi}^H\right)\mathbf{\Psi}$$

Journal of Physics: Conference Series **330** (2011) 012010 doi:10.1088/1742-6596/330/1/012010

$$= -\int\int\int dxdydz\boldsymbol{\Psi}^T\hat{p}_k\boldsymbol{\Psi}^* = \int\int\int dxdydz\boldsymbol{\Psi}^H\hat{p}_k\boldsymbol{\Psi}. \quad (61)$$

Therefore Eq. (39), with the normalization condition Eq. (50), is the first quantized equation of the photon.

5. Lagrangian for Maxwell's equations without sources

One can write a Lagrangian for Maxwell's equations simply by substituting $\boldsymbol{\Psi} = \mathcal{N}(\mathbf{E} + i\mathbf{B})$ into the photon Lagrangian Eq. (47). One obtains,

$$\frac{L}{c\mathcal{N}} = -\mathbf{E}^T(\hat{\mathbf{p}}\cdot\mathbf{S})\mathbf{E} - \mathbf{B}^T(\hat{\mathbf{p}}\cdot\mathbf{S})\mathbf{B} + \mathbf{i}\mathbf{E}^T\frac{\hat{E}}{c}\mathbf{B} - i\mathbf{B}^T\frac{\hat{E}}{c}\mathbf{E} + \frac{\lambda^*}{2}\hat{\mathbf{p}}\cdot\boldsymbol{\Psi} - \frac{\lambda}{2}\hat{\mathbf{p}}\cdot\boldsymbol{\Psi}^* \quad (62)$$

$$= -\hbar\left(\mathbf{E}\cdot(\nabla\times\mathbf{E}) + \mathbf{B}\cdot(\nabla\times\mathbf{B}) + \frac{1}{c}\mathbf{E}\cdot(\partial_t\mathbf{B}) - \frac{1}{c}\mathbf{B}\cdot(\partial_t\mathbf{E})\right)$$

$$-\hbar\left(\mathrm{Im}(\lambda)\,\nabla\cdot\mathbf{E} - \mathrm{Re}(\lambda)\,\nabla\cdot\mathbf{B}\right). \quad (63)$$

The conserved probability current Eq. (49) takes the form,

$$\partial_t\left(\boldsymbol{\Psi}^H\boldsymbol{\Psi}\right) + c\boldsymbol{\nabla}\cdot\left(\boldsymbol{\Psi}^H\mathbf{S}\boldsymbol{\Psi}\right) = \mathcal{N}^2\left[\partial_t\left(\mathbf{E}^2 + \mathbf{B}^2\right) + 2c\boldsymbol{\nabla}\cdot(\mathbf{E}\times\mathbf{B})\right] = 0, \quad (64)$$

which coincides with the Poynting theorem. The normalization coefficient $\mathcal{N}$ can be found from the normalization requirement,

$$\int\int\int dxdydz\boldsymbol{\Psi}^H\boldsymbol{\Psi} = |\mathcal{N}|^2\int\int\int dxdydz\left(\mathbf{E}^2 + \mathbf{B}^2\right) = 1, \quad (65)$$

hence,

$$|\mathcal{N}|^{-2} = \int\int\int dxdydz\left(\mathbf{E}^2 + \mathbf{B}^2\right). \quad (66)$$

Thus the probability density for one photon is,

$$\boldsymbol{\Psi}^H\boldsymbol{\Psi} = \frac{\mathbf{E}^2 + \mathbf{B}^2}{\int\int\int dxdydz\left(\mathbf{E}^2 + \mathbf{B}^2\right)}, \quad (67)$$

and the normalized photon wavefunction is,

$$\boldsymbol{\Psi} = \frac{\mathbf{E} + i\mathbf{B}}{\sqrt{\int\int\int dxdydz\left(\mathbf{E}^2 + \mathbf{B}^2\right)}}. \quad (68)$$

6. Relativistic quaternions

The α_μ matrices of the photon equation Eqs.(34-37) form a representation of what we will call "relativistic quaternions". They satisfy the following relations,

$$(\alpha_1)^2 = (\alpha_2)^2 = (\alpha_3)^2 = (\alpha_0)^2 = \alpha_0, \quad (69)$$

$$\alpha_1\alpha_2 = i\alpha_3 \quad \text{(and cyclic permutation)}, \quad (70)$$

$$\alpha_k\alpha_l = -\alpha_l\alpha_k, \quad \text{for } k \neq l, \quad k,l = 1,2,3. \quad (71)$$

The α_μ matrices factorize the d'Alembertian, therefore also the relativistic quaternions will factorize it. The relativistic quaternions are a particular case of complex quaternions. Let $\mathbf{Q}$ be a quaternion, which traditionally is written as,

$$\mathbf{Q} = c_0\mathbf{1} + c_1\mathbf{i} + c_2\mathbf{j} + c_3\mathbf{k}, \tag{72}$$

$$\mathbf{i}^2 = \mathbf{j}^2 = \mathbf{k}^2 = -\mathbf{1}, \quad \mathbf{ij} = -\mathbf{ji}, \quad \mathbf{ik} = -\mathbf{ki}, \quad \mathbf{kj} = -\mathbf{jk}, \quad \mathbf{ijk} = -\mathbf{1}, \tag{73}$$

where the c_μ are real coefficients. Let us rewrite Eq. (72) as,

$$\mathbf{Q} = c_0\mathbf{q}_0 + c_1\mathbf{q}_1 + c_2\mathbf{q}_2 + c_3\mathbf{q}_3, \tag{74}$$

$$\mathbf{q}_0 \equiv \mathbf{1}, \quad \mathbf{q}_1 \equiv \mathbf{i}, \quad \mathbf{q}_2 \equiv \mathbf{j}, \quad \mathbf{q}_3 \equiv \mathbf{k}. \tag{75}$$

The conjugate quaternion is defined as,

$$\mathbf{Q}^C = c_0\mathbf{q}_0 - c_1\mathbf{q}_1 - c_2\mathbf{q}_2 - c_3\mathbf{q}_3. \tag{76}$$

Using Eqs. (72-76) we obtain,

$$\mathbf{Q}^C\mathbf{Q} = \mathbf{q}_0\left(c_0^2 + c_1^2 + c_2^2 + c_3^2\right). \tag{77}$$

The quaternions are not suitable for factorizing the d'Alembertian. The factorization can be done with complex quaternions, for which the coefficients c_μ in Eq. (72) or Eq. (74) can be complex. The relativistic quaternion with complex coefficients is defined as,

$$\mathbf{R} = c_0\mathbf{r}_0 + c_1\mathbf{r}_1 + c_2\mathbf{r}_2 + c_3\mathbf{r}_3, \tag{78}$$

$$\mathbf{r}_0 = \mathbf{q}_0 = \mathbf{1}, \quad \mathbf{r}_1 = i\mathbf{q}_1, \quad \mathbf{r}_2 = i\mathbf{q}_2, \quad \mathbf{r}_3 = i\mathbf{q}_3, \tag{79}$$

and its conjugate as,

$$\mathbf{R}^C = c_0\mathbf{r}_0 - c_1\mathbf{r}_1 - c_2\mathbf{r}_2 - c_3\mathbf{r}_3, \tag{80}$$

The multiplication table of the $\mathbf{r}_\mu$ base (and its representations) is,

$\times$	$\mathbf{r}_0$	$\mathbf{r}_1$	$\mathbf{r}_2$	$\mathbf{r}_3$
$\mathbf{r}_0$	$\mathbf{r}_0$	$\mathbf{r}_1$	$\mathbf{r}_2$	$\mathbf{r}_3$
$\mathbf{r}_1$	$\mathbf{r}_1$	$\mathbf{r}_0$	$i\mathbf{r}_3$	$-i\mathbf{r}_2$
$\mathbf{r}_2$	$\mathbf{r}_2$	$-i\mathbf{r}_3$	$\mathbf{r}_0$	$i\mathbf{r}_1$
$\mathbf{r}_3$	$\mathbf{r}_3$	$i\mathbf{r}_2$	$-i\mathbf{r}_1$	$\mathbf{r}_0$

. (81)

and,

$$\mathbf{R}^C\mathbf{R} = \mathbf{R}\mathbf{R}^C = \mathbf{r}_0\left(c_0^2 - c_1^2 - c_2^2 - c_3^2\right), \tag{82}$$

which allows factorization of the d'Alembertian.

Another important advantage of the relativistic quaternions is that they have a representation in terms of Pauli matrices Eq. (3),

$$\mathbf{R}_\sigma = c_0\boldsymbol{\sigma}_0 + c_1\boldsymbol{\sigma}_x + c_2\boldsymbol{\sigma}_y + c_3\boldsymbol{\sigma}_z, \tag{83}$$

Journal of Physics: Conference Series **330** (2011) 012010 doi:10.1088/1742-6596/330/1/012010

i.e. $\mathbf{R}_{\boldsymbol{\sigma}}$ is a second rank mixed spinor. In this way we have shown the connection between the relativistic quaternions and spinors. Now we can rewrite the photon (and Maxwell's) equations Eq. (39) as spinor equation,

$$\left(\frac{\hat{E}}{c}\boldsymbol{\sigma}_0 - \hat{p}_x\boldsymbol{\sigma}_x - \hat{p}_y\boldsymbol{\sigma}_y - \hat{p}_z\boldsymbol{\sigma}_z\right)(\lambda\boldsymbol{\sigma}_0 + \boldsymbol{\Psi}_x\boldsymbol{\sigma}_x + \boldsymbol{\Psi}_y\boldsymbol{\sigma}_y + \boldsymbol{\Psi}_z\boldsymbol{\sigma}_z) = 0. \tag{84}$$

In the section devoted to the quantum photon (spin 1) other representation of the relativistic quaternion in terms of 4×4 matrices is given in Eqs. (124) and (125). Similarly, in the section devoted to the quantum graviton (spin 2) other representation of the relativistic quaternion in terms of 8×8 matrices is given in Eqs. (145) and (147).

The 4×4 matrices Eqs. (34), (37), (124) and (125), which are representations of the relativistic quaternions, have eigenvalues -1 and 1 (twice). This is consistent with the condition of having helicity in the forward and backward directions only. The above matrices contain implicitly the subsidiary conditions, which is consistent with their role in suppressing the helicity components in other than forward and backward directions. Other properties are:

$$\det(\alpha_i) = \det(\beta_i) = 1, \quad i = x, y, z, \tag{85}$$

$$\mathrm{tr}(\alpha_i) = \mathrm{tr}(\beta_i) = 0. \tag{86}$$

The Pauli matrices σ_i have eigenvalues -1 and 1 and,

$$\det(\sigma_i) = -1, \quad \mathrm{tr}(\sigma_i) = 0. \tag{87}$$

Taking into account that the α_i and β_i have the same eigenvalues as the σ_i and the validity of Eqs. (85-87), we can conclude that α_i and β_i are equivalent to the matrices

$$\Sigma_i^{(4)} = \begin{pmatrix} \sigma_i & \\ & \sigma_i \end{pmatrix}, \quad i = 0, x, y, z, \tag{88}$$

In a similar way one can show that the 8×8 γ_i matrices of Eqs. (145) and (147) are equivalent to the 8×8 matrices

$$\Sigma_i^{(8)} = \begin{pmatrix} \sigma_i & & & \\ & \sigma_i & & \\ & & \sigma_i & \\ & & & \sigma_i \end{pmatrix}. \tag{89}$$

The above results can be generalized. Thus the relativistic quaternions have representations in terms of 4s×4s matrices

$$\Sigma_i^{(4s)} = \begin{pmatrix} \sigma_i & & & \\ & \sigma_i & & \\ & & \ddots & \\ & & & \sigma_i \end{pmatrix}. \tag{90}$$

An example of the above representation can be obtained by replacing Eq. (84) with,

$$\left(\frac{\hat{E}}{c}\Sigma_0^{(4)} + \hat{p}_x\Sigma_x^{(4)} + \hat{p}_y\Sigma_y^{(4)} + \hat{p}_z\Sigma_z^{(4)}\right)\begin{pmatrix} \lambda + \Psi_z \\ \Psi_x + i\Psi_y \\ \Psi_x - i\Psi_y \\ \lambda - \Psi_z \end{pmatrix} = 0, \tag{91}$$

or to use,

$$\begin{pmatrix} \lambda + \Psi_z \\ \Psi_x + i\Psi_y \\ \Psi_x - i\Psi_y \\ \lambda - \Psi_z \end{pmatrix} = \begin{pmatrix} \lambda + \Psi_0 \\ -\sqrt{2}\Psi_1 \\ \sqrt{2}\Psi_{-1} \\ \lambda - \Psi_0 \end{pmatrix}. \tag{92}$$

7. Dirac's equations for massless particles of any spin

Dirac has derived equations for massless particles with spin s, which in the ordinary vector notation[4] are,

$$\{s\hat{p}_0 + S_x\hat{p}_x + S_y\hat{p}_y + S_z\hat{p}_z\}\,\psi = 0, \tag{93}$$

$$\{s\hat{p}_x + S_x\hat{p}_0 - iS_y\hat{p}_z + iS_z\hat{p}_y\}\,\psi = 0, \tag{94}$$

$$\{s\hat{p}_y + S_y\hat{p}_0 - iS_z\hat{p}_x + iS_x\hat{p}_z\}\,\psi = 0, \tag{95}$$

$$\{s\hat{p}_z + S_z\hat{p}_0 - iS_x\hat{p}_y + iS_y\hat{p}_x\}\,\psi = 0, \tag{96}$$

where ψ a $(2s+1)$ component wave function and S_n are the spin $(2s+1)\times(2s+1)$ matrices which satisfy,

$$[S_x, S_y] = iS_z, \quad [S_z, S_x] = iS_y, \quad [S_y, S_z] = iS_x, \quad S_x^2 + S_y^2 + S_z^2 = s(s+1)I^{(s)}. \tag{97}$$

Above the $\hat{p}_n$ are the momenta, $\hat{p}_0 = \hat{E}/c$, $\hat{E}$ the energy, and $I^{(s)}$ is a $(2s+1)\times(2s+1)$ unit matrix. As we shall see below, for the case $s = 1$, Eq. (93) will lead to the Faraday and Ampere-Maxwell laws. The Gauss laws can be derived from Eqs. (93-96) in a way which will be described below. Eqs. (93-96) were analyzed extensively by Bacry[5], who derived them using Wigner's condition [6] on the Pauli-Lubanski vector W^μ for massless fields,

$$W^\mu = s\hat{p}^\mu, \quad \mu = x_0, x, y, z. \tag{98}$$

In the next section Eqs. (93-96) will be retrieved by factorizing the d'Alembertian.

8. Wave equations for massless particles of any spin

Using the metric convention $g_{00} = -g_{11} = -g_{22} = -g_{33} = 1$, we will employ a special form of the Pauli-Lubanski vector operator W^μ suggested by H. Bacry[5] for massless particles,

$$W^\mu = \frac{-i}{2}\varepsilon^{\mu\nu\rho\lambda}S_{\rho\lambda}\hat{p}_\nu, \quad S_{\rho\lambda} = \begin{pmatrix} 0 & S_x & S_y & S_z \\ -S_x & 0 & -iS_z & iS_y \\ -S_y & iS_z & 0 & -iS_x \\ -S_z & -iS_y & iS_x & 0 \end{pmatrix} \equiv (\mathbf{S}, i\mathbf{S}), \tag{99}$$

where $S_{\rho\lambda}$ is the antysymmetric spin tensor, $\mathbf{S}$ is the spin vector matrix operator, s is the spin of the particle with zero mass and the components of W^μ are $(2s+1)\times(2s+1)$ matrices. We shall use the following relation[7],

$$W_\mu W^\mu = -s\,(s+1)\,\hat{p}_\mu\hat{p}^\mu I^{(s)}, \tag{100}$$

where $I^{(s)}$ is the $(2s+1)\times(2s+1)$ unit matrix. One can easily see that,

Journal of Physics: Conference Series **330** (2011) 012010 doi:10.1088/1742-6596/330/1/012010

$$(W_\mu - s\hat{p}_\mu)(W^\mu + s\hat{p}^\mu) = -s(2s+1)\hat{p}_\mu\hat{p}^\mu I^{(s)}, \tag{101}$$

or

$$(W_\mu + s\hat{p}_\mu)(W^\mu - s\hat{p}^\mu) = -s(2s+1)\hat{p}_\mu\hat{p}^\mu I^{(s)}. \tag{102}$$

Eq. (102) or Eq. (101) will be the basis for all our derivations. From Eq. (102) one can see that if ψ, the wavefunction, which is a $(2s+1)$ component spherical tensor operator of rank s satisfies,

$$(W^\mu + s\hat{p}^\mu)\psi = 0, \quad \mu = 0,1,2,3, \tag{103}$$

also the basic one-particle requirement,

$$\hat{p}_\mu\hat{p}^\mu I^{(s)}\psi = 0, \tag{104}$$

will be satisfied. Explicitly we have,

$$W^0 = W_0 = -S_x\hat{p}_x - S_y\hat{p}_y - S_z\hat{p}_z, \tag{105}$$

$$-W^1 = W_x = -S_x\hat{p}_0 - iS_y\hat{p}_z + iS_z\hat{p}_y, \tag{106}$$

$$-W^2 = W_y = -S_y\hat{p}_0 - iS_z\hat{p}_x + iS_x\hat{p}_z, \tag{107}$$

$$-W^3 = W_z = -S_z\hat{p}_0 - iS_x\hat{p}_y + iS_y\hat{p}_x. \tag{108}$$

Eqs. (105) and (108) can be presented in a nonmanifestly covariant form[5] as,

$$(\hat{\mathbf{p}}\cdot\mathbf{S} - s\hat{p}^0)\psi = 0, \tag{109}$$

$$(\mathbf{S}\hat{p}^0 + i\mathbf{S}\times\hat{\mathbf{p}} - s\hat{\mathbf{p}})\psi = 0, \tag{110}$$

where $\hat{\mathbf{p}}$ the particle momentum and $\mathbf{S}$ is the spin vector matrix operator with the properties given in Eq. (97). Dirac[4] has suggested to use Eq. (109) as the basic one and to substitute from it $s\hat{p}^0 = -\hat{\mathbf{p}}\cdot\mathbf{S}$ into Eqs. (110). In this way one obtains,

$$\left(\frac{1}{s}\mathbf{S}(\hat{\mathbf{p}}\cdot\mathbf{S}) + i\mathbf{S}\times\hat{\mathbf{p}} - s\hat{\mathbf{p}}\right)\psi = 0, \tag{111}$$

which are subsidiary conditions on Eq. (109). But the three Eqs. (111) are not independent. If one multiplies Eqs. (111) by $\mathbf{S}$ one finds,

$$\mathbf{S}\cdot\left(\frac{1}{s}\mathbf{S}(\hat{\mathbf{p}}\cdot\mathbf{S}) + i\mathbf{S}\times\hat{\mathbf{p}} - s\hat{\mathbf{p}}\right) = 0, \tag{112}$$

and only one of the three Eqs. (111) is independent. According to our experience the simplest one is the z-component equation,

$$\left(\frac{1}{s}S_z(\hat{\mathbf{p}}\cdot\mathbf{S}) + i(S_x\hat{p}_y - S_y\hat{p}_x) - s\hat{p}_z\right)\psi = 0. \tag{113}$$

Now we can collect our results and make the quantum transition substitutions,

$$\hat{\mathbf{p}} \rightarrow -i\hbar\boldsymbol{\nabla}, \quad \hat{p}^0 \rightarrow i\hbar\frac{\partial}{c\partial t}, \tag{114}$$

and obtain the final result,

$$\left(si\hbar\frac{\partial}{c\partial t} + i\hbar\mathbf{S}\cdot\boldsymbol{\nabla}\right)\psi = 0, \tag{115}$$

with the subsidiary conditions,

$$-i\hbar\left[\frac{1}{s}S_z\left(\mathbf{S}\cdot\boldsymbol{\nabla}\right) + i\left(S_x\frac{\partial}{\partial y} - S_y\frac{\partial}{\partial x}\right) - s\frac{\partial}{\partial z}\right]\psi = 0. \tag{116}$$

In the equations we derived above, the wavefunctions are $(2s+1)$ component spherical tensor operators of rank s. In order to be consistent with the Cartesian representation, we will have to work in terms of spherical tensors. A consistent formalism is worked out in Appendix D.

9. The photon equation revisited

We will present the photon equation in the angular momentum representation using the formalism given by Eqs.(99-D.12) of Appendix D. The wavefunction and the spin matrices are,

$$\Psi = \begin{pmatrix} \Psi_1 \\ \Psi_0 \\ \Psi_{-1} \end{pmatrix}, \tag{117}$$

$$\tilde{S}_x = \frac{-1}{\sqrt{2}}\begin{pmatrix} 0 & 1 & 0 \\ 1 & 0 & 1 \\ 0 & 1 & 0 \end{pmatrix}, \quad \tilde{S}_y = \frac{i}{\sqrt{2}}\begin{pmatrix} 0 & -1 & 0 \\ 1 & 0 & -1 \\ 0 & 1 & 0 \end{pmatrix}, \quad \tilde{S}_z = \begin{pmatrix} -1 & 0 & 0 \\ 0 & 0 & 0 \\ 0 & 0 & 1 \end{pmatrix}, \tag{118}$$

$$I^{(3)} = \begin{pmatrix} 1 & 0 & 0 \\ 0 & 1 & 0 \\ 0 & 0 & 1 \end{pmatrix},$$

We have 4 equations for $s = 1$. The main equation,

$$\left(\hat{\mathbf{p}}\cdot\tilde{\mathbf{S}} - \frac{\hat{E}}{c}I^{(3)}\right)\Psi = 0, \tag{119}$$

and the subsidiary equations,

$$\left(\tilde{\mathbf{S}}\left(\hat{\mathbf{p}}\cdot\tilde{\mathbf{S}}\right) + i\tilde{\mathbf{S}}\times\hat{\mathbf{p}} - \hat{\mathbf{p}}I^{(3)}\right)\Psi = 0. \tag{120}$$

The z- subsidiary condition is,

$$\left[\tilde{S}_z\left(\tilde{S}_x\hat{p}_x + \tilde{S}_y\hat{p}_y + \tilde{S}_z\hat{p}_z\right) + i\tilde{S}_x\hat{p}_y - i\tilde{S}_y\hat{p}_x - \hat{p}_z I^{(3)}\right]\Psi$$

$$= \begin{pmatrix} 0 & 0 & 0 \\ \frac{1}{2}\sqrt{2}\hat{p}_x - \frac{1}{2}i\sqrt{2}\hat{p}_y & -\hat{p}_z & -\frac{1}{2}\sqrt{2}\hat{p}_x - \frac{1}{2}i\sqrt{2}\hat{p}_y \\ 0 & 0 & 0 \end{pmatrix}\begin{pmatrix} \Psi_1 \\ \Psi_0 \\ \Psi_{-1} \end{pmatrix}$$

Journal of Physics: Conference Series **330** (2011) 012010 doi:10.1088/1742-6596/330/1/012010

$$= \begin{pmatrix} 0 & 0 & 0 \\ \hat{p}_{-1} & -\hat{p}_0 & \hat{p}_1 \\ 0 & 0 & 0 \end{pmatrix} \begin{pmatrix} \Psi_1 \\ \Psi_0 \\ \Psi_{-1} \end{pmatrix} = 0, \tag{121}$$

which yields,

$$\hat{p}_{-1}\Psi_1 - \hat{p}_0\Psi_0 + \hat{p}_1\Psi_{-1} = -\hat{p}_x\Psi_x - \hat{p}_y\Psi_y - \hat{p}_z\Psi_z = -\hat{\mathbf{p}}\cdot\mathbf{\Psi} = \mathbf{0},$$

so that the subsidiary condition becomes,

$$\hat{\mathbf{p}}\cdot\mathbf{\Psi} = \mathbf{0}, \tag{122}$$

as before in Eq. (11). The main equation and the subsidiary condition form the photon equation in the angular momentum representation. In order to get an equation which include the subsidiary condition implicitly, we proceed in a similar way to the derivation of Eqs.(40-43). One introduces the wavefunction,

$$\Phi = \begin{pmatrix} \lambda \\ \Psi_1 \\ \Psi_0 \\ \Psi_{-1} \end{pmatrix}, \tag{123}$$

where λ is an arbitrary constant, and complements the $\tilde{S}_x, \tilde{S}_y, \tilde{S}_z$ matrices to 4×4 matrices $\beta_x, \beta_y, \beta_z$,which include implicitly the subsidiary condition. The required matrices are,

$$\beta_x = \frac{-1}{\sqrt{2}} \begin{pmatrix} 0 & -1 & 0 & 1 \\ -1 & 0 & 1 & 0 \\ 0 & 1 & 0 & 1 \\ 1 & 0 & 1 & 0 \end{pmatrix}; \; \beta_y = \frac{i}{\sqrt{2}} \begin{pmatrix} 0 & -1 & 0 & -1 \\ 1 & 0 & -1 & 0 \\ 0 & 1 & 0 & -1 \\ 1 & 0 & 1 & 0 \end{pmatrix}; \tag{124}$$

$$\beta_z = \begin{pmatrix} 0 & 0 & -1 & 0 \\ 0 & -1 & 0 & 0 \\ -1 & 0 & 0 & 0 \\ 0 & 0 & 0 & 1 \end{pmatrix}; \; \beta_0 = \begin{pmatrix} 1 & 0 & 0 & 0 \\ 0 & 1 & 0 & 0 \\ 0 & 0 & 1 & 0 \\ 0 & 0 & 0 & 1 \end{pmatrix}. \tag{125}$$

One can check that these matrices provide a representation of the relativistic quaternions Eqs. (69) - (79), and therefore factorize the d'Alembertian,

$$\left(\beta_0\partial_0 + \beta_x\partial_x + \beta_y\partial_y + \beta_z\partial_z\right)\left(\beta_0\partial_0 - \beta_x\partial_x - \beta_y\partial_y - \beta_z\partial_z\right)$$

$$= \left(\partial_0^2 - \partial_x^2 - \partial_y^2 - \partial_z^2\right)\beta_0. \tag{126}$$

The (free) photon equation takes the form,

$$\left(\hat{E}\beta_0 - c\hat{\mathbf{p}}\cdot\boldsymbol{\beta}\right)\Phi = 0, \tag{127}$$

where Φ is given in Eq.(123). In a similar way to the derivation of the photon Lagrangian Eq.(45) one can construct the Lagrangian L,

$$L = \frac{\Phi^H}{2}\left(\hat{E}\beta_0 - c\hat{\mathbf{p}}\cdot\boldsymbol{\beta}\right)\Phi - \frac{\Phi^T}{2}\left(\hat{E}\beta_0 - c\hat{\mathbf{p}}\cdot\boldsymbol{\beta}^*\right)\Phi^*, \tag{128}$$

where Φ^T is the transpose of Φ,

$$\Phi^T = \begin{pmatrix} \lambda & \Psi_1 & \Psi_0 & \Psi_{-1} \end{pmatrix}, \tag{129}$$

and Φ^H is the Hermitian conjugate of Φ which, according to Appendix B, is,

$$\Phi^H = \begin{pmatrix} \lambda^* & -\Psi_{-1}^* & \Psi_0^* & -\Psi_1^* \end{pmatrix}. \tag{130}$$

By varying the Lagrangian L with respect to the components of Φ^*, Eq.(127) is obtained. By varying L with respect to the components of Φ, the equation with the opposite helicity,

$$\left(\hat{E}\beta_0 - c\hat{\mathbf{p}}\cdot\boldsymbol{\beta}^*\right)\Phi^* = 0, \tag{131}$$

is obtained.

10. The quantum graviton

We now turn to a derivation of the graviton (spin 2) equation in the angular momentum representation following the formalism of Eqs.(99-D.10). The wavefunction $\boldsymbol{\Psi}$ is,

$$< 2, m|\boldsymbol{\Psi} >= \begin{pmatrix} \Psi_2 \\ \Psi_1 \\ \Psi_0 \\ \Psi_{-1} \\ \Psi_{-2} \end{pmatrix}, \tag{132}$$

and the spin matrices are,

$$\tilde{S}_x = \begin{pmatrix} 0 & -1 & 0 & 0 & 0 \\ -1 & 0 & -\sqrt{\frac{3}{2}} & 0 & 0 \\ 0 & -\sqrt{\frac{3}{2}} & 0 & -\sqrt{\frac{3}{2}} & 0 \\ 0 & 0 & -\sqrt{\frac{3}{2}} & 0 & -1 \\ 0 & 0 & 0 & -1 & 0 \end{pmatrix}, \tag{133}$$

$$\tilde{S}_y = i\begin{pmatrix} 0 & -1 & 0 & 0 & 0 \\ 1 & 0 & -\sqrt{\frac{3}{2}} & 0 & 0 \\ 0 & \sqrt{\frac{3}{2}} & 0 & -\sqrt{\frac{3}{2}} & 0 \\ 0 & 0 & \sqrt{\frac{3}{2}} & 0 & -1 \\ 0 & 0 & 0 & 1 & 0 \end{pmatrix}, \tag{134}$$

$$\tilde{S}_z = \begin{pmatrix} -2 & 0 & 0 & 0 & 0 \\ 0 & -1 & 0 & 0 & 0 \\ 0 & 0 & 0 & 0 & 0 \\ 0 & 0 & 0 & 1 & 0 \\ 0 & 0 & 0 & 0 & 2 \end{pmatrix}, \quad I^{(5)} = \begin{pmatrix} 1 & 0 & 0 & 0 & 0 \\ 0 & 1 & 0 & 0 & 0 \\ 0 & 0 & 1 & 0 & 0 \\ 0 & 0 & 0 & 1 & 0 \\ 0 & 0 & 0 & 0 & 1 \end{pmatrix}. \tag{135}$$

We have 4 equations for $s = 2$, the main equation,

$$\left(\hat{\mathbf{p}}\cdot\tilde{\mathbf{S}} - 2\frac{\hat{E}}{c}I^{(5)}\right)\Psi = 0, \tag{136}$$

Journal of Physics: Conference Series **330** (2011) 012010 doi:10.1088/1742-6596/330/1/012010

and the subsidiary conditions,

$$\left(\frac{1}{2}\tilde{\mathbf{S}}\left(\hat{\mathbf{p}}\cdot\tilde{\mathbf{S}}\right)+i\tilde{\mathbf{S}}\times\hat{\mathbf{p}}-2\hat{\mathbf{p}}I^{(5)}\right)\Psi=0. \tag{137}$$

The z- subsidiary condition is,

$$\left[S_z\left(S_x\hat{p}_x+S_y\hat{p}_y+S_z\hat{p}_z\right)/2+iS_x\hat{p}_y-iS_y\hat{p}_x-2\hat{p}_zI^{(5)}\right]\Psi=K\Psi=0, \tag{138}$$

where,

$$K=\begin{pmatrix} 0 & 0 & 0 & 0 & 0\\ \frac{3}{\sqrt{2}}\hat{p}_{-1} & -\frac{3}{2}\hat{p}_0 & \sqrt{3}\hat{p}_1 & 0 & 0\\ 0 & \sqrt{3}\hat{p}_{-1} & -2\hat{p}_0 & \sqrt{3}\hat{p}_1 & 0\\ 0 & 0 & \sqrt{3}\hat{p}_{-1} & -\frac{3}{2}\hat{p}_0 & \frac{3}{\sqrt{2}}\hat{p}_1\\ 0 & 0 & 0 & 0 & 0 \end{pmatrix}. \tag{139}$$

The 3 subsidiary conditions are each valid up to a multiplicative constant. We will choose the multiplicative constants so that the following decomposition is obtained,

$$-4\hat{p}_\mu\hat{p}^\mu I^{(s)}\Psi=\left(W_0-2\frac{\hat{E}}{c}I^{(5)}\right)\left(W_0+2\frac{\hat{E}}{c}I^{(5)}\right)\Psi+P^HP\Psi, \tag{140}$$

where P^H is the Hermitian conjugate of P. The normalized subsidiary condition will fulfil the equation,

$$P\Psi=0, \tag{141}$$

where,

$$P=\begin{pmatrix} 0 & 0 & 0 & 0 & 0\\ \sqrt{6}\hat{p}_{-1} & -\sqrt{3}\hat{p}_0 & \hat{p}_1 & 0 & 0\\ 0 & \sqrt{3}\hat{p}_{-1} & -2\hat{p}_0 & \sqrt{3}\hat{p}_1 & 0\\ 0 & 0 & \hat{p}_{-1} & -\sqrt{3}\hat{p}_0 & \sqrt{6}\hat{p}_1\\ 0 & 0 & 0 & 0 & 0 \end{pmatrix}. \tag{142}$$

In order to get an equation with the subsidiary conditions included in it implicitly, one has to introduce an eight-component wavefunction,

$$\Phi=\begin{pmatrix} \lambda_1\\ \lambda_0\\ \lambda_{-1}\\ \Psi_2\\ \Psi_1\\ \Psi_0\\ \Psi_{-1}\\ \Psi_{-2} \end{pmatrix}, \tag{143}$$

where $\lambda_1,\lambda_0,\lambda_{-1}$ are arbitrary constants, and to complement the 5×5 $\tilde{S}_x,\tilde{S}_y,\tilde{S}_z$ matrices into 8×8 matrices $\gamma_x,\gamma_y,\gamma_z$, so that the first 3 raws and 3 columns generate the subsidiary conditions. In addition we require that these matrices factorize the d'Alembertian. The required matrices are:

$$\gamma_x = \frac{1}{2}\begin{pmatrix} 0 & \frac{1}{\sqrt{2}} & 0 & \sqrt{3} & 0 & -\frac{1}{\sqrt{2}} & 0 & 0 \\ \frac{1}{\sqrt{2}} & 0 & \frac{1}{\sqrt{2}} & 0 & \sqrt{\frac{3}{2}} & 0 & -\sqrt{\frac{3}{2}} & 0 \\ 0 & \frac{1}{\sqrt{2}} & 0 & 0 & 0 & \frac{1}{\sqrt{2}} & 0 & -\sqrt{3} \\ \sqrt{3} & 0 & 0 & 0 & -1 & 0 & 0 & 0 \\ 0 & \sqrt{\frac{3}{2}} & 0 & -1 & 0 & -\sqrt{\frac{3}{2}} & 0 & 0 \\ -\frac{1}{\sqrt{2}} & 0 & \frac{1}{\sqrt{2}} & 0 & -\sqrt{\frac{3}{2}} & 0 & -\sqrt{\frac{3}{2}} & 0 \\ 0 & -\sqrt{\frac{3}{2}} & 0 & 0 & 0 & -\sqrt{\frac{3}{2}} & 0 & -1 \\ 0 & 0 & -\sqrt{3} & 0 & 0 & 0 & -1 & 0 \end{pmatrix}, \tag{144}$$

$$\gamma_y = \frac{i}{2}\begin{pmatrix} 0 & \frac{1}{\sqrt{2}} & 0 & -\sqrt{3} & 0 & -\frac{1}{\sqrt{2}} & 0 & 0 \\ -\frac{1}{\sqrt{2}} & 0 & \frac{1}{\sqrt{2}} & 0 & -\sqrt{\frac{3}{2}} & 0 & -\sqrt{\frac{3}{2}} & 0 \\ 0 & -\frac{1}{\sqrt{2}} & 0 & 0 & 0 & -\frac{1}{\sqrt{2}} & 0 & -\sqrt{3} \\ \sqrt{3} & 0 & 0 & 0 & -1 & 0 & 0 & 0 \\ 0 & \sqrt{\frac{3}{2}} & 0 & 1 & 0 & -\sqrt{\frac{3}{2}} & 0 & 0 \\ \frac{1}{\sqrt{2}} & 0 & \frac{1}{\sqrt{2}} & 0 & \sqrt{\frac{3}{2}} & 0 & -\sqrt{\frac{3}{2}} & 0 \\ 0 & \sqrt{\frac{3}{2}} & 0 & 0 & 0 & \sqrt{\frac{3}{2}} & 0 & -1 \\ 0 & 0 & \sqrt{3} & 0 & 0 & 0 & 1 & 0 \end{pmatrix}, \tag{145}$$

$$\gamma_z = \frac{1}{2}\begin{pmatrix} 1 & 0 & 0 & 0 & -\sqrt{3} & 0 & 0 & 0 \\ 0 & 0 & 0 & 0 & 0 & -2 & 0 & 0 \\ 0 & 0 & -1 & 0 & 0 & 0 & -\sqrt{3} & 0 \\ 0 & 0 & 0 & -2 & 0 & 0 & 0 & 0 \\ -\sqrt{3} & 0 & 0 & 0 & -1 & 0 & 0 & 0 \\ 0 & -2 & 0 & 0 & 0 & 0 & 0 & 0 \\ 0 & 0 & -\sqrt{3} & 0 & 0 & 0 & 1 & 0 \\ 0 & 0 & 0 & 0 & 0 & 0 & 0 & 2 \end{pmatrix}, \tag{146}$$

$$\gamma_0 = \begin{pmatrix} 1 & 0 & 0 & 0 & 0 & 0 & 0 & 0 \\ 0 & 1 & 0 & 0 & 0 & 0 & 0 & 0 \\ 0 & 0 & 1 & 0 & 0 & 0 & 0 & 0 \\ 0 & 0 & 0 & 1 & 0 & 0 & 0 & 0 \\ 0 & 0 & 0 & 0 & 1 & 0 & 0 & 0 \\ 0 & 0 & 0 & 0 & 0 & 1 & 0 & 0 \\ 0 & 0 & 0 & 0 & 0 & 0 & 1 & 0 \\ 0 & 0 & 0 & 0 & 0 & 0 & 0 & 1 \end{pmatrix}. \tag{147}$$

One can check that the $\gamma_0\gamma_x, \gamma_y, \gamma_z$, form a representation of the relativistic quaternion, and therefore factorize the d'Alembertian, i.e.,

$$\begin{aligned} &\left(\gamma_0\partial_0 + \gamma_x\partial_x + \gamma_y\partial_y + \gamma_z\partial_z\right)\left(\gamma_0\partial_0 - \gamma_x\partial_x - \gamma_y\partial_y - \gamma_z\partial_z\right) \\ = \;&\left(\partial_0^2 - \partial_x^2 - \partial_y^2 - \partial_z^2\right)\gamma_0. \end{aligned} \tag{148}$$

With this in mind, the (free) quantum graviton equation takes the form,

$$\left(2\hat{E}\gamma_0 - c\hat{\mathbf{p}}\cdot\boldsymbol{\gamma}\right)\Phi = 0, \tag{149}$$

where Φ is given in Eq. (143). More details about the properties of the $\gamma_x, \gamma_y, \gamma_z$ matrices can be found in Appendix C. In a similar way to the derivation of the photon Lagrangian, Eq. (128), one can construct also a Lagrangian for the quantum graviton,

$$L = \frac{\Phi^H}{2}\left(2\hat{E}\gamma_0 - c\hat{\mathbf{p}}\cdot\boldsymbol{\gamma}\right)\Phi - \frac{\Phi^T}{2}\left(2\hat{E}\gamma_0 - c\hat{\mathbf{p}}\cdot\boldsymbol{\gamma}^*\right)\Phi^*, \tag{150}$$

where Φ^T is the transpose of Φ, and Φ^H is the Hermitian conjugate of Φ which, according to Appendix B, is,

$$\Phi^H = \left(\begin{array}{cccccccc} \lambda_{-1}^* & \lambda_0^* & \lambda_1^* & \Psi_{-2}^* & -\Psi_{-1}^* & \Psi_0^* & -\Psi_1^* & \Psi_2^* \end{array}\right). \tag{151}$$

By varying the Lagrangian L with respect to the components of Φ^* Eq. (149) is obtained. By varying L with respect to the components of $\boldsymbol{\Phi}$, the equation with the opposite helicity,

$$\left(2\hat{E}\gamma_0 - c\hat{\mathbf{p}}\cdot\boldsymbol{\gamma}^*\right)\Phi^* = 0, \tag{152}$$

is obtained. The probability current, evaluated in a similar way as in Appendix A, is,

$$\partial_t\left(\boldsymbol{\Psi}^H\boldsymbol{\Psi}\right) + \frac{c}{2}\boldsymbol{\nabla}\cdot\left(\boldsymbol{\Psi}^H\tilde{\mathbf{S}}\boldsymbol{\Psi}\right) = 0. \tag{153}$$

Here $\boldsymbol{\Psi}^H\boldsymbol{\Psi}$ is interpreted as the density of probability which should be normalized to unity,

$$\int\int\int dxdydz\boldsymbol{\Psi}^H\boldsymbol{\Psi} = 1, \tag{154}$$

under the condition that the wavefunction (as a wave packet) vanishes properly at infinity. The constants in Eq. (143) do not appear in the equations that determine the normalization, namely Eqs. (132),(153) and (154).Using the Lagrangian Eq. (150), the expectation values of the energy and momenta are,

$$\int\int\int dxdydzT^{00} = \int\int\int dxdydz\boldsymbol{\Psi}^H\hat{E}\boldsymbol{\Psi} =<\hat{E}>, \tag{155}$$

$$\int\int\int dxdydzT^{0k} = \int\int\int dxdydz\boldsymbol{\Psi}^H\left(c\hat{p}_k\right)\boldsymbol{\Psi} =< c\hat{p}_k >, \quad k = 1,2,3, \tag{156}$$

in agreement with the definitions of quantum operators. Therefore Eq. (149) with the normalization condition Eq. (154) is the first quantized equation of the graviton.

11. Covariance of the equations

The quantum equations for massless particles of spin s were derived from the covariant equations Eqs. (100-104) with the result

$$\left(s\hat{E}\eta_0 - c\hat{\mathbf{p}}\cdot\boldsymbol{\eta}\right)\Phi = 0, \tag{157}$$

where $\eta_0, \eta_1, \eta_2, \eta_3$ are $4s \times 4s$ matrices representing the relativistic quaternions and Φ is a $4s$ component wave function with $2s-1$ zeros. Equations similar to Eq. (157) were analyzed by Lomont[[8]] (including the problem of the zeros in the wave functions) and were shown to be covariant. Moreover, as $\left(s\hat{E}\eta_0 - c\hat{\mathbf{p}}\cdot\boldsymbol{\eta}\right)$ factorizes the d'Alembertian, each component of the wave function satisfies the relativistic wave equation.

As an example let us take Eq. (39) for $s = 1$, which was derived by combining two covariant Maxwell equations Eq. (28), therefore it should be also covariant. The wave function Eq. (38) has a zero component which results from the sum of two covariant equations, therefore it will remain zero in any frame. The representation under which the wave function Eq. (38) transforms is $D^{(1/2,1/2)}$ which has basis of spin 0 (one component) and spin 1 (3 components). As the photon is of spin 1, the spin zero component has to be eliminated. This is the reason of the zero component in Eq. (38) and it has to stay this way in all frames. The spin one 3 component wave function Eq. (18), Eq. (19) or Eq. (48) (proportional to $\mathbf{E} + \mathbf{iH}$) is also covariant as was shown by Laporte and Uhlenbeck[[9]], the 3 components belong to a second rank symmetric spinor.

Similar situation appears in handling the "graviton". The 8 component wave function transforms under $D^{(3/2,1/2)}$ which has basis of spin 1 (3 components) and spin 2 (5 components). In order to eliminate the spin one contribution the wave function has to have 3 zeros.

12. Summary and conclusions

We have shown above, how the quantum photon equation and the resulting Maxwell's equations can be derived from first principles, similar to those which have been used to derive the Dirac relativistic electron equation. We have worked out a general method of deriving quantum equations for massless particles of any spin, based on the factorization of the d'Alembertian. We have applied this procedure for the spin one (photon) and the spin two (graviton) in two ways, in order to write free particle wave equations with explicit and implicit subsidiary conditions. To this aim the factorization was achieved by using "relativistic quaternions" and their matrix representations (See Sec.6).

The unexplained substitution of the photon wavefunction Ψ in terms of the electric $\mathbf{E}$ and the magnetic field $\mathbf{B}$,

$$\mathbf{\Psi}=\mathcal{N}\left(\mathbf{E}+i\mathbf{B}\right), \tag{158}$$

leads to Maxwell's equations. Thus Maxwell's equations can be considered as the first quantized one-photon quantum equation. This fact was not realized for a long time, because Maxwell's equations do not contain the Planck constant $\hbar$. The Planck constant $\hbar$ and the normalization constant $\mathcal{N}$ are cancelled out in Eqs. (18) and (19).

Maxwell's equations were derived using different representations of relativistic quaternions, i.e., in terms of α and β matrices, Eqs. (39) and (127), respectively. It was demonstrated that these equations can be obtained from scalar Lagrangians and each one of them leads to Maxwell's equations. In addition a conserved probability current Eq. (49) was derived, which allowed definition of the probability density and wavefunction normalization. The energy-momentum tensors were calculated and found to be consistent with the quantum expectation values of the energy and momentum operators.

The quantum graviton equation was derived first with explicit subsidiary conditions, Eqs. (136) and (139). Rather than using the z subsidiary condition, Eqs. (137), we defined normalized subsidiary conditions, Eqs. (141, 142), which allowed construction of a single quantum graviton equation Eq. (149) in terms of 8×8 γ matrices Eqs. (144-147). The 8×8 γ matrices form a representation of the relativistic quaternions and in this way allow the factorization of the d'Alembertian. Moreover the subsidiary conditions are implicitly included in the new equation. We have included in the 8-component wavefunction Eq. (143) 3 constant components $\lambda_1, \lambda_0, \lambda_{-1}$ which allowed us to construct a scalar Lagrangian from which all the equations can be derived. A conserved probability current Eq. (153) was derived, which allowed to define the probability density and wavefunction normalization. The energy-momentum tensor was found to be consistent with the quantum expectation values of the energy and momentum operators.

Journal of Physics: Conference Series **330** (2011) 012010 doi:10.1088/1742-6596/330/1/012010

In this way we have shown that the quantum graviton equation Eq. (149) satisfies the basic requirements of a quantum theory.

In summary, we have developed a formalism of quantum equations for massless particles of arbitrary spin and applied it to spin one (photon) and spin two (graviton). The formalism allowed us to formulate scalar Lagrangians, derive probability currents and wavefunction normalization and to understand Maxwell's equations as first quantized photon equation. The same procedures were applied to the quantum graviton. It remains to understand the connection between the quantum graviton and gravitation.

Appendix A. Conserved currents of probability

The photon equation Eq.(10) and Eq.(27), satisfies,

$$\left(\frac{\hat{E}}{c}I^{(3)} - \hat{\mathbf{p}}\cdot\mathbf{S}\right)\mathbf{\Psi} = 0, \tag{A.1}$$

$$\left(\frac{\hat{E}}{c}I^{(3)} + \hat{\mathbf{p}}\cdot\mathbf{S}\right)\mathbf{\Psi}^* = 0. \tag{A.2}$$

From Eqs.(A.1-A.2) we have,

$$\mathbf{\Psi}^H\left(\frac{\hat{E}}{c}I^{(3)} - \hat{\mathbf{p}}\cdot\mathbf{S}\right)\mathbf{\Psi} + \mathbf{\Psi}^T\left(\frac{\hat{E}}{c}I^{(3)} + \hat{\mathbf{p}}\cdot\mathbf{S}\right)\mathbf{\Psi}^*$$

$$= \frac{\hat{E}}{c}\left(\mathbf{\Psi}^H\mathbf{\Psi}\right) - \mathbf{\Psi}^H\left(\hat{\mathbf{p}}\cdot\mathbf{S}\right)\mathbf{\Psi} + \mathbf{\Psi}^T\left(\hat{\mathbf{p}}\cdot\mathbf{S}\right)\mathbf{\Psi}^*. \tag{A.3}$$

Let us show that,

$$\mathbf{\Psi}^H\left(\hat{\mathbf{p}}\cdot\mathbf{S}\right)\mathbf{\Psi} - \mathbf{\Psi}^T\left(\hat{\mathbf{p}}\cdot\mathbf{S}\right)\mathbf{\Psi}^* = \hat{\mathbf{p}}\cdot\left(\mathbf{\Psi}^H\mathbf{S}\mathbf{\Psi}\right). \tag{A.4}$$

Indeed,

$$\mathbf{\Psi}^H\left(\hat{\mathbf{p}}\cdot\mathbf{S}\right)\mathbf{\Psi} - \mathbf{\Psi}^T\left(\hat{\mathbf{p}}\cdot\mathbf{S}\right)\mathbf{\Psi}^* =$$

$$\left[\Psi_y^*\left(-i\hat{p}_x\Psi_z\right) + \Psi_z^*\left(i\hat{p}_x\Psi_y\right) + \Psi_y\left(i\hat{p}_x\Psi_z^*\right) + \Psi_z\left(-i\hat{p}_x\Psi_y^*\right)\right]$$

$$+\left[\Psi_x^*\left(i\hat{p}_y\Psi_z\right) + \Psi_z^*\left(-i\hat{p}_y\Psi_x\right) + \Psi_x\left(-i\hat{p}_y\Psi_z^*\right) + \Psi_z\left(i\hat{p}_y\Psi_x^*\right)\right]$$

$$+\left[\Psi_x^*\left(-i\hat{p}_z\Psi_y\right) + \Psi_y^*\left(i\hat{p}_z\Psi_x\right) + \Psi_x\left(i\hat{p}_z\Psi_y^*\right) + \Psi_y\left(-i\hat{p}_z\Psi_x^* + i\hat{p}_x\Psi_z^*\right)\right]$$

$$= \hat{p}_x\left[i\left(\Psi_z^*\Psi_y\right) - i\left(\Psi_y^*\Psi_z\right)\right] + \hat{p}_y\left[i\Psi_x^*\Psi_z - i\Psi_z^*\Psi_x\right] + \hat{p}_z\left[i\Psi_y^*\Psi_x - i\Psi_x^*\Psi_y\right]$$

$$= \hat{p}_x\left(\mathbf{\Psi}^H\mathbf{S}_x\mathbf{\Psi}\right) + \hat{p}_y\left(\mathbf{\Psi}^H\mathbf{S}_y\mathbf{\Psi}\right) + \hat{p}_z\left(\mathbf{\Psi}^H\mathbf{S}_z\mathbf{\Psi}\right) = \hat{\mathbf{p}}\cdot\left(\mathbf{\Psi}^H\mathbf{S}\mathbf{\Psi}\right).$$

Finally from Eq.(A.3) and Eq.(A.4) one obtains,

$$\frac{\hat{E}}{c}\left(\mathbf{\Psi}^H\mathbf{\Psi}\right) - \hat{\mathbf{p}}\cdot\left(\mathbf{\Psi}^H\mathbf{S}\mathbf{\Psi}\right) = 0, \tag{A.5}$$

or

$$\partial_t\left(\mathbf{\Psi}^H\mathbf{\Psi}\right) + c\mathbf{\nabla}\cdot\left(\mathbf{\Psi}^H\mathbf{S}\mathbf{\Psi}\right) = 0. \tag{A.6}$$

Eq.(A.6) is the conserved current of probability and $\mathbf{\Psi}^H\mathbf{\Psi}$ is interpreted as the density of probability which should be normalized to unity.

Appendix B. Complex conjugation of spherical tensors

In this appendix we relate spherical tensors to Cartesian ones and evaluate the outcome of complex conjugation of spherical tensors. For spherical vectors $\mathbf{V}^{(1)}$ (tensors of rank 1) we have,

$$\mathbf{V}_m^{(1)} \equiv \begin{pmatrix} V_1^{(1)} \\ V_0^{(1)} \\ V_{-1}^{(1)} \end{pmatrix} = \begin{pmatrix} -\frac{1}{\sqrt{2}}\left(V_x + iV_y\right) \\ V_z \\ \frac{1}{\sqrt{2}}\left(V_x - iV_y\right) \end{pmatrix}, \tag{B.1}$$

where V_x, V_y, V_z are the Cartesian components of the vector $\mathbf{V}$,

$$\mathbf{V}_i \equiv \begin{pmatrix} V_x \\ V_y \\ V_z \end{pmatrix} = \begin{pmatrix} -\frac{1}{\sqrt{2}}\left(V_1^{(1)} - V_{-1}^{(1)}\right) \\ \frac{i}{\sqrt{2}}\left(V_1^{(1)} + V_{-1}^{(1)}\right) \\ V_z \end{pmatrix}. \tag{B.2}$$

Let us construct the spherical tensor $\mathbf{V}_m^{C(1)}$ out of the complex conjugates of $\mathbf{V}_i$,

$$\mathbf{V}_m^{C(1)} \equiv \begin{pmatrix} -\frac{1}{\sqrt{2}}\left(V_x^* + iV_y^*\right) \\ V_z^* \\ \frac{1}{\sqrt{2}}\left(V_x^* - iV_y^*\right) \end{pmatrix}. \tag{B.3}$$

Taking the complex conjugate of Eq.(B.1) and the definition Eq.(B.3) one finds,

$$\left(V_m^{(1)}\right)^* = (-1)^m V_{-m}^{C(1)}, \quad m = 1, 0, -1. \tag{B.4}$$

The Hermitian conjugate of $\mathbf{V}^{(1)}$ can be found from Eqs.(B.1-B.4),

$$\left(\mathbf{V}_m^{(1)}\right)^H \equiv \begin{pmatrix} V_1^{(1)} \\ V_0^{(1)} \\ V_{-1}^{(1)} \end{pmatrix}^H = \begin{pmatrix} \left(V_1^{(1)}\right)^* & \left(V_0^{(1)}\right)^* & \left(V_{-1}^{(1)}\right)^* \end{pmatrix} \tag{B.5}$$

$$= \begin{pmatrix} -V_{-1}^{C(1)} & V_0^{C(1)} & -V_1^{C(1)} \end{pmatrix}.$$

We also find that,

$$\mathrm{Re}\left(V_m^{(1)}\right) = \frac{1}{2}\left(V_m^{(1)} + (-1)^m V_{-m}^{C(1)}\right), \tag{B.6}$$

$$\mathrm{Im}\left(V_m^{(1)}\right) = \frac{-i}{2}\left(V_m^{(1)} - (-1)^m V_{-m}^{C(1)}\right), \quad m = 1, 0, -1. \tag{B.7}$$

For tensors of rank 2 $\mathbf{T}_m^{(2)}$ the procedure is similar. Their relation to a traceless symmetric Cartesian tensors ($S_{11} + S_{22} + S_{33} = 0$) is,

$$\mathbf{T}_m^{(2)} \equiv \begin{pmatrix} T_2^{(2)} \\ T_1^{(2)} \\ T_0^{(2)} \\ T_{-1}^{(2)} \\ T_{-2}^{(2)} \end{pmatrix} = \begin{pmatrix} \sqrt{\frac{3}{2}}\left(S_{11} - S_{22} + 2iS_{12}\right) \\ -\sqrt{6}\left(S_{13} + iS_{23}\right) \\ 2S_{33} - S_{11} - S_{22} = -3\left(S_{11} + S_{22}\right) \\ \sqrt{6}\left(S_{13} - iS_{23}\right) \\ \sqrt{\frac{3}{2}}\left(S_{11} - S_{22} - 2iS_{12}\right) \end{pmatrix}, \tag{B.8}$$

Journal of Physics: Conference Series **330** (2011) 012010 doi:10.1088/1742-6596/330/1/012010

where S_{ij} are the components of the corresponding traceless Cartesian tensor. The inverse relations are,

$$S_{11} = \frac{1}{\sqrt{24}}\left(T_2^{(2)} + T_{-2}^{(2)}\right) - \frac{1}{6}T_0^{(2)},$$

$$S_{22} = -\frac{1}{\sqrt{24}}\left(T_2^{(2)} + T_{-2}^{(2)}\right) - \frac{1}{6}T_0^{(2)},$$

$$S_{12} = S_{21} = -i\frac{1}{\sqrt{24}}\left(T_2^{(2)} - T_{-2}^{(2)}\right),$$

$$S_{13} = S_{31} = -\frac{1}{\sqrt{24}}\left(T_1^{(2)} - T_{-1}^{(2)}\right),$$

$$S_{23} = S_{32} = i\frac{1}{\sqrt{24}}\left(T_1^{(2)} + T_{-1}^{(2)}\right),$$

$$S_{33} = -S_{11} - S_{22} = \frac{1}{3}T_0^{(2)}.$$

Let us construct the spherical tensor $\mathbf{T}_m^{C(2)}$ out of the complex conjugates of $\mathbf{S}_{ij}$,

$$\mathbf{T}_m^{C(2)} \equiv \begin{pmatrix} T_2^{C(2)} \\ T_1^{C(2)} \\ T_0^{C(2)} \\ T_{-1}^{C(2)} \\ T_{-2}^{C(2)} \end{pmatrix} = \begin{pmatrix} \sqrt{\frac{3}{2}}\left(S_{11}^* - S_{22}^* + 2iS_{12}^*\right) \\ -\sqrt{6}\left(S_{13}^* + iS_{23}^*\right) \\ 2S_{33}^* - S_{11}^* - S_{22}^* \\ \sqrt{6}\left(S_{13}^* - iS_{23}^*\right) \\ \sqrt{\frac{3}{2}}\left(S_{11}^* - S_{22}^* - 2iS_{12}^*\right) \end{pmatrix}, \tag{B.9}$$

Taking the complex conjugate of Eq.(B.8) and the definition Eq.(B.9) one finds,

$$\left(T_m^{(2)}\right)^* = (-1)^m T_{-m}^{C(2)}, \quad m = 2, 1, 0, -1, -2. \tag{B.10}$$

and,

$$\mathrm{Re}\left(T_m^{(2)}\right) = \frac{1}{2}\left(T_m^{(2)} + (-1)^m T_{-m}^{C(2)}\right), \tag{B.11}$$

$$\mathrm{Im}\left(T_m^{(2)}\right) = \frac{-i}{2}\left(T_m^{(2)} - (-1)^m T_{-m}^{C(2)}\right), \quad m = 2, 1, 0, -1, -2. \tag{B.12}$$

The Hermitian conjugate of $\mathbf{T}_m^{(2)}$ can be found from Eqs.(B.8-B.10),

$$\begin{pmatrix} T_2^{(2)} \\ T_1^{(2)} \\ T_0^{(2)} \\ T_{-1}^{(2)} \\ T_{-2}^{(2)} \end{pmatrix}^H = \left(\left(T_2^{(2)}\right)^* \; \left(T_1^{(2)}\right)^* \; \left(T_0^{(2)}\right)^* \; \left(T_{-1}^{(2)}\right)^* \; \left(T_{-2}^{(2)}\right)^* \right)$$

$$= \left(T_{-2}^{C(2)} \; -T_{-1}^{C(2)} \; T_0^{C(2)} \; -T_1^{C(2)} \; T_2^{C(2)} \right). \tag{B.13}$$

Appendix C. Properties of the γ matrices

The matrix P Eq. (142) can be presented as a scalar product of the matrices Π_x, Π_y, Π_z, with the vector $\hat{p}_x, \hat{p}_y, \hat{p}_z$,

$$P = \hat{\mathbf{p}} \cdot \mathbf{\Pi}, \tag{C.1}$$

where,

$$\Pi_x = \begin{pmatrix} 0 & 0 & 0 & 0 & 0 \\ \sqrt{3} & 0 & -\frac{1}{\sqrt{2}} & 0 & 0 \\ 0 & \sqrt{\frac{3}{2}} & 0 & -\sqrt{\frac{3}{2}} & 0 \\ 0 & 0 & \frac{1}{\sqrt{2}} & 0 & -\sqrt{3} \\ 0 & 0 & 0 & 0 & 0 \end{pmatrix}, \tag{C.2}$$

$$\Pi_y = i \begin{pmatrix} 0 & 0 & 0 & 0 & 0 \\ -\sqrt{3} & 0 & -\frac{1}{\sqrt{2}} & 0 & 0 \\ 0 & -\sqrt{\frac{3}{2}} & 0 & -\sqrt{\frac{3}{2}} & 0 \\ 0 & 0 & -\frac{1}{\sqrt{2}} & 0 & -\sqrt{3} \\ 0 & 0 & 0 & 0 & 0 \end{pmatrix}, \tag{C.3}$$

$$\Pi_z = \begin{pmatrix} 0 & 0 & 0 & 0 & 0 \\ 0 & -\sqrt{3} & 0 & 0 & 0 \\ 0 & 0 & -2 & 0 & 0 \\ 0 & 0 & 0 & -\sqrt{3} & 0 \\ 0 & 0 & 0 & 0 & 0 \end{pmatrix}. \tag{C.4}$$

The $\gamma_x, \gamma_y, \gamma_z$,matrices are constructed according to,

$$\gamma_k = \frac{1}{2} \begin{pmatrix} \begin{pmatrix} -\tilde{S}_k^{(1)} \end{pmatrix} & \begin{pmatrix} \Pi_k \end{pmatrix} \\ \begin{pmatrix} \Pi_k^H \end{pmatrix} & \begin{pmatrix} \tilde{S}_k^{(2)} \end{pmatrix} \end{pmatrix}, \quad k = x, y, z \tag{C.5}$$

where $\tilde{S}_k^{(2)}$ are spin 2 matrices Eqs. (133), 135) and $\tilde{S}_k^{(1)}$ are spin 1 matrices Eqs. (118).

Appendix D. Spherical tensor formalism

We follow the definition of the scalar product for irreducible spherical tensors of rank J. Let $A_m \equiv \langle A|Jm\rangle$, and $B_m \equiv \langle B|Jm\rangle$, $-J \leq m \leq J$, be such tensors, the scalar product is defined[10] according to,

$$\langle A|B\rangle \equiv (A \cdot B) = \sum_{m=-J}^{J} \langle A|Jm\rangle\langle Jm|B\rangle \tag{D1}$$

$$= \sum_{m=-J}^{J} (-1)^m \langle A|Jm\rangle\langle B| -m, J\rangle = \sum_{m=-J}^{J} (-1)^m A_m B_{-m}, \tag{D2}$$

Journal of Physics: Conference Series **330** (2011) 012010 doi:10.1088/1742-6596/330/1/012010

from which we infer that,

$$\sum (-1)^m \left|J,m\right> < J,-m| = \sum (-1)^m \left|J,-m\right> < J,m| = I^{(J)}, \tag{D.1}$$

is the identity operator in the subspace of given J. Let us denote,

$$(-1)^m \left|J,-m\right> = \left|J,-\tilde{m}\right>, \tag{D.2}$$

hence,

$$\sum \left|J,-\tilde{m}\right> < J,m| = I^{(J)}. \tag{D.3}$$

Within this formalism,

$$< J,-\tilde{m}_2|(W^\mu + s\hat{p}^\mu I^{(s)})|\Psi > \tag{D.4}$$

$$= \sum_{m_1} < J,-\tilde{m}_2|(W^\mu + s\hat{p}^\mu I^{(s)})|J,-\tilde{m}_1 >< J,m_1|\Psi >= 0,$$

where $< J,m_1|\Psi >$ is a column vector. The matrix elements of the spin operators satisfy,

$$\tilde{S}_{m_2,m_1} =< J,-\tilde{m}_2|S_x|J,-\tilde{m}_1 >= - < J,m_2|S_x|J,m_1 >, \ i.e.\ \tilde{S}_x = -S_x. \tag{D.5}$$

In a similar way we obtain,

$$\tilde{S}_y = S_y, \quad and \quad \tilde{S}_z = -S_z. \tag{D.6}$$

For practical calculations we introduce,

$$S_+ = S_x + iS_y,\ S_- = S_x - iS_y, S_x = \frac{1}{2}\left(S_+ + S_-\right),\ S_y = \frac{-i}{2}\left(S_+ - S_-\right), \tag{D.7}$$

with their matrix elements given by,

$$\left\langle s,m+1\right| S_+ \left|sm'\right\rangle = \sqrt{(s-m)\,(s+m+1)}\delta_{mm'}, \tag{D.8}$$

$$\left\langle s,m-1\right| S_- \left|sm'\right\rangle = \sqrt{(s+m)\,(s-m+1)}\delta_{mm'}, \tag{D.9}$$

$$\left\langle sm\right| S_z \left|sm'\right\rangle = m\delta_{mm'}, \quad -s \le m \le s. \tag{D.10}$$

The momentum vector $\hat{\mathbf{p}}$ in this representation is,

$$< 1,m|\hat{\mathbf{p}} >= \begin{pmatrix} \hat{p}_1 \\ \hat{p}_0 \\ \hat{p}_{-1} \end{pmatrix} = \begin{pmatrix} -\frac{1}{\sqrt{2}}\left(\hat{p}_x + i\hat{p}_y\right) \\ \hat{p}_z \\ \frac{1}{\sqrt{2}}\left(\hat{p}_x - i\hat{p}_y\right) \end{pmatrix}. \tag{D.11}$$

The inverse relation is,

$$\begin{pmatrix} \hat{p}_x \\ \hat{p}_y \\ \hat{p}_z \end{pmatrix} = \begin{pmatrix} -\frac{1}{\sqrt{2}}\hat{p}_1 + \frac{1}{\sqrt{2}}\hat{p}_{-1} \\ \frac{i}{\sqrt{2}}\hat{p}_1 + \frac{i}{\sqrt{2}}\hat{p}_{-1} \\ \hat{p}_0 \end{pmatrix}. \tag{D.12}$$

References

[1] Gersten A 2001 *Found. Phys.* **31** 1211-1231 and references therein.
[2] Gersten A 1999 *Found. Phys. Lett.* **12** 291-298.
[3] Jackson J D 1999 *Classical Electrodynamics 3rd ed.* John Wiley & Sons New York 556.
[4] Dirac P A M 1936 *Proc. Roy. Soc.* **A155** 447-59.
[5] Bacry H 1976 *Nuovo Cimento* **A32** 448-60.
[6] Wigner E P 1939 *Ann. Math.* **40** 149.
[7] Gersten A 2000 *Found. Phys. Lett.* **13** 185-192.
[8] Lomont J S 1958 *Phys. Rev.* **111** 1710-1716.
[9] Laporte O and Uhlenbeck G E 1931 *Phys. Rev.* **37** 1380-1397.
[10] de-Shalit A and Talmi I 1963 *Nuclear Shell Theory* Academic Press New York.

7th Biennial Conf. on Classical and Quantum Relativistic Dynamics of Particles and Fields IOP Publishing
Journal of Physics: Conference Series **330** (2011) 012011 doi:10.1088/1742-6596/330/1/012011

On the Stueckelberg-like generalization of general relativity

Matej Pavšič
Jožef Stefan Institute, Jamova 39, SI-1000, Ljubljana, Slovenia
E-mail: matej.pavsic@ijs.si

Abstract. We first consider the Klein-Gordon equation in the 6-dimensional space $M_{2,4}$ with signature $+----+$ and show how it reduces to the Stueckelberg equation in the 4-dimensional spacetime $M_{1,3}$. A field that satisfies the Stueckelberg equation depends not only on the four spacetime coordinates x^μ, but also on an extra parameter τ, the so called evolution time. In our setup, τ comes from the extra two dimensions. We point out that the space $M_{2,4}$ can be identified with a subspace of the 16-dimensional Clifford space, a manifold whose tangent space at any point is the Clifford algebra Cl(1,3). Clifford space is the space of oriented r-volumes, $r = 0, 1, 2, 3$, associated with the extended objects living in $M_{1,3}$. We consider the Einstein equations that describe a generic curved space $M_{2,4}$. The metric tensor depends on six coordinates. In the presence of an isometry given by a suitable Killing vector field, the metric tensor depends on five coordinates only, which include τ. Following the formalism of the canonical classical and quantum gravity, we perform the 4 + 1 decomposition of the 5-dimensional general relativity and arrive, after the quantization, at a generalized Wheeler-DeWitt equation for a wave functional that depends on the 4-metric of spacetime, the matter coordinates, and τ. Such generalized theory resolves some well known problems of quantum gravity, including "the problem of time".

1. Introduction

1.1. The problem of time in quantum gravity

Despite being a very successful theory at the classical level, general relativity has turned out to be problematic when attempted to be consistently quantized. Amongst others, there is the so called 'problem of time' (for a recent review see [1]). This can be seen if we perform the canonical quantization. If we start from the Einstein-Hilbert action, and perform the 1 + 3 Arnowith-Deser-Misner (ADM) decomposition of spacetime, $M_{1,3} = \mathbb{R} \times \Sigma$, then the action of general relativity can be cast into the 'phase space' form [2, 3]

$$I[q_{ij}, p^{ij}, N, N^i] = \int dt\, d^3x\, \left[p^{ij}\dot{q}_{ij} - N\, H(q_{ij}, p^{ij}) - N_i H^i(q_{ij}, p^{ij})\right]. \tag{1}$$

Here q_{ij}, $i, j = 1, 2, 3$, is a 3-metric on a space hypersurface Σ, and p^{ij} is the corresponding canonically conjugate momentum, whilst N and N_i are, respectively, laps and shift functions having the role of Lagrange multipliers leading to the constraints

$$\mathcal{H}(q_{ij}, p^{ij}) \approx 0, \quad \text{and} \quad \mathcal{H}^i(q_{ij}, p^{ij}) \approx 0, \tag{2}$$

Journal of Physics: Conference Series **330** (2011) 012011 doi:10.1088/1742-6596/330/1/012011

which are associated with the diffeomorphism invariance of the original Einstein-Hilbert action. The Hamiltonian is a linear combination of constraints and the evolution is a pure gauge. There is no physical evolution time in such a theory.

Upon quantization, the above constraints become the wave functional equations. For instance, the first constraint becomes the Wheeler-DeWitt equation

$$\mathcal{H}(q_{ij}, -i\tfrac{\delta}{\delta q_{ij}})\,\Psi[q_{ij}] = 0 \tag{3}$$

whilst the second set of constraints become

$$\mathcal{H}^i(q_{ij}, -i\tfrac{\delta}{\delta q_{ij}})\,\Psi[q_{ij}] = 0. \tag{4}$$

We see that in quantum theory there is no spacetime, but only space Σ, , because the wave function(al) depends only on 3-geometry, represented by q_{ij}. Thus, in addition to the absence of an external time, we have also the problem of the disappearance of spacetime.

1.2. A possible remedy: the Stueckelberg theory

In the Stueckelberg theory [4], besides the four spacetime coordinates x^μ, there is an extra parameter τ. The coordinate $x^0 \equiv t$ is not the 'evolution parameter'. The evolution parameter is τ, considered to be a universal "world time".

In quantum theory of a 'point particle', the wave function is

$$\psi(\tau, x^\mu), \tag{5}$$

and is normalized according to $\int \mathrm{d}^4x\,\psi^*\psi = 1$. We will show how τ arises from extra two dimensions, one space like and one time like dimension.

Then we will show that 'extra dimensions' need not be the 'true' extra dimensions, i.e., some extra dimensions in addition to four spacetime dimensions, but can be associated with the space of matter configurations. A particular case of such configuration space is Clifford space, a manifold whose tangent space at any point is the Clifford algebra Cl(1,3). Clifford space is the space of oriented r-volumes, $r = 0, 1, 2, 3$, associated with the extended objects living in $M_{1,3}$. In this paper we focus our attention to a 6-dimensional subspace, $M_{2,4}$, of Clifford space. We consider the Einstein equations that describe a generic curved space $M_{2,4}$. Then we perform the ADM-like 1+4 decomposition of a 5-dimensional subspace $M_{2,3}$ of $M_{2,4}$, our argument being that the additional dimension can be neglected in the presence of an isometry given by a suitable Killing vector field, because then the metric tensor depends on five coordinates only.

We will show how in the quantized theory the problems of time and of spacetime disappear in such a generalized theory. The latter problem does not occur, because the wave functional now depends on spacetime 4-geometry, represented by the metric $g_{\mu\nu}$. The problem of time we resolve by adding a suitable matter part to the action.

2. Klein-Gordon equation in 6D

Let us consider the action for the massless Klein-Gordon field in 6-dimensions:

$$I[\phi, \phi^*] = \int \mathrm{d}^6x\; \partial_M\phi^*\,\partial^M\phi \tag{6}$$

where $\phi = \phi(x^M)$, $M = 0, 1, 2, 3, 5, 6$. Let us split the index M into a 4-dimensional part and the part due to the extra two dimensions according to $M = (\mu, \bar{M})$, $\mu = 0, 1, 2, 3$, $\bar{M} = 5, 6$, and let us assume that the metric has the following form:

$$G_{MN} = \begin{pmatrix} g_{\mu\nu} & 0 & 0 \\ 0 & 0 & -1 \\ 0 & -1 & 0 \end{pmatrix}. \tag{7}$$

The latter metric can be transformed into

$$G'_{MN} = \begin{pmatrix} g_{\mu\nu} & 0 & 0 \\ 0 & -1 & 0 \\ 0 & 0 & 1 \end{pmatrix}, \tag{8}$$

which is the pseudo euclidean metric with signature $(+ - - - - +)$. By inserting the metric (7) into the action (6), we obtain

$$I[\phi, \phi^*] = \int \mathrm{d}^6 x \, (g^{\mu\nu} \, \partial_\mu \phi^* \partial_\nu \phi \; - \; \partial_\tau \phi^* \partial_\lambda \phi \; - \; \partial_\lambda \phi^* \partial_\tau \phi), \tag{9}$$

where we have denoted $x^5 \equiv \tau$, $x^6 \equiv \lambda$.

Taking the ansatz

$$\phi = e^{i\Lambda\lambda} \psi(\tau, x^\mu), \tag{10}$$

where Λ is a constant, we have

$$I[\psi, \psi^*] = \int \mathrm{d}\tau \, \mathrm{d}^4 x \;\; [\partial_\mu \psi^* \partial^\mu \psi \; + \; i\Lambda \, (\psi^* \partial_\tau \psi \; - \; \partial_\tau \psi^* \psi)] \, , \tag{11}$$

which is the well known Stueckelberg action. We have omitted the integration over λ, because it gives a constant factor which can be absorbed into the definition of the fields.

Considering the corresponding equations of motion, we find that from a massless Klein-Gordon equation in 6D

$$\partial^M \partial_M \phi = 0 \tag{12}$$

we obtain the Stueckelberg equation

$$i \, \partial_\tau \psi \;\; = \;\; \frac{1}{2\Lambda} \, \partial^\mu \partial_\mu \psi \tag{13}$$

The constant Λ comes from the 6th dimension $x^6 \equiv \lambda$. More precisely, Λ is an eigenvalue of the canonical momentum conjugate to λ. By ansatz (10), coordinate λ is eliminated from the action, whilst the eigenvalue Λ remains.

To sum up, if the signature of two extra dimensions is $(-+)$, and if, instead in the coordinates x'^5, x'^6 in which the metric is diagonal (Eq. (8)), we work in the coordinates $x^5 \equiv \tau = \frac{1}{2}(x'^5 + x'^6)$, $x^6 \equiv \lambda = \frac{1}{2}(x'^5 - x'^6)$ in which the metric is non diagonal (Eq. (7)), then we obtain the Stueckelberg equation for a wave function that depends on τ and x^μ. The coordinates τ, λ are analogous to the light cone coordinates $(x^0 + x^1)/2$, $(x^0 - x^1)/2$. Notice that, because τ is like a 'light cone' coordinate, we have the first derivative of ψ with respect to τ.

2.1. More formal considerations: Point particle in 6D and its quantization

Let us consider a classical action for a point particle in 6-dimensional space:

$$I[X^M] = M_p \int d\sigma (\dot{X}^M \dot{X}_M)^{1/2}, \tag{14}$$

where $M = 0, 1, 2, 3, 5, 6$, and M_p is the particle's mass in 6D. Here σ is a parameter, denoting a point on the worldline, and $\dot{X}^\mu = \mathrm{d}X^\mu / \mathrm{d}\sigma$.

An equivalent action is a functional of the coordinates X^M, the canonically conjugate momenta P_M, and a Lagrange multiplier α:

$$I[X^M, P_M, \alpha] \; = \; \int d\sigma \left(P_M \dot{X}^M \; - \; \frac{\alpha}{2} (P_M P^M \; - \; M_p^2) \right). \tag{15}$$

Journal of Physics: Conference Series **330** (2011) 012011 doi:10.1088/1742-6596/330/1/012011

Varying the latter action with respect to P_M, we obtain the relation between velocities and momenta, $\dot{X}^M = \alpha P^M$.

If we split the coordinates according to

$$X^M = (X^\mu, X^{\bar{M}}) \,, \qquad \bar{M} = 5, 6, \tag{16}$$

and express four momenta in terms of velocities, $P^\mu = \dot{X}^\mu/\alpha$, then the action (15) becomes

$$I[X^\mu] = \int d\sigma \left(\frac{\dot{X}^\mu \dot{X}_\mu}{2\alpha} + P_{\bar{M}} \dot{X}^{\bar{M}} - \frac{\alpha}{2}(P_{\bar{M}} P^{\bar{M}} - M_p^2) \right). \tag{17}$$

The second term in the latter action can be omitted, because by partial integration it can be transformed into the form

$$\int \mathrm{d}\sigma \left(\frac{\mathrm{d}}{\mathrm{d}\sigma}(P_{\bar{M}} X^{\bar{M}}) - \dot{P}_{\bar{M}} X^{\bar{M}} \right), \tag{18}$$

and if we use the equations of motion, $\dot{P}_{\bar{M}} = 0$, then only the total derivative term remains.

The third term in eq. (17) can be rewritten in terms of the 4D mass, m. Namely, by varying (15) we obtain the mass shell constrain in 6D:

$$\delta\alpha : \; G^{MN} P_M P_N - M_p^2 = 0, \tag{19}$$

which can be decomposed according to

$$g^{\mu\nu} P_\mu P_\nu + G^{\bar{M}\bar{N}} P_{\bar{M}} P_{\bar{N}} - M_p^2 = 0. \tag{20}$$

From $M_p^2 = P^M P_M = P^\mu P_\mu + P_{\bar{M}} P^{\bar{M}}$, we have

$$m^2 = M_p^2 - P_{\bar{M}} P^{\bar{M}}, \tag{21}$$

where $m^2 \equiv P^\mu P_\mu$. If 6D mass M_p is equal to zero, then the 4D mass is due to the 5th and the 6th component of momentum only:

$$m^2 = -G^{\bar{M}\bar{N}} P_{\bar{M}} P_{\bar{N}} = 2P_5 P_6. \tag{22}$$

So, from eq. (17), using (18), and (21), we obtain the well known Howe-Tucker action for a massive particle in 4-dimensional spacetime:

$$I[X^\mu] = \tfrac{1}{2} \int d\sigma \left(\frac{\dot{X}^\mu \dot{X}_\mu}{\alpha} + \alpha m^2 \right). \tag{23}$$

Upon quantization, the classical constraint (19) becomes the Klein-Gordon equation

$$(G^{MN} \hat{P}_M \hat{P}_N - M_p^2)\phi = 0, \tag{24}$$

where $\hat{P}_M = -i\hbar\partial/\partial X^M$ is the momentum operators. We will use unit in which $\hbar = c = 1$, and write $\partial_M \equiv \partial/\partial X^M$.

We can decompose eq. (24) into a 4D and a 2D part:

$$(-g^{\mu\nu}\partial_\mu\partial_\nu - G^{\bar{M}\bar{N}}\partial_{\bar{M}}\partial_{\bar{N}} - M_p^2)\phi = 0, \tag{25}$$

Journal of Physics: Conference Series **330** (2011) 012011 doi:10.1088/1742-6596/330/1/012011

which gives

$$(-g^{\mu\nu}\partial_\mu\partial_\nu + 2\partial_5\partial_6 - M_p^2)\phi = 0. \tag{26}$$

By the ansatz

$$\phi = e^{iP_6x^6}\psi(x^5, x^\mu) \tag{27}$$

and by denoting $x^5 \equiv \tau$, $P_6 \equiv \Lambda$, eq. (26) gives

$$i\frac{\partial}{\partial\tau}\psi = \frac{1}{2\Lambda}\left(g^{\mu\nu}\partial_\mu\partial_\nu + M_p^2\right)\psi \tag{28}$$

If, in particular, the 6D mass M_p is zero, then we have the usual Stueclkelberg equation (13). Alternatively, M_p^2 in Eq. (28) can be eliminated by the phase change $\psi \to \exp[-i\frac{M_p^2}{2\Lambda}\tau]\psi$.

We have seen that the Stueckelberg equation in which the wave function depends not only on four spacetime coordinates x^μ, but also on an extra parameter τ (evolution parameter), is embedded in the 6D theory with one time-like and one space-like extra dimension:

6D space $M_{2,4}$ signature $+ - - - - +$	$\longrightarrow$	Stueckelberg theory in $M_{1,3}$ with invariant evolution parameter τ	(29)

At this point it is interesting to observe, that an extra time like and an extra space like dimension are necessary in the "two time" physics [5], based on the extended phase space action that is invariant under local Sp(2) transformations. A special case of the latter action is the phase space action for a relativistic point particle in 4-dimensions. Since our phase space action (15) is in six, and not in four dimensions, this means that in the considered 6-dimensional theory we do not not impose the Sp(2) constraints on X^M and P^M. We can envisage that such constraints are imposed in a space of a higher dimensionality, and that a particular, gauge fixed, case of the higher dimensional, Sp(2) invariant, action, is the phase space action (15). Thus, our approach differes from that in refs. [5]. We do not impose the Sp(2) constraints in $M_{2,4}$, but we admit the possibility that such constraints hold in a higher dimensional space within the context of a theory that is a modification of the two time physics [5]. In such a way it is possible to embed the Stueckelberg theory into the theory based on the local Sp(2) invariance.

What about ghosts? It is usually taken for granted that time like dimensions imply ghosts. But there is another, not so well known, possibility that is based on an alternative definition of vacuum [6], in which case no ghosts are associated with time like dimensions. How this works within the context of string theory and quantum field theory, and how this can resolve the cosmological constant problem, was shown in Refs. [7, 8, 9].

A question arises as to what is a physical meaning of the extra dimensions. This will be discuss in next section.

3. The space $M_{2,4}$ as a subspace of Clifford space

Clifford space, C, is the space of oriented r-volumes, $r = 0, 1, 2, 3$, associated with extended objects, such as strings/branes living in spacetime $M_{1,3}$. The concept of Clifford space—a manifold whose tangent space at any point is the Clifford algebra $Cl(1,3)$—has been discussed in refs. [10]–[18]. It was found that a curved C, since being a higher dimensional space, enables the unifications of interactions à la Kaluza-Klein without introducing the extra dimensions of spacetime. The 'extra dimensions' of C are due to the fundamental extended nature of physical objects, they are the dimensions of a configurations space. In principle, those degrees of freedom are not hidden from our direct observation, therefore we do not need to compactify such 'internal' space. Here we will exploit the fact that the space $M_{2,4}$, used in previous section, can be identified with a subspace of C.

3.1. Clifford space: a quenched configuration space of extended objects–branes

Strings and branes have infinitely many degrees of freedom. But at first approximation we can consider just the center of mass, x^μ, $\mu = 0,1,2,3$. Next approximation is in considering the holographic coordinates, $x^{\mu\nu}$, of the oriented area enclosed by the string. We may go even further and search for eventual thickness of the object. If the string has finite thickness, i.e., if actually it is not a string, but a 2-brane, then there exist the corresponding volume degrees of freedom, $x^{\mu\nu\rho}$.

In general, for an extended object in $M_{1,3}$, we have 16 coordinates [17, 18]

$$x^M \equiv x^{\mu_1 \ldots \mu_r}\,, \quad r = 0,1,2,3,4. \tag{30}$$

They are the projections of r-dimensional volumes (areas) onto the coordinate planes. Although branes have infinitely many degrees of freedom, we can sample them by a finite set of coordinates x^M that denote position in a 16-dimensional space. Let us first assume that the latter space is flat. Then the position can be described by a vector $x = x^M \gamma_M$, where x^M are components, and γ_M basis vectors. For the basis vectors we will take the basis elements of the Clifford algebra $Cl(1,3)$, thus $\gamma_M \equiv \gamma_{\mu_1} \wedge \gamma_{\mu_2} \wedge \ldots \wedge \gamma_{\mu_r}$, $r = 0,1,2,3,4$. The vector $x \in Cl(1,3)$, picturesquely called 'polyvector', is an aggregate of r-vectors, i.e., of scalars, vectors, bivectors, threevectors (pseudovectors) and fourvectors (pseudo scalars). We can now assume that the Clifford algebra $Cl(1,3)$ is a tangent space of a 16-dimensional manifold, called Clifford space C. If the manifold C is flat, then it is isomorphic to the Clifford algebra $Cl(1,3)$, which is the tangent space at a chosen point $P \in C$, say the "origin". In general, C can have non vanishing curvature, in which case it is not isomorphic to $Cl(1,3)$.

Coordinates x^M of Clifford space C can be used to model extended objects, whatever they are. The latter coordinates, the so called 'polyvector coordinates', are a generalization of the concept of center of mass [17]. Instead of describing extended objects in "full detail", we can describe them in terms of the center of mass, area and volume coordinates. Namely, a configuration of an extended object, such as a brane, has infinitely many degrees of fredom, and the space $\mathcal{M}$ of all possible brane configurations [9] is infinite dimensional. A full description of a brane corresponds to a point in $\mathcal{M}$ that requires infinitely many "coordinates", i.e., the brane embedding functions $X^\mu(\xi^a)$. A "quenched" description of a brane corresponds to a point in C that needs sixteen coordinates only, i.e., the coordinates x^M. Therefore, the Clifford space, C, is a quenched configuration space for extended objects [19].

Instead of the usual relativity, formulated in spacetime, in which the interval is

$$\mathrm{d}s^2 = g_{\mu\nu}\,\mathrm{d}x^\mu \mathrm{d}x^\nu \tag{31}$$

let us consider the theory in which the interval is extended to Clifford space:

$$\mathrm{d}S^2 = G_{MN}\,\mathrm{d}x^M \mathrm{d}x^N \tag{32}$$

where $\mathrm{d}x^M \equiv \mathrm{d}x^{\mu_1 \ldots \mu_r}$, $r = 0,1,2,3,4$.

In particular, extended objects can be fundamental branes.

The line element (32) can be written as the scalar product of the Clifford number

$$\mathrm{d}X = \mathrm{d}x^M \gamma_M \equiv \mathrm{d}x^{\mu_1\mu_2\ldots\mu_r}\gamma_{\mu_1\mu_2\ldots\mu_r}\,, \qquad r = 0,1,2,3,4 \tag{33}$$

with its reverse $\mathrm{d}X^\ddagger$:

$$\mathrm{d}S^2 \equiv |\mathrm{d}X|^2 \equiv \mathrm{d}X^\ddagger * \mathrm{d}X = \mathrm{d}x^M \mathrm{d}x^N G_{MN} \equiv \mathrm{d}x^M \mathrm{d}x_M. \tag{34}$$

The metric is given by the scalar product of the basis Clifford numbers:

$$G_{MN} = \gamma_M^\ddagger * \gamma_N \equiv \langle \gamma_M^\ddagger \gamma_N \rangle_0. \tag{35}$$

Journal of Physics: Conference Series **330** (2011) 012011 doi:10.1088/1742-6596/330/1/012011

Reversion, denoted by ‡, is an operation that reverses the order of vectors in a Clifford product: $(\gamma_{\mu_1}\gamma_{\mu_2}...\gamma_{\mu_r})^\ddagger = \gamma_{\mu_r}...\gamma_{\mu_2}\gamma_{\mu_1}$. In flat Clifford space, $\gamma_M \equiv \gamma_{\mu_1\mu_2...\mu_r}$ is the wedge product of basis vectors, $\gamma_M = \gamma_{\mu_1} \wedge \gamma_{\mu_2} \wedge ... \wedge \gamma_{\mu_r}$, at every point of C. This is not the case in curved C.

With the definition (35) of the metric, signature of C is $(8,8)$. Therefore, $M_{2,4}$ is a subspace of C.

3.2. Dynamics

The following action generalizes the action for a point particle of the ordinary special relativity:

$$I = M_p \int d\sigma\, (\eta_{MN}\dot{X}^M\dot{X}^N)^{1/2}, \tag{36}$$

where σ is an arbitrary continuous parameter. From the latter action we obtain the following equations of motion:

$$\ddot{X}^M \equiv \frac{\mathrm{d}^2X^M}{\mathrm{d}\sigma^2} = 0 \tag{37}$$

Here η_{MN} is the analogue of Minkowski metric with signature $(8,8)$.

Since X^M are interpreted as r-volume coordinates, the equations of motion (37) imply that the volume (in particular the area) changes linearly with σ. If the coordinates X^M sample a brane, then the above dynamics can only hold for a tensionless brane. For a brane with tension one has to introduce curved Clifford space and generalize eqs. (36),(37) to arbitrary metric with non vanishing curvature [12, 13, 14].

A worldline $X^M(\sigma)$ in C represents the evolution of a 'thick particle' in spacetime $M_{1,3}$. In C we have a line, a worldline $X^M(\sigma)$, whilst in spacetime $M_{1,3}$, we have a thick line whose centroid line is $X^\mu(\sigma)$. It describes a thick particle, i.e., an extended object, in spacetime. The thick particle can be an aggregate of p-branes for various $p = 0, 1, 2, \ldots$. But such interpretation is not obligatory. A thick particle may be a conglomerate of whatever extended objects that can be sampled by 'polyvector' coordinates $X^M \equiv X^{\mu_1\mu_2...\mu_r}$.

4. Einstein's equations in $M_{2,4}$

Let x^M, $M = 0,1,2,3,5,6$ be coordinates, and $G_{MN} = G_{MN}(x^M)$ a metric tensor in $M_{2,4}$. The Einstein-Hilbert action in the presence of a point like source[1] reads

$$I[X^M, G_{MN}] = M_p \int \mathrm{d}\sigma\, (\dot{X}^M\dot{X}^N G_{MN})^{1/2} + \frac{1}{16\pi\mathcal{G}} \int \mathrm{d}^6x\, \sqrt{-G}R^{(6)} \tag{38}$$

If we vary the latter action with respect to $X^M(\sigma)$, we obtain the geodesic equation,

$$\frac{1}{\sqrt{\dot{X}^2}}\frac{\mathrm{d}}{\mathrm{d}\sigma}\left(\frac{\dot{X}^M}{\sqrt{\dot{X}^2}}\right) + \Gamma^M_{JK}\frac{\dot{X}^J\dot{X}^K}{\dot{X}^2} = 0, \tag{39}$$

and if we vary it with respect to $G_{MN}(x^M)$, we obtain the Einstein equations,

$$R^{MN} - \frac{1}{2}G^{MN}R = 8\pi\,\mathcal{G} \int \mathrm{d}\sigma\, \delta^6(x - X(\sigma))\dot{X}^M\dot{X}^N \tag{40}$$

We can use eqs. (38),(40) as an approximation to a physical situation in which instead of the δ-distribution we have a distribution due to an extended source.

[1] In our interpretation of the space $M_{2,4}$ as a subspace of Clifford space C, which is a configuration space associated with an extended object, a point like source in $M_{2,4}$ is a thick source in 4D spacetime $M_{1,3}$.

Journal of Physics: Conference Series **330** (2011) 012011 doi:10.1088/1742-6596/330/1/012011

The 6D Ricci scalar can be written as

$$R^{(6)} = R^{(4)} + \text{ extrinsic curvature term}_{5,6}, \tag{41}$$

where the subscripts 5,6 mean that the extrinsic curvature is due to the presence of the 5th and 6th dimension. Instead of performing such ADM-like $2+4$ decomposition, we will follow an easier procedure. We will consider a $1+5$ decomposition in which case we have

$$R^{(6)} = R^{(5)} + \text{ extrinsic curvature term}_6 \tag{42}$$

If there exist suitable isometries in the 6D space $M_{2,4}$, and if we choose a suitable 5D subspace $M_{2,3}$, then the extrinsic curvature terms in eq. (42) can vanish. Namely, the extrinsic curvature term tells how the hypersurface is bended with respect to the emebdding space, and it can be bended so that the extrinsic curvature is zero[2]. We will assume that this is the case.

The 5D Ricci scalar, in turn, can also be decomposed in an analogous way:

$$R^{(5)} = R^{(4)} + \text{ extrinsic curvature term}_5 \tag{43}$$

In particular, let us consider the ADM-like 1+4 decomposition $M_{2,3} = \mathbb{R} \times M_{1,3}$, where $M_{1,3}$ is spacetime. Then the 5D metric can be decomposed as

$$G_{MN} = \begin{pmatrix} \mathcal{N}_\mu \mathcal{N}^\mu + \mathcal{N}^2, & \mathcal{N}_\mu \\ \mathcal{N}_\nu\,, & g_{\mu\nu} \end{pmatrix}, \qquad \begin{matrix} M,N = 0,1,2,3,5 \\ \mu,\nu = 0,1,2,3 \end{matrix} \tag{44}$$

where the indices M, N now assume five values only, and $\mathcal{N} = 1/\sqrt{G^{55}}$. The inverse metric is

$$G^{MN} = \begin{pmatrix} 1/\mathcal{N}^2, & \mathcal{N}^\mu/\mathcal{N}^2 \\ \mathcal{N}^\nu/\mathcal{N}^2, & g^{\mu\nu} + \mathcal{N}^\mu \mathcal{N}^\nu/\mathcal{N}^2 \end{pmatrix} \tag{45}$$

The extrinsic curvature is

$$K_{\mu\nu} = \mathcal{D}_\nu n_\mu = \frac{1}{2\mathcal{N}} \left(D_\nu \mathcal{N}_\mu + D_\mu \mathcal{N}_\nu - \frac{\partial g_{\mu\nu}}{\partial \tau} \right), \qquad \tau \equiv x^5. \tag{46}$$

Here $\mathcal{D}_\nu$ is the 5D covariant derivative, D_ν the 4D covariant derivative, and n_M the normal to $M_{1,3}$.

The 4D metric $g_{\mu\nu}$ depends not only on four spacetime coordinates x^μ, but also on an extra parameter τ.

Introducing

$$p^{\mu\nu} = \kappa \sqrt{-g}\,(K g^{\mu\nu} - K^{\mu\nu}), \tag{47}$$

where $\kappa = 1/(16\pi\mathcal{G})$, $g \equiv \det g_{\mu\nu}$, and $K \equiv g^{\mu\nu} K_{\mu\nu}$, we can write the 5D action in the 'phase space' form:

$$I_G[g_{\mu\nu}, p^{\mu\nu}, \mathcal{N}, \mathcal{N}^\mu] = \int d\tau\, d^4x\, [p^{\mu\nu}\, \dot{g}_{\mu\nu} - \mathcal{N}\mathcal{H}(g_{\mu\nu}, p^{\mu\nu}) - \mathcal{N}_\mu \mathcal{H}^\mu(g_{\mu\nu}, p^{\mu\nu})], \tag{48}$$

where

$$\mathcal{H} = 2\kappa \sqrt{-g}\,\mathcal{N}^2 G^{55} = \kappa \sqrt{-g}\,(R^{(4)} + K^2 - K^{\mu\nu} K_{\mu\nu}) \tag{49}$$

$$\mathcal{H}^\mu = 2\kappa \sqrt{-g}\,\mathcal{N}\, G^{5\mu} = 2\, D_\nu p^{\mu\nu} \tag{50}$$

[2] In flat embedding space this means that the hypersurface is not bended at all.

Journal of Physics: Conference Series **330** (2011) 012011 doi:10.1088/1742-6596/330/1/012011

The terms with extrinsic curvature in eq. (49) can be expressed in terms of $p^{\mu\nu}$:

$$K^2 - K^{\mu\nu} K_{\mu\nu} = \frac{1}{\kappa^2(-g)} \left(\frac{p^2}{D-1} - p^{\mu\nu} p_{\mu\nu} \right), \tag{51}$$

where $D = g_{\mu\nu} g^{\mu\nu} = 4$ and $p \equiv g_{\mu\nu}\, p^{\mu\nu} = \sqrt{-g}\,(D-1)K$.

Here $p^{\mu\nu}$ are the canonical momenta conjugated to the 4D metric $g_{\mu\nu}$, whilst $\mathcal{N}$ and $\mathcal{N}_\mu$ are Lagrange multipliers for the constraints

$$\mathcal{H} = 0, \tag{52}$$

$$\mathcal{H}^\mu = 0. \tag{53}$$

Upon quantization, $g_{\mu\nu}$ and $p^{\mu\nu}$ become operators that can be represented as

$$g_{\mu\nu} \to g_{\mu\nu}, \quad p_{\mu\nu} \to -i \frac{\delta}{\delta g_{\mu\nu}} \tag{54}$$

More precisely, momentum operator has to satisfy the condition of Hermiticity, therefore the above definition is not quite correct in curved spaces, and has to be suitably modified. There also exists the factor ordering ambiguity that has to be adequately delt with. We are not interested here into such issues, therefore the expressions with $-i\delta/\delta g_{\mu\nu}$ have symbolic meaning only.

The 'Hamiltonian' constraint, $\mathcal{H} \approx 0$, becomes the Wheeler–DeWitt equation:

$$\left[-\frac{1}{2\kappa\sqrt{-g}} \left(\frac{g_{\mu\nu}\, g_{\alpha\beta}}{D-1} - g_{\mu\alpha}\, g_{\nu\beta} \right) \frac{\delta^2}{\delta g_{\mu\nu} \delta g_{\alpha\beta}} + \kappa \sqrt{-g} R^{(4)} \right] \Psi[g_{\mu\nu}] = 0\,, \quad D = 4. \tag{55}$$

Now the wave function(al) depends on 4-geometry, represented by a spacetime metric $g_{\mu\nu}\,(x^\mu)$. In this theory we have no problem of spacetime. We also have no problem of time, if by 'time' we understand the coordinate time $t \equiv x^0$.

However, the evolution parameter τ has disappeared from the quantized theory. There is no τ in the wave functional equation (55). Now we have the problem of τ. One possibility is to take the position that this is not a problem. It is important that we do not have the problem of $t \equiv x^0$, whereas missing τ is not a problem at all.

Another possibility is to bring τ into the game by considering matter degrees of freedom. In our approach the latter degrees of freedom are described by coordinates of Clifford space, one of them being interpreted as τ. To describe matter configurations, we have to consider also the matter part of the action.

As a model we consider the action (38) in which $R^{(6)}$ is replaced with $R^{(5)}$, and $\mathrm{d}^6 x$ with $\mathrm{d}^5 x$, the indices being now $M, N = 0, 1, 2, 3, 5$. The gravitational part we then replace by the equivalent phase space action (48). The matter part of the action we also replace by the phase space form:

$$I_{\rm m} = \int d\sigma \left(P_M \dot{X}^M - \frac{\alpha}{2} (G_{MN} P^M P^N - M_p^2) \right) \tag{56}$$

Splitting the metric according to (44), we have

$$I_{\rm m} = \int d\sigma \left(P_M \dot{X}^M - \frac{\alpha}{2} \left[g_{\mu\nu} (P^\mu + \mathcal{N}^\mu P^5)(P^\nu + \mathcal{N}^\nu P^5) + \mathcal{N}^2 P^5 P^5 - M_p^2 \right] \right) \tag{57}$$

To cast the matter part into a form comparable to the gravitational part of the action, we insert the integration over $\delta^5(x - X(\sigma)) d^5 x$, which gives identity. In both parts of the action, I_m and I_G, now stands the integration over $\mathrm{d}^5 x$. Recall that we identified $x^5 \equiv \tau$.

Varying the total action

$$I = I_G + I_m \tag{58}$$

with respect to α, $\mathcal{N}$ and $\mathcal{N}^\mu$, we obtain the constraints

$$\delta\alpha: \quad g_{\mu\nu}(P^\mu + \mathcal{N}^\mu P^5)(P^\nu + \mathcal{N}^\nu P^5) + \mathcal{N}^2 P^5 P^5 - M_p^2 = 0, \tag{59}$$

$$\delta\mathcal{N}: \quad \mathcal{H} + \int \mathrm{d}\sigma\alpha\mathcal{N}\delta^5(x - X(\sigma))P^5 P^5 = 0, \tag{60}$$

$$\delta\mathcal{N}^\mu: \quad \mathcal{H}_\mu - \int \mathrm{d}\sigma\alpha\, \delta^5(x - X(\sigma))\, g_{\mu\nu}(P^\nu + \mathcal{N}^\nu P^5)P^5 = 0. \tag{61}$$

where $\mathcal{H}$ and $\mathcal{H}^\mu$ are given in eqs. (49),(50), and $\kappa \equiv 16\pi\mathcal{G}$. We can write $\mathcal{H}$ compactly as

$$\mathcal{H} = \frac{1}{\kappa}\mathcal{G}_{\mu\nu\,\alpha\beta}\, p^{\mu\nu} p^{\alpha\beta} + \kappa\sqrt{-g}R^{(4)}, \tag{62}$$

with the metric

$$\mathcal{G}_{\mu\nu\,\alpha\beta} = \frac{1}{2\sqrt{-g}}\left[\frac{g_{\mu\nu}\, g_{\alpha\beta}}{D-1} - \frac{1}{2}(g_{\mu\alpha}\, g_{\nu\beta} + g_{\mu\beta}\, g_{\nu\alpha})\right], \qquad D = 4. \tag{63}$$

In a quantized theory, the constraints (59)–(61) become operator equations acting on a state vector. The constraint (59) can be put straightforwardly into its quantum version by replacing $P_\mu \to \hat{P}_\mu = -i\partial_\mu$, $P_5 \to \hat{P}_5 = -i\partial_5$. The latter definition of momentum operator holds in flat space only. In curved space we have to take a modified definition. For instance, a possible definition [20] that renders P_M hermitian, and also resolves the factor ordering ambiguity, is $\hat{P}_M = -i[\partial_M + (-G)^{-1/4}\partial_M(-G)^{-1/4}]$. An alternative procedure was proposed in Ref. [21].

So we have

$$\left[g_{\mu\nu}(\hat{P}^\mu + \mathcal{N}^\mu \hat{P}^5)(\hat{P}^\nu + \mathcal{N}^\nu \hat{P}^5) + \mathcal{N}^2 \hat{P}^5 \hat{P}^5 - M_p^2\right]\Psi = 0. \tag{64}$$

But the constraints (60),(61), because of the δ-distribution, are not practical for a direct translation into their corresponding quantum equivalents. Usually, for a quantum description of gravity in the presence of matter, one does not take the matter action in the form (57). Instead, one takes for I_m an action for, e.g., a scalar or spinor field, and then attempts to quantize the total action (58) following the established procedure of quantum field theory. Here I would like to point out that one can nevertheless start from the action (57) and use all the constraints (59)–(61).

Let us consider the Fourier transform of the constraint (60), the zero mode being given by the integral

$$\int \mathrm{d}^5x\, \mathcal{H} = -\int \alpha \mathrm{d}\sigma\, \mathcal{N}(P^5)^2 \tag{65}$$

Writing $\mathrm{d}^5x = \mathrm{d}^4x\, \mathrm{d}x^5$ and introducing $H = \int \mathrm{d}^4x\mathcal{H}$, we have

$$\int \mathrm{d}x^5\, H = -\int \alpha \mathrm{d}\sigma\, \mathcal{N}(P^5)^2, \tag{66}$$

or

$$\mathrm{d}x^5\, H = -\,\alpha \mathrm{d}\sigma\, \mathcal{N}(P^5)^2, \tag{67}$$

from which it follows

$$\frac{1}{\alpha}\frac{\mathrm{d}X^5}{\mathrm{d}\sigma} H = -\mathcal{N}(P^5)^2. \tag{68}$$

Here we have replaced the coordinate x^5, denoting a point in the $5D$ manifold, with the coordinate X^5, denoting a point on the worldline. Using the equation of motion (resulting from varying the action (56) with respect to P^M),

$$P^M = \frac{\dot{X}^M}{\alpha}, \tag{69}$$

where $\dot{X}^M \equiv \mathrm{d}X^M/\mathrm{d}\sigma$, we find that $P^5 = \dot{X}^5/\alpha$. Using the latter expression in eq. (68), we obtain

$$H = -\mathcal{N}P^5. \tag{70}$$

Similarly, from the constraint (61) we obtain

$$H_\mu = g_{\mu\nu}(P^\nu + \mathcal{N}^\nu P^5), \tag{71}$$

where $H^\mu = \int \mathrm{d}^4x\,\mathcal{H}^\mu$. Let us now use the relations $P^M = G^{MN}P_N$ and $P_M = G_{MN}P^N$ with the metrics (44),(45), and rewrite eqs. (70),(71) into the form with covariant components of momenta P_μ, P_5:

$$H = -\frac{1}{\mathcal{N}}(P_5 + \mathcal{N}^\mu P_\mu), \tag{72}$$

$$H_\mu = P_\mu. \tag{73}$$

The above result is nothing but a manifestation of the fact that the integration of a stress-energy tensor over a certain hypersurface gives momentum. Here momentum is $P_M = (P_\mu, P_5)$. Using (50), eq. (73) can be rewritten as

$$2\int_\Omega \mathrm{d}^4x\,\mathrm{D}_\nu p^{\mu\nu} = 2\int_B \mathrm{d}\Sigma_\nu p^{\mu\nu} = P^\mu\,, \qquad P^\mu \equiv g^{\mu\nu}P_\nu, \tag{74}$$

where B is the boundary of a region Ω in the 4-space, and $\mathrm{d}\Sigma_\nu$ is an element of the boundary surface. The relation (74) is analogous to the Gauss law in electrodynamics. Bear in mind that the momentum P^M points along a worldline $X^M(\sigma)$, $M = (\mu, 5)$, which intersects 4D spacetime in one point. Therefore, the integral in eq. (74) is different from zero only when the 3-surface B embraces the intersection point.

For the Lagrange multipliers we can choose $\mathcal{N} = 1$ and $\mathcal{N}^\mu = 0$, which simplifies eqs. (72) and (59) into

$$H = -P_5, \tag{75}$$

$$g^{\mu\nu}P_\mu P_\nu + P_5P_5 - M_p^2 = 0. \tag{76}$$

It is now straightforward to consider the quantum versions of the constraints (75),(73) together with the constraint (76). We have[3]

$$\left(g^{\mu\nu}\frac{\mathrm{D}^2}{\mathrm{D}X^\mu \mathrm{D}X^\nu} + \frac{\partial^2}{\partial\tau^2} + M_p^2\right)\Psi = 0 \tag{77}$$

$$\int \mathrm{d}^4x\left(-\frac{1}{\kappa}G_{\mu\nu\alpha\beta}\frac{\delta^2}{\delta g_{\mu\nu}\delta g_{\alpha\beta}} + \kappa\sqrt{-g}R^{(4)}\right)\Psi = i\frac{\partial}{\partial\tau}\Psi\,, \qquad \tau \equiv X^5. \tag{78}$$

$$\int \mathrm{d}^4x\,\mathrm{D}_\nu\left(-i\frac{\delta}{\delta g_{\mu\nu}}\right)\Psi = -i\frac{\partial}{\partial X^\mu}\Psi \tag{79}$$

[3] See the texts after Eqs. (54) and (63). We chose the factor ordering in order to achieve covariance in the space comprised of X^μ. Therefore, in Eq. (77) we have the covariant derivative $\mathrm{D}/\mathrm{D}X^\mu$. In an analogous way should be interpreted Eq. (78).

Journal of Physics: Conference Series **330** (2011) 012011 doi:10.1088/1742-6596/330/1/012011

The latter equations impose the operator constraints on a quantum state that is represented by $\Psi[\tau, X^\mu, g_{\mu\nu}(x^\mu)]$ which depends on the particle's coordinates X^μ, the fifth coordinate $X^5 \equiv \tau$, and the spacetime metric $g_{\mu\nu}(x^\mu)$. In other words, Ψ is a function of τ, X^μ, and a functional of $g_{\mu\nu}(x^\mu)$. Eq. (78) is just like the Schrödinger equation, with τ as evolution parameter. Therefore, the "problem of τ" does not exist in this quantum model for a point particle coupled to a gravitational field. Had we performed a split from six to four dimensions (and not from five to four as we did in this section), then in eq. (77), instead of ∂_τ^2, we would have $\partial_\lambda \partial_\tau \sim \Lambda \partial_\tau$ (see sec. 2), so that eq. (77) would become the Stueckelberg equation.

The system (77)–(79) describes at once a Klein-Gordon wave function for a relativistic particle, and the wave functional for a gravitational field. It is only an incomplete description of the physical system. A complete description would require to take into account the infinite set of constraints due to all Fourier modes of the the constraints (59)–(61).

5. Discussion and conclusion

We have shown how the Stueckelberg equation for a relativistic point particle comes from a 6-dimensional space, $M_{2,4}$, with signature (2,4), that is $(+ - - - - +)$. Two extra dimensions, one time like and one space like, are necessary, because then in the equation we obtain the first derivative of the wave function with respect to a Lorentz, SO(1,3), invariant parameter τ which is identified with the fifth coordinate X^5.

An argument in favor of such 6D space comes from the works on the two time (2T) physics [5] that is invariant under local Sp(2) transformations between coordinates and momenta. In such theory there are three Lagrange multipliers associated with three constraints, which cannot be satisfied in 4D spacetime $M_{1,3}$. They can be satisfied in 6D space $M_{2,4}$, or in a suitable higher dimensional space. Since the theory by Bars et al. [5] is based on very strong foundations, we can conclude that the 6D space is a reasonable subsitute for 4D spacetime. It enables to formulate the 2T physics on the one hand, and the Stueckelberg theory on the other hand, but not both at once. A relationship between the two theories has yet to be explored. A clue is to consider a higher than six dimensional space and to impose the Sp(2) constraints on the variables entering the phase space action, and thus obtain a generalization of the 2T physics. The phase space action (15) in six dimensions—that the Stueckelberg theory in embedded in—is a particular, gauge fixed, case of the Sp(2) invariant action in higher than six dimensions. According to such view, the local Sp(2) invariance holds in a higher dimensional space, whereas in the 6-dimensional subspace $M_{2,4}$, it is broken. But, in $M_{2,4}$ one might expect the problem with ghosts due to the extra time like dimension. Concerning ghosts, it was shown in Refs. [7, 8, 9] that they do not necesarily occur in spaces with time like dimensions, if one defines vacuum in an alternative way, as proposed by Jackiw et al. [6].

There exists another direction of research, which is based on the concept of configuration space, i.e., the space of possible matter configurations. An example of such space is the 16D space of oriented r-volumes, associated with extended objects, e.g., branes. We call it Clifford space, C, because it is a manifold whose tangent space at any point is a Clifford algebra $Cl(1,3)$. If we define the metric according to eq. (35), then the signature of C is $(8,8)$. A subspace of C is $M_{2,4}$. Therefore, if we adopt the concept of Clifford space, C, we do not need to postulate extra dimensions of spacetime, in order to have the 6D space formulation of the Stueckelberg theory, or of the 2T physics. Four dimensions of C can be identified with the four dimensions of spacetime, whilst the remaining 12 dimensions of C are associated with the intrinsic configurations of matter living in the 4-dimensional spacetime.

We have considered the general relativity in Clifford space, more precisely in the 6D subspace with signature (2,4). The action contains the Einstein-Hilbert term which is a functional of the metric only, and a matter term, which is a functional of matter degrees of freedom coupled to the metric. As a model we have considered a point like source. We have performed the ADM decomposition of a 5D subspace into the spacetime $M_{1,3}$ and a part due to the 5th dimensions, x^5. The action gives the mass shell constraint in 5-dimensions, and the constraints that generalize the Hamiltonian and momentum constraints of the canonical gravity, with the extra terms due to the presence of the point particle source. After quantization those constraints become the operator constraints acting on a state that can be represented as a functional of the spacetime metric $g_{\mu\nu}$, $\mu, \nu = 0, 1, 2, 3$, a function of the particle coordinates X^μ, and the fifth coordinates, $X^5 \equiv \tau$, which has the role of the Stueckelberg evolution parameter. In the Stueckelberg theory the 'true' time is the Lorentz, SO(1,3), invariant evolution parameter τ, and not the coordinate $x^0 \equiv t$. Since such parameter occurs in the wave function(al) for the gravitational field, we conclude that there is no 'problem of time' in this theory.

Acknowledgments

This research was supported by the Ministry of High Education, Science and Technology of Slovenia.

References

[1] Anderson E 2011 The Problem of Time in Quantum Gravity *Preprint* 1009.2157 [gr-qc]
[2] DeWitt B S 1967 *Phys. Rev.* **160** 1113
[3] Arnowitt R, Deser S and Misner C W 1962 The Dynamics of General Relativity, published in *Gravitation: An Introduction to Current Research* Ed. Witten L (New York: Wiley), p. 227
[4] Fock V 1937 *Phys. Z. Sowj.* **12** 404
Stueckelberg E C G 1941 *Helv. Phys. Acta*
Stueckelberg E C G 1942 **15** 23
Horwitz L P and C. Piron C 1973 *Helv. Phys. Acta* **46** 316
Horwitz L P and Rohrlich F 1981 *Physical Review* D **24** 1528
Horwitz L P, Arshansky R I and Elitzur A C 1988*Found. Phys* **18** 1159
Feynman R P 1951 *Phys. Rev* **84** 108
Fanchi J R 1993 *Found. Phys.* **23** 287, and many references therein
Fanchi J R 1993 *Parametrized Relativistic Quantum Theory* (Dordrecht: Kluwer) Pavšič M 1991 *Found. Phys.* **21** 1005
[5] Bars I, Deliduman C, and Andreev O 1998 *Phys. Rev.* D **58** 066004
Bars I 1998 *Phys. Rev.* D **58** 066006
Bars I 2001 *Class. Quant. Grav.* **18** 3113
Bars I 2006 *Phys. Rev.* D **74** 085019
[6] Cangemi D, Jackiw R and Zwiebach B 1996 *Annals of Physics* **245** 408
Benedict E, Jackiw R and Lee H J 1996 *Phys. Rev.* D **54** 6213
[7] Pavšič M 2005 *Found. Phys.* **35** 1617 (*Preprint* hep-th/0501222)
[8] Pavšič M 1999 *Phys. Lett.* A **254** 119 (*Preprint* hep-th/9812123)
[9] Pavšič M 2001 *The Landscape of Theoretical Physics: A Global View; From Point Particle to the Brane World and Beyond, in Search of Unifying Principle* (Dordrecht: Kluwer) *Preprint* gr-qc/0610061
[10] Castro C 2000 *Chaos Solitons Fractals* **11** 1721 (*Preprint* hep-th/9912113)
[11] Pavšič M 2001 *Found. Phys.* **31** 1185 (*Preprint* hep-th/0011216)
[12] Pavšič M 2005 *Phys. Lett.* B **614** 85 (Preprint hep-th/0412255)
[13] Pavšič M 2006 *Int. J. Mod. Phys.* A **21** 5905 (*Preprint* gr-qc/0507053)
[14] Pavšič M 2009 On the Relativity in Configuration Space: A Renewed Physics In Sight *Preprint* 0912.3669 [gr-qc]
[15] Pavšič M 2008 *J. Phys.* A **41** 332001 (*Preprint* 0806.4365 [hep-th])
[16] Castro C and Pavšič M 2005 *Prog. Phys.* **1** 31
[17] Pavšič M 2003 *Found. Phys.* **33** 1277 (*Preprint* gr-qc/0211085)
[18] Pavšič M 2007 *Found. Phys.* **37** 1197 (*Preprint*hep-th/0605126)
[19] Ansoldi S, Aurilia A, Castro C and Spallucci E 2001 *Phys. Rev.* D **64** 026003 (*Preprint* hep-th/0105027)
Aurilia A, Ansoldi S and Spallucci E 2002 *Class. Quant. Grav.* **19** 3207 (*Preprint* hep-th/0205028)
[20] DeWitt B S 1957 *Rev. Mod. Phys* **29** 377
DeWitt B S 1952 *Phys. Rev.* **85** 653
[21] Pavšič M 2003 *Class. Quant. Grav.* **20** 2697 (*Preprint* gr-qc/0111092)

Journal of Physics: Conference Series **330** (2011) 012012 doi:10.1088/1742-6596/330/1/012012

Electric and magnetic fields: do they need Lorentz covariance?

Zbigniew Oziewicz

Universidad Nacional Autónoma de México, Facultad de Estudios Superiores Cuautitlán
C.P. 54714 Cuautitlán Izcalli, Estado de México

E-mail: `oziewicz.zbigniew@gmail.com`, `oziewicz@unam.mx`

Abstract. Electric and magnetic fields are relative. They depend not only on a choice of electromagnetic sources via Maxwell equations, but also on a choice of observer, a choice of material reference-system. In 1908 Minkowski defined electric and magnetic fields on a **four**-dimensional spacetime, as tensorial concomitants of observer. Minkowski defined Lorentz-group-covariance of concomitant tensor field as group-action that commute with contractions. Present-day textbooks interpret Lorentz-group-covariance of concomitant tensor differently than Minkowski in 1908. In 2003-2005 Tomislav Ivezíc re-invented Minkowski's group-covariance. Different interpretations of group-covariance, lead to different relativity transformations of electric and magnetic fields.

An objective of present article is to explore third possibility, implicit in [Minkowski 1908, §11.6], where a set of all relativity transformations of all material observers forms a groupoid category, which is *not* a group.

Keywords: history of relativity theory, group-covariance, material reference frame, observer, tetrad versus monad, electric and magnetic fields, groupoid category, relativity groupoid as a category

Physics and Astronomy Classification Scheme (PACS) 1999/2000. 03. special relativity; 01.65.+g History of science;

2000 Mathematics Subject Classification. 51B20 Minkowski geometry, 53A17 Kinematics, 53A35 Non-Euclidean geometry, 53B30 Lorentz metric, 83A05 Special relativity.

1. Introduction: symmetry group of Maxwell equations

In 1904 Lorentz and Poincaré showed that a system of four Maxwell's differential equations (observer-dependent) has a symmetry group the same as a symmetry group of metric tensor. This was the Lorentz and Poincaré isometry group. In 1909/1910 Bateman and independently Cunningham, showed that Lorentz and Poincaré were wrong, because they discovered only a **subgroup** of actual symmetry group of Maxwell equations. In fact symmetry group of Maxwell equations of electromagnetic field is conformal group of dimension 15, this is actual symmetry group of massless radiation.

What does it mean to be a symmetry of differential equation? Symmetry act on solutions of differential equation and maps solutions into solutions. Therefore a symmetry is iso-morphism of a manifold of solutions.

Journal of Physics: Conference Series **330** (2011) 012012 doi:10.1088/1742-6596/330/1/012012

As we know perfectly today symmetry group of massless Maxwell's equations, the conformal group, was never considered as a group permuting material reference systems. Isometry group of metric tensor was considered incorrectly as the *same* as a symmetry group of Maxwell's equations. If this could be made plausible then a hypothesis, accepted in 1905 by Albert Einstein, and in 1908 explicitly by Hermann Minkowski, that this non-Euclidean metric tensor of empty spacetime and this Lorentz isometry-group is a group that can permute not only solutions of massless Maxwell's equations, but can permute also material reference-systems of relativity theory?

In present paper we argue that light-like massless radiation must not be considered as a reference system, and therefore must be outside of domain of transformations of material reference-systems. From the other hand, isometry of metric, as it is the Lorentz group, must possess in domain *all* vectors, including also light-like vectors in Minkowski space. This implies that a set of all transformations among all material reference systems do not need to be postulated a priori to be the Lorentz isometry-group.

We propose here an alternative for the Einstein isometric special relativity. We suggest that a set of all relativity transformations (between material reference-systems, among the time-like vector fields) could be a groupoid viewed as a *category*, that it is *not* a group[1]. One consequence of groupoid relativity, that it is implicit in Minkowski last publication in 1908, in §11.6, is examined here for groupoid relativity transformation of the electric and magnetic fields, and groupoid transformation of the electromagnetic sources Equation (12.7)–Equation (12.8). This consequence could eventually be tested experimentally.

We wish to show, following Minkowski's implicit idea in §11.6, that Maxwell theory of electromagnetic fields does not need the concept of Lorentz group, nor Lorentz-covariance, and can be formulated alternatively in terms of groupoid relativity, in terms of groupoid category that is not a group.

1.1. What is groupoid?

We owe to readers brief explication of the difference among concept of a group, a very special groupoid category, and a groupoid viewed as a category. Groupoid generalize a concept of a group. Every group is a groupoid category, but not every groupoid category is a group.There are two differences interrelated. We refer to:

http://mathworld.wolfram.com/Groupoid.html.

The first property of a group distinguishing them from a general groupoid category, is that a group possess *unique* neutral element. In the case of Lorentz relativity group, the zero velocity $\mathbf{0}_{\text{Earth}}$ of Earth relative to Earth, and zero velocity $\mathbf{0}_{\text{Sun}}$ of Sun relative to Sun, are identified as the unique neutral Lorentz boost, $\mathbf{0}_{\text{Sun}} \equiv \mathbf{0}_{\text{Earth}} \equiv \mathbf{0} \in$ Lorentz group. In case of groupoid relativity, instead, these are different elements,

$$\mathbf{0}_{\text{Sun}} \neq \mathbf{0}_{\text{Earth}} \in \text{Groupoid}. \tag{1.1}$$

Every two elements of a group must be composed, group binary operation is *global*. For example, if L_1 and L_2 are two elements of Lorentz group, hence their group compositions, $L_1 L_2$ and $L_2 L_1$, belong to Lorentz group. This is not the case in a groupoid category, where not every pair of elements of a groupoid (*i.e.* not every pair of morphism of groupoid category) must be composable. The groupoid binary operation is not necessarily global. In our example of relativity groupoid, for a set of four different material reference systems denoted by $\mathbf{P}, \mathbf{Q}, \mathbf{R}, \mathbf{S}$,

[1] Frequently groupoid is understood as binary operation, renamed by Bourbaki as *magma*. Groupoid viewed as a cateory is not magma.

Journal of Physics: Conference Series **330** (2011) 012012 doi:10.1088/1742-6596/330/1/012012

consider two groupoid transformations,

$$\mathbf{P} \xrightarrow{g_1} \mathbf{Q} \qquad \text{and} \qquad \mathbf{R} \xrightarrow{g_2} \mathbf{S}. \tag{1.2}$$

Each of these transformations is invertible in the case of a group and also necessarily invertible in a case of groupoid. However, if $\mathbf{Q} \neq \mathbf{R}$ and $\mathbf{S} \neq \mathbf{P}$, these transformations, g_1 and g_2, can be constructed in such way that a priori (e.g. $\mathbf{Q}$ not in domain of g_2, etc) this would *not* allow us to compose them. In this case, a set of all such invertible transformations without of global composition is said to be a *groupoid category*, briefly groupoid.

Einstein's second postulate, Postulate II state: light velocity is independent of motion of emitting source [Einstein 1905]. It is rather known that this second Einstein's postulate is a consequence of **only one** Minkowski's postulate, that a group of all relativity transformations among material reference systems must coincide with Lorentz isometry group. One can deduce the second Einstein's postulate also as a consequence of groupoid relativity [Oziewicz 2005]. Therefore as far as light velocity is concerned, there is **no** difference among Lorentz-group relativity and groupoid relativity. It is, therefore, impossible to infer Lorentz relativity transformations from second Einstein's postulate alone. In fact Lorentz-*group* relativity transformations were silently *assumed* by Einstein. Einstein postulated *group* structure. Equivalently, one can deduce group structure by postulating reciprocity of relative velocity (also postulated by Einstein in 1905), *i.e.* that inverse of relative velocity $\mathbf{v}$ is $-\mathbf{v}$. Reciprocity axiom of relative velocity does not hold in groupoid relativity [Świerk 1988, Matolcsi 1994, Oziewicz 2005].

2. Notation: basis-free four-dimensional space-time

It is worth to mention that idea of spacetime was implicit in Galileo's observation of relativity of space (thus absolute spacetime), and that *four*-dimensional space-time, with 'fourth time coordinate', was proposed by Jean Le Rond D'Alembert (1717–1783) in articles about the partial differential wave equation, published in Encyclopedia of Denis Diderot around 1772.

Starting in 1884, Gregorio Ricci-Curbastro developed tensor analysis, and, working jointly with his student Levi Civita, they made clear in 1901 the difference among 'contravariant' and 'covariant' vectors. Élie Cartan (1869–1951) used a name a differential form as a synonym for Ricci's covariant vector field. Category theory born in 1945, clarify this distinction in terms of dual spaces, and in terms of dual pair of vector and covector fields. It was clarified that Ricci's 'covariant' vector must acquire categorical meaning as contravariant vector.

In 1908 Hermann Minkowski introduced the following terminology, a *space vector*, as a synonym of present terminology *space-like vector in a four-dimensional Minkowski spacetime*, and a *space-time vector*, that presently is understood as a vector in Minkowski spacetime.

Minkowski in 1908	Presently
space vector	space-like vector
space-time vector	vector, or '4-vector'

The Minkowski terminology was, and is, misinterpreted, in particular, when considering observer-dependent product-structure or

$$\text{a splitting of spacetime into: 'space'} \oplus \text{'time'}.$$

The name *spacetime*, introduced in 1908 by Minkowski, is misleading, suggesting incorrectly that this concept is derived from two primitive concepts of 'space' and 'time'. It is just opposite, the very primitive concept is space-time of events, and space is a *derived* concept that needs an artificial *choice* of material body, e.g. Earth or Sun, as a reference system, see Equation (2.1) below. But any such *choice* is irrelevant for physical phenomena, it is no more then for example a convenience for a computer program. If a material reference system is chosen, Earth or Sun,

Journal of Physics: Conference Series **330** (2011) 012012 doi:10.1088/1742-6596/330/1/012012

then corresponding space of *this* material body is *not* a fiber in space-time, but it is rather a quotient-space = spacetime/material-body,

$$\text{Space} \equiv \frac{\text{Spacetime}}{\text{material body}} \qquad \text{Time} \equiv \frac{\text{Spacetime}}{\text{Convention of simultaneity}}, \tag{2.1}$$

$$\text{Proper-Time} \equiv \frac{\text{Spacetime}}{\text{Metric simultaneity of material body}}. \tag{2.2}$$

Misinterpretation of Minkowski's terminology grows into three-dimensional and four-dimensional quantities, 3-dimensional vector, 3-space vectors, 3D versus 4D quantities, 4-vectors, four-tensors, etc. (e.g. Rindler 1969, §5.4 Four-vector, §5.10 Three-Force and Four-Force; Landau & Lifshitz, 1975, §6. Four-Vectors). A vector does not have dimension. It is a manifold and a vector-space that possess dimension.

Spacetime splitting does not mean that there then appear some strange '3-dimensional quantities'. Like Minkowski in 1908, we exclusively use **four**-dimensional space-time manifold only. In all our expressions, a boldface electric field $\mathbf{E}$, and a boldface magnetic field $\mathbf{B}$, denote 'contravariant' vector fields (derivations, see next section) on *four*-dimensional spacetime manifold. All tensor fields, scalar fields, vector fields, bivector fields, differential multiforms, etc., are tensor fields on four-dimensional spacetime manifold only.

Our boldface notation distinguishes 'contravariant' multi**vector** fields, from the 'covariant' differential forms denoted by italics letters, or by Greek alphabet. Thus, $\mathbf{F}$ denotes a bivector field, whereas F denotes a differential biform. A material observer is a time-like vector field, and is denoted by boldface $\mathbf{P}$ for an observer **P**aul, and, by boldface $\mathbf{R}$ for an observer **R**ose.

The source of misinterpretation is a concept of a vector field. Many authors denote by $\mathbf{E}$ time dependent electric field strength , *i.e. not static* electric field, $\mathbf{E}(\mathbf{r}, t)$, and of course such electric field is a vector field on four-dimensional space-time, independently of a choice of a basis. This notation assume silently that derivative of some time coordinate (a parameter t) in direction of $\mathbf{E}$ is zero, $\mathbf{E}t \equiv (dt)\mathbf{E} = 0$. This means that this 4D vector field $\mathbf{E}$ on spacetime is tangent to three-dimensional super-surfaces t = const. This condition could be equivalent to stressed many times by Minkowski, that electric field $\mathbf{E}$ must by orthogonal to time-like observer-vector field $\mathbf{P}$, *i.e.* $\mathbf{E} \cdot \mathbf{P} \equiv g(\mathbf{P} \otimes \mathbf{E}) = 0$, for some Minkowski's metric tensor g. To be Minkowski' s metric tensor g does not means here that curvature and torsion of connection must be absent, we emphasize only that this non-Euclidean spacetime metric was recognized by Minkowski (and early by Henri Poincaré).

$$\text{If} \quad g\mathbf{P} = dt \quad \text{then:} \quad \mathbf{E} \cdot \mathbf{P} = 0 \quad \Longleftrightarrow \quad \mathbf{E}t = 0. \tag{2.3}$$

Nobody call $\mathbf{E}$ to be 'nonrelativistic' because exists another vector $\mathbf{P}$ orthogonal to $\mathbf{E}$!

How to understand apparently 'nonrelativistic' electric vector field $\mathbf{E}(\mathbf{r}, t)$ in textbooks by Sommerfeld [1948], or by Jackson [1962] ? Somebody interpret this as infinite-family of vector fields on three-dimensional space, family that is parameterized by time parameter. But of course, such family of vector fields is not a vector field on space, it is not a '3D' vector. Calling time-dependent vector field as 'nonrelativistic' bring objection: nonrelativistic one can associate to limit $c \Rightarrow \infty$, but this field $\mathbf{E}$ is a solution of Maxwell differential equations. Authors call $\mathbf{E}(\mathbf{r}, t)$ incorrectly as '3D' vector, but in fact this is a vector field on space-time, *i.e.* this is '4D' vector field.

Some readers would prefer as more clear an explanation in terms of mathematical bases. When nowhere-vanishing vector field, $\mathbf{E} \neq \mathbf{0}$, is chosen to be a part of mathematical basis, e.g. if a basis of vector fields (a frame) is, $\{e_0, \mathbf{E}, e_2, e_3\}$, then, this vector field $\mathbf{E}$ in such basis has exactly one non-vanishing scalar component, i.e. $\mathbf{E} \simeq (0, 1, 0, 0)$. Each non-zero vector field on a four-dimensional spacetime possesses a set of three adopted (or associated) differential one-forms

Journal of Physics: Conference Series **330** (2011) 012012 doi:10.1088/1742-6596/330/1/012012

annihilating this vector field. Each non-zero electric vector field, $\mathbf{E} \neq \mathbf{0}$, in such adopted cobasis (coframe), has only one non-zero scalar component. For example, let there exist the differential one-forms, $\alpha_1 \wedge \alpha_2 \wedge \alpha_3 \neq 0$, such that, $\alpha_1\mathbf{E} = 0$, $\alpha_2\mathbf{E} = 0$ and $\alpha_3\mathbf{E} = 0$. In a co-frame basis, $\alpha_1 \wedge \alpha_2 \wedge \alpha_3 \wedge \alpha_4 \neq 0$, a vector field $\mathbf{E} \neq \mathbf{0}$ possesses only one non-vanishing scalar component, $\alpha_4\mathbf{E} \neq 0$.

In this basis-dependent meaning, should this vector field $\mathbf{E}$ *on spacetime* to be '1D' or '3D quantity'? Phrases '3D **E**', and '3D **B**', as well as three-dimensions, could be understood in several different and misleading ways:

- 'Time-**independent**' tensor on spacetime, i.e. tensor with one Lie-symmetry.
- Tensor on three-dimensional sub-manifold.
- Tensor on three-dimensional quotient-manifold.
- Basis-dependent concept that in an accidental (badly chosen) mathematical basis, '3D **E**' has at most three non-vanishing scalar components.

In the present paper we do not need terminology like 3D, 3-dimensional vector and no '4-vectors', '4-velocities', - all misleading. We do not accept the distinction between '3D' and '4D quantities', because if '4' means a dimension of a space-time, then *all* time-dependent tensor fields are on spacetime, and in the present paper, and in *all* textbooks, in Jackson's **F** and **E**, etc., are always '4D', independently of how many components are not vanishing.

Two different material observers, call them **P**aul and **R**ose, each have their own material reference-system. In each material reference-system, concomitant-compound electric and magnetic vector fields are defined in 1908 by Minkowski, see Definition 7.1 and expressions Equation (7.1) below,

$$\mathbf{E}(\mathbf{P}\mathrm{aul}), \quad \mathbf{B}(\mathbf{P}\mathrm{aul}), \quad \mathbf{E}(\mathbf{R}\mathrm{ose}), \quad \mathbf{B}(\mathbf{R}\mathrm{ose}). \tag{2.4}$$

In a generic basis of four independent vector fields, each of these vector fields Equation (2.4) possesses *four* non-zero scalar components; never more than four. A differential one-form of an electric field, $E(F, \mathbf{P}\mathrm{aul})$, is a tensor field on exactly the same manifold on which differential bi-form F, and an observer-monad field **P**aul, are living.

One-hundred years after duality was realized, and important distinction among a concept of a (coordinate) vector field, ∂_x, and a concept of a coordinate-free differential form, dx, was realized, still it is very hard to find this distinction in present University textbooks. Another story is to understand that every vector field is coordinate-free, and that also coordinate-vector-field ∂_x that is completely fixed by one coordinate co-frame, is also coordinate-free as every tensor field must be.

Differential one-form, known also as Pfaff form, like differential of a scalar field, dx, is not Leibniz's small infinitesimal increment; instead, following an idea of Isaac Barrow, it is a real-valued function on vectors (vector is synonym of a process), associating to a vector (process) the directional derivative (of scalar field) along this vector, see for example [Bishop and Goldberg 1968, page 58]. This paper distinguishes contravariant vectors and tensors on four-dimensional spacetime, denoted by boldface fonts, from covariant vectors and tensors, and in particular differential forms, denoted by italic or Greek letters.

A differential one-form is a synonym of covariant vector field, it is a scalar-field-valued evaluation-map on vector fields. For example, $(dx)\mathbf{P} = \mathbf{P}x$ is a (directional) derivative of a scalar field x along a vector field $\mathbf{P}$. *Every* derivative is directional, however this most important point is missed in all Calculus texbooks. A vector field annihilated by a given differential one-form is said to be in kernel of this differential form. Therefore, a value of differential one-form on vector field is said to be evaluation, and does not need a concept of a scalar product applied to pair of vector fields of the same covariance. For example, $\mathrm{ev}(df \otimes \mathbf{P}) \equiv (df)\mathbf{P} \equiv \mathbf{P}f$, is evaluation, and *not* scalar product.

Journal of Physics: Conference Series **330** (2011) 012012 doi:10.1088/1742-6596/330/1/012012

The concomitant of compound electric field, Paul-dependent electric field as Pfaff differential form, $E(F, \mathbf{Paul})$, has a time-like vector field $\mathbf{P}$ in his kernel, $\{E(F, \mathbf{P})\}\mathbf{P} = 0$. For $E = g\mathbf{E}$, equivalently, $\mathbf{E} \cdot \mathbf{P} \equiv 0$, [Minkowski 1908 §11.6, Eq. (49)]. Hence, whenever a vector field $\mathbf{P}$ is chosen to be a basis vector field, then in such mathematical basis electric field on spacetime possesses no more than three non-vanishing scalar components.

The last objective of the present note is to advocate a basis-free and coordinate-free approach, and therefore, basis-dependent classification into 4D, 3D, 2D and 1D 'quantities', is totally irrelevant for mathematics and for physics.

3. Each vector field is a derivation

Most textbooks of XXI century separate Calculus's concept of a derivative or a derivation, from algebraic concept of a vector field. The source of this separation lay also in sorrowful history of Calculus, a war of Newton against Leibniz. Students learn ancient historical Pierre de Fermat's definition of derivative of a function (derivative of a scalar field) that do not allow to imagine that derivative of a scalar field in reality is *not* unique, and needs always a choice of a vector field along which such derivative one can calculate. *Every* derivative is directional, however initial chapters of Calculus textbooks insists incorrectly that no one vector field is involved in the concept of derivative. Derivative of a not constant scalar field x along a coordinate vector field ∂_x, is, by definition, $\partial_x x = 1$, but this is *not* the only derivative possible. The same scalar field x, has a different derivative along another coordinate vector field, say $\partial_{(x^2+2)}$, namely, $\partial_{(x^2+2)} x = \frac{1}{2x}$, this is the chain rule. There are so many derivatives of a given not constant scalar field as many there are vector fields. No vector field chosen, it is not possible to calculate derivative of a function! The most frequent Calculus-books problem 'Calculate derivative of '$2x + 1$', is meaningless, because $\partial_{(2x+1)}(2x + 1) = 1$.

Each partial derivative, like ∂_x, is a coordinate vector field (a derivation of an algebra of the scalar fields), and, as a partial derivative, is not given uniquely by a given scalar field x. Notation ∂_x is misleading because this coordinate vector field ∂_x depends on a choice of coordinate system, on a choice of an integrals of motion, for example we need to choose a scalar field t for ∂_x such that $\partial_x t = 0$. This information is missing in notation ∂_x. Textbooks of thermodynamics use a correct notation,

$$(\partial_T)_P \quad \neq \quad (\partial_T)_V. \tag{3.1}$$

When writing ∂_x it is most important a choice of a coordinate chart to which a scalar field x belongs; for example, $x \in \{x, t\}$. If $\partial_x t = 0$, then $\partial_x \equiv (\partial_x)_t$.

Each electric field and each magnetic field, is a vector field, therefore these fields are *derivations* of algebra of scalar fields. Therefore coordinate expressions of these vector fields must be as follows

$$\begin{aligned}
\mathbf{E} &= (\mathbf{E}x^\mu)\partial_\mu = (\mathbf{E}x^\mu)\frac{\partial}{\partial x^\mu}, \\
\mathbf{B} &= (\mathbf{B}x^\mu)\partial_\mu = (\mathbf{B}x^\mu)\frac{\partial}{\partial x^\mu}, \\
\mathbf{P} &= (\mathbf{P}x^\mu)\partial_\mu = (\mathbf{P}x^\mu)\frac{\partial}{\partial x^\mu}, \\
\mathbf{R} &= (\mathbf{B}x^\mu)\partial_\mu = (\mathbf{P}x^\mu)\frac{\partial}{\partial x^\mu}, \quad \text{etc.}
\end{aligned} \tag{3.2}$$

Evidently, in Equation (3.2), scalar components, like $\mathbf{E}x^\mu$, is a derivative of a scalar field x^μ, along electric field $\mathbf{E}$. This scalar component is denoted historically as $\mathbf{E}^\mu \equiv \mathbf{E}x^\mu$. In what follows we will not need and not use neither coordinates nor bases. Physics is coordinate-free and bases-free.

Journal of Physics: Conference Series **330** (2011) 012012 doi:10.1088/1742-6596/330/1/012012

4. Herman Minkowski

In 1908 Hermann Minkowski published his last paper, entitled "The foundations for electromagnetic phenomena in the moving bodies" (Die Grundlagen für die electromagnetischen Vorgänge in bewegten Körpen). In 1910, after Minkowski's death, two other papers were published under Minkowski's name. The 1910-paper, of almost the same title, was written by his pupil Max Born, and sometimes is referred as Minkowski and Born paper, although was published under the name of Hermann Minkowski alone. When comparing the Minkowski 1908-paper with Born's 1910-paper, it is clear that Born's 1910-interpretation was different from 1908-paper by Minkowski. Born put full emphasis on Lorentz-group covariance, whereas Minkowski in Part II §11.6 of his 1908-paper defined electric and magnetic fields in a covariance-free way.

Figure 1. Hermann Minkowski 1864-1909
Portrait provided at http://library.thinkquest.org/05aug/01273/whoswho.html

Minkowski's 1908 paper deserves commemoration more than Einstein's 1905 paper did, for several following reasons.

- Minkowski defined Lorentz group as isometry group of a metric and this definition do *not* involve a concept of reference system. Relativity is formulated by Minkowski as **one axiom-postulate only**: a group of all transformations among material reference systems coincide with Lorentz isometry group. See exactly the same definition for example in [Schrödinger

1931, Bergmann 1942, Einstein 1949, Bargmann 1957]. This postulate (subject of revision in present paper) reduced the relativity principle to Lorentz group-covariance.

- A tensor is said to be a *concomitant* tensor, if it is build-from/dependent-on other primary tensors. Minkowski defined group-covariance, and in particular at the beginning of §11.6, defined Lorentz-covariance of concomitant tensors. Minkowski's definition of Lorentz-covariance does not appears in contemporary University textbooks on electromagnetism. Tomislav Ivezíc rediscovered the Minkowski group-covariance in 2003-2005.
- The entire §11.6 of Minkowski's 1908-paper is devoted to ingenious invention of definition of electric and magnetic fields, as concomitants of absolute electromagnetic field. These concepts are covariance-free, and do not need at all Lorentz-isometry group.
- Relativity axiom, **only one** Minkowski's axiom, does not involve a concept of relative velocity among material reference systems. Within Lorentz-group axiom, relative velocity belongs to Lobachevski factor space, see [Varićak 1924], and this factor space is not unique. This implies that relative velocity among reference systems is also not unique. Contrary to relativity axiom (not unique relative velocity), in §11.6 of 1908-paper, Minkowski defined the **unique** relative velocity among material reference systems identified with normalized time-like vectors.
  When relative velocity among pair of reference systems is **not** deduced from Lorentz isometry transformation, *i.e.* if it is not an element of Lobachevsky factor space, but it is postulated a priori, then a set of all transformations among material reference-monads is not a group, but it is a groupoid.
- The Minkowski 1908-paper attracted most attention for constitutive equations that include electric permittivity $\mathbf{D} = \varepsilon\mathbf{E}$, and magnetic permeability $\mathbf{H} = \frac{1}{\mu}\mathbf{B}$, extended by Minkowski to moving material reference system.
  Rousseaux [2008] derived Galilean limit $\gamma \Rightarrow 1$, of constitutive equations, considering that only this limit can be tested experimentally in present day practice.

5. Observer: physical material reference system

Each measurement needs a choice of a **material** observer [Brillouin 1970]. What is a precise mathematical model for a physical material reference system?

5.1. Einstein in 1905: observer is a coordinate-basis = tetrad

Lorentz transformation is 'defined' frequently as transformation of scalar coordinate system, from $\{x^\mu\}$ to $\{x^{\mu\prime}\}$. Here x^μ is interpreted as a scalar field, $\mathbb{R}$-valued function on a manifold of spacetime events. Each coordinate system determine coordinate co-frame, $\{x^\mu\} \Longrightarrow \{dx^\mu\}$. In 1905 Einstein considered that a physical material reference system can be identified with a mathematical coordinate-basis in spacetime, an observer $= \{dx^\mu\}$. Such observer is known also as a (holonomic, integrable) frame, or as a (holonomic) tetrad. Whereas tensor fields are basis-free, i.e., observer-free within identification of an observer with a basis, therefore, every tensor field is absolute (not relative) within observers-bases. When observer is a basis, then, physically meaningful (experimentally measurable) quantities are scalar components of absolute-tensors, relative to a chosen coordinate-basis-observer. This is because these scalar components of absolute-tensors are observer-dependent, being dependent on a choice of an observer-basis.

Relativity transformation does not acts on events, does not acts on Minkowski space-time. Therefore does not acts on scalar fields, not on functions, not on coordinates. To say that a scalar field is Lorentz invariant (or GL-invariant) is empty phrase because Lorentz group does not acts on scalar fields. A phrase 'Lorentz-invariant parameter' e. g. in [Trump and Schieve 1999 p. 13] raise eyebrows. Domain of Lorentz group are all vectors, and this induce transformation of

Journal of Physics: Conference Series **330** (2011) 012012 doi:10.1088/1742-6596/330/1/012012

all tensors, except of scalar fields. Lorentz group commute with contraction of tensors, cf with [Minkowski 1908 §11.6 formula (43) and ff.].

Each observer-frame-tetrad, $\{dx^\mu\}$, defines a unique Lorentz metric tensor g, such that relative to g, this observer-frame is Lorentz orthonormal basis. Then, Lorentz isometry group is defined as a group permuting orthonormal observer-bases, leaving derived Lorentz metric-tensor invariant. The Lorentz-covariance, then, is identified with transformation of scalar components with respect to change of observers-frames. This is the philosophy shared by all present textbooks, see, for example [Hehl and Obukhov 2003, §D.5.4, pages 296–297]. The transformation of observer-frames and of scalar components, is considered to be only relativity transformations. All tensors that are basis-free are considered to be absolute = Lorentz-invariant.

A transformation is said to be passive if domain of transformation is a non-linear manifold of all bases of a vector space. 'Passive' means **active** action on coordinate-basis, but not on individual absolute-vectors. Some authors claim that tensor fields (like for example differential biform of electromagnetic field F), are invariant with respect to passive Lorentz transformations. This claim is incorrect, because every tensor field is basis-free and coordinate-free, and can be presented in arbitrary non-orthogonal basis, not necessarily in Lorentz-basis. Therefore, in fact, each tensor is invariant relative to passive general linear group, GL-group, acting on bases.

Of course, every tensor is GL-covariant, and also Lorentz-covariant, relative to GL and Lorentz transformations acting on individual vectors (and not on bases).

5.2. Euler's fluid and Minkowski in 1908: observer = monad

An objection against Einstein's identification of physical material reference system with mathematical coordinate-basis or with a frame, is that neither coordinates nor bases need physical concept of a material body. How distinguish coordinates of material "point" from not material point? No inertial mass in involved in a mathematical concept of a frame as a basis [Brillouin 1970]. In fact kinematics does not exists without a concept of inertial mass and center of mass, however in textbooks concept of a mass is considered as a part of dynamics, and not of kinematics.

An alternative point of view, different from Einstein's in 1905, is axiom that physics (observers, measurements, etc.) is coordinate-free, and basis-free. Because several physical concepts are observer-dependent, like electric field, hence an observer, within this axiom, cannot be identified with a mathematical basis.

If physics is coordinate-free and basis-free, then, mass-irrelevant old coordinate kinematics as presented for example in [Whittaker 1952], is useless, and must be replaced by Leonard Euler's material fluid introduced in 1754 as a vector field in spacetime. Euler's fluid was reincarnated by Minkowski in 1908: physical material reference system is identified with a time-like fluid vector field in a space-time, known also as a monad, and abbreviated in the present note as observer-monad, or as observer-vector. Within this view mathematical bases are irrelevant for physics. Clearly, a time-like vector field cannot describe a massless radiation, and therefore is related to some non-zero mass-density as in Euler's approach. Mass-density 'ρ' enter to energy-momentum tensor of a perfect fluid, and it is indispensable for two-body kinematics of center-of-mass and for reduced mass.

The monad-observers, time-like fluids, were re-invented independently by Eckart in 1940, Ehlers in 1961, and by Abraham Zelmanov (1913-1987) in his PhD Dissertation in 1944, and in his publication in 1976. Minkowski's first invention of monad-observer in 1908 went to oblivion. For discussion of Einstein's tetrad, versus Minkowski's monad, we refer to [Mitskievich 2006, Chapter 2].

Consider two material bodies, reference systems, **P** for **P**aul, and **R** for **R**ose. Each material reference system is a monad, *i.e.* it is g-normalized time-like vector field. All material bodies

Journal of Physics: Conference Series **330** (2011) 012012 doi:10.1088/1742-6596/330/1/012012

(inertial or non-inertial) in spacetime must be treated on the same footing and must therefore be not-distinguishable by normalization [Minkowski 1908 Part I §4 Eq. (19); Part II §8 Eq. (27) and §11.6 Eq. (46)],

$$\mathbf{P}^2 \equiv \mathbf{P}\cdot\mathbf{P} = \mathbf{R}\cdot\mathbf{R} = -1. \tag{5.1}$$

Vector field is coordinate-free, therefore coordinate expressions, like, $\mathbf{P} = \mathbf{P}^\mu \partial_\mu = \partial_{ct}$, are not so important. The scalar magnitude of light velocity in vacuo is denoted by c.

Let $\mathfrak{w} = \mathfrak{w}_x\partial_x + \mathfrak{w}_y\partial_y + \mathfrak{w}_z\partial_z$, be space-like velocity of Rose relative to Paul, i.e. a relative velocity as measured by Paul, $\mathfrak{w}\cdot\mathbf{P} = 0$. Two observers-monads are as follows [Minkowski 1908 §4 before time-like normalization (19); §11.6, before equation (46)],

$$\text{Rose's monad}\quad \mathbf{R} \equiv \gamma\left\{\mathfrak{w} + \mathbf{P}\right\}, \qquad \text{Paul's monad}\quad \mathbf{P} = \partial_{ct}. \tag{5.2}$$

Minkowski's observer is denoted by a letter 'w' in [Minkowski 1908 before Eq. (46)], and here is denoted by **R**ose **R**.

Minkowski in 1908 before Eq. (46)	Our notation in Equation (5.2)
$w \equiv \{w_1, w_2, w_3.w_4\}$	$\mathbf{R}$
i	$\mathbf{P}$
$\mathfrak{w} \equiv \{\mathfrak{w}_x, \mathfrak{w}_y, \mathfrak{w}_z\}$	$\mathfrak{w} = \mathfrak{w}_x\partial_x + \mathfrak{w}_y\partial_y + \mathfrak{w}_z\partial_z$

Minkowski's line before equation (46), is the definition of relative velocity $\mathfrak{w}$, and can be read as follows

$$w \equiv \{w_1, w_2, w_3, w_4\} \equiv w_1\partial_x + w_2\partial_y + w_3\partial_z + w_4\partial_{ict} = \gamma\{\mathfrak{w} + i\partial_{ict}\} \tag{5.3}$$

This pair of Minkowski's observers, Equation (5.2), **P** and **R**, is Eq. (8a) in [Hamdan 2006].

Heaviside [1888, 1889] introduced scalar factor γ, and his definition that follows, is equivalent to Minkowski's normalization $\mathbf{R}^2 = -1$ in [Minkowski Eq. (46)]

$$\mathbf{P}\cdot\mathbf{P} = \mathbf{R}\cdot\mathbf{R} = -1 \quad\Longrightarrow\quad \mathbf{P}\cdot\mathbf{R} = -\gamma, \tag{5.4}$$

$$\mathbf{R}^2 = -1 \quad\Longleftrightarrow\quad \left(\frac{\mathbf{u}}{c}\right)^2 = 1 - \frac{1}{\gamma^2}, \tag{5.5}$$

$$\gamma \equiv \frac{1}{\sqrt{1-\mathbf{u}^2/c^2}} \simeq 1 + \frac{1}{2}\left|\frac{\mathbf{u}}{c}\right|^2 + \frac{3}{8}\left|\frac{\mathbf{u}}{c}\right|^4 + \frac{5}{16}\left|\frac{\mathbf{u}}{c}\right|^6 + \ldots \tag{5.6}$$

6. Cross product of vectors in spacetime

Relativity transformation of electric and magnetic fields, see Equation (11.3) and Equation (12.15) below, need orientation-dependent Gibbs-like cross product of vectors, $\times$, in spacetime of dimension four. Gibbs's cross product of vectors, $\times$, needs Hodge-star isomorphisms (or duality). One Hodge-star acting on Grassmann's multi-vectors, and another Hodge-star acting on Grassmann multi-forms. Hodge star '$\star$' was invented by Hermann Grassmann under the name Ergänzung, and denoted by vertical dash |, [Grassmann 1862, Chapter 3, §4-5]. In 1908 Minkowski denoted dual of F by F^*. Star-notation, $\star$, dual of differential form F by $\star F$, introduced Hermann Weyl in 1945. Hodge-star isomorphism is a tensorial concomitant of metric tensor-field g, and depends on a choice of orientation [e.g. duality in Misner, Thorne & Wheeler 1973, §3.5; Kocik 1998 and WEB page; Oziewicz 1994; Cruz & Oziewicz 2003]. Cross-product of vectors gives an orientation-dependent *pseudo*-vector, *i.e.* strictly speaking this is *not* a binary operation.

In dimension three, binary cross product of vectors, $\mathbf{u}\times\mathbf{v} \equiv \star(\mathbf{u}\wedge\mathbf{v})$, was invented by Clifford, and was popularized by Heaviside in monograph *Electromagnetic Theory* [1893], and by Gibbs

Journal of Physics: Conference Series **330** (2011) 012012 doi:10.1088/1742-6596/330/1/012012

in *Vector Analysis* [1901]. However, Maxwell differential equations needs physical intensive and extensive vector fields on **four**-dimensional spacetime, and spacetime was 'not known' explicitly before Minkowski in 1908. Clifford-Gibbs cross product in dimension three is totally irrelevant for electromagnetism, have nothing to do with the subject matter of electric and magnetic fields, and it darkens ideas. Electromagnetic laws, four Maxwell's differential equations, ponderomotive force (called the Lorentz force), and relativity transformation of electric and magnetic fields, with Gibbs cross in *three* dimensions, become a thoughtless mechanical set of strange formulas, and this is mortal for electromagnetism and radiation in spacetime. Relativity transformation of electric and magnetic fields, see Equation (11.3) below in §11, ponderomotive force , **need** cross product of vectors in **four**-dimensional spacetime.

Generalization of cross product in arbitrary dimension was considered by Eckmann in 1942. Plebański with Przanowski in 1988 defined binary cross product of vectors in arbitrary dimension, in terms of augmented quaternion-like algebra of para-vectors. However both these attempts we consider not satisfactory because, among other things, orientation-dependence is either lost or it is not explicit.

Minkowski's definition of magnetic pseudo-field B needs Hodge-star isomorphism $\star$ acting on Grassmann's multiforms. Hodge-star intertwine the Grassmann exterior multiplication, the exterior wedge product acting on right, denoted by $e_{\mathbf{P}}\mathbf{R} \equiv \mathbf{P} \wedge \mathbf{R}$, with interior product that is dual to exterior, denoted by $i_{\mathbf{P}} \equiv (e_{\mathbf{P}})^*$,

$$\star \circ e_P = i_{\mathbf{P}} \circ \star, \qquad e_P \circ \star = \star \circ i_{\mathbf{P}}, \tag{6.1}$$

$$\mathbf{P}^2 \cdot \mathrm{id} = e_P \circ i_{\mathbf{P}} + i_{\mathbf{P}} \circ e_P. \tag{6.2}$$

For interior product we use also following abbreviation, $i_{\mathbf{P}}F \equiv \mathbf{P} \cdot F$.

6.1 Definition (Cross product in spacetime)**.** Let $\mathbf{A}$, $\mathbf{B}$ and $\mathbf{P}$ be vector fields on four-dimensional space-time. Gibbs-like 'binary' cross-product, $\times_{\mathbf{P}}$, of vector fields, $\mathbf{A}$ and $\mathbf{B}$, is orientation-dependent, and $\mathbf{P}$-dependent, and is defined in terms of the Hodge star map as follows,

$$\mathbf{A} \times_{\mathbf{P}} \mathbf{B} \equiv \star(\mathbf{A} \wedge \mathbf{P} \wedge \mathbf{B}) = -\mathbf{B} \times_{\mathbf{P}} \mathbf{A}. \tag{6.3}$$

Hence, cross-product of vectors in dimension four, is a 'ternary' operation. The same definition Equation (6.3) applies for covariant differential one-forms. In four-dimensional spacetime, binary cross $\times_{\mathbf{P}}$ depends on a choice of an auxiliary vector field $\mathbf{P}$. This vector-field-dependence of binary-cross in a spacetime, is of crucial importance for understanding. It is either not realized or thoughtlessly suppressed, when presenting Lorentz transformations of electric and magnetic fields, and when presenting ponderomotive (Lorentz) force as a tensorial concomitant of the electromagnetic field and electromagnetic spin-charge density, see subsection 12.1.1 and expression Equation (12.6) below.

6.2 Exercise. Let $\mathbf{P}^2 = -1$, $\mathfrak{w} \cdot \mathbf{P} = 0$, and $\mathbf{P} \cdot \mathbf{E} = 0$. Then

$$\mathfrak{w} \times_{\mathbf{P}} \{\mathfrak{w} \times_{\mathbf{P}} \mathbf{E}\} = \mathfrak{w}^2\, \mathbf{E} - (\mathfrak{w} \cdot \mathbf{E})\, \mathfrak{w}. \tag{6.4}$$

7. Minkowski in 1908: electric and magnetic fields are concomitants

In 1908 Minkowski introduced electromagnetic field as a differential bi-form F on spacetime. A differential biform of electromagnetic field, F (or a bivector field $\mathbf{F}$, $F \equiv g^{\wedge}\mathbf{F}$), are often called the Faraday tensors, however they were introduced in 1908 by Minkowski (1864-1909), and not by Michael Faraday (1791-1867).

It is convenient to consider electric and magnetic fields also as differential one-forms on four-dimensional spacetime, denoted by italics letters, E and B, instead of corresponding vector fields, $\mathbf{E}$ and $\mathbf{B}$, where $E \equiv g\mathbf{E}$, etc.

7.1 Definition (Electric and magnetic fields). [Minkowski 1908, Part II: Electromagnetic Phenomena, §11.6] Let observer Paul be given by time-like vector field $\mathbf{P}$, and can also be given as a time-like differential one-form, denoted by italic letter $P \equiv g\mathbf{P}$. These are fields on four-dimensional Lorentzian space-time with a metric tensor g.

Minkowski defined electric and magnetic fields on *four-dimensional spacetime* as differential one-forms being the following observer-dependent concomitants,

$$\begin{aligned} E(F, \mathbf{Paul}) &\equiv \quad \mathbf{P} \cdot F, \\ B(F, \mathrm{Paul}) &\equiv \star(P \wedge F) = \mathbf{P} \cdot (\star F), \qquad \star B(F, P) = P \wedge F. \end{aligned} \tag{7.1}$$

Electric and magnetic fields are concomitant of two variables, they are F-dependent (depends on electromagnetic sources via Maxwell equations), and they are observer-dependent (depends on time-like Paul- reference system).

7.2 Exercise. Set $E \equiv E(F, \mathbf{P})$ and $B \equiv B(F, P)$. As a corollary we have in particular the following relation,

$$E\mathbf{P} = 0 = \mathbf{E} \cdot \mathbf{P}, \qquad B\mathbf{P} = 0 = \mathbf{B} \cdot \mathbf{P}, \tag{7.2}$$

$$(\star F)^2 \simeq -F^2, \qquad -F = P \wedge E + \star(P \wedge B), \tag{7.3}$$

$$\begin{aligned} E^2 &= (\mathbf{P} \cdot F)^2 = P^2 F^2 - (P \wedge F)^2, \\ B^2 &= (\mathbf{P} \cdot \star F)^2 = P^2 (\star F)^2 - \{\star(\mathbf{P} \cdot F)\}^2. \end{aligned} \tag{7.4}$$

Minkowski's definition 7.1 was re-invented by number of authors, e.g. [Fecko 1997; Kocik 1998; Ivezić 2003; Cruz and Oziewicz 2003; Hehl and Obukhov 2003, §B.2.2 page 123, Definition (B.2.10)]. Hehl and Obukhov denoted observer vector field by 'n' in Hehl's (B.1.22), see pages 115-117, and depart from Minkowski Definition 7.1 in two respects. Firstly, Hehl and Obukhov are metric-free except of Chapter E.4. Their observer n is restricted by transverse condition for a given 'would-be-time σ' scalar field on page 115. This is similar to an idea that an observer is an idempotent endomorphism field, $n \otimes d\sigma$, with trace$(n \otimes d\sigma) = 1$ [Świerk 1988; Kocik 1998]. Many-observer, many-body metric-free kinematics generated by trace-class idempotents we considered in [Oziewicz 2007].

Secondly, Hehl and Obukhov stress that electric and magnetic fields are in **three-dimension** and not on spacetime [Hehl and Obukhov 2003, after (B.2.9) on page 123]. In Minkowski definition 7.1 electric field is on four-dimensional spacetime.

We must emphasize essential difference of Minkowski's 1908-Definition 7.1 when compared with other textbooks. According to Minkowski, electric field $\mathbf{E}$ is F-dependent and observer-dependent, depends on electromagnetic field, $\mathbf{E} = \mathbf{E}(F, \mathbf{P})$. Electromagnetic field F is absolute (observer-free), and electric field is **derived** concept that is observer-relative.

7.3 Misleading 'definition' (Electromagnetic field strength). In some textbooks, one can find the following 'definition'

$$F^{\alpha\beta} \equiv \begin{pmatrix} 0 & -E_x & -E_y & -E_z \\ E_x & 0 & -B_z & B_y \\ E_y & B_z & 0 & -B_x \\ -E_z & -B_y & B_x & 0 \end{pmatrix} \tag{7.5}$$

See for example [Sommerfeld 1948, 1964 §26 B and C; Landau and Lifshitz since 1951, edition 1975, formula (23.5) on page 61; Møller 1952 §53 page 141; Fock 1955, 1961 §24; Tonnelat 1959 Chapter 9; Jackson, since 1962, last edition 1999 formula (11.137); Barut 'Electrodynamics' 1964, 1980, page 96].

Journal of Physics: Conference Series **330** (2011) 012012 doi:10.1088/1742-6596/330/1/012012

What means 'definition' Equation (7.5)? Authors explain: on the right there are 'well known time-dependent three-dimensional' electric and magnetic fields, denoted by $\mathbf{E}(\mathbf{r}, t)$ and $\mathbf{B}(\mathbf{r}, t)$, as primary concepts given in terms of basis-dependent components, $\mathbf{E} = \{E_x, E_y, E_z\}$ and $\mathbf{B} = \{B_x.B_y, B_z\}$. Whereas we consider that electric field is basis-free. On the left is new 'tensor' F **defined** in terms of 'well known time-dependent three-dimensional' electric and magnetic fields, *i.e.* $F = F(\mathbf{E}, \mathbf{B})$, so electromagnetic field is defined as $\mathbf{E}$-dependent and $\mathbf{B}$-dependent.

Time-dependent electric field $\mathbf{E}$ must be vector field on four-dimensional spacetime, independently of how many basis-dependent components vanishes. Definition 7.5 suggest that time-component E_t vanishes or it is not known, or maybe $\mathbf{E}$ is in fact a time-dependent field on three-dimensional space? But in fact this never appeared time-component is nothing but $E_t = \mathbf{E} \cdot \mathbf{P}$ a scalar product with suppressed observer $\mathbf{P}$. We have the Minkowski conditions $\mathbf{E} \cdot \mathbf{P} = 0 = \mathbf{B} \cdot \mathbf{P}$, *i.e.* electric and magnetic fields are $\mathbf{P}$-constrained. Such observer-constrained vector field on four-dimensional space-time Ivezić call '3D' quantity, and we consider that this name is misleading.

One can try to interpret 'definition' Equation (7.5), as well as Sommerfeld's six-vector, as follows. Primary concept must be observer, electric and magnetic fields needs observer as the constraint, and then electromagnetic field in Equation (7.5) must be also observer-constrained, $F_{\mathbf{P}}$,

$$\text{observer} \quad \hookrightarrow \quad \text{electric and magnetic} \quad \hookrightarrow \quad \text{electromagnetic}, \tag{7.6}$$

$$\mathbf{P} \quad \hookrightarrow \quad \mathbf{E} \cdot \mathbf{P} = 0 = \mathbf{B} \cdot \mathbf{P} \quad \hookrightarrow \quad F_{\mathbf{P}}(\mathbf{E}, \mathbf{B}) \simeq F(\mathbf{P}, \mathbf{E}, \mathbf{B}). \tag{7.7}$$

Let look for explicit $\mathbf{P}$-dependence of electromagnetic field $F_{\mathbf{P}}$ in Equation (7.5). This task has the well known analogy in basis-dependent theorem.

Using identity Equation (6.2) [Minkowski 1908 §11.6 identity (45)] and the Minkowski definitions Equation (7.1), Minkowski deduced the following **theorem** [Minkowski formulas (55)-(56)],

$$F = \text{id}\ F \simeq E(F) \wedge P - i_{\mathbf{P}} \star B(F). \tag{7.8}$$

Minkowski's **theorem** Equation (7.8), some textbooks takes naively as the **definition** 7.5 of observer-dependent electromagnetic field,

$$F(P, E, B) \equiv E \wedge P - i_{\mathbf{P}} \star B, \quad \text{with} \quad E \cdot P = 0 = B \cdot P. \tag{7.9}$$

Coordinate-free and basis-free definition Equation (7.9) is equivalent to matrix 'definition' Equation (7.5).

In textbooks that accept 'definition' Equation (7.5), electric field $\mathbf{E}$ is observer-constrained vector field on *four*-dimensional space-time $\mathbf{E} \cdot \mathbf{P} = 0$. Scalar components, $E_x \equiv (dx)\mathbf{E} = \mathbf{E}x$, $E_t \equiv (dt)\mathbf{E} = \mathbf{E}t$, etc, are $\mathbf{E}$-dependent. But we must be careful here, because these textbooks consider that vector depends on his scalar components! For example, some authors consider that expression $\mathbf{E} = E_x \partial_x + \ldots$, is a *definition* (whereas it is a theorem!). These textbooks consider that on the right there are obvious and clear physical basis vector $\mathbf{i} \equiv \partial_x$ (dogma that basis is physical and not mathematical), and measurable scalar field $E_x \equiv (dx)\mathbf{E}$, whereas on the left there is an 'artificial' symbol, $\mathbf{E}$, that must be understood as a function of E_x, *i.e.* it is insisting that vector $\mathbf{E}$ is nothing but a set of components $\mathbf{E} \equiv \{E_x, E_y, E_z, E_t\}$ and vector $\mathbf{E} = \mathbf{E}(E_x, \ldots)$ is E_x-dependent. When we change E_x then such vector $\mathbf{E}$ must change, $\mathbf{E}(E_x, \ldots)$. It is hard to find worse case.

In fact, theorem, $\mathbf{E} = E_x \mathbf{i} + E_y \mathbf{j} + E_z \mathbf{k} + \ldots$, must be understand that a vector $\mathbf{E}$ and a basis $\{\mathbf{i}, \mathbf{j}, \mathbf{k}, \ldots\}$, are independent concepts, and scalar component E_x is $\mathbf{E}$-dependent, and is basis-dependent, $E_x(\mathbf{E}, \mathbf{i}, \mathbf{j}, \mathbf{k}, \ldots)$,

$$\text{Theorem:} \quad \mathbf{E} = E_x(\mathbf{E}, \mathbf{i}, \mathbf{j}, \mathbf{k}, \ldots)\, \mathbf{i} + E_y(\mathbf{E}, \mathbf{i}, \mathbf{j}, \mathbf{k}, \ldots)\, \mathbf{i} + \ldots \tag{7.10}$$

Journal of Physics: Conference Series **330** (2011) 012012 doi:10.1088/1742-6596/330/1/012012

The worst textbooks of algebra consider Equation (7.10) as the definition of a vector, without aware that every vector of a basis is also a vector, and is impossible to define a vector **E** in terms of another vector **i**, etc.

What means 'definition' Equation (7.5)? In order to define electromagnetic field F, we need first to have primary electric and magnetic fields, that probably must have never more than three components maybe, because component E_t is absent in 'definition' Equation (7.5). Time-dependence of such electric field is explicit in Maxwell equation, e.g. $\mathbf{E} = \partial_t \mathbf{A} + \ldots$, that must read as Lie-Ślebodziński derivative of potential in direction of observer field, $E = \mathcal{L}_{\mathbf{P}} A - d(A\mathbf{P})$. Such electric field with strange $E_t = 0$ or $E_t \neq 0$ must be known without of knowing E_t, before we can start to define electromagnetic field F Equation (7.5). According to some lecturers I knew personally F is artificial mathematical constructs that does not exists in Nature.

In present paper electromagnetic field F is primary concept, and it is **E**-independent, as state by Minkowski in 1908.

Textbook presentations of four Maxwell differential equations hidden reference system. Implicit reference system is a source of **wrong** interpretation of Maxwell equations by Dirac suggesting magnetic monopol. The explicit observer-dependence is given in [Minkowski 1908; Fecko 1997; Kocik 1998; Cruz Guzman & Oziewicz 2003].

8. Isometric relativity theory (not about gravity)

8.1 Definition (Minkowski 1908, Introduction and §11.6)**.** Theory where material reference systems are allowed to be connected by isometric Lorentz-group transformations **only** is said to be the *isometric relativity theory*. Isometric relativity, or Lorentz-group relativity, is defined by this **only one** postulate.

8.2 Clarification. The same definition of relativity theory was given by many authors: Schrödinger 1931; Peter Bergmann 1942, page 159; Einstein 1949; V. Bargmann 1957 p. 161; Barut 1964 Chapter I.6; Hehl and Obukhov 2003, §C.2, page 211.

It is worthy to emphasize that Albert Einstein in 1949 accepted Minkowski **one-axiom** definition, and forget about his two 1905-postulates including constancy of light velocity. Einstein separated special from general relativity in terms of properties of metric tensor. However the very concept of *relativity of observers* has nothing to do with properties of metric tensor, observers are *relative* independently of covariant derivative ∇g, independently of connections $\{\nabla\}$ and their curvatures and torsions.

8.3 Clarification. Often it is stressed incorrectly that relativity theory deals with *inertial* material systems only. This dogma insist that between non-inertial material bodies does not exist physically equivalent reference systems subject of relativity theory. Do Nature offer inertial reference systems at all? or they are not-natural mathematical idealizations only? After all condition to be inertial is *differential* condition on material system and it is hard to find in abundant 'special relativity' bibliography at least one paper where this differential condition is assumed and used. A material system **P** is said to be inertial if covariant derivative vanishes, $\nabla \mathbf{P} = 0$. Minkowski definition 8.1 hold for all material reference systems, including also non-inertial ones, as already noted in [Logunov 1990, 2004]. Logunov define (special) relativity theory as spacetime with absolute pseudo-Euclidean metric tensor g, and this implicitly implies that only allowed transformations between material reference systems are g-isometries, as in Minkowski Definition 8.1

8.4 Clarification. Definition 8.1 does **not** use a concept of relative velocity. Within such axiomatic isometric relativity theory, relative velocity must lay in Lobachevsky factor space. It was Vladimir Varićak in 1924 who developed non-Euclidean geometry as most natural structure for relativity theory. We stress here important fact (overlooked by Varićak) that Lobachevsky factor space is not unique, and this imply that relative velocity between given reference systems is also not unique, because every such relative velocity needs a priori a choice of Lobachevsky factor space, and each choice is equivalent to privileged Æther.

Lorentz group is in fact equivalent to reciprocity principle: each velocity of reference system **R** relative to an observer **P**, not unique $\mathbf{v}_{\mathbf{P},\mathbf{R}}$ in Lobachevsky factor-space, satisfies reciprocity condition, $\mathbf{v}_{\mathbf{P},\mathbf{R}} = -\mathbf{v}_{\mathbf{R},\mathbf{P}}$. For not uniqueness of relative velocity $\mathbf{v}_{\mathbf{P},\mathbf{R}}$ we refer to [Oziewicz 2007]. One can assume reciprocity of relative velocity, and deduce Lorentz group transformations, see, for example [Silagadze 2008].

8.5 Clarification. One can define Lorentz isometry group either as:

- Acting on observers-frames-tetrads with Lorentz 'passive' action on scalar components [Einstein 1905], for postulated absolute-vectors.
- Acting on individual covariant-vectors in Minkowski spacetime (when a basis is physically meaningless), with a Lorentz 'active'-action on the Lorentz-covariant tensor fields.
- Lorentz isometry group, as every isometry, does not act on manifold of events. This means that *all* scalar fields are GL-invariant, and in particular they are Lorentz-invariant.

8.6 Clarification. In Section 12 we define an alternative to above group-relativity. This alternative relativity theory is based on **only one** axiom that each ordered pair of material reference system possesses **only one unique** relative velocity. This unique relative velocity was introduced by Minkowski [Minkowski 1908 §11.6, before equation (46)], as the velocity of matter. As the consequence of this axiom a set of all relativity transformations among material reference systems is a groupoid that is not a group. This alternative relativity theory we call the *groupoid relativity*. Domain of groupoid transformation does not include light-like and space-like vectors. Therefore groupoid transformation do not extends to tensor algebra: it can not be neither isometry nor violate isometry. Metric tensor need not to be postulated as absolute.

9. Minkowski in 1908: Lorentz-covariance

Hermann Minkowski in Part II Electromagnetic Phenomena [Minkowski 1908, §11.6, after expression (44)], explains what Lorentz-covariance of a concomitant vector field, like Equation (7.1), means on a four-dimensional spacetime manifold. Let **F** denotes bi-vector field, and α is a differential Pfaff one-form (all tensor fields are on spacetime). Hence inner product can be interpreted in many ways,

$$(\mathbf{F}_*)\alpha \equiv \mathbf{F} \circ e_\alpha \equiv \alpha \cdot \mathbf{F} \equiv i_\alpha \mathbf{F}. \tag{9.1}$$

We repeat Minkowski's text in our notation. Let GL denote general-linear group, and let $L \in GL$. This means that L is an invertible endomorphism. GL-covariance, according to Minkowski, means the following set of transformations,

$$\begin{aligned} i_\alpha \mathbf{F} &\longmapsto L(i_\alpha \mathbf{F}), \\ \alpha &\longmapsto L^{*-1}\alpha, \\ \mathbf{F} &\longmapsto L^\wedge \mathbf{F}. \end{aligned} \tag{9.2}$$

Reader could ask why transformation of form must be 'contragradient'? We assume that every vector transform as $\mathbf{P} \longmapsto L\mathbf{P}$, and that GL-action commute with contraction. All scalar fields are GL-invariant, therefore

$$\alpha\mathbf{P} \longmapsto L(\alpha\mathbf{P}) = (L^?\alpha)(L\mathbf{P}) = (L^?\alpha \circ L)\mathbf{P} = \{(L^* \circ L^?)\alpha\}\mathbf{P} = \alpha\mathbf{P}. \tag{9.3}$$

$$\begin{aligned} (\mathbf{F}_*)\alpha \equiv i_\alpha \mathbf{F} \quad \xrightarrow{\text{transformation}} \quad & L(i_\alpha \mathbf{F}) = i_{L^{*-1}\alpha}\{L^\wedge \mathbf{F}\} = (L^\wedge \mathbf{F})_*(L^{*-1}\alpha) \\ & = (L \circ \mathbf{F}_* \circ L^*) \circ (L^{*-1}\alpha), \end{aligned} \tag{9.4}$$

$$(\mathbf{F}_*)\alpha \quad \longmapsto \quad \{(\mathbf{F}_*)\alpha\}' \equiv L((\mathbf{F}_*)\alpha) = (\mathbf{F}'_*)\,\alpha'. \tag{9.5}$$

This is, up to notation, Minkowski's definition of GL-covariance [Minkowski 1908, §11.6, after expressions (44)]. Minkowski is considering Lorentz-covariance (and notation A for our L^{*-1}), however, a concept of group-covariance is exactly the same for general linear group GL, as well as for any subgroup of GL, for example for Lorentz-isometry subgroup, $O_g \subset GL$, $O(3,1) \subset GL(4)$.

In 1908 Minkowski realized 'pure mathematical formality' [quotation from the first page of his Introduction], that Lorentz transformation is an isomorphism of a vector space, then entire algebra of tensor fields must be Lorentz-covariant. Every vector is Lorentz-covariant, and an observer time-like vector field, also must be Lorentz-covariant. All tensor fields, F and $\mathbf{P}$, must be Lorentz-covariant. Lorentz transformation must act on all tensor fields, including time-like vector fields. Hence electromagnetic field F, potential A, and **P**aul $\mathbf{P}$, must be Lorentz-covariant.

Lorentz-covariance of concomitant tensor fields that are observer-dependent, like Lorentz covariance of electric and magnetic fields, Equation (7.1), Lorentz covariance of charge and current densities, Equation (12.7)-Equation (12.8), is misunderstood in textbooks on electromagnetism. First definition of Lorentz-covariance of electric and magnetic fields, is Einstein's definition by means of transformation of *coordinate basis* in a system of four differential Maxwell equations, $\partial_t \mathbf{E} = \operatorname{rot} \mathbf{B}$, etc., [Einstein 1905, Part II. Electrodynamischer Teil, pages 907-909; Bergmann 1942, page 106; Hamdan 2006]. A fact that four Maxwell equations depend on a choice of observer, depend on a choice of a product structure (space $\oplus$ time)-split, was realized by Fecko in 1997 and by Kocik in 1998, see also [Cruz Guzman and Oziewicz 2003].

For how Lorentz-covariance and Lorentz-invariance is understood one hundred years after Minkowski we refer to [Arunasalam 2001].

10. Ivezić in 2005: Lorentz-covariance

In 2005 Ivezić re-discovered Minkowski's 'pure mathematical formality' of Lorentz-covariance. Ivezić's version of relativity is coined invariant special relativity = ISR, where invariant is synonym to be basis-free, as in [Misner, Thorne and Wheeler 1973, end of Chapter 2]. A tensor field is basis-free, therefore with respect to a 'passive' action of Lorentz group on non-linear manifold of bases-frames-tetrads, every tensor field is Lorentz-invariant. Such Lorentz-invariance is misleading, if a mathematical basis has no physical interpretation: basis-free tensor is trivially GL-invariant passively, and Lorentz isometry subgroup is here irrelevant. An alternative is to try to interpret a mathematical basis as a physical-experimental concept; however, this is outside of the philosophy of the present discussion.

Ivezić defined Lorentz-covariance of compound electric and magnetic fields, Equation (7.1), exactly as defined by Minkowski in 1908 in §11.6, just before formula (46). We stress that Minkowski in 1908 does **not** use in practice his definition of Lorentz-covariance. Lorentz transformation of electric and magnetic concomitant vector fields, according to Ivezić definition of Lorentz-covariance, is as follows.

Consider two reference systems, labeled **P** for **P**aul, and **R** for **R**ose, identified, following Minkowski in 1908, with time-like vector fields on four-dimensional Lorentzian space-time with a metric tensor g.

Space-like electric and magnetic fields (time dependent) as measured by **P**aul are denoted respectively by, $\mathbf{E}(\mathbf{P})$ and $\mathbf{B}(\mathbf{P})$. Analogously, we denote electric and magnetic fields as measured by **R**ose, by $\mathbf{E}(\mathbf{R})$ and $\mathbf{B}(\mathbf{R})$. It was *assumed* by Minkowski in [1908, §11.6], see Exercise 7.2 above, that

$$\mathbf{P}\cdot\mathbf{E}(\mathbf{P}) = 0 = \mathbf{P}\cdot\mathbf{B}(\mathbf{P}) \qquad \text{and} \qquad \mathbf{R}\cdot\mathbf{E}(\mathbf{R}) = 0 = \mathbf{R}\cdot\mathbf{B}(\mathbf{R}). \tag{10.1}$$

In Eq. Equation (10.1), all vector fields are on a **four**-dimensional spacetime manifold: in order to satisfy the differential Maxwell equations, time-dependence must be allowed.

Journal of Physics: Conference Series **330** (2011) 012012 doi:10.1088/1742-6596/330/1/012012

The question is: how are these electric and magnetic fields, Equation (10.1), which are due to the same sources, but are measured by two *different* observers in relative motion, how they are related?

The electric and magnetic fields are *relative*, they depend not only on electromagnetic sources, via the Maxwell differential equations, but also on the choice of a material reference system, e.g. the transformation Equation (10.4) or Equation (11.3) or Equation (12.15), that we are going to discuss in details.

Let $\mathbf{u}$ be space-like velocity of Rose relative to Paul, $\mathbf{u}\cdot\mathbf{P}=0$, [Minkowski 1908, §11.6, before Eq. (46)].

10.1 Assumption. $\mathbf{u}\cdot\mathbf{P}=0$,

$$\left.\begin{array}{l}\mathbf{E}\equiv\mathbf{E}(F,\mathbf{Paul})\\ \mathbf{B}\equiv\mathbf{B}(F,\mathbf{Paul})\end{array}\right\}\xrightarrow{\text{Lorentz-boost}}\left\{\begin{array}{l}\mathbf{E}^I\equiv\mathbf{E}(F^J,\mathbf{Rose})\\ \mathbf{B}^I\equiv\mathbf{B}(F^J,\mathbf{Rose})\end{array}\right. \tag{10.2}$$

$$\mathbf{E}\cdot\mathbf{P}=0=\mathbf{B}\cdot\mathbf{P},\qquad \mathbf{E}^I\cdot\mathbf{R}=0=\mathbf{B}^I\cdot\mathbf{R}. \tag{10.3}$$

10.2 Theorem (Ivezić 2005 page 307, formulas (8-9-10).

$$\begin{aligned}\mathbf{E}^I&=\mathbf{E}+\gamma\left(\mathbf{E}\cdot\frac{\mathbf{u}}{c}\right)\left\{\boldsymbol{P}+\frac{\gamma}{\gamma+1}\frac{\mathbf{u}}{c}\right\},\\ \mathbf{B}^I&=\mathbf{B}+\gamma\left(\mathbf{B}\cdot\frac{\mathbf{u}}{c}\right)\left\{\boldsymbol{P}+\frac{\gamma}{\gamma+1}\frac{\mathbf{u}}{c}\right\}.\end{aligned} \tag{10.4}$$

There are **no** formulae like, Equation (10.2)-Equation (10.4), in Minkowski's paper in 1908. In Equation (10.2)-Equation (10.16), a superscript I is for Ivezić. Minkowski never used his definition of group-covariance in practice. Ivezić-transformed electric field is no longer orthogonal to first observer, $\mathbf{E}^I\cdot\mathbf{P}=-\gamma\,\mathbf{E}\cdot\mathbf{u}/c$, according to relativity condition Equation (10.3).

Proof-Hint. Ivezić's transformation Equation (10.4) is specified case of well known Lorentz isometry transformation of vector, see for example [Fock 1955, 1961 §24], with only one notable difference, that in contrast to textbooks presentations here time-like observer $\mathbf{P}$ is explicit *variable*, Lorentz-covariant.

In 1937 Élie Cartan noted that isometry can be generated by Grassmann bivector. Consider the following bivector,

$$b\equiv\mathbf{P}\wedge\gamma\frac{\mathbf{u}}{c},\quad\text{with}\quad \mathbf{P}\cdot\mathbf{u}=0\quad\text{and}\quad\mathbf{P}^2=-1. \tag{10.5}$$

Let $\mathbf{E}$ be arbitrary vector. Isometry transformation of $\mathbf{E}$, generated by above bivector, L_b, is as follows [Oziewicz 2005, 2006, 2007, 2009],

$$L_b\mathbf{E}=\mathbf{E}+\gamma\left(\mathbf{E}\cdot\frac{\mathbf{u}}{c}\right)\left\{\mathbf{P}+\frac{\gamma}{\gamma+1}\frac{\mathbf{u}}{c}\right\}-(\mathbf{E}\cdot\mathbf{P})\left\{\gamma\frac{\mathbf{u}}{c}+(\gamma-1)\mathbf{P}\right\} \tag{10.6}$$

$$\Longrightarrow\qquad \mathbf{R}\equiv L_b\mathbf{P}=\gamma\left(\frac{\mathbf{u}}{c}+\mathbf{P}\right). \tag{10.7}$$

Lorentz transformation of vector Equation (10.6) is nothing but, for example, Fock's transformation in [Fock 1955 or 1961, formulas (24.39)-(24.40)], or Jackson's formulas (11.19) in [Jackson 1962]. Namely, vector $\mathbf{E}$ can be $\mathbf{P}$-decomposed as the sum of $\mathbf{E}^\perp$ that is orthogonal to time-like observer $\mathbf{P}$, and $\mathbf{E}^\parallel$, that is parallel to time-like observer,

$$\begin{aligned}\mathbf{E}\simeq\mathbf{E}^\perp+\mathbf{E}^\parallel,&\qquad\mathbf{E}^\perp\cdot\mathbf{P}\equiv0,\\ (L_b\mathbf{E})^\parallel&=\gamma\left\{\mathbf{E}^\parallel+\left(\mathbf{E}^\perp\cdot\frac{\mathbf{u}}{c}\right)\mathbf{P}\right\},\\ (L_b\mathbf{E})^\perp&=\mathbf{E}^\perp+\frac{\gamma^2}{\gamma+1}\left(\mathbf{E}^\perp\cdot\frac{\mathbf{u}}{c}\right)\frac{\mathbf{u}}{c}-\gamma(\mathbf{E}^\parallel\cdot\mathbf{P})\frac{\mathbf{u}}{c}.\end{aligned} \tag{10.8}$$

Journal of Physics: Conference Series **330** (2011) 012012 doi:10.1088/1742-6596/330/1/012012

Let compare Equation (10.4) with derivation by Ivezić in [2005, page 307]. Ivezić used rotor introduced in Clifford algebra by David Hestenes, for a simple bivector Equation (10.5),

$$R \equiv \frac{\gamma + 1 + b}{\sqrt{2(\gamma+1)}}, \qquad b \equiv \mathbf{P} \wedge \gamma \frac{\mathbf{u}}{c}. \tag{10.9}$$

Denoting space-like relative velocity by, $\beta \equiv \mathbf{u}/c$, and time-like observer by, $\gamma_0 \equiv \mathbf{P}$, Ivezić in 2005 arrived to following expression

$$\mathbf{E}^I = \tfrac{1}{2}(\gamma+1)\mathbf{E} + \gamma(\mathbf{E}\cdot\beta)\,\mathbf{P} \pm \tfrac{1}{2}\tfrac{\gamma^2}{\gamma+1}\,\beta\,\mathbf{E}\,\beta. \tag{10.10}$$

Last term in Equation (10.10) is Clifford product of three one-vectors,

$$\beta\,\mathbf{E}\,\beta = 2\,(\mathbf{E}\cdot\beta)\,\beta - \beta^2\,\mathbf{E}, \qquad \beta^2 = \frac{(\gamma+1)(\gamma-1)}{\gamma^2}. \tag{10.11}$$

Plus sign in Equation (10.10) leads to Ivezić's relativity transformation Equation (10.4). □

10.3 Corollary.

$$(\mathbf{E}^I)^2 = \mathbf{E}^2, \qquad (\mathbf{B}^I)^2 = \mathbf{B}^2, \qquad \mathbf{E}^I\cdot\mathbf{B}^I = \mathbf{E}\cdot\mathbf{B}, \tag{10.12}$$

$$(\mathbf{P}+\mathbf{u}/c)\cdot\mathbf{E}^I = 0, \qquad (\mathbf{P}+\mathbf{u}/c)\cdot\mathbf{B}^I = 0, \tag{10.13}$$

$$\mathbf{P}\cdot\mathbf{E}^I = -\mathbf{E}\cdot\gamma\mathbf{u}/c, \qquad \mathbf{P}\cdot\mathbf{B}^I = -\mathbf{B}\cdot\gamma\mathbf{u}/c. \tag{10.14}$$

10.4 Corollary.

$$\mathbf{R} \equiv \gamma\left(\mathbf{P} + \frac{\mathbf{u}}{c}\right) \qquad \Longrightarrow \qquad \mathbf{R}\cdot\mathbf{E}^I \equiv 0. \tag{10.15}$$

10.5 Corollary.

$$\begin{aligned} \mathbf{u}\cdot\mathbf{E} = 0 \quad &\Longleftrightarrow \quad \mathbf{E}^I = \mathbf{E}, \\ \mathbf{u}\cdot\mathbf{B} = 0 \quad &\Longleftrightarrow \quad \mathbf{B}^I = \mathbf{B}. \end{aligned} \tag{10.16}$$

11. Absolute observer

Different electromagnetic fields, variable electromagnetic fields can be registered in the same fixed reference system. Transformation of electromagnetic field dos not imply that an observer must also be transformed.

11.1 Side remark. In 1905 and again in 1907 Albert Einstein derived relativity transformation of electric and magnetic fields, transformations Equation (11.3) below, applying Lorentz isometry-group to a system of four differential Maxwell equations. Landau and Lifshitz re-derived transformations, Equation (11.3) below, without Maxwell equations [Landau and Lifshitz 1975, §24]. Whereas Hamdan re-deduced transformations Equation (11.3) using Maxwell equations as Einstein did [Hamdan 2006]. We note that four Maxwell's equations have already a fixed observer assumed a priori [Fecko 1997; Kocik 1998].

Observer need not to be inertial, and isometry need not commute with differential d, therefore Lorentz-covariance of **four** Maxwell differential equations that are observer-dependent is not obvious.

Journal of Physics: Conference Series **330** (2011) 012012 doi:10.1088/1742-6596/330/1/012012

In this section we assume that differential biform of electromagnetic field F 'would to be Lorentz-covariant'. Lorentz group is subgroup of symmetry group of massless Maxwell equations, thus maps solutions to solutions $F \longmapsto F^L$. But we will depart here from Lorentz-covariance that action of Lorentz group on tensor algebra and on Grassmann algebra, and in particular Lorentz-action on differential biforms, must be induced from action on vector fields. We assume that Lorentz group does **not acts** on vector fields. Observers are assumed **neither** Lorentz-covariant **nor** Lorentz-invariant. One can fix observer for variable electromagnetic fields and variable electromagnetic sources. No such group-covariance is logically allowed. In this meaning we do not need here Lorentz-covariance at all. After all one can consider a priori transformations of electromagnetic fields not induced by isometries, but induced by transformations of electromagnetic sources.

A change of electromagnetic field, $F \longmapsto F^L$, is interpreted as due to variable sources, due to moving electric charges and magnetic spins, a source motion relative to a **fixed** observer or relative to a fixed source. Electromagnetic source, charge-current density, need not to be time-like, can be light-like as well as space-like, therefore it is a different concept then a material time-like observer.

We accept Minkowski's definition of concomitant electric and magnetic fields Equation (7.1), where we assume that observer-monad **Paul** is fixed. Be fixed is not the same as to be Lorentz-invariant. We depart from Lorentz-covariance, because presented below transformation of biforms is independent of a choice of a fixed observers. Lorentz group is assumed do not acts on observers.

11.2 Assumption.

$$\begin{aligned} &F \xrightarrow{\text{transformation}} F^L \equiv L \circ F \circ L^*,\\ &\left.\begin{array}{r}\mathbf{E}(F,\mathbf{Paul})\\ \mathbf{B}(F,\mathbf{Paul})\end{array}\right\} \xrightarrow{\text{transformation}} \left\{\begin{array}{l}\mathbf{E}^L \equiv \mathbf{E}(F^L,\mathbf{Paul})\\ \mathbf{B}^L \equiv \mathbf{B}(F^L,\mathbf{Paul})\end{array}\right., \end{aligned} \tag{11.1}$$

$$\mathbf{P}\cdot\mathbf{E} = 0 = \mathbf{P}\cdot\mathbf{B}, \quad \text{and} \quad \mathbf{P}\cdot\mathbf{E}^L = 0 = \mathbf{P}\cdot\mathbf{B}^L. \tag{11.2}$$

In present section $\mathbf{u}$ is interpreted as a space-like velocity of moving charge-current relative to Paul; i.e., a relative velocity as measured by a fixed time-like Paul, $\mathbf{u}\cdot\mathbf{P} = 0$.

11.3 Theorem. *Above assumptions leads to following transformation of electric and magnetic fields.*

$$\begin{aligned} \mathbf{E}^L &= \gamma\left\{\mathbf{E}(F) + \frac{\mathbf{u}}{c}\times_{\mathbf{P}}\mathbf{B}(F)\right\} - \frac{\gamma^2}{\gamma+1}\left\{\frac{\mathbf{u}}{c}\cdot\mathbf{E}(F)\right\}\frac{\mathbf{u}}{c},\\ \mathbf{B}^L &= \gamma\left\{\mathbf{B}(F) - \frac{\mathbf{u}}{c}\times_{\mathbf{P}}\mathbf{E}(F)\right\} - \frac{\gamma^2}{\gamma+1}\left\{\frac{\mathbf{u}}{c}\cdot\mathbf{B}(F)\right\}\frac{\mathbf{u}}{c}, \end{aligned} \tag{11.3}$$

11.4 Clarification. Transformation Equation (11.3) must not be called the *relativity* transformation because reference system, an observer vector field $\mathbf{P}$ is here fixed. There is **only one** material reference system $P = g\mathbf{P}$ and two different massless electromagnetic fields $F^L \neq F$,

$$P\wedge E + \star(P\wedge B) = -F \quad \neq \quad -F^L = P\wedge E^L + \star(P\wedge B^L). \tag{11.4}$$

Transformed electric field with absolute (not covariant) observer is necessarily orthogonal to fixed Æther-like observer, $\mathbf{E}^L\cdot\mathbf{P} \equiv 0$. Absolute Æther-observer $\mathbf{P}$ and his proper-time are intact.

11.5 Corollary. $(\mathbf{E}^L)^2 - (\mathbf{B}^L)^2 = \mathbf{E}^2 - \mathbf{B}^2$.

11.6 Clarification. Transformation Equation (11.1)-Equation (11.3) coincides *formally* with given by observers-tetrads: Pauli 1921; Bergmann 1942, 1976; Sommerfeld 1948 §28; Fock 1955, 1961 §24, formulas 24.37-24.38 on page 102; Tonnelat 1959 Chapter 9; Jackson 1962, 1999, formula (11.149); Landau and Lifshitz 1975 §24; Tisza 1976, Chapter 4 (4.1.31); Ingarden and Jamiołkowski 1979, 1985 §18.2; Dvoeglazov & Quintanar Gonzlez 2006; Rousseaux 2008.

Jammer in monograph [1961, Chapter XI], interpreted relativity transformation as coordinate-change that transform the D'Alembert wave differential equation on **four**-dimensional spacetime into Poisson equation in **three**-dimensional space. We disagree with such interpretation [Oziewicz 2008].

John Field in 2006, considering a **moving** charge, derived another scalar factor at $\mathbf{u}\cdot\mathbf{E}$, just γ, instead of $\frac{\gamma^2}{\gamma+1}$ as it is in Equation (11.3):

$$\begin{aligned}\mathbf{B}(\text{static charge}) \equiv 0, \quad \mathbf{E} \equiv \mathbf{E}(\text{static charge}),\\ \mathbf{E}(\text{moving charge}) = \gamma\,\mathbf{E} - \gamma\left\{\frac{\mathbf{u}}{c}\cdot\mathbf{E}\right\}\frac{\mathbf{u}}{c}.\end{aligned} \tag{11.5}$$

Group-free derivation of transformation by Heaviside, by Thomson, and by Field in 2006, need not be the same as due to isometry.

11.7 Clarification. In 2005 Ivezić observed logical and mathematical inconsistency of textbook treatments of Lorentz-covariance. He noted that it is illogical to consider a closed differential biform F to be Lorentz-covariant, and at the same time, keep observer's time-like vector field, a '4-velocity', $\mathbf{P} \simeq (1,0,0,0)) \simeq \gamma_0$, to be Lorentz-absolute (not Lorentz-invariant and not Lorentz-covariant). For example, compare how absolute observer is hidden in calculations presented in [Misner, Thorne and Wheeler 1973, Chapter 3].

Proof of Theorem 11.3. Our 'proof' is not conceptually correct, because we use covariance as Grassmann algebra map, Equation (11.7)-Equation (11.8) below, and we only hope that such transformation of differential biform of electromagnetic field $F \longmapsto F^L$ (without transforming vectors) can be deduced from transformation of electromagnetic sources.

Let L be endomorphism of a module of vector fields. L extends to Grassmann algebra morphism $L^\wedge$. L-covariance means exactly that L extends to tensor-algebra morphism, Grassmann-algebra morphism and that L commute with evaluation. Let 'e' denotes left regular (adjoint) representation of Grassmann algebra, 'e' is for Grassmann's historical extension, exterior, creator. Grassmann algebra morphism means the following covariance-rule

$$\forall \quad \text{multivectors} \quad \mathbf{P},\mathbf{R}, \qquad e_{\mathbf{P}}\mathbf{R} \equiv \mathbf{P}\wedge\mathbf{R}, \tag{11.6}$$

$$L^\wedge \circ e_{\mathbf{P}} = e_{L^\wedge\mathbf{P}} \circ L^\wedge \quad \Longleftrightarrow \quad i_{\mathbf{P}} \circ L^{*\wedge} = L^{*\wedge} \circ i_{L^\wedge\mathbf{P}}. \tag{11.7}$$

Consider electric field as defined by Minkowski in 1908, Definition 7.1,

$$E \equiv i_{\mathbf{P}}F \quad \longmapsto \quad E^L \equiv i_{\mathbf{P}}F^L = i_{\mathbf{P}}\,L^{*\wedge}F = L^*\{i_{L\mathbf{P}}F\}. \tag{11.8}$$

The decomposition Equation (11.4) for $\mathbf{u}\cdot\mathbf{P} = 0$ gives

$$i_{\mathbf{u}}F = i_{\mathbf{u}}B(P) + \{E(P)\mathbf{u}\}\,g\mathbf{P}. \tag{11.9}$$

We now use Equation (10.5)-Equation (10.7). This gives transformation Equation (11.3). □

Journal of Physics: Conference Series **330** (2011) 012012 doi:10.1088/1742-6596/330/1/012012

12. Groupoid relativity is isometry-free

Since 2005 we are trying oppose dogmatic trend, that set of all relativity transformations acting on material observers, must coincide with Lorentz isometry group. Lorentz isometry group must act on *all* vectors, also on light-like vector, that can not be reference system. We argue, hence, that material reference systems need not to be necessarily connected by isometry [Oziewicz 2005, 2006, 2007]. Among others, Poincaré group is symmetry group of metric tensor of *empty* energy-less space-time.

12.1 Definition (Groupoid category)**.** A category consists of family of objects and family of arrows/morphisms. Category with every morphism being isomorphism, with two-sided inverse, is said to be *groupoid category*. In particular, a group is a groupoid one-object-category.

12.2 Definition (Relativity groupoid)**.** Let each object of a groupoid be a material body, not necessarily inertial, given in terms of future directed g-normalized time-like vector field, for instance as in Minkowski's Definition Equation (5.2). Two material bodies in a relative rest are considered to be the same one body - one reference system. Let each morphism be **unique** binary relative velocity as follows,

$$\text{If}\quad \mathbf{P}^2 = -1 \quad\text{and}\quad \mathbf{P}\cdot\mathbf{u} = 0 \quad\text{then:}\quad \mathbf{P} \xrightarrow{\ \mathrm{u}\ } \mathbf{R} = \gamma\left(\mathbf{P} + \frac{\mathbf{u}}{c}\right), \tag{12.1}$$

$$\frac{\mathbf{u}(\mathbf{P}\mapsto\mathbf{R})}{c} \equiv -\mathbf{P} - \frac{\mathbf{R}}{\mathbf{P}\cdot\mathbf{R}}, \tag{12.2}$$

This groupoid category is said to be the *relativity groupoid.*

Each pair of elements of a group is composable. In contrast, this is not the case for morphisms/arrows in a groupoid with more than one object. Consider four massive observers-monads, objects, $\mathbf{P}, \mathbf{R}, \mathbf{Q}$, and $\mathbf{S}$, with $\mathbf{R} \neq \mathbf{Q}$ and $\mathbf{S} \neq \mathbf{P}$.The groupoid arrows, $\mathbf{P} \longmapsto \mathbf{R}$, and, $\mathbf{Q} \longmapsto \mathbf{S}$, are not composable.

Minkowski's basis-dependent expressions [Minkowski 1908 Part II §8 (27)] and [Minkowski 1908 Part II §11.6 (46)], see Equation (5.1), must be understand as basis-free groupoid transformation Equation (12.1). Groupoid transformation, Equation (12.1)-Equation (12.2), we propose in Thesis by Świerk [1988]; and these expressions are foundation of monograph by Matolcsi [1994, Part II, §4.2, page 191]. Hehl and Obukhov's expression (E.4.10) in [2003, page 349], is essentially the same as Equation (12.1)-Equation (12.2). See also [Oziewicz since 2005].

Groupoid kinematics is basis-free, and postulate that electromagnetic field is absolute, i.e. that closed differential biform F, is observer-free for arbitrary non-inertial observer [Minkowski 1908, §11.6; Cruz & Oziewicz 2003; Oziewicz 2005, 2006, 2007]. Axiom of absolute F, is within the principle of absolute reality by Giovanni P. Gregori [2005, page 12], stating that physical reality of Nature is observer-free, that laws of Physics and Mathematics are absolute and do not need existence of observers = humans. Postulate of absolute electromagnetic field F is contrary to basis-free relativity, where closed differential bi-form of electromagnetic field, F, must be Lorentz-covariant (not absolute).

In a groupoid kinematics electromagnetic field is postulated to be absolute, *i.e.* observer-free, and we have,

$$\left.\begin{matrix}\mathbf{E}(F,\text{Paul})\\ \mathbf{B}(F,\text{Paul})\end{matrix}\right\} \xrightarrow[\qquad\qquad\qquad]{\text{groupoid action}\atop\text{not isometry}} \left\{\begin{matrix}\mathbf{E}^M \equiv \mathbf{E}(F,\text{Rose})\\ \mathbf{B}^M \equiv \mathbf{B}(F,\text{Rose})\end{matrix}\right. . \tag{12.3}$$

In Equation (12.3)-Equation (12.15), a superscript 'M' is for Minkowski.

Minkowski's implicit assumption, that F is observer-free, implies that relativity transformation of observer-dependent electric and magnetic fields is different when compared with the following cases:

- Lorentz-transformation of electromagnetic field with a fixed observer, Theorem 11.3.
- Minkowski's Lorentz-covariance in Ivezić's 'invariant special relativity' ISR-theory Equation (10.2)-Equation (10.16), [Ivezić 2003, 2005]. Transformation of electromagnetic field is induced from transformation of observer. Both transformations are related. One can chose any observer for a given electromagnetic field, but **group-covariance** require that every group action on observer must be together with induced action on electromagnetic field Equation (9.2).

12.1. Sources of electric and magnetic fields: spin

12.1.1. Ponderomotive force as a sum of three terms Observer-independent source vector field $\mathbf{J}$ is a sum of observer-dependent vector magnetic spin $\mathbf{s}$ and a scalar punctual charge density ρ as follows

$$\mathbf{J} = \mathbf{s}(\mathbf{R}) - \mathbf{R} \wedge \rho(\mathbf{R}) = \mathbf{s}(\mathbf{P}) - \mathbf{P} \wedge \rho(\mathbf{P}). \tag{12.4}$$

Analogously observer-free electromagnetic field, a differential bi-form F is a sum of observer-dependent magnetic field B and electric field E [Minkowski 1908 §11.6 identity (45)],

$$P \equiv g\mathbf{P} \quad \text{and} \quad R \equiv g\mathbf{R} \quad \Longrightarrow \quad F = B(\mathbf{R}) - R \wedge E(\mathbf{R}) = B(\mathbf{P}) - P \wedge E(\mathbf{P}). \tag{12.5}$$

Therefore observer-free ponderomotive differential form (identified with the Lorentz force) consists of **three** terms, because magnetic field does not interact with a scalar charge density, $\rho\, i_{\mathbf{P}} B \equiv 0$, whereas an electric field E interact both with charge and with magnetic spin giving also time-like contribution

$$\mathbf{J} \cdot F = -\rho\, E + \mathbf{s} \cdot B + (E\, \mathbf{s})\, P. \tag{12.6}$$

12.1.2. Convection weaker within groupoid. Let $\mathbf{u}$ be a space-like velocity of **R**ose relative to **P**aul, *i.e.* a relative velocity as measured by **P**aul, $\mathbf{u} \cdot \mathbf{P} = 0$.

12.3 Theorem. *Electric punctual charge-densities, scalar fields $\rho(\boldsymbol{P})$ and $\rho(\mathbf{R})$, and vectors magnetic spin- and current-densities, $\mathbf{s}(\boldsymbol{P})$ and $\mathbf{s}(\mathbf{R})$, as measured by these two observers are related by means of groupoid transformation as follows*

$$\rho(\mathbf{R}) = \gamma \left\{ \rho(\boldsymbol{P}) + \frac{\mathbf{u}}{c} \cdot \mathbf{s}(\boldsymbol{P}) \right\}, \tag{12.7}$$

$$\mathbf{s}(\mathbf{R}) = \mathbf{s}(\boldsymbol{P}) + \rho(\boldsymbol{P}) \left\{ \gamma^2 \frac{\mathbf{u}}{c} + (\gamma^2 - 1)\boldsymbol{P} \right\} + \gamma^2 \left(\frac{\mathbf{u}}{c} \cdot \mathbf{s}(\boldsymbol{P}) \right) \left(\boldsymbol{P} + \frac{\mathbf{u}}{c} \right). \tag{12.8}$$

The relativity transformation of the scalar charge-densities Equation (12.7), coincide with the Lorentz-group covariant transformation, compare for example with [Tonnelat 1959, Chapter 9, formulas (9.3) and (9.4).] Whereas the transformation Equation (12.8) differs from the Lorentz-group transformation, and follows from the groupoid kinematics considered here.

When $\mathbf{s}(\mathbf{P}) = 0$, then $\mathbf{s}(\mathbf{R})$ is purely convection current density,

$$\text{Electric convection} = \rho \left\{ \gamma^2 \frac{\mathbf{u}}{c} + (\gamma^2 - 1)\mathbf{P} \right\}, \tag{12.9}$$

$$|\text{Electric convection}| = \begin{cases} (\gamma^2 - 1)\,\rho & \text{within groupoid,} \\ \sqrt{\gamma^2 - 1}\,\rho & \text{within Lorentz group.} \end{cases} \tag{12.10}$$

Formula Equation (12.9) is within groupoid relativity prediction. We see that the electric convection current due to the charge in the motion, within groupoid relativity, contains only

even powers of relative velocity, Equation (12.12). This must be compared with the Lorentz transformations within the isometric special relativity, where the electric convection is predicted to be stronger
for $|\frac{\mathbf{u}}{c}| << 1$,

$$\sqrt{\gamma^2-1} = \left|\frac{\mathbf{u}}{c}\right|\gamma, \tag{12.11}$$

$$(\gamma^2-1) \simeq \left|\frac{\mathbf{u}}{c}\right|^2 + \left|\frac{\mathbf{u}}{c}\right|^4 + \ldots, \qquad \sqrt{\gamma^2-1} \simeq \left|\frac{\mathbf{u}}{c}\right|^1 + \tfrac{1}{2}\left|\frac{\mathbf{u}}{c}\right|^3 + \ldots \tag{12.12}$$

$$\sqrt{\gamma^2-1} = \gamma^2 - 1 \quad\Longleftrightarrow\quad \left|\frac{\mathbf{u}}{c}\right| = 0 \quad\text{or}\quad \left|\frac{\mathbf{u}}{c}\right| = \frac{1}{\sqrt{2}}. \tag{12.13}$$

12.2. Lorentz-covariance-free magnetic and electric fields

12.4 Theorem (Minkowski 1908, §11.6). *Let* $\mathbf{E}(\mathbf{R})$ *and* $\mathbf{B}(\mathbf{R})$ *be electric and magnetic fields measured by* $\boldsymbol{Rose} = \gamma(\boldsymbol{P}+\mathbf{u}/c)$*, Definition Equation (5.2). Let,* $\mathbf{E} \equiv \mathbf{E}(\boldsymbol{P})$*, and* $\mathbf{B} \equiv \mathbf{B}(\boldsymbol{P})$*, be the electric and magnetic fields as measured by Paul. According to Minkowski definition Equation (7.1)*

$$\boldsymbol{P}\cdot\mathbf{E}(\boldsymbol{P}) = 0 = \boldsymbol{P}\cdot\mathbf{B}(\boldsymbol{P}) \quad\Longrightarrow\quad \mathbf{R}\cdot\mathbf{E}(\mathbf{R}) = 0 = \mathbf{R}\cdot\mathbf{B}(\mathbf{R}). \tag{12.14}$$

Then, these fields are related by means of the following covariance-free transformation,

$$\begin{aligned}\mathbf{E}^M &\equiv \mathbf{E}(\mathbf{R}) = \gamma\left\{\mathbf{E}(\boldsymbol{P}) + \frac{\mathbf{u}}{c}\times_{\boldsymbol{P}}\mathbf{B}(\boldsymbol{P})\right\} + \gamma\left\{\frac{\mathbf{u}}{c}\cdot\mathbf{E}(\boldsymbol{P})\right\}\boldsymbol{P},\\ \mathbf{B}^M &\equiv \mathbf{B}(\mathbf{R}) = \gamma\left\{\mathbf{B}(\boldsymbol{P}) - \frac{\mathbf{u}}{c}\times_{\boldsymbol{P}}\mathbf{E}(\boldsymbol{P})\right\} + \gamma\left\{\frac{\mathbf{u}}{c}\cdot\mathbf{B}(\boldsymbol{P})\right\}\boldsymbol{P}.\end{aligned} \tag{12.15}$$

Note that transformations of electromagnetic sources, electric charge and spin-current densities, as given early in Equation (12.7)-Equation (12.8), are covariance-free, in the same manner as Equation (12.15).

The groupoid transformation, Equation (12.2) → Equation (12.3) → Equation (12.15), is induced on concomitants from the primary action on massive observers-monads, on time-like vector-fields only, Equation (12.2). The transformation of concomitant-fields Equation (12.15) were derived by Minkowski in [1908, §11.6, his Eqs. (47-48) and (51-52)]. Minkowski did not make it clear that such transformation among massive observers, Equation (12.2)-Equation (12.15), is *not* an isometry, because, among other, the domain do not include light-like vectors. The Minkowski groupoid transformations, Equation (12.2)-Equation (12.15), are *not* the Lorentz transformations.

Hamdan, University of Aleppo in Syria, died tragically in 2008. Hamdan similarly, considered in 2006 that a transformation of monads, Equation (12.2), is a Lorentz transformation, [Hamdan 2006, Eqs. (8a-8b)]. In fact this is a groupoid transformation, because a vector $\mathbf{P}$ *must* be time-like, and not all such transformations are composable. Besides, one can show directly that the groupoid transformation is not an isometry. For example, consider a pair of monads, $\mathbf{P} \neq \mathbf{Q}$, as follows,

$$\mathbf{P}\cdot\mathbf{u} = 0 = \mathbf{Q}\cdot\mathbf{u}, \quad \text{together with the groupoid isomorphisms,}$$

$$\mathbf{P} \xrightarrow{\ \mathbf{u}\ } \gamma\left(\mathbf{P}+\frac{\mathbf{u}}{c}\right), \quad\text{and,}\quad \mathbf{Q} \xrightarrow{\ \mathbf{u}\ } \gamma\left(\mathbf{Q}+\frac{\mathbf{u}}{c}\right), \tag{12.16}$$

$$\mathbf{P}\cdot\mathbf{Q} \xrightarrow{\ \mathbf{u}\ } \gamma^2\left(\mathbf{P}+\frac{\mathbf{u}}{c}\right)\cdot\left(\mathbf{Q}+\frac{\mathbf{u}}{c}\right) = \gamma^2\mathbf{P}\cdot\mathbf{Q} + \gamma^2 - 1 \neq \mathbf{P}\cdot\mathbf{Q}. \tag{12.17}$$

This shows that a groupoid transformation of monads is not an isometry.

Journal of Physics: Conference Series **330** (2011) 012012 doi:10.1088/1742-6596/330/1/012012

Let's compare the Lorentz-covariant electromagnetic field, $F \longmapsto F^L$, with **fixed** observer-monad-Æther, Equation (11.1)-Equation (11.3), and non-isometric groupoid transformation of monads, $P \xrightarrow{\mathbf{u}} R$, with an observer-free electromagnetic field F, Equation (12.2) → Equation (12.3) → Equation (12.15), where transformation of observers is not related with transformation of electromagnetic fields. The only visual difference are terms containing the scalar products, $\mathbf{u} \cdot \mathbf{E}$ and $\mathbf{u} \cdot \mathbf{B}$. These terms within 'Æther' Equation (11.3), are proportional to space-like relative velocity $\mathbf{u}$. Within relativity groupoid these terms give time-like contributions.

If spacelike relative velocity $\mathbf{u}$ is orthogonal in spacetime to electric and magnetic vector fields, then Equation (11.3) coincides with prediction of groupoid relativity in Theorem 12.4,

$$\text{if,} \quad \mathbf{u}\cdot\mathbf{E}(F) = 0 = \mathbf{u}\cdot\mathbf{B}(F) \quad \Longrightarrow \quad \mathbf{u}\cdot\mathbf{E}(F^J) = 0 = \mathbf{u}\cdot\mathbf{B}(F^J). \tag{12.18}$$

12.3. Experimental consequences

There are the following consequences,

$$\mathbf{u} \times_P \mathbf{B} = 0 \quad \Longrightarrow \tag{12.19}$$

$$\text{For fixed 'Æther':} \qquad \mathbf{E}^L \cdot \mathbf{E} - \gamma \mathbf{E}^2 = -\frac{\gamma^2}{\gamma+1}\left(\frac{\mathbf{u}}{c}\cdot\mathbf{E}\right)^2. \tag{12.20}$$

$$\text{Within relativity groupoid:} \quad \mathbf{E}^M \cdot \mathbf{E} - \gamma \mathbf{E}^2 = \qquad 0. \tag{12.21}$$

Whereas within Lorentz-covariance á la Minkowski 1908, ∀ $\mathbf{u}$, and ∀ $\mathbf{B}$, Equation (10.4), there is [Ivezić 2005],

$$\mathbf{E}^I \cdot \mathbf{E} - \mathbf{E}^2 = +\frac{\gamma^2}{\gamma+1}\left(\frac{\mathbf{u}}{c}\cdot\mathbf{E}\right)^2. \tag{12.22}$$

The differences among three theories are of the second order β^2, for $\beta \equiv \frac{\mathbf{u}}{c}$,

$$\gamma \simeq 1 + \tfrac{1}{2}\beta^2 + \ldots, \qquad \tfrac{\gamma^2}{\gamma+1} \simeq \tfrac{1}{2} + \tfrac{3}{8}\beta^2 + \ldots, \tag{12.23}$$

$$\mathbf{E}(\mathbf{u})\cdot\mathbf{E} - \mathbf{E}^2 = \tfrac{1}{2}\begin{cases} \beta^2\,\mathbf{E}^2 - (\mathbf{E}\cdot\beta)^2 & \text{fixed 'Æther' Equation (11.3)} \\ (\mathbf{E}\cdot\beta)^2 & \text{Ivezić's group-covariance} \\ \beta^2\,\mathbf{E}^2 & \text{groupoid relativity.} \end{cases} \tag{12.24}$$

We need to stress again that in expressions Equation (12.20)-Equation (12.22), the electric field measured by an observer 'at rest', and electric field measured by a moving observer-monad, within the three different relativity theories, all are the vector fields on **four**-dimensional space-time manifold,

$$\mathbf{E}, \quad \mathbf{E}^J, \quad \mathbf{E}^I, \quad \mathbf{E}^M \equiv \mathbf{E}(\text{Rose}), \quad \text{and} \quad \mathbf{u}. \tag{12.25}$$

In a randomly chosen not-adopted mathematical basis, each of these vector fields Equation (12.25) possesses **four non-zero** scalar components. In terminology used by Ivezić, the fields, $\mathbf{E}$ and $\mathbf{E}^J, \mathbf{E}^I, \mathbf{E}^M$, etc, in Equation (12.20)-Equation (12.22) and in Equation (12.25), are '4D-quantities'.

Note that there are the following implications

$$\left\{\begin{array}{l} \mathbf{u}\cdot\mathbf{E}(P) = 0 \\ \mathbf{u}\cdot\mathbf{B}(P) = 0 \end{array}\right\} \Longrightarrow \left\{\begin{array}{l} \mathbf{u}\cdot\mathbf{E}(R) = 0 \\ \mathbf{u}\cdot\mathbf{B}(R) = 0 \end{array}\right\} \Longrightarrow \left\{\begin{array}{l} P\cdot\mathbf{E}(R) = 0 \\ P\cdot\mathbf{B}(R) = 0 \end{array}\right\}$$
$$\Longrightarrow \quad \mathbf{E}(P)\wedge\mathbf{B}(P)\wedge\mathbf{E}(R)\wedge\mathbf{B}(R) = 0. \tag{12.26}$$

13. Conclusion

We propose an alternative for Einstein's special relativity. We suggest that the set of all relativity transformations between material reference systems (between time-like vector fields) could be a groupoid that it is *not* a group. One consequence of this groupoid-relativity, that is implicit in the Minkowski last publication in 1908 [Minkowski 1908 §11.6], is examined here for groupoid transformation of electric and magnetic fields, Theorem 12.4 and Equation (12.15), and for groupoid transformation of electromagnetic sources Equation (12.7)–Equation (12.8). This consequence could eventually be tested experimentally.

In the present paper we also repeated essentially what in 1908 Hermann Minkowski explained on Lorentz isometry group acting on vectors and on vector fields on space-time. The domain of the Lorentz isometry are vectors, and Lorentz isometry transformation of vectors induce Lorentz transformation of all tensor algebra except of scalar fields. The electric field is a vector field on space-time (is time-dependent), and under Lorentz isometry must transforms as every vector field. This (trivial) fact was independently re-discovered by Tomislav Ivezić [Ivezić 2003, 2005].

According to Minkowski and Ivezić, and according to present author, an electric field, as a vector field on space-time is defined as dependent on the electromagnetic field tensor F, *i.e.* $\mathbf{E} = \mathbf{E}(F)$. In 1908 Minkowski explained how Lorentz transformation of vectors induce transformation of tensors, and how transformation of a tensor F induce the transformation of a vectors built from other tensors.

Some present-day textbooks consider electric and magnetic fields as *primary* 'time-dependent three-dimensional vectors', versus the primary electromagnetic field F introduced in 1908 by Minkowski, *i.e.* some textbooks consider that $F \equiv F(\mathbf{E}, \mathbf{B})$, versus Minkowski's $\mathbf{E} = \mathbf{E}(F)$ and $\mathbf{B} = \mathbf{B}(F)$.

There are lecturers of electromagnetism (under name 'electrodynamics' with 'magnetism' removed by Ampère in 1828), convinced by definition of $F(\mathbf{E}, \mathbf{B})$ Equation (7.5) that electromagnetic field F is artificial mathematical construction without physical contents, and that in physical reality there are only electric $\mathbf{E}$ and magnetic $\mathbf{B}$ fields. Even this $\mathbf{E}$ must be a strange vector field on space-time with permanent amputee component E_t for not known scalar field t. Such strange 'amputee' vector field **on space-time** is called by many textbooks a '3D' quantity. This was called up by Tomislav Ivezić [Ivezić 2003, 2005, among other].

Acknowledgments

The present version was essentially inspired by extensive Skype and email discussions during many last years with Tomislav Ivezić from Zagreb. I am most cordially thankful to Tomislav for these important discussions. We agree with Tomislav about the meaning of Lorentz covariance and if Lorentz-group covariance is postulated then the only mathematically correct Lorentz-covariant transformations of electric and magnetic fields are transformations derived by Tomislav in 2003-2005. On the other hand we do not agree with interpretation of a concept of a vector in physical literature. I do not like $3D$- or $4D$-vector unlucky terminology, whereas Tomislav is constantly stressing importance of these names and the hidden meaning they keep.

We also disagree about interpretation of Minkowski publication in 1908. Minkowski's well known short 14-pages lecture in Cologne *Raum und Zeit* was published in 1909, and in fact it is a short introduction to and summary of 1908 long paper of 59 pages we are referring. With Tomislav we agree that Minkowski defined Lorentz-covariance as Tomislav is advocating since 2005, however Tomislav vigorously disagree that Minkowski introduced the modern concept of observer as a time-like vector field and a relative velocity among two such observers.

References

Arunasalsam V 2001 Lorentz covariance versus invariance: deeper insight, Physics Essays **14** (4) 329–340
Bargmann V 1957 Relativity, Reviews of Modern Physics **29** (2) 161–174

Barut A. O 1964 1980 Electrodynamics and Classical Theory of Fields and Particles, New York: Dover

Bateman H 1909/1910 The transformation of the electrodynamical equations, Proceedings of London Mathematical Society, Volume **8** (February 8), pages 223–264

Bergmann Peter Gabriel 1942 Introduction to the Theory of Relativity, Prentice-Hall, Inc. 1942; New York: Dover Publications, Inc. 1976

Bishop Richard L., and Samuel I. Goldberg 1968 1980 *Tensor Analysis on Manifolds* (New York: Dover Publications)

Brillouin Léon 1970 Relativity Reexamined, Academic Press, New York

Celakoska Emilija G 2008 On isometry links between 4-vectors of velocity, Novi Sad Journal of Mathematics **38** (3) 165–172

Celakoska Emilija G., Kostadin Trencevski and Dusan Cakmakov 2010 On Lorentz links between 4-vectors in Minkowski space, in progress in 2010

Cruz Guzmán José de Jesus, and Zbigniew Oziewicz 2003 Four Maxwell's equations for non-inertial observer, Bulletin de la Société de Sciences et des Lettres de Łodź, Volume LIII, Série: Recherches sur les Déformations, Volume XXXIX pp. 107-140; PL ISSN 0459-6854

Cunningham E 1909/1910 The principle of relativity in electrodynamics and extension thereof, Proceedings of London Mathematical Society, Volume **8** (February 8), pages 77–98

Dvoeglazov Valeri V., and J. L. Quintanar González 2006 A note on the Lorentz transformations for photon, Foundations Physics Letters **19** (2) 195–200; http://www.arxiv.org, arXiv physics/0410169.

Eckart Carl 1940 The thermodynamics of irreversible processes. Part III. Relativistic theory of simple fluid, The Physical Review **58** 919–924

Eckmann B 1942/1943 Comment. Math. Helvetica **17** page 318

Ehlers J 1961 Akad. Wiss. Lit. Mainz, Abhandl. Math.–Nat. Kl. **11**

Einstein Albert 1905 Zur Electrodynamik bewegter Körper, Annalen der Physik (Leipzig) **17** 891–921

Einstein Albert 1949 Autobiographisches, In: *Albert Einstein, Philosopher-Scientist* (Illinois)

Fecko Marián 1997 On 3+1 decomposition with respect to an observer field via differential forms, Journal of Mathematical Physics **38** (9) 4542–4560

Field John H 2006 Physica Scripta **74** 702, http://www.arxiv.org, arXiv physics/0501130

Fock Vladimir A 1955 1961 Space, Time and Gravitation, Moskva: GITTL. Nauka 1964. New York: Pergamon Press - MacMillan 1959, 1964. MR21#7042

Gibbs Josiah Willard 1901 *Vector Analysis*

Grassmann Hermann 1862 *Die Ausdehnungslehre* (Berlin: Verlag 1862; Extension Theory, American Mathematical Society 2000)

Gregori Giovanni P 2005 *Relativity, Quanta, Gravitation and Cosmology* (Rome: Science Edition 2005, ISSN 1615-2824)

Hajra Sankar 2011 On the history and status of relativity theory *Galilean Electrodynamics* **22** (3) 57–59

Hamdan Nizar 2006 On the invariance of Maxwell's field equations under Lorentz transformations, Galilean Electrodynamics **17** (6) 115–117

Heaviside Oliver 1888 The electro-magnetic effects of a moving charge, The Electrician **22** 147–148

Heaviside Oliver 1889 On the electromagnetic effects due to the motion of electrification through a dielectric, Philosophical Magazine Ser. **27** 324–339

Heaviside Oliver 1893 *Electromagnetic Theory* (London: The Electrician, Volume 1 in 1893, Volume 3 in 1912, 466 pages)

Hehl Friedrich W., and Yuri N. Obukhov 2003 *Foundations of Classical Electrodynamics*, (Boston, Berlin: Birkhäuser)

Hehl Friedrich W 2008 Maxwell's equations in Minkowski's world: their premetric generalization and the electromagnetic energy-momentum tensor, Annalen der Physik (Berlin) **17** 691-704, ArXiv gr-qc 0807.4249v1

Ingarden Roman Stanisław, and Andrzej Jamiołkowski 1979 1985 *Classical Electrodynamics* Warszawa: Państwowe Wydawnictwo Naukowe 1979, 1980, 1981. Amsterdam: Elsevier 1985

Itin Y. and Y. Friedman 2008 Annalen der Physik (August 2008) ArXiv gr-qc 0807.2625v1

Ivezić Tomislav 2003 The proof that the standard transformations of E and B are not the Lorentz transformations, Foundations of Physics **33** 1339

Ivezić Tomislav 2005 The proof that Maxwell equations with the 3D E and B are not covariant upon the Lorentz transformations but upon the standard transformations: the new Lorentz invariant field equations, Foundations of Physics **35** 1585

Ivezić Tomislav 2005 The difference between the standard and the Lorentz transformations of the electric and the magnetic fields, Foundations of Physics Letters **18** (4) 301–324

Ivezić Tomislav 2005 Axiomatic geometric formulation of electromagnetism with only one axiom: the field equation for the bivector field F with an explanation of the Trouton-Noble experiment, Foundations of Physics Letters **18** 401

Ivezić Tomislav 2006 Lorentz invariant Majorana formulation of the field equations and Dirac-like equation for the free photo, Electronic Journal of Theoretical Physics **10** 131–142

Ivezić Tomislav 2006 Four-dimensional geometric quantities versus the usual three-dimensional quantities: the resolution of Jackson's paradox

Ivezić Tomislav 2009 Generalized Uhlenbeck-Goudsmit hypothesis. Magnetic's and electric's spins

Ivezić Tomislav 2010 Lorentz transformations of the electric and magnetic fields according to Minkowski, Physica Scripta **82** 1–6; preprint arXiv: physics.gen-ph 0906.3166

Ivezić Tomislav 2011 The electromagnetic field equations for moving media. Preprint arXiv:1101.3292 v1 physics.gen-ph 17 January 2011

Ivezić Tomislav 2011 The constitutive relations and the magnetoelectric effect for moving media with homogeneous and isotropic electric and magnetic properties

Jackson John David 1962 1975 1999 *Classical Electrodynamics* (New York: John Wiley & Sons ISBN 83-01-00309-X)

Jackson John David 2002 From Lorenz to Coulomb and other explicit gauge transformations, American Journal of Physics **70** (9) 917–928; http://ojps.aip.org/ajp/

Jammer Max 1961 *Concepts of Mass* (Harvard University Press)

Jancewicz Bernard 2008 Pre-metric electrodynamics, Annalen der Physik (August 2008) arXv: physics.class-ph 0807.2989v1

Kocik Jerzy 1998 Relativistic observer and Maxwell's equations: an example of a non-principal Ehresmann connection
http://www.math.siu.edu/jkocik.htm, University of Illinois at Urbana preprint P-98-10-029

Landau Lev Davidovich, and Evgenii Mikchailovich Lifshitz 1951 1975 *The Classical Theory of Fields* (Oxford - New York: Pergamon Press)

Logunov A. A 1990 Lectures in Relativity and Gravitation, Pergamon Press

Logunov A. A 2004 Henri Poincaré and Relativity Theory,
arXiv:physics/0408077 v3 23 August 2004

Matolcsi Tamás 1994 *Spacetime without Reference Frames* (Budapest: Akadémiai Kiadó ISBN 963 05 6433 5)

Minkowski Hermann 1908 Die Grundlagen für die electromagnetischen Vorgänge in bewegten Körpen, Nachrichten von der König. Gesellschaft der Wissenschaften zu Göttingen, Mathematisch-Physikalische Wiss. Klasse, 53–111.

- Reprinted in: Minkowski Hermann 1911 *Gesammelte Abhandlungen* Unter Mitwirkung A. Speiser, H. Weyl, Hrsg. David Hilbert (Leipzig, Berlin: B. G. Teubner 1911, Bd. 2. S. 352–404).
- The French translation by Paul Langevin, manuscript at:
http://hal.archives-ouvertes.fr/hal-00321285/fr/.
Paul Langevin, inventor of the twin paradox, paradox known also as the Langevin paradox.
- 1920 The fundamental equations for electromagnetic processes in moving bodies, http://en.wikisource.org, in: The Principle of Relativity, Calcutta University Press pp. 1–69. Translated by Megh Nad Saha.
- 1983The Russian translation in: *Einsteinian Sbornik 1978-1979* Moskva: Nauka pages 5-63
- 2010 Hermann Minkowski, The fundamental equations for electromagnetic processes of mobil bodies, pages 1–45, translated by A. F. Kracklauer

Minkowski Hermann 1910 (paper written by Max Born after Minkowski's death in 1909) Eine Ableitung der Grundgleichungen für die electromagnetischen Vorgänge in bewegten Körpern vom Standpunkt der Electronentheorie, Math. Ann. **68** 526–556

Misner Charles W., Kip S. Thorne, John Archibald Wheeler 1973 *Gravitation* (San Francisco: W. H. Freeman and Company)

Mitskievich Nikolai V 2006 *Relativistic Physics in Arbitrary Reference Frames* (Huntington, New York: Nova Science Publishers, Inc.)

Møller C 1952 ... 1969 The Theory of Relativity, Oxford at the Clarendon Press

Nunez Yepez H. N., A. L. Salas Brito, and C. A. Vargas 1988 Revista Mexicana de Física **34** (4) 636–644

Oziewicz Zbigniew 1994 Classical field theory and analogy between Newton's and Maxwell's equations", Foundation of Physics **24** (10) 1379–1402. http://www.arxiv.org, arXiv hep-th/9312009

Oziewicz Zbigniew 2005 How do you add relative velocities? In: Pogosyan George S., Luis Edgar Vicent and Kurt Bernardo Wolf, Editors, *Group Theoretical Methods in Physics* (Bristol: Institute of Physics, Conference Series Number 185, ISBN 0-7503-1008-1). Full version http://www.worldnpa.org/pdf/group25.pdf

Oziewicz Zbigniew 2006 The Lorentz boost-link is not unique http://www.arxiv.org, arXiv math-ph/0608062
http://www.worldnpa.org/pdf/Merida.pdf

Journal of Physics: Conference Series **330** (2011) 012012 doi:10.1088/1742-6596/330/1/012012

Oziewicz Zbigniew 2006 Electric field and magnetic field in a moving reference system, short version, In: M. R. Adhikari et al. , Editors, *Physical Interpretations of Relativity Theory* (Kolkata: International Symposium on Recent Advances in Mathematics, Calcutta Mathematical Society) pp. 47–53

Oziewicz Zbigniew 2007 Relativity groupoid instead of relativity group, International Journal of Geometric Methods in Modern Physics **4** (4) 1-11; http://www.arxiv.org, arXiv math.CT/0608770 http://www.worldnpa.org/pdf/ijgmmpOziewicz.pdf

Oziewicz Zbigniew 2007 *T*ernary relative velocity. Astonishing conflict of the Lorentz group with relativity (Physical Interpretatiions of Relativity Theory) ed M. C. Duffy and V. O. Gladyshev (Moscow: Bauman Moscow State Technical University ISBN 978-5-7038-3178-6) pp 292–303 (arXiv:1104.0682 v1 physics.gen-ph 29 March 2011) http://www.worldnpa.org/pdf/ternaryvelocity.pdf

Oziewicz Zbigniew 2008 Electric field and magnetic field in a moving reference system, Review Bulletin of the Calcutta Mathematical Society **16** (1) 49–66

Oziewicz Zbigniew, and Cynthia Kolb Whitney 2008 Electric and magnetic fields according to Hermann Minkowski, Proceedings of the Natural Philosophy Alliance NPA, Albuquerque 2008, Volume **5** No. 2, pages 183–194, http://www.worldnpa.org/php/

Oziewicz Zbigniew 2009 Isometry from reflections versus isometry from bivector, Advances in Applied Clifford Algebras **19** (3-4) 793–817

Oziewicz Zbigniew, and William S. Page 2011 Concepts of relative velocity preprint arXiv:1104.0684 v1 physics.gen-ph 30 March 2011

Plebański Jerzy Franciszek, and Maciej Przanowski 1988 Notes on a cross product of vectors, Journal of Mathematical Physics **29** (11) 2334–2337

Rindler Wolfgang 1969 1977 *Essential Relativity, Special, General, and Cosmological* (New York: Springer-Verlag; Oxford University Press 2001)

Rousseaux Germain 2008 On the electrodynamics of Minkowski at low velocities, EPL - A Letters Journal Exploring the Frontiers of Physics **84** 20002–p1–p4, www.epljournal.org

Schrödinger Ernst 1931 Specielle Relativitätstheorie und Quantenmechanik", Sitzungsber. Preuss. Akademie Wissen. phys.-math. **Bd. 12** 238–247

Silagadze Z. K 2008 Relativity without tears, Acta Physica Polonica **B 39** (No. 4) 811–885

Sommerfeld Arnold 1948 *Electrodynamics* (Wiesbaden 1948, Academic Press 1964, §28.C, pp. 239–241)

Świerk Dariusz 1988 *Relativity theory and product structures*, Master Thesis, Uniwersytet Wrocławski, Instytut Fizyki Teoretycznej, Poland, 1988, pages 1–42

Thomson J. J. (Lord Kelvin) 1889 The Phil. Mag. pages 1–14

Tisza László 1976 http://ocw.mit.edu/ans7870/resources/tisza/index.htm, Chapter 4, (4.1.31)

Tonnelat Marie-Antoinette 1959 *Les Principes de la Théorie Électromagnétique et de la Relativité* (Paris, Masson et Cie)

Trump Matthew A., and W. C. Schieve 1999 *Classical Relativistic Many-Body Dynamics* (Dordrecht: Kluwer Academic Publishers, Fundamental Theories of Physics Volume **103**)

Varićak Vladimir 1924 *Darstellung Der Relativitätstheorie Im Dreidimensionalen Lobatschefskijschen Räume* (Zagreb 1924); *Relativity in three dimensional Lobachevsky space*, translated by A. F. Kracklauer (Lulu 2007, ISBN: 978-1-84753-364-7)

Weyl Hermann 1945 Annals of Mathematics **44** (1945) pp. 1–6

Whittaker Edmund T 1952 *A treatise on the analytical dynamics of particles and rigid bodies with an introduction to the problem of three bodies* (Cambridge University Press)

Zel'manov A. L 1945 Monad formalism, (Ph. D. Thesis), Doklady Akademii Nauk SSSR **227** (1979) page 78

Journal of Physics: Conference Series **330** (2011) 012013 doi:10.1088/1742-6596/330/1/012013

Equations of motion in general relativity and quantum mechanics

Paul O'Hara

Dept. of Mathematics, Northeastern Illinois University, 5500 North St. Louis Avenue, Chicago, Illinois 60625-4699

E-mail: `pohara@neiu.edu`

Abstract. In a previous article a relationship was established between the linearized metrics of General Relativity associated with geodesics and the Dirac Equation of quantum mechanics. In this paper the extension of that result to arbitrary curves is investigated. A generalized Dirac equation is derived and shown to be related to the Lie derivative of the momentum along the curve. In addition,the equations of motion are derived from the Hamilton-Jacobi equation associated with the metric and the wave equation associated with the Hamiltonian is then shown not to commute with the Dirac operator. Finally, the Maxwell-Boltzmann distribution is shown to be a consequence of geodesic motion.

KEY WORDS: non-geodesic motion, Dirac equation, equations of motion, Maxwell-Boltzmann distribution.

1. Introduction

In a recent paper Frank Tipler [10] gives a derivation of the Schrodinger equation using the Hamilton-Jacobi principle of action. In doing so, he is able to transpose Erwin Schrodinger's "*purely* formal procedure" [1] of replacing $\frac{\partial W}{\partial t}$ in the Hamilton-Jacobi equation with $\pm\frac{h}{2\pi i}\frac{\partial}{\partial t}$ with a more direct derivation of the wave equation. Likewise, in a paper by Marie-Noelle Celerier and Laurent Nottale [4] in 2003, the Dirac equation is derived by taking the square root of the Klein-Gordan equation using the "bi-quaternionic action" associated with geodesic motion and the Hamilton-Jacobi principle of action. Other approaches not directly based on the use of the action principle have also been used. For example, Ng and Dam used a geometrical derivation of the Dirac equation by exploiting "rotational invariance" and "the explicit use of the spin-$\frac{1}{2}$ property of ψ" [5], while Martin Rivas quantized a Poincare invariant Hamiltonian in which the spin angular momentum of the particle is constant with respect to the center of mass observer to derive the Dirac equation.[9]

This paper contains parallels with the approach of Tipler, and Celerier and Nottale in that it makes use of the Hamilton Jacobi equation but it also differs in that it emphasizes the fundamental role of the metric in enabling us to derive two equations associated with non-geodesic motion: one a generalized Dirac equation, which captures the kinematics, and the other an equation of motion describing the dynamics.

Throughout the paper, $(\mathcal{M}, g)$ will denote a connected four dimensional Hausdorff manifold, with metric g of signature -2. At every point p on the space-time manifold $\mathcal{M}$ we erect a local tetrad $e_0(p), e_1(p), e_2(p), e_3(p)$ such that a point x has coordinates $x = (x^0,\ x^1,\ x^2,\ x^3) = x^a e_a$ in this tetrad coordinate system, while the spinor Ψ can be written as $\Psi = \psi^i e_i(p)$, where ψ^i represent the coordinates of the spinor with respect to the tetrad at p. This is permissable since a spinor is also an element of a vector space. In particular, when each ψ^i is equal for each i then we can write $\Psi = \psi\xi$ where the spinor $\xi = (e_i)$ and $||\Psi|| = 1$. When this occurs, we say that ψ is a scalar field. Such a scalar field will occur when Ψ is parallel transported along a geodesic. For example, in the case of the Dirac equation, the plane wave solution can be written as

$$\Psi(x) = e^{-ipx/\hbar} u_+(p), \tag{1}$$

where $\psi = e^{-ipx/\hbar}$ and $u_+(p) = \begin{pmatrix} u_1 \\ u_2 \end{pmatrix}$ and allows for two independent solutions for each momentum p. Also in this case, the spinor $u(p)$ remains constant along a geodesic. This reflects Rivas result in that the derivation of the Dirac equation presupposes the motion of a particle "in a plane orthogonal to S, which is constant in this frame," [9] where S refers to spin-angular momentum. In terms of quantum mechanics each constant can be identified with a quantum number. Also at each point p we can establish a tangent vector space $T_p(\mathcal{M})$, with basis $\{\partial_0, \partial_1, \partial_2, \partial_3\}$ and a dual 1-form space, denoted by T_p^* with basis $\{dx_0, dx_1, dx_2, dx_3\}$ at p, defined by

$$dx^\mu \partial_\nu \equiv \partial_\nu x^\mu = \delta^\mu_\nu. \tag{2}$$

We refer to the basis $\{dx^0, dx^1, dx^2, dx^3\}$ as the basis of one forms dual to the basis $\{\partial_0, \partial_1, \partial_2, \partial_3\}$ of vectors at p.

With notation clarified, we note that in a previous article [8], the Dirac equation was derived as a dual of a linearized metric tensor for geodesics. The linkage was accomplished in a natural way by associating a generalized Dirac equation with those operators which are duals of differential one-forms, obtained by linearizing the metric tensors of General Relativity (expressed locally as a Minkowski metric). Specifically, in that paper, if

$$ds^2 = g_{\mu\nu} dx^\mu dx^\nu = \eta_{ab} dx^a dx^b \tag{3}$$

where a and b refer to local tetrad coordinates and η to a rigid Minkowski metric tensor of signature -2, it was shown that a matrix $\tilde{ds} \equiv \gamma_a dx^a$ with eigenvalue ds could be associated with the metric tensor $\tilde{ds} \otimes \tilde{ds}$ (which from now on we will denote by ds^2) such that $\{\gamma_a, \gamma_b\} = 2\eta_{ab}$.

In addition, by noting that the $\tilde{ds}$ matrix is the dual of the expression $\tilde{\partial}_s \equiv \gamma^a \frac{\partial}{\partial x^a}$ this enabled us to associate a generalized Dirac equation

$$\tilde{\partial}_s \psi\xi \equiv \gamma^a \frac{\partial \psi}{\partial x^a} \xi = \frac{\partial \psi}{\partial s} \xi, \tag{4}$$

with the motion of a particle along a geodesic, with a fixed spinor ξ.

This paper is an extension of that previous work. First, we re-investigate this result by showing that the existence of a Hamilton-Jacobi function associated with a family of space filling curves is equivalent to defining the Dirac equation at almost every point on the manifold. Indeed, this association also suggests a deeper understanding of the term "wave-particle duality." Secondly, we show that the existence of such functions also determines the existence of both quantum and classical ideal gases defined in terms of the Bose-Einstein and Boltzmann statistics respectively. Thirdly, we introduce accelerations and then show that the integrability of the Hamilton-Jacobi function determines a coherent set of natural motions and also implies the non-commutativity of the Hamiltonian and action operators associated with non-geodesic motion.

Journal of Physics: Conference Series **330** (2011) 012013 doi:10.1088/1742-6596/330/1/012013

2. Generalized Dirac equation

The generalized Dirac equation defined above relies on the definition of the four-momentum in special relativity and upon the fact that $\tilde{\partial}_s\psi$ and $\tilde{ds}$ are parallel along geodesics, and consequently by the chain rule their product is $\frac{d\psi}{ds}ds$. In contrast, when accelerations are introduced we will find that in general

$$\frac{\tilde{ds}}{ds}.\tilde{\partial}_s\psi = \frac{1}{2}\left\{\frac{\tilde{ds}}{ds},\tilde{\partial}_s\psi\right\} + \frac{1}{2}\left[\frac{\tilde{ds}}{ds},\tilde{\partial}_s\psi\right] \tag{5}$$

$$= \frac{d\psi}{ds} + \frac{\vec{ds}}{ds} \wedge \frac{\partial\vec{\Psi}}{\partial s}. \tag{6}$$

In this regard, note that equation (5) represents an operator equation and the four γ matrices used to define $\tilde{ds} = \gamma^a dx_a$ and $\tilde{\partial}_s\psi = \gamma^a \frac{\partial\psi}{\partial x_a}$ are a spinor representation of four unit vectors constituting a local Minkowski tetrad at any point on the curve. Equation (6), on the other hand, is written in the usual vector notation. The two equations can be identified by noting that the anti-commutator and commutator relationships in (5) associated with the spinor tetrad, define a dot product and a cross product respectively, which can can also be expressed in standard vector notation as in equation (6). Direct calculation of the dot product of $\tilde{ds}$ and its dual $\tilde{\partial}_s\psi$ or equivalently of $d\vec{x}^a$ and $\frac{\vec{\partial\psi}}{dx_a}$ gives $\frac{d\psi}{ds}ds$.

This latter term can also be directly related to the Hamilton-Jacobi characteristic function, which in turn can be associated with a coherent set of natural motions [3]. Indeed, to remove any ambiguity, we begin with the following definitions:

Definition 1 *A function* $W = \int_{\sigma(\lambda)}\left(\mathbf{p}\frac{d\mathbf{x}}{d\lambda} - H\frac{dt}{d\lambda}\right)d\lambda$ *is called a Hamilton-Jacobi function if the integral is path independent for all curves in* $\{\sigma(\lambda) \in (\mathcal{M},g)\}$ *and* $\frac{\partial W}{\partial t} = -H(x_1,x_2,x_3,t,\frac{\partial W}{\partial x_1},\frac{\partial W}{\partial x_2},\frac{\partial W}{\partial x_3})$ *defined with respect to a local tetrad .*

Equivalently, we can say that $dW = \mathbf{pdx} - Hdt$ is an exact differential.

We begin by focusing on the kinematics associated with the function W and by showing how it can be related to the Dirac equation. This leads to the following lemmas and corollaries:

Lemma 1 *Let* $\psi(W(x^a))$ *be a differentiable function of* W *and let* $\psi' = \frac{d\psi}{dW}$. $W = -\int \eta^{ab}p_a dx_b = \int \boldsymbol{p}d\boldsymbol{x} - Hdt$ *is the Hamilton-Jacobi function, iff* $\psi(W) = \int \boldsymbol{p}^*d\boldsymbol{x} - H^*dt$ *is also a Hamilton-Jacobi function.*

Proof:($\Rightarrow$) First note $p_a^* = \psi' p_a$, from the differentiability of $\psi(W)$, and $p_o = H$. Since W is a Hamilton-Jacobi function:

$$\begin{aligned}
\frac{\partial\psi(W)}{\partial t} &= \psi'\frac{\partial W}{\partial t} \\
&= -\psi' H(x_1,x_2,x_3,t,p_1,p_2,p_3) \\
&= -\psi' H(x_1,x_2,x_3,t,p_1^*/\psi',p_2^*/\psi',p_3^*/\psi') \\
&= -H^*(x_1,x_2,x_3,t,p_1^*,p_2^*,p_3^*).
\end{aligned}$$

Therefore, $\psi(W)$ is a Hamilton-Jacobi function.

($\Leftarrow$) Conversely if $\psi(W)$ is a Hamilton-Jacobi function then

$$\begin{aligned}
\frac{\partial W}{\partial t} &= \frac{\partial\psi(W)}{\partial t}/\psi' \\
&= -H^*(x_1,x_2,x_3,t,p_1^*,p_2^*,p_3^*)/\psi' \\
&= -H(x_1,x_2,x_3,t,p_1^*/\psi',p_2^*/\psi',p_3^*/\psi') \\
&= -H(x_1,x_2,x_3,t,p_1,p_2,p_3).
\end{aligned}$$

Journal of Physics: Conference Series **330** (2011) 012013 doi:10.1088/1742-6596/330/1/012013

Therefore, W is a Hamilton-Jacobi function.

In addition, from equations (5) and (6) we obtain the following Lemma which gives a necessary and sufficient condition for the existence of such Hamilton-Jacobi functions.

Lemma 2 *Let $\psi(W(\mathbf{x},t))$ be a differentiable function and $\{\sigma(\lambda)\}$ a family of curves on the manifold with unit tangent vectors $\frac{\tilde{ds}}{ds}$ with respect to a local tetrad then $\tilde{ds}.\tilde{\partial}_s\psi(W)$ is an exact differential iff $\psi(W)$ is a Hamilton-Jacobi function such that $p^*_a = \frac{d\psi}{ds}\frac{dx_a}{dt} = \frac{\partial\psi_{\|}(W)}{\partial x^a}$, where $\psi(W) = \psi_{\|}(W) + \psi_{\perp}(W)$.*

Proof: From equations (5), if $\frac{\tilde{ds}}{ds}.\tilde{\partial}_s\psi(W)$ is an exact differential then $[\frac{\tilde{ds}}{ds}, \frac{\partial\tilde{\psi(W)}}{\partial s}]$ is also an exact differential. This means that $\psi(W) = \psi_{\|}(W) + \psi_{\perp}(W)$ where $[\frac{\tilde{ds}}{ds}, \frac{\partial\tilde{\psi_{\|}(W)}}{\partial s}] = 0$ and $\psi_{\perp}(W) = c_o t + c_1 x_1 + c_2 x_2 + c_3 x_3$, with c_o, c_1, c_2, c_3 being constants.

Also, $[\frac{\tilde{ds}}{ds}, \frac{\partial\tilde{\psi_{\|}(W)}}{\partial s}] = 0$ means $\frac{dx^a}{d\lambda}$is parallel (for any parameter λ including the curve length s) to $\frac{\partial\vec{\psi_{\|}(W)}}{\partial s}$, and $\exists g(\lambda)$ such that $p^*_{\|a} \equiv g(\lambda)\frac{dx_a}{d\lambda} = \frac{\partial\psi_{\|}(W)}{\partial x^a}$.

Also given $\frac{\tilde{ds}}{ds}.\tilde{\partial}_s\psi_{\|}(W)$ is an exact differential and denoting $x_0 = t$, gives

$$\begin{aligned}\frac{\tilde{ds}}{ds}.\tilde{\partial}_s\psi_{\|}(W) &= \frac{\partial\psi_{\|}(W)}{\partial x^1}\frac{dx^1}{ds} + \frac{\partial\psi_{\|}(W)}{\partial x^2}\frac{dx^2}{ds} + \frac{\partial\psi_{\|}(W)}{\partial x^3}\frac{dx^3}{ds} + \frac{\partial\psi_{\|}(W)}{\partial t}\frac{dt}{ds} \\ &= \frac{d\psi_{\|}(W)}{ds}.\end{aligned}$$

On substituting $g(s)\frac{dx_a}{ds} = \frac{\partial\psi_{\|}(W)}{\partial x^a}$ and noting that $\frac{dx^a}{ds}\frac{dx_a}{ds} = 1$ gives $g(s) = \frac{d\psi_{\|}(W)}{ds}$. It follows that

$$\left(\frac{\partial\psi_{\|}(W)}{\partial t}\right)^2 = (p^{*1}_{\|})^2 + (p^{*2}_{\|})^2 + (p^{*3}_{\|})^2 + \left(\frac{d\psi_{\|}(W)}{ds}\right)^2 = \left(H^*_o\left(x^a, p^{*1}_{\|}, p^{*2}_{\|}, p^{*3}_{\|}\right)\right)^2. \quad (7)$$

Therefore $\psi_{\|}(W)$ is a Hamilton-Jacobi function, as also is $\psi(W) = \psi_{\|}(W) + \psi_{\perp}(W)$ with $p^*_a = p^*_{\|a} + c_a$.

Conversely, given $p^*_a = \frac{d\psi}{ds}\frac{dx_a}{dt} + c_a = \frac{\partial\psi_{\|}(W)}{\partial x^a} + \frac{\partial\psi_{\perp}(W)}{\partial x^a} = \frac{\partial\psi(W)}{\partial x^a}$ then $\left[\tilde{ds}, \tilde{\partial}_s\psi_{\|}\right] = 0$ and since $\psi(W)$ is a Hamilton-Jacobi function, it follows from Equation (5) and the definition of $\psi_{\perp}$ that

$$\tilde{ds}.\tilde{\partial}_s\psi_{\|}(W) = d\psi_{\|}(W) \qquad \text{and} \qquad \left[\frac{\tilde{ds}}{ds}, \frac{\partial\tilde{\psi_{\perp}(W)}}{\partial s}\right]$$

are both integrable. Therefore,$\tilde{ds}.\tilde{\partial}_s\psi(W)$ is an exact differential. The result follows.

Corollary 1 *If $\psi(W) = W = -m_o s + k$ is a Hamilton-Jacobi function for each fixed k then $g(s) \equiv m_o$ is called the rest mass associated with linear motion in the plane $W(s,k)$, and*

$$H^2 = p_1^2 + p_2^2 + p_3^2 + m_o^2.$$

Proof: Follows from Equation (7).

We now show that for the family of curves defined in Lemma 2, we can always construct a Hamilton-Jacobi function locally at each point on the curve.

Corollary 2 *Let $ds^2 = dx^a dx_a$ define a metric tensor locally along a piece-wise smooth congruence of curves, $\{\sigma(\lambda) = x^a(\lambda)\}$, which fill coordinate space with only one curve passing through each point, then a Hamilton-Jacobi function can be constructed such that*

$$W = \int_\sigma m\frac{ds}{d\lambda}ds = \int \mathbf{p}d\mathbf{x} - m\frac{dt}{d\lambda}dt, \qquad where \quad p^a = m\frac{dx^a}{d\lambda}, \; m = m_o\frac{d\lambda}{ds}. \tag{8}$$

Note $\frac{\partial W}{ds} = m_o$ is called the rest mass.

Proof: At every space-time point in a gravitational field, by the Principle of Equivalence, an inertial frame can be constructed. Now, define $p^a = m\frac{dx^a}{d\lambda}$, $m = m(s)\frac{d\lambda}{ds}$, where $m(s) = m_o$ is a constant. Then $dW \equiv m\left(\frac{ds}{d\lambda}\right)^2 d\lambda = m_o ds$ is clearly an exact differential, such that

$$W = m_o s + k = m_o\sqrt{x_1^2 + x_2^2 + x_3^2 - t^2} + k, \qquad k \text{ constant}$$

in a neighborhood of any point. Therefore $\tilde{\partial}_s W = \gamma^a \frac{\partial W}{\partial x^a}$ exists almost everywhere and $dW = \tilde{ds}.\tilde{\partial}_s\psi(W)$. It follows from Lemma 2 that W is a Hamilton-Jacobi function.

Remark: The corollary has shown the existence of W at every point on the manifold by showing that W can be constructed in terms of a local tetrad as a linear function of the arc length, s. In effect, this means that we are defining the Hamilton-Jacobi function along a geodesic passing through the point x_o. In practice, it is sufficient to integrate $\frac{\partial W}{\partial x^a} = m_o\frac{dx^a}{ds}$ to obtain W, as in the example below. Moreover, although $p_o = p(x_o)$ is a constant along the geodesic passing through x_o, this does not mean that $p^a = \frac{\partial W}{\partial x^a}$ is constant along an arbitrary curve. Indeed, this latter condition only follows, if motion is along a geodesic.

Example: As an example of the corollary, consider a motion of a particle of rest mass m_o fired into the air without resistance. The Equation of Motion for such a projectile in Minkowski space are given by

$$\mathbf{F} = -m_o g\mathbf{j}, \tag{9}$$

or equivalently, denoting $\frac{dx}{ds} = \dot{x}$,

$$m_o\ddot{x} = 0 \qquad m_o\ddot{y} = -m_o g. \tag{10}$$

Solving the equations gives

$$x = x_0 + u_x s \quad \text{and} \qquad y = y_0 + u_y s - \frac{1}{2}gs^2. \tag{11}$$

However, in terms of the corollary above, we begin not with equations of motion (9) but with the line increments associated with the kinematics described by equation (11). Specifically,

$$\begin{aligned} -ds^2 &= dx^2 + dy^2 - dt^2 \\ \text{iff} \quad - m_o ds &= m_o\dot{x}dx + m_o\dot{y}dy - m_o\dot{t}dt \\ \text{iff} \quad - m_o ds &= m_o u_x dx + m_o(u_y - gs)dy - m_o\dot{t}dt. \end{aligned}$$

If we now let $dW = -m_o ds$ along a geodesic and require that W be an integrable function passing through (x_0, y_0, t_0), we obtain $W = m_o u_x x + m_o(u_y - gs)y - m_o c^2\dot{t}t + w_o$. Indeed, on taking partial derivatives, we find

$$\frac{\partial W}{\partial x} = m_o u_x = p_x, \qquad \frac{\partial W}{\partial y} = m_o(u_y - gs) = p_y \qquad \left(\frac{\partial W}{\partial t}\right) = -m_o\dot{t}$$

and

$$\left(\frac{\partial W}{\partial t}\right)^2 = m_o^2 + p_1^2 + p_2^2 + p_3^2 = H^2.$$

Therefore, H is a Hamilton-Jacobi function.

In effect, Lemma 2 establishes a relationship between Hamilton-Jacobi functions and the commutator relationship, $[\tilde{ds}, \tilde{\partial}_s\psi(W)]$ applied to Equation (5) (or equivalently Equation (6)). We now use the same commutator relationship to establish another important property relating the dual operator $\tilde{\partial}_s$ and the metric operator $\tilde{ds}$ associated with the increments along a curve. Indeed, Equation (5) could be described as the most general form of a "wave-equation" associated with a curve in space-time. However, we put the expression "wave-equation" in quotes to emphasize that $\psi(W)$ is not necessarily a wave-function of quantum mechanics. For the moment, ψ they can be any C_1 function defined on the manifold. In a previous paper, we have noted that ψ can only be interpreted as a quantum wave function when further restrictions are imposed on the function space, such as requiring that it be an L^2 function. Intuitively, we could think of $\tilde{\partial}\psi(W(s))$ as a wave associated with the vibration of a curve $\sigma(s)$ in space-time, whose tangent is $\tilde{ds}$, with respect to a local tetrad coordinate system. This leads to the following lemma:

Lemma 3 *If $\psi(W)$ is a Hamilton-Jacobi function such that $[\tilde{\partial}_s W, \tilde{ds}] = 0$ then there exists a simultaneous eigenfunction ξ such that*

$$(\tilde{\partial}_s\psi)\xi(p) = \partial\psi_s\xi(p), \tag{12}$$

where $\partial\psi_s = \frac{\partial\psi}{\partial s}$, which in the case of geodesic motion reduces to

$$\tilde{\partial}_s\Psi = \frac{d\Psi}{ds}, \qquad \textit{where} \qquad \Psi = \psi\xi. \tag{13}$$

Remark: *$(\tilde{\partial}_s\psi)\xi = \tilde{\partial}_s\Psi$ in general, since x is independent of p in phase space. However, $\frac{d\Psi}{ds} \neq \psi'(p)\xi$ unless motion is along a geodesic.*

Proof: First note that $[\tilde{\partial}_s W, \tilde{ds}] = 0$ implies $[\tilde{\partial}_s\psi, \tilde{ds}] = [\psi'(W)\tilde{\partial}_s W, \tilde{ds}] = 0$. Therefore, there exists simultaneous eigenvectors $\xi = \xi(p)$ such that $\tilde{ds}\xi = ds\xi$ and $(\tilde{\partial}_s\psi)\xi = \gamma^a p_a^*\xi(p) = \gamma^a p_a\psi'\xi(p) = mc\psi'(p)\xi(p) = (\partial_s\psi)\xi(p)$.
Also, $\xi(p)$ is constant along a geodesic and therefore

$$\tilde{\partial}_s\Psi = \frac{d\Psi}{ds}, \qquad \text{where} \qquad \Psi = \psi\xi.$$

The result follows.

Corollary 3 *Let $ds^2 = dx^a dx_a$ define a metric tensor on a manifold and $p^a = m(s)\frac{dx^a}{ds}$ along a curve. Then there exists a Hamilton-Jacobi function $\psi(W)$, and a vector $\xi(p)$ such that*

$$(\tilde{\partial}_s\psi)\xi(p) = (\partial_s\psi)\xi(p).$$

Proof: By Cor.2 it is sufficient to define $p_a = \frac{\partial W}{\partial x^a}$, and solve for W to obtain a Hamilton-Jacobi equation. It now follows that

$$[\tilde{\partial}_s W, \tilde{ds}] = [\gamma^a \frac{\partial W}{\partial x^a}, \gamma^a dx_a] = [\gamma^a p_a, \gamma^a dx_a] = 0. \tag{14}$$

Journal of Physics: Conference Series **330** (2011) 012013 doi:10.1088/1742-6596/330/1/012013

But from Lemma (3) if $[\tilde{\partial}_s W, \tilde{ds}] = 0$, there exists a $\psi = \psi(W)$ such that

$$(\tilde{\partial}_s\psi)\xi(p) = (\partial_s\psi)\xi(p).$$

The result follows.

Remark: We refer to Equation (12) as a generalized Dirac equation associated with a curve, and Equation(13) as a generalized Dirac equation associated with geodesics. It reduces to the usual form of the Dirac equation if we let $\psi = Ae^{\kappa W}$, where A is an arbitrary constant and $\kappa = \frac{i}{\hbar}$:

Corollary 4 *Let $\psi = Ae^{\kappa W}$ where $W = -ms + k$ as defined in Cor. 1 and $\kappa = \frac{i}{\hbar}$ then*

$$\gamma^a \frac{\partial \Psi}{\partial x^a} = -\frac{i}{\hbar} m\Psi. \tag{15}$$

Proof: Direct substitution for W in Equation (13) reduces to the conventional Dirac equation

$$\gamma^a \frac{\partial \Psi}{\partial x^a} = -\frac{i}{\hbar} m\Psi.$$

The above equation can be rewritten in the conventional form, if we multiply across by $-i\hbar\gamma^0$, define $\alpha^0 = \gamma^0$, $\alpha^a = \gamma^0\gamma^a$ and let $E = i\hbar H^* = i\hbar\frac{\partial\psi}{\partial t}$ to get

$$\left[-i\hbar\left(\alpha_1\frac{\partial}{\partial x_1} + \alpha_2\frac{\partial}{\partial x_2} + \alpha_3\frac{\partial}{\partial x_3}\right) + \alpha_0 m\right]\Psi = E\Psi \qquad \text{where} \qquad \Psi = \psi\xi. \tag{16}$$

Corollary 5 *Consider motion in a plane with Hamilton-Jacobi equation $W = -m_o s + c_1x^1 + c_2x^2 + c_3x^3 + c_0x^0 + d = W_\parallel + W_\perp$. Let $S_a \equiv \frac{\partial W_\perp}{\partial x^a} = c_a$ (called the spin), $\tilde{d\sigma} = \gamma_a dx^a + \gamma_a(c^a/m_o)ds$, $\tilde{\partial}_\sigma W \equiv \gamma^a\frac{\partial W}{\partial x^a} = \gamma^a(p_a + c_a)$, then there exists an eigenfunction ξ such that*

$$(\tilde{\partial}_\sigma\psi)\xi(p) = (\partial_\sigma\psi)\xi(p). \tag{17}$$

Proof: By construction $[\gamma^a\frac{\partial W}{\partial x^a}, \tilde{d\sigma}] = 0$ therefore by Lemma 3

$$(\tilde{\partial}_\sigma\psi)\xi(p) = (\partial_\sigma\psi)\xi(p).$$

Remark: (1) This corollary confirms Martin Rivas' observation that the Dirac equation presupposes the motion of a particle "in a plane orthogonal to S, which is constant in this frame," [9] where S is the spin.

(2) The duality relating line increments and the Dirac equation brings to the foreground philosophical issues regarding the difference between quantum and classical mechanics, and in particular how the two might be related. A more detailed discussion of this point will be found in the next section and also in [8].

Next, we prove a theorem that relates the Lie derivative and the Dirac equation for a particle in a closed system.

Theorem 1 *Consider a family of curves $\{\sigma(s)\}$ parameterized by the curve length s with unit tangent vectors $u^a = \frac{dx^a}{ds}$ associated with the increments $ds^2 = dx^a dx_a$ then the Lie derivative $\mathcal{L}_u(\mathbf{p}) = 0$ iff there exists a Hamilton-Jacobi function $\psi(W)$ such that $[\gamma^a\frac{\partial(W)}{\partial x^a}, \tilde{ds}] = 0$ and*

$$(\tilde{\partial}_s\psi)\xi(p_\parallel) = (\partial_s\psi)\xi(p_\parallel).$$

Journal of Physics: Conference Series **330** (2011) 012013 doi:10.1088/1742-6596/330/1/012013

Proof:($\Rightarrow$) Since $\mathcal{L}_u(\mathbf{p}) = 0$ there exists a coordinate system (in fact the rest frame) [2] such that $u^a = \delta^a_0$ and $p^a_{,0} = 0$. Therefore, in a general Lorentz frame $p^a = m_o u^a + m_o v^a$ where m_o is constant and $\frac{\partial v^a}{\partial x^b}\frac{\partial x^b}{\partial s} = 0$. Denote $m_o u^a$ by $p^a_{\|}$. Then $p^a_{\|} = \frac{\partial W}{\partial x^a}$ defines a Hamilton-Jacobi function by Cor. 2 and consequently for $\psi = \psi(W)$, $[\tilde{\partial}_s\psi, \tilde{ds}] = 0$. It follows from Lemma 3 that

$$(\tilde{\partial}_s\psi)\xi(p_{\|}) = (\partial_s\psi)\xi(p_{\|}).$$

($\Leftarrow$) Recall $\psi = \psi(W)$. Therefore $\frac{\partial\psi}{\partial x^a} = \psi'\frac{\partial W}{\partial x^a}$. Now $[\gamma^a\frac{\partial W}{\partial x^a}, \tilde{ds}] = 0$ implies $\gamma^a\frac{\partial W}{\partial x^a} = m(s)\frac{\tilde{ds}}{ds}$ for some function $m(s)$. But $(\tilde{\partial}_s\psi)\xi(p_{\|}) = (\partial_s\psi)\xi(p_{\|})$ implies $m(s) = \frac{\partial W}{\partial s}$, while Cor. (2) implies that $m(s) = m_o$. Now let $p^a_{\|} = \frac{\partial W}{\partial x^a} = m_o\frac{dx^a}{ds}$. Define $p^a = p^a_{\|} + m_o v^a$, where u^a is a Killing vector for v^a (i.e.$\mathcal{L}_u(\mathbf{v}) = 0$). It now follows that

$$\begin{aligned}
\mathcal{L}_u(p^a) &= \mathcal{L}_u(p_{\|}) + \mathcal{L}_u(mv^a) \\
&= (p^a_{\|})_{;b}u^b - u^a_{;b}p^b_{\|} \\
&= mu^a_{;b}u^b - mu^a_{;b}u^b \\
&= 0.
\end{aligned}$$

The theorem has been proven.

Remark: The theorem states that the Dirac equation associated with a particle exists and is defined locally if and only if the wave function is Lie transported along the curve whose action is W.

By way of concluding this section, we make some final observations:

- In general for any Hamilton-Jacobi function $\psi(W)$ it is possible to define $\tilde{\partial}_s\psi = \tilde{\partial}_s\psi_{\|} + \tilde{\partial}_s\psi_{\perp}$ such that $[\tilde{\partial}_s\psi_{\|}, \tilde{ds}] = 0$ and $\{\tilde{\partial}_s\psi_{\|}, \tilde{ds}\} = 0$. $\tilde{\partial}_s\psi_{\|}$ is the projected cosine along $\tilde{ds}$ and satisfies
$$(\tilde{\partial}_s\psi)_{\|}\xi(p) = (\partial_s\psi)_{\|}\xi(p), \qquad \text{with } p_a = \frac{\partial W}{\partial x^a},$$
where $\tilde{ds}\xi(p) = ds\xi(p)$. Also, $\tilde{\partial}_s\psi_{\perp}$ defines the spin along $\tilde{ds}$.
- The Hamilton-Jacobi function can be re-written in covariant form for a general coordinate system as follows:
$$dW = g^{\mu\nu}p_\mu dx_\nu, \tag{18}$$
with the corresponding wave operator
$$\tilde{\gamma}^\mu\frac{\partial\psi}{\partial x^\mu}\xi = \tilde{\gamma}^\mu p_\mu\psi'\xi \tag{19}$$
associated with the action along a curve, provided $2g^{\mu\nu} = \tilde{\gamma}^\mu\tilde{\gamma}^\nu + \tilde{\gamma}^\nu\tilde{\gamma}^\mu$, where $\tilde{\gamma}^\mu = \frac{\partial x^\mu}{\partial x^a}\gamma^a$.
- The generalized Dirac equation
$$\tilde{\gamma}^\mu\frac{\partial\psi}{\partial x^\mu}\xi = \frac{\partial\psi}{\partial s}\xi. \tag{20}$$
can be defined along an arbitrary curve and is always covariant.
- Gauge potentials of the form A^μ can be introduced into the system by defining
$$p_\mu = m_o\frac{dx_\mu}{d\tau} = \partial_\mu W - eA_\mu. \tag{21}$$
However, in general $\oint A^\mu dx_\mu \neq 0$ and therefore is not an exact differential and $\tilde{ds}\partial_s\tilde{W} \neq \frac{d\psi}{ds}ds$ but does obey Eqn (5). An analysis of this will be given elsewhere.

Journal of Physics: Conference Series **330** (2011) 012013 doi:10.1088/1742-6596/330/1/012013

- As already noted, the above approach deepens our understanding of the Principle of Complementarity. The particle properties should be directly associated with the path increments. The wave properties emerge from Equation (13).
- It should be clear that the strict form of the Dirac equation (13) pertains to the kinematics and not the dynamics of the motion. It describes the kinematics with respect to a local tetrad. The dynamics requires further work, which we will do in the next section. Indeed, the restriction of the motion to geodesics also explains why we obtain distinct energy and momentum levels. Geodesic motion presupposes constant momentum and energy, which can either be discrete or form a continuum depending on the boundary conditons. Quantum mechanics associates these constants with quantum numbers.

2.1. Relationship between quantum and classical mechanics

From a strictly mathematical perspective a particle of rest mass m_o moving in Minkowski space could equally be a cannon ball (a classical object) or an electron being ejected by a neutron in beta decay (a quantum mechanics object). In effect, the difference between the two objects are determined by the initial boundary conditions. In the classical case the line increment from which the Hamilton-Jacobi function is derived gives rise to a dual non-quantum "wave-function," which is a point mass, given by $\psi(s) = A\delta(ms)$. In contrast, in the case of quantum mechanics the same line increment is dual to a family of L^2 functions in such way that the initial boundary conditions coming from the physics are statistical, non-deterministic in nature and incorporates quantization. In other words, in the case of a strictly classical particle, the mechanics can be determined (in principle) from the initial conditions applied directly to the properties of the line increments, with the non-quantum "wave-equation" representing a point-mass and not contributing any additional information. In the case of a quantum particle the opposite appears to be true. It is precisely the "wave-equation" that encapsulates the dynamics of the particle, although the solution to the "wave-equation" is dependent upon the line increments associated with the classical particle. This also gives a new insight into the Principle of Complementarity.

Example: Consider a particle of rest mass m_o moving with uniform velocity u_o with respect to proper time along the x-axis in Minkowski space. The Hamilton-Jacobi function is given by $W = m_o u_x x - m_o \dot{t} t = -m_o s$ such that $p_x = m_o u_x$ and $H = m_o \dot{t}$. Indeed as a classical particle with $x = 0$ when $s = 0$, then $x = u_o s$, where $u_o = \frac{dx}{ds}$. In terms of the coordinate system (x, t) of the laboratory frame this can be written as $x = u_x t$, where $u_x = (u_o \frac{ds}{dt})$ is constant. Moreover, this information can also be encapsulated in a family of Dirac delta functionals defined by $\psi(W) = \delta_s(W) \equiv W(s)$ which in term of the laboratory frame is equivalent to $\psi(W(x,t) = \delta(x - u_x t)$. In other words the wave function as a point mass indicates that at any time t we can find the particle in the position $x = u_x t$ with probability 1. The wave functional per se adds no new information. It essentially encapsulates the identical information already obtained by analyzing the Hamilton-Jacobi equation derived directly from the line increments $ds^2 = u_o dx - \dot{t} dt$.

On the other hand, when we turn to the problem of decay and an electron moving along the x-axis, we have no way of knowing where it will be unless we place a detector somewhere enroute, which then measures position at a particular instant. Unlike the classical problem, such a measurement does not allow us to predict subsequent motion in a deterministic way. Rather, we assume that its motion is described by the generalized Dirac equation with eigenvector solution $\psi(W) = A\exp(k(u_o x - \dot{t}t))$, k a dimension full constant. This in turn prompts the question of interpretation. The author claims that such interpretations are multiple and depends on the question being asked.

For example, for a detector placed at x_o the arrival times of a beta-decay source can be modeled with an exponential distribution $\theta\exp(-\theta t)$ with mean $\frac{1}{\theta}$. Moreover, for $x = x_o$,

Journal of Physics: Conference Series **330** (2011) 012013 doi:10.1088/1742-6596/330/1/012013

$\psi^2_{x_o}(W) = A\exp(2ku_o x_o)\exp(-2k\dot{t}t)$ defines a probability distribution in t. This can be identified with the exponential distribution by letting $\theta = 2k\dot{t} = A\exp(2ku_o x_o)$. In practise, θ can be measured experimentally by considering an ensemble of n beta particles and measuring the mean time of detection of each one of them at x_o. Specifically if the ith decay is detecting at time t_i, then

$$\frac{1}{\bar{\theta}} = \frac{1}{n}\sum_{i=0}^{n}(t_i - t_{i-1}) = \frac{t_n - t_o}{n}.$$

As a second example, consider n-identical and independent particles in Minkowski space with Hamilton-Jacobi function $W = \sum_{i=1}^{n} m_o s_i$, (each i representing a different particle). Then the function defined by

$$\psi(s_1, s_2, \ldots, s_n) = \prod_{i=1}^{n}\psi(s_i) \tag{22}$$

$$= A\exp(k\sum_{i=1}^{n}(t^2 - p_1^2 - p_2^2 - p_3^2) \tag{23}$$

$$= A\exp(knt^2)\exp\left(-k\sum_{i=1}^{n}(p_1^2 + p_2^2 + p_3^2)_i\right) \tag{24}$$

$$= A\exp(knt^2)\exp\left(-k\sum_{i=1}^{n}\mathbf{p_i^2}\right), \tag{25}$$

where $\mathbf{p}_i = (p_1^2 + p_2^2 + p_3^2)_i$, is a solution to the generalized Dirac equation (13), and may be loosely referred to as a "wave-function." As a function in Minkowski Space it cannot be normalized, as seen from Equations (23)-(25). However, for any fixed t, it can be normalized as an L^p function, $p > 0$ in the Euclidean space E^3. Moreover, as defined it can be used to represent either classical or quantum ideal gases according to the initial boundary conditions imposed on it.

For example, for a suitable choice of A and k, with $\epsilon_i \equiv \frac{\mathbf{p}_i^2}{2m_o}$, the squared-"wave function" can be written as

$$\psi^2(W) = A(N)\psi^2(t)\exp\left(-\frac{\beta}{2m_o}\sum_i \mathbf{p}_i^2\right) = A(N)\psi^2(t)\exp\left(-\beta\sum_i \epsilon_i\right)$$

such that

$$\psi^2(W(\mathbf{p}_1 \ldots \mathbf{p}_n) \equiv A(N)\exp\left(-\frac{\beta}{2m_o}\sum_i \mathbf{p}_i^2\right) = A(N)\exp\left(-\beta\sum_i \epsilon_i\right)$$

is independent of t and defines a Maxwell-Boltzmann distribution for an ensemble of n classical particles defined in the center of mass frame.

In addition, if we also require that the values of $\mathbf{p}_i^2$ are such that

$$\epsilon_i = \mathbf{p}_i^2 \in \{\hbar^2 k_l^2/2m_o | l = 0, 1, 2, 3, \ldots\}, \qquad E = \sum n_l \epsilon_l, \ \sum n_l = n,$$

and impose indistinguishability conditions (encapsulated in the probability amplitude term A) then

$$\psi^2(W(\mathbf{p}_1 \ldots \mathbf{p}_n)) = A(N, n_1, n_2, \ldots, n_\infty)\exp\left(-\beta\sum n_i \epsilon_i\right)$$

Journal of Physics: Conference Series **330** (2011) 012013 doi:10.1088/1742-6596/330/1/012013

defines a Bose-Einstein statistic or Fermi-Dirac statistic depending on whether the occupation numbers n_l range over the set $\{0, 1, 2, \ldots\}$ or the set $\{0, 1\}$.

The above formalism also begs the question as to why L^2 functions are needed. Indeed, for the Maxwell-Boltzmann statistics as derived above, it would be sufficient to work with any L^p function, provided it is also a solution to the Generalized Dirac equation. However, when we refer to the Dirac equation proper (16) the appearance of i and the imposition of periodic boundary conditions to obtain standing waves, requires that the normalization process be restricted to L^2 functions. In reality, quantum mechanics is an empirical science with measurements and observations being made in real time in the laboratory frame, and any probability interpretation should be made with this in mind. In this regard, the above examples highlight the importance of boundary conditions when interpreting the significance of a wave function ψ. In the case of the beta decay problem, the wave function $\psi(x,t) \equiv \psi(x)\psi(t|x)$ such that $||\psi(t|x) = 1||$. In other words, for each x we can associate a (conditional) probability distribution with $\psi(t|x)$. Similarly, in the case of both the Maxwell-Boltzmann and Bose-Einstein statistics, $||\psi(W(\mathbf{p}_1 \ldots \mathbf{p}_n)) = \psi(E) = 1||$.

3. Non-geodesic motion and the Hamiltonian

In the previous section we related the Hamilton-Jacobi characteristic function directly to the solution of the generalized Dirac equation and noted that it can represent both classical and quantum solutions of the equation, dependent upon the initial conditions. Moreover, from a mathematical perspective ψ can be a functional on a space of compact support C_c^∞, or an L^p function defined on some domain. However, within the context of General Relativity, Theorem 1 shows that the general form of the solutions are determined only locally and not globally, especially when we consider motion along a non-geodesic. Indeed, the existence of non-geodesics suggests that other factors other than gravity may be involved. For the purpose of quantum mechanics, we will take $\psi \in L^2(E^3)$, where E is a Euclidean space and $\Psi \in L^2 \otimes H$ where H is a finite dimensional Hilbert Space associated with the spin. For example, $\psi \in L^2$ but $\psi \otimes \xi \in L^2 \otimes H$. Ψ and ψ will also be referred to as defining the state of the system, with and without spin respectively.

In this regard, we need also to be aware that from a strict physics perspective, things are more nuanced. Dirac, for example, defines the state as the maximum knowledge that we may have about the system. However, when one formulates a mathematical theory one is always limited by the definitions and restrictions imposed by the "space" within which the theory is formulated. For example, in the previous section W was restricted to being a Hamilton-Jacobi function which required that momentum be defined as $p_a = \frac{\partial W}{\partial x^a}$. However, from Lemma 1, we know that $\psi(kW)$ is also a Hamilton-Jacobi function and in particular $\frac{\partial \psi(kW)}{\partial x^a} = kp_a\psi(kW)$, if ψ is an eigenfunction. Moreover, if k is then chosen to be the complex number i, quantum bound states result. Similarly, the Hamiltonian H was defined strictly in terms of position and momentum. Consequently, in this context the energy states of the system depend only on position and momentum and on the imposed boundary conditions. In other branches of physics, the state, may also depend on temperature, potential, electric and magnetic fields.

3.1. Hamilton's equations of motion

The Hamilton-Jacobi equation is given by $W = \int(\mathbf{p}dx - Hdt)$ with the understanding that the integration is independent of the path. Using this, we now derive Hamilton's equations of motion directly from this definition, without making any explicit recourse to the Calculus of Variations.

First, for each differentiable function W, we write $\frac{dW}{ds} = \dot{W}$. Also note that $||\frac{dx^a}{ds}|| = 1$

requires that $\dot{W} = \frac{\partial W}{\partial s} = -H(s)$, and on taking the derivative of the Hamilton-Jacobi function with respect to s we obtain

$$-H(s) = \mathbf{p}\dot{\mathbf{x}} - H\dot{t}, \tag{26}$$

with the understanding that for $a \in \{1, 2, 3\}$

$$\dot{W} = \frac{\partial W}{\partial s} = -H(s), \qquad \frac{\partial W}{\partial x^a} = p_a \qquad \text{and} \qquad \frac{\partial W}{\partial t} = -H. \tag{27}$$

Note that these can be written in covariant form $p^\mu = g^{\mu\nu}\frac{\partial W}{\partial x^\nu}$. However, for simplicity and clarity we will continue to work with local tetrads.

Differentiating (27) with respect to s gives

$$\frac{\partial H}{\partial s} = \dot{H}(s) \qquad \text{and} \qquad \dot{p}^a = -\frac{\partial H(s)}{\partial x^a}. \tag{28}$$

To derive the remaining equation of motion we follow a method introduced by Synge and Griffith(see [3]):

$$\dot{p}_a = \frac{\partial}{\partial x^a}\left(\frac{dW}{ds}\right) \tag{29}$$

$$= \frac{\partial}{\partial x^a}\left(\frac{\partial W}{\partial x^b}\dot{x}^b + \frac{\partial W}{\partial t}\dot{t}\right) \tag{30}$$

$$= \frac{\partial^2 W}{\partial x^a \partial x^b}\dot{x}^b + \frac{\partial^2 W}{\partial x^a \partial t}\dot{t}. \tag{31}$$

Also,

$$\frac{\partial}{\partial x^a}\left(\frac{\partial W}{\partial t}\right)\dot{t} = -\frac{\partial}{\partial x^a}\left(H\left(x^b, \frac{\partial W}{\partial x^b}, t\right)\right)\dot{t} \tag{32}$$

$$= -\frac{\partial H}{\partial x^a}\dot{t} - \frac{\partial H}{\partial p^b}\frac{\partial^2 W}{\partial x^a \partial x^b}\dot{t} \tag{33}$$

$$= -\frac{\partial H(s)}{\partial x^a} - \frac{\partial H}{\partial p^b}\frac{\partial^2 W}{\partial x^a \partial x^b}\dot{t}. \tag{34}$$

Combining equations (31) and (34) and substituting from (28) yields

$$\frac{\partial^2 W}{\partial x^a \partial x^b}\left(\dot{x}^b - \frac{\partial H(s)}{\partial p^b}\right) = 0. \tag{35}$$

Therefore,

$$\dot{x}^b = \frac{\partial H(s)}{\partial p^b} \qquad \text{provided } det\left(\frac{\partial W}{\partial x^a \partial x^b}\right) \neq 0. \tag{36}$$

This completes the derivation of the canonical equations of motion.

In terms of tetrad summation notation these can be rewritten as

$$\frac{dx^a}{ds} = \eta^{ab}\frac{\partial H(s)}{\partial p^b}, \qquad \frac{dp^a}{ds} = -\eta^{ab}\frac{\partial H(s)}{\partial x^b}. \tag{37}$$

These are the same equations assumed by Horwitz, Schieve and Piron in their work on Stueckelberg theory applied to the Gibb's ensemble[6]. It should also be noted, in reference to Eqn. (37) that in tetrad coordinates by the principle of equivalence $\frac{dp^a}{ds} = \frac{Dp^a}{ds}$, where

D represents the covariant derivative. This follows because the affine connection vanishes on a geodesic. In terms of a generalized coordinate system both equations of motion can be subsumed into the covariant form:

$$\frac{dx^\mu}{d\tau} = g^{\mu\nu}\frac{\partial K}{\partial p^\nu}, \qquad \frac{Dp^\mu}{d\tau} = -g^{\mu\nu}\frac{\partial K}{\partial x^\nu}, \tag{38}$$

where $\frac{Dp^\mu}{d\tau} = p^\mu_{;}u^\nu = \dot{p} + \Gamma^\mu_{\nu\lambda}\dot{x}^\mu\dot{x}^\lambda$. Note that in general the covariant derivative is necessary on the left hand side, since $\frac{dp^\mu}{ds}$ is not a tensor and transforms like acceleration under a change of coordinate systems. On the other hand the right hand side of the second equation is already covariant.

3.2. Non-commutative operators

We begin this section by recalling that for this article $\Psi = \psi^i e_i$ represents a spinor on a manifold, while ψ^i are scalar functions (cf. page 2). Also if $\Psi = \psi\xi$, where ξ is a spinor then ψ is said to define a scalar field. It then follows that $\tilde{\partial}_s\psi(W) \equiv \gamma^a\frac{\partial\psi(W)}{\partial x^a}$ and $\tilde{\partial}_s\psi(H) = \gamma^a\frac{\partial\psi(H)}{\partial x^a}$ are operators, because of the presence of the γ matrices, where W is the Hamilton-Jacobi function and H is the Hamiltonian of the system. In particular,

$$[\tilde{\partial}_s\psi(W), \tilde{\partial}_s\psi(H)] \neq 0. \tag{39}$$

To see this, note that from equations (27) and (28) applied to $\psi(W)$ and $\psi(H)$ respectively, we obtain

$$\tilde{\partial}\psi(W) = \gamma^a\frac{\partial\psi(W)}{\partial x^a} = \gamma^a p_a\psi'(W) \tag{40}$$

and

$$\tilde{\partial}\psi(H) = \gamma^a\frac{\partial\psi(H)}{\partial x^a} = \gamma^a\dot{p}_a\psi'(H). \tag{41}$$

Now unless $\dot{p}^a = g(t)p^a$ for some function $g(t)$ then they cannot commute. Indeed, the condition $\dot{p}^a = g(t)p^a$ is equivalent to $x^a = \exp(\int^t g(\omega)d\omega)$ independently of a, which defines a non affine parameter along a geodesic (cf. [2] Prob., 1.13, n2). Consequently, there do not exist simultaneous eigenvectors, and both operators cannot be simultaneously measured, except along a geodesic. This in itself explains some of the difficulty with dynamics in quantum mechanics. It is also clear from Equation (5) that the non-commutativity is very much related to the presence of the γ matrices and the Dirac Algebra, and although $\psi(W)$ or $\psi(H)$ may themselves be L^p functions nevertheless $\Psi = \psi\xi$, ξ a spinor, may be interpreted as an element of the finite four dimensional Hilbert space associated with the spin of the particle. Indeed, the non-commutativity indicates that spin only becomes manifest in accelerations, and also suggests a second quantization procedure associated with particles moving off geodesics.

In practice, both equations can be seen as useful depending on the circumstances. When the dynamical system corresponds to motion on geodesics then $\dot{p} = 0$ and has no role to play. In this case the generalized Dirac equation (13) can be used. In the event that the dynamical system is undergoing accelerations with respect to the laboratory frame then we could solve the eigenvector equation

$$\tilde{\partial}\psi(H)\xi = \gamma^a\frac{\partial\psi(H)}{\partial x^a}\xi = \frac{dm}{ds}\xi \tag{42}$$

where $\frac{dm}{ds} = ||f^a||$ is the norm of the four-force f^a. It should also be pointed out that since the Principle of Equivalence guarantees the existence of a coordinate system x^a such that $\ddot{x}^a = 0$, the existence of absolute acceleration presupposes the existence of a non-gravitational force such that $a^a = \ddot{x} \neq 0$. This also means that in the presence of such an acceleration the rest mass will not

Journal of Physics: Conference Series **330** (2011) 012013 doi:10.1088/1742-6596/330/1/012013

be a constant but will change according to whether its accelerating or decelerating. Accordingly, depending on whether $\frac{dm}{ds} = 0, > 0, < 0$, three cases arise which could be interpreted as defining a null, time-like, and space-like event in momentum space. This, together, with the second quantization properties will be investigated in another research paper.

3.3. Statistical mechanics and ideal gases

Another interesting application is to theory of Ideal gases. Indeed, as noted in Section 2.1 both classical and quantum statistics distributions can be derived from solving the generalized Dirac equation, with the difference between the two being related to the boundary conditions. In fact, in equations (23)-(25) we already assumed the form of the "wave-function" for an ideal gas, both classical and quantum. However, it now remains to derive this fundamental Hamiltonian function as an application of Hamilton's equation's of motions.

First, consider a single particle with a Hamilton-Jacobi function given by $W = -ms$ associated with motion along a geodesic in Minkowski space, then $H = \frac{\partial W}{\partial s} = m$. In terms of the equations of motion there exists a differentiable function $\psi = \psi(H)$ such that

$$\frac{\partial \psi}{\partial x^a} = -\psi' \dot{p}^a, \quad \text{where } \dot{p} = \frac{dp}{ds}, \tag{43}$$

which can be re-written in spinor notation as

$$\gamma^a \frac{\partial \psi}{\partial x^a} = -\gamma^a \psi' \dot{p}^a. \tag{44}$$

Taking the dot product of (43) with $m\frac{dx^a}{ds}$ and using the chain rule, (or equivalently by taking the inner product $\frac{1}{2}\{\gamma^a \frac{\partial \psi(H)}{\partial x^a}, \gamma^a p_a\}$ in (44)) gives

$$\frac{d\psi(H)}{ds} = -\psi'(H)\dot{p}^a p_a. \tag{45}$$

To solve, recall that $\psi'(H) = \frac{d\psi(H)}{dH}$ and $2\dot{p}_a p^a = \frac{d}{ds}(p_a p^a)$, and consequently $H = \frac{1}{2}p_a p^a$. In particular when $\psi' = k\psi$, an eigenvector, and k an eigenvalue, solving for ψ gives

$$\psi = Ae^{\frac{k}{2}p^a p_a}. \tag{46}$$

If k is real and time dependent then

$$\psi(t, x) = A\psi(t)\psi(x|t) = A\exp(kt^2)\exp(-k(p_1^2 + p_2^2 + p_3^2)),$$

is not normalizable. However, for each t, $\psi(x|t)$ is an $L^p(E^3)$ function and can be normalized. Indeed, the normalized set of functions $\{\psi(x|t)\}$ defines a Markov process in t. Moreover, if k is a constant as it is for geodesic motion, then $\psi(x_1, x_2, x_3|t) = \psi(x_1, x_2, x_3)$. On the other hand, if we take $k = \frac{i}{\hbar}$ then $\psi(p|t) = \psi(p)$ is an L^2 function with periodic boundary conditions.

With this in mind, we now extend this result to a system of n independent particles as would occur in some statistical systems, and define the joint wave function as an independent product of n single particle eigenfunctions with real k, or equivalently as an ensemble of n independent eigenfunctions. Written as an L^2 function (although for a classical particle we could equally work with L^1)

$$\psi = e^{\frac{k}{2}\sum_1^n p^a p_a}. \tag{47}$$

Journal of Physics: Conference Series **330** (2011) 012013 doi:10.1088/1742-6596/330/1/012013

Moreover, if the motion is along a geodesic then $H = constant$ and $p^a p_a = m$ for each particle Consequently,

$$\psi = c\exp\left[\frac{k}{2}m\sum_n(\dot{t}^2 - \dot{x}_1^2 - \dot{x}_2^2 - \dot{x}_3^2)\right] = e^{\frac{k}{2}nm}. \tag{48}$$

Defining $T = \frac{1}{k_B k}$, to be the temperature, where k_B is Boltzmann's constant, and using the separation of variables for time independent states to write $\psi = \psi(t)\psi(x_1, x_2, x_3)$, where

$$\psi(x_1, x_2, x_3|t) = c\exp\left(\frac{-\frac{1}{2}m}{k_B T}\Sigma_n(\dot{x}_1^2 + \dot{x}_2^2 + \dot{x}_3^2)\right) \tag{49}$$

determines a Maxwell-Boltzmann statistics for free particles at any time t. Note also that $|\psi(x_1, x_2, x_3|t)|^2$ defines a normal distribution provided the variance is given by $\sigma_x^2 = \sigma_y^2 = \sigma_z^2 = \sigma^2 = \frac{k_B T}{2m}$. Equations (47) and (48) can be interpreted to mean that the system is in equilibrium, with total conserved energy $\sum_n m$. Indeed, for each (local) t the same Maxwell distribution occurs. Moreover, if T is not constant then $k = k(T)$ varies and in this case, on solving for ψ in (43) one obtains

$$\psi = e^{\int k(T)p_a\dot{p}^a dt}. \tag{50}$$

Also from equation (44) we should note that $\frac{\psi}{\psi'}$ is always an exact differential, and incorporates $k(T)$. If $k = k(T)$ has no explicit time dependence then ψ will define a stationary state. Note that in the case of a closed system, T is proportional to the population variance of the velocities $\sum_n \frac{(u_i-\mu)^2}{n}$.

4. Conclusion

The article has attempted to establish a relationship between the metrics of General Relativity and Quantum Mechanics. This has been achieved by first relating the metric structure of spacetime to the Hamilton-Jacobi function and then using this relationship to derive a Generalized Dirac equation. In addition we have derived Hamilton's equations of motion directly from the Hamilton-Jacobi equation and then used these equations to determine a dynamical equation for the evolution of the system. The dynamical equations derived do not commute with the generalized Dirac equation and consequently cannot be measured simultaneously. The dynamical equations also permit the derivation of the statistical mechanics of the system. Indeed, in this paper the Maxwell-Boltzmann distribution was derived directly from the equations of motion.

Finally, it should be noted that we have restricted ourselves to scalar fields as defined in the introduction. However, in its most general form, we can write

$$\gamma^a\frac{\partial\Psi}{\partial x^a} = \Phi \tag{51}$$

where Φ would be defined by the physics of the problem. For example, Maxwell's equations in Minkowski space can be written in spinor form as

$$i\alpha^a\frac{\partial\Psi}{\partial x^a} = -4\pi\Phi, \tag{52}$$

where $\phi_0 = \rho$ is charge density, and $\phi_a = j_a$, $a \in \{1, 2, 3\}$ is a current density. Also in this case, $\phi_0 = 0$ and $\phi_a = H_a - iE_a$, where H_a and E_a are the magnetic and electric fields respectively [7]. Such cases would need further study.

Acknowledgments
I would like to thank Prof. Lamberto Rondoni from the *Politecnico di Torino* and also the referees for their invaluable suggestions and input while I was writing this paper.

References
[1] O'Raifeartaigh L 1997 *The Dawning of Gauge Theory* Princeton University Press: Princeton New Jersey 92.
[2] Poisson E 2004 *A Relativistist's Toolkit* Cambridge University Press: Cambridge 9.
[3] Synge J and Griffith G 1959 *Principles of Mechanics* New York and Tokyo: McGraw Hill 448-450.
[4] Celerier M-N and Nattale L 2003 *Electromagn. Phenom* **3** 70.
[5] Ng Y and van Dam H 2003 *Phys.Lett* **A309** 335.
[6] Horwitz L and Schieve W. 1981 *Ann. of Phys.* **137** 307.
[7] Moses H 1958 *Physical Rev.* **6** 1670-1679.
[8] O'Hara P 2005 *Found. Phys.* **35** 1563-1584.
[9] Rivas M 1994 *J. Math. Phys.* **35** No. 7.
[10] Tipler F and Dupre M 2010 General Relativity As an Aether Theory *Preprint* arXiv:1007.4572.

7th Biennial Conf. on Classical and Quantum Relativistic Dynamics of Particles and Fields IOP Publishing
Journal of Physics: Conference Series **330** (2011) 012014 doi:10.1088/1742-6596/330/1/012014

Harmonic oscillator states with integer and non-integer orbital angular momentum

Martin Land

Department of Computer Science, Hadassah College, 37 HaNeviim Street, Jerusalem

E-mail: `martin@hadassah.ac.il`

Abstract. We study the quantum mechanical harmonic oscillator in two and three dimensions, with particular attention to the solutions as basis states for representing their respective symmetry groups — O(2), O(1,1), O(3), and O(2,1). The goal of this study is to establish a correspondence between Hilbert space descriptions found by solving the Schrodinger equation in polar coordinates, and Fock space descriptions constructed by expressing the symmetry operators in terms of creation/annihilation operators. We obtain wavefunctions characterized by a principal quantum number, the group Casimir eigenvalue, and one group generator whose eigenvalue is $m + s$, for integer m and real constant parameter s. For the three groups that contain O(2), the solutions split into two inequivalent representations, one associated with $s = 0$, from which we recover the familiar description of the oscillator as a product of one-dimensional solutions, and the other with $s > 0$ (in three dimensions, solutions are found for $s = 0$ and $s = 1/2$) whose solutions are non-separable in Cartesian coordinates, and are hence overlooked by the standard Fock space approach. The O(1,1) solutions are singlet states, restricted to zero eigenvalue of the symmetry operator, which represents the boost, not angular momentum. For O(2), a single set of creation and annihilation operators forms a ladder representation for the allowed oscillator states for any s, and the degeneracy of energy states is always finite. However, in three dimensions, the integer and half-integer eigenstates are qualitatively different: the former can be expressed as finite dimensional irreducible tensors under O(3) or O(2,1) while the latter exhibit infinite degeneracy. Creation operators that produce the allowed integer states by acting on the non-degenerate ground state are constructed as irreducible tensor products of the fundamental vector representation. However, the half-integer eigenstates are infinite-dimensional, as expected for the non-compact O(2,1), and no Fock space description is found for this case. For all $s \neq 0$ solutions, the SU(N) symmetry of the harmonic oscillator Hamiltonian recently discovered by Bars is spontaneously broken by the ground state. The connection of this symmetry breaking to the non-separability into one-dimensional Cartesian solutions is demonstrated. We conclude with a discussion of the O(2,1) solutions, and the problem of timelike ghost excitations.

1. Introduction

1.1. Background

Along with its classical counterpart, the quantum harmonic oscillator is a well-studied model with exact solutions and connections to many physical systems for which it serves as foundation or approximation. Beyond its application to atomic and molecular spectra, statistical mechanics, and by way of various relativistic generalizations, to quark dynamics, several general techniques

Journal of Physics: Conference Series **330** (2011) 012014 doi:10.1088/1742-6596/330/1/012014

associated with the harmonic oscillator, including Dirac's factorization of the Hamiltonian into creation/annihilation operators and the resulting Fock space of uncoupled modes, serve as conceptual building blocks in areas ranging from blackbody radiation to canonical quantization and string theory. Yet, despite the subject's long history, fundamental new insights continue to emerge.

In modeling the hadron mass spectrum on the excitation spectrum of quark bound states [1], Feynman, Kislinger, and Ravndal (FKR) stimulated interest in finding a consistent relativistic generalization of the quantum harmonic oscillator. This model, which the authors themselves characterized as "naive and obviously wrong in its simplicity"[1] raises a number of questions, including difficulty with ground state normalization, negative norm ghost states, and consequent problems with unitarity. Nevertheless, their generalization of the nonrelativistic Hamiltonian to a Lorentz-invariant Stueckelberg form [2], and the interpretation of its eigenvalues as the mass of a bound state, determined dynamically and not subject to *a priori* constraint, were not seen as shortcomings of the theory. In obtaining the relativistic path integral, Feynman had previously treated mass as a dynamical quantity that takes fixed eigenvalues for asymptotic particle states [3], but in that case argued that mass is conserved dynamically by the electromagnetic field (see also [4, 5] on this point). In the FKR model, the harmonic oscillator was taken as a phenomenological binding force, providing a Lorentz-invariant ground state and a positive definite spectrum. Because states with timelike excitations can have negative norm (and perhaps because the interpretation of such excitations had not been entirely clarified), only spacelike excitations are admitted in the model. This *ad hoc* exclusion of timelike modes can lead to violations of unitarity, because sums of states that omit negative norm contributions may exceed unity.

Seeking to address the difficulties in the FKR model, Kim and Noz (KN) proposed to replace the scalar ground state with a Lorentz covariant solution to the eigenvalue equation [6], and study excited states through the action of Lorentz boosts on this solution. Although this ground state is normalizable and all states have positive norm, timelike modes contribute negative energy. By dealing with boosts rather than excitation modes of the Hamiltonian, the model sidesteps the negative spectrum problem.

Another model for the relativistic oscillator was found by Horwitz and Arshansky [7] who applied a many-body extension [8] of Stueckelberg's covariant Hamiltonian quantum mechanics to obtain relativistic generalizations of the classical central force bound state problems. According to a virial theorem, presented in appendix A, restriction to spacelike dynamics guarantees a positive spectrum, but since there is no obvious way to realize this nonholonomic constraint in Cartesian coordinates, the eigenvalue equation was posed in a hyperspherical parameterization. In order to obtain bound states solutions with the correct spectrum and symmetry properties, it was necessary to first solve the eigenvalue equation in an O(2,1)-invariant subspace of the spacelike sector, and then induce a representation of O(3,1) over the O(2,1)-covariant wavefunctions. The resulting wavefunctions are eigenstates of mass, the Casimir operators for O(3,1) and its O(2,1) and O(3) subgroups, and one angular momentum component. Under the raising/lowering combinations of the boost and angular momentum operators, the wavefunctions transform as O(3,1) tensors, providing an infinite dimensional representation of boosts, as expected for a non-compact group, and a finite, half-integer representation of angular momentum. Splitting the degeneracy through the Zeeman [9] and Stark [10] effects leads to finite, even multiplicity. The

[1] Such disclaimers appear in a number of Feynman's papers, including his derivation of the relativistic path integral. He evidently found the technique useful in deflecting premature criticism of certain of his more original ideas.

Journal of Physics: Conference Series **330** (2011) 012014 doi:10.1088/1742-6596/330/1/012014

half-integer representation seems to be related to a lowering of the ground state mass eigenvalue. In the FKR and KN models, despite the supression of timelike excitations, the ground state mass is found to be $2\hbar\omega$ as one may expect for a relativistic system by associating $\frac{1}{2}\hbar\omega$ with each of 4 degrees of freedom. The Horwitz-Arshansky solution, on the other hand, exhibits the lower ground state level $\frac{3}{2}\hbar\omega$ one would like to obtain for an oscillator with 3 space dimensions. The common origin of the half-integer angular momentum and lowered ground state mass eigenvalue appears in the solution in hyperspherical coordinates. Kastrup [11] has shown that the choice of Cartesian coordinates overlooks the singularity at the origin of polar coordinates, implicitly choosing one solution from a family of harmonic oscillators with different ground state levels. Only in certain cases can solutions in hyperspherical coordinates be identified with the standard Cartesian wavefunctions.

In this paper, we study the harmonic oscillator in two and three dimensions, with emphasis on the solutions as basis states for representations of their respective symmetry groups: O(2), O(1,1), O(3), and O(2,1). The overall goal of this study was to solve the Schrodinger equation in polar coordinates, adapting the method of Horwitz and Arshansky to each group respectively, in order to establish a correspondence between the resulting Hilbert space descriptions and Fock space descriptions, constructed by writing the symmetry operators in terms of creation/annihilation operators. To provide a faithful representation, the states in both descriptions must be eigenstates of the Hamiltonian and a maximal set of mutually commuting symmetry operators. The states should also provide ladder representations — for the mode number in two and three dimensional Fock space, for the symmetry eigenvalues associated with the non-abelian groups in $D = 3$, and for a unitary dynamical symmetry recently discovered by Bars [12]. The results of these attempts reflect physical differences among the oscillator models. Wavefunctions with integer and non-integer angular momentum levels were found that provide faithful representations of O(2), as well as the groups O(3) and O(2,1) that contain O(2) as a subgroup. Combinations of non-diagonal operators were found that raise and lower eigenvalues of mode number and angular momentum. On the other hand, the O(1,1) states are Lorentz invariant, rather than covariant, permitting only the zero eigenvalue for the symmetry generator, which in this case represents the boost, rather than angular momentum. These scalar states, which turn out to be non-normalizable for any mode number, effectively represent a one dimensional oscillator, providing an essentially trivial representation of the Lorentz symmetry.

For all groups, integer angular momentum wavefunctions can be expressed as linear combinations of the familiar Cartesian wavefunctions, while non-integer wavefunctions do not admit separation in Cartesian coordinates, and are therefore not unitarily connected to the familiar integer solutions. The non-integer solutions for O(2) are shown to be nonunitarity on algebraic grounds. Nevertheless, complete solutions are obtained in Fock space, for the integer and non-integer representations of O(2), as well as the integer representations of O(3) and O(2,1), by positing states as representations of the polar ladder operators. However, the Dirac factorization of the Hamiltonian does not lead to a Fock space representation of any kind for the non-integer states of either group in $D = 3$.

Generally, irreducible unitary representations of a compact group are finite dimensional, while unitary representations of a non-compact group are either singlet or infinite dimensional. As expected, the integer and non-integer representations of O(2) are finite dimensional, and the representation of O(1,1) is a singlet. While an infinite dimensional representation of O(1,1) may exist, it was not found either in Hilbert space or Fock space. On algebraic grounds, the infinite dimensional, non-integer representation of O(3) must be nonunitary, and a similar conclusion must hold for the finite dimensional, integer representation of O(2,1), which is equivalent to the FKR model.

Following a brief discussion of the oscillator formalism, we present the $D = 2$ representations in section 2 and the non-abelian $D = 3$ representations in section 3. Further approaches to a unitary, infinite dimensional Fock space representation of O(2,1) will be discussed in a subsequent paper.

1.2. Cartesian formalism

We write a Schrodinger equation of the type

$$K\psi = i\partial_\tau\psi = \kappa\psi \tag{1}$$

with harmonic oscillator Hamiltonian

$$K = \frac{1}{2}\left(p^2 + \omega^2 x^2\right) = \frac{1}{2}\eta_{\mu\nu}\left(p^\mu p^\nu + \omega^2 x^\mu x^\nu\right) \tag{2}$$

to describe either an $\mathrm{O}(D)$ nonrelativistic oscillator with Euclidean metric

$$\eta_{\mu\nu} = \delta_{\mu\nu}, \quad \mu,\nu = 1,...,D \tag{3}$$

or an $\mathrm{O}(D-1,1)$ relativistic oscillator with Lorentz metric

$$\eta_{\mu\nu} = \mathrm{diag}\left(-1,1,...,1\right), \quad \mu,\nu = 0,...,D-1. \tag{4}$$

For the relativistic case, (1) is a Stueckelberg equation with Lorentz-invariant evolution parameter τ, which as in the Euclidean case, admits a virial theorem (see appendix A) showing that restriction of the motion to spacelike dynamics guarantees a positive spectrum. The standard approach to Fock space proceeds by separation of Cartesian variables and subsequent application of Dirac's factorization of the one-dimensional Hamiltonian for each degree of freedom. Assuming a product solution of one-dimensional oscillators

$$\psi\left(x\right) = \prod_\mu \psi\left(x^\mu\right) \qquad\qquad \kappa = \sum_\mu \kappa^\mu \tag{5}$$

the Hamiltonian separates into a sum of D mode-number terms as

$$K = \omega\eta_{\mu\nu}\left(\bar{a}^\mu a^\nu + \frac{1}{2}\eta^{\mu\nu}\right) = \omega\sum_\mu \eta_{\mu\mu}\left(N^\mu + \frac{1}{2}\eta^{\mu\mu}\right) \tag{6}$$

with creation/annihilation operators

$$a^\mu = \frac{1}{\sqrt{2}}\left(x^\mu + ip^\mu\right) \qquad\qquad \bar{a}^\mu = \frac{1}{\sqrt{2}}\left(x^\mu - ip^\mu\right) \tag{7}$$

that satisfy

$$\left[a^\mu, \bar{a}^\nu\right] = \eta^{\mu\nu} \tag{8}$$

and mode number operators $N^\mu = \bar{a}^\mu a^\mu$ (no summation) that satisfy

$$\left[N^\mu, \bar{a}^\nu\right] = \eta^{\mu\nu}\bar{a}^\mu \qquad \left[N^\mu, a^\nu\right] = -\eta^{\mu\nu}a^\mu \qquad \left[N^\mu, N^\nu\right] = 0. \tag{9}$$

The products $|n\rangle = \prod_\mu |n^\mu\rangle$ of N^μ eigenstates form a Fock space of orthogonal oscillator modes with the ladder property

$$\bar{a}^\mu\left|n\right\rangle = e^{i\phi_+^\mu}\sqrt{n^\mu + \eta^{\mu\mu}}\left|n + \eta^{\mu\mu}\mathbf{e}_\mu\right\rangle \qquad a^\mu\left|n\right\rangle = e^{i\phi_-^\mu}\sqrt{n^\mu}\left|n - \eta^{\mu\mu}\mathbf{e}_\mu\right\rangle, \tag{10}$$

Journal of Physics: Conference Series **330** (2011) 012014 doi:10.1088/1742-6596/330/1/012014

where the $\mathbf{e}_\mu$ are unit vectors in the occupation number space

$$(\mathbf{e}_\mu)^\lambda = \delta_\mu^\lambda \tag{11}$$

and the particular choice of phases $e^{i\phi_+^\mu}$ and $e^{i\phi_-^\mu}$, all taken to be 1 for the nonrelativistic Euclidean oscillator, has non-trivial consequences for the relativistic oscillator.

The FKR model [1] can be recovered by choosing the phases

$$e^{i\phi_+^0} = e^{i\phi_-^0} = i \qquad e^{i\phi_+^\mu} = e^{i\phi_-^\mu} = 1,\ \mu > 0, \tag{12}$$

which preserves the roles of $\bar{a}^0$ as creation operator and a^0 as annihilation operator for the timelike mode, under the requirement that $n^0 \leq 0$ so that

$$\bar{a}^0 \left|n\right\rangle = \sqrt{1-n^0}\ \left|n - \mathbf{e}_0\right\rangle \quad a^0 \left|n\right\rangle = \sqrt{-n^0}\ \left|n + \mathbf{e}_0\right\rangle . \tag{13}$$

Although these states have positive spectrum

$$\left\langle n^0\ \mathbf{n}\ \middle|\ K\ \middle|\ n^0\ \mathbf{n}\right\rangle = \omega\left(\left|n^0\right| + \sum_{\mu>0} n^\mu + \frac{1}{2}\eta_{\mu\nu}\eta^{\mu\nu}\right) \tag{14}$$

they have negative norm for odd values of n^0

$$\left\langle n^0\ \mathbf{n}\middle|\ n^0\ \mathbf{n}\right\rangle = \frac{1}{(-n^0)!}\left\langle 0\right| \left(a^0\right)^{-n^0} \left(\bar{a}^0\right)^{-n^0} \left|0\right\rangle = \left\langle 0\right| \left(\eta^{00}\right)^{-n^0} \left|0\right\rangle = (-1)^{-n^0} \tag{15}$$

requiring that the negative norm states (ghosts) be suppressed. These authors chose to exclude all excited timelike modes by applying the covariant condition

$$(p \cdot a)\, \psi = 0 \tag{16}$$

which forces ψ into the ground state along the momentum p. Although this approach is formally comparable to Gupta-Bleuler quantization [13] of the electromagnetic field, where (16) expresses the Lorentz gauge condition as an operator equation and eliminates negative norm states along the field's lightlike momentum, there is no general gauge freedom for the relativistic oscillator that justifies this procedure. The first order differential equations

$$a^\mu \psi_0\,(x) = 0 \tag{17}$$

lead to the ground state solution proposed in [1],

$$\psi_0\,(x) = e^{-x^2/2} = e^{-\left(\mathbf{x}^2 - t^2\right)/2} \tag{18}$$

which clearly requires some regularization procedure for normalization. In the absence of such a proper normalization, difficulties arise in the interpretation of all Fock space procedures, in particular the evaluation of the norm. To see this, we suppose that having no prior knowledge of Dirac factorization, we procede naively from separation of Cartesian variables (5) to formulate the eigenvalue problem as

$$\frac{1}{2}\left(\mathbf{p}^2 + \mathbf{x}^2 - p_0^2 - t^2\right)\psi_{\mathbf{n}}\,(\mathbf{x})\,\psi_{n_t}\,(t) = \left(\kappa_x + \kappa_y + \kappa_z + \kappa_t\right)\psi_{\mathbf{n}}\,(\mathbf{x})\,\psi_{n_t}\,(t)\,. \tag{19}$$

Journal of Physics: Conference Series **330** (2011) 012014 doi:10.1088/1742-6596/330/1/012014

The space part has the familiar O(3) solution

$$\psi_{\mathbf{n}}(\mathbf{x}) = A_{\mathbf{n}} H_{n_x}(x) H_{n_y}(y) H_{n_z}(z) e^{-\mathbf{x}^2/2} \tag{20}$$

where the Hermite polynomials satisfy

$$H''(x) - 2xH'(x) + (\kappa - 1) H(x) = 0 \qquad H_n(x) = (-1)^n e^{x^2} \frac{d^n}{dx^n} e^{-x^2} \tag{21}$$

while the time part is

$$\psi_{n_t}(t) = A_{n_t} \tilde{H}_{n_t}(t) e^{t^2/2} \tag{22}$$

with polynomials satisfying

$$\tilde{H}''(t) + 2t\tilde{H}'(t) - (\kappa - 1) \tilde{H}(t) = 0 \qquad \tilde{H}_n(t) = e^{-t^2} \frac{d^n}{dt^n} e^{t^2}. \tag{23}$$

Writing the norm for these states as

$$\|\psi_{n_t \mathbf{n}}(t, \mathbf{x})\| = |A_{\mathbf{n}} A_{n_t}|^2 \int d^3x \left|H_{n_x}(x) H_{n_y}(y) H_{n_z}(z)\right|^2 e^{-\mathbf{x}^2} \int dt \left|\tilde{H}_{n_t}(t)\right|^2 e^{t^2} \tag{24}$$

the divergence of the second integral seems clear enough, but we would not suspect that the norm could be negative. Specifically, for the first excited timelike state, which by (15) should be negative, we might cleverly (but still naively) observe that the state may be written

$$\psi_1(t) = 2A_1 t e^{t^2} = A_1 (t + \partial_t) e^{t^2/2} \tag{25}$$

so that integration by parts leads from

$$\|\psi_{1\mathbf{n}}(t, \mathbf{x})\| = 4|A_1|^2 \int dt\ t^2 e^{t^2} = A_1^2 \int dt \left[(t + \partial_t) e^{t^2/2}\right] (t + \partial_t) e^{t^2/2} \tag{26}$$

to

$$\begin{aligned} \|\psi_{1\mathbf{n}}(t, \mathbf{x})\| &= A_1^2 \int dt\ e^{t^2/2} (t - \partial_t)(t + \partial_t) e^{t^2/2} &(27)\\ &= A_1^2 \int dt\ e^{t^2/2} \Big\{[(t - \partial_t), (t + \partial_t)] + (t - \partial_t)\Big\} e^{t^2/2} &(28)\\ &= -A_1^2 \int dt\ e^{t^2}. &(29) \end{aligned}$$

While this obvious error follows from the improper integration by parts for the divergent ground state, it is just the Hilbert space expression of the transposition of operators performed in (15). In the absence of a proper normalization for the ground state, the identification of negative normed states in this theory through (15) remains questionable, even when the states are restricted to spacelike support.

The KN model [6] can be recovered by choosing $e^{i\phi^\mu_+} = e^{i\phi^\mu_-} = 1$ for all phases, so it follows from (10) that $\bar{a}^0$ acts as an annihilation operator and a^0 is the creation operator for the timelike mode, such that the ground state mode must have $n^0 \geq 1$. This role reversal between $\bar{a}^0$ and a^0 insures that timelike excitations have positive norm,

$$\langle n^0\ \mathbf{n}|\ n^0\ \mathbf{n}\rangle = \frac{1}{(n^0-1)!} \langle 1\ \mathbf{0}| \left(\bar{a}^0\right)^{n^0-1} \left(a^0\right)^{n^0-1} |1\ \mathbf{0}\rangle = \langle 1\ \mathbf{0}| \left(-\eta^{00}\right)^{n^0-1} |1\ \mathbf{0}\rangle = 1 \tag{30}$$

but leads to an indefinite spectrum

$$\langle n^0\ \mathbf{n}|K|n^0\ \mathbf{n}\rangle = \omega\left(-\left|n^0\right| + \sum_{\mu>0} n^\mu + \frac{1}{2}\eta_{\mu\nu}\eta^{\mu\nu}\right). \tag{31}$$

The requirements on the ground state

$$\bar{a}^0\left|1\ \mathbf{0}\right\rangle = 0 \qquad a^\mu\left|1\ \mathbf{0}\right\rangle = 0\ ,\ \mu > 0 \tag{32}$$

lead to a set of first order differential equations that reproduce the normalizable KN ground state solution

$$\psi_0(x) = A_0 e^{-t^2} e^{-x^2/2} = A_0 e^{-\left(t^2+\mathbf{x}^2\right)/2}. \tag{33}$$

Although this model suggests a negative dynamic contribution to the mass of the bound state, the states are normalizable and no ghosts are present.

In order to solve the Hamiltonian eigenvalue problem following the approach of Horwitz and Arshansky [7], we restrict the wavefunctions to spacelike support, imposed by solving the harmonic oscillator problem in polar coordinates of the generic form

$$x = \rho\hat{n} \tag{34}$$

where ρ is the invariant length

$$\rho = \sqrt{\eta_{\mu\nu}x^\mu x^\nu} \tag{35}$$

$\hat{n}$ is a spacelike unit vector given for the respective symmetry groups by

$$\begin{array}{cc} \mathrm{O}(2) & \mathrm{O}(1,1) \\ \begin{bmatrix}\hat{n}^1\\ \hat{n}^2\end{bmatrix} = \begin{bmatrix}\cos\phi\\ \sin\phi\end{bmatrix} & \begin{bmatrix}\hat{n}^0\\ \hat{n}^1\end{bmatrix} = \begin{bmatrix}\sinh\beta\\ \frac{\rho}{|\rho|}\cosh\beta\end{bmatrix} \\ \mathrm{O}(3) & \mathrm{O}(2,1) \\ \begin{bmatrix}\hat{n}^1\\ \hat{n}^2\\ \hat{n}^3\end{bmatrix} = \begin{bmatrix}\begin{pmatrix}\cos\phi\\ \sin\phi\end{pmatrix}\sin\theta\\ \cos\theta\end{bmatrix} & \begin{bmatrix}\hat{n}^0\\ \hat{n}^1\\ \hat{n}^2\end{bmatrix} = \begin{bmatrix}\sinh\beta\\ \begin{pmatrix}\cos\phi\\ \sin\phi\end{pmatrix}\cosh\beta\end{bmatrix} \end{array} \tag{36}$$

where for O(1,1)

$$-\infty < \rho < \infty \qquad -\infty < \beta < \infty \tag{37}$$

and for O(2), O(3), O(2,1)

$$0 \le \rho < \infty \qquad -\infty < \beta < \infty. \tag{38}$$

The associated gradient operators are given in appendix B. In these coordinates, the Schrodinger equation takes the form

$$\left[-\partial_\rho^2 - \frac{D-1}{\rho}\partial_\rho + \frac{1}{\rho^2}\mathbf{M}^2 + \rho^2 - \varepsilon\right]\psi = 0 \tag{39}$$

where the energy/mass eigenvalue is $\kappa = \frac{\omega}{2}\varepsilon$ and $\mathbf{M}^2$ is the differential operator formed as the Casimir invariant from the generators

$$M^{\mu\nu} = x^\mu p^\nu - x^\nu p^\mu \tag{40}$$

of the relevant symmetry group. Separation of radial and (hyper)angular variables leads to replacement of $\mathbf{M}^2$ by its eigenvalue, so that the Schrodinger equation becomes a radial equation and the (hyper)angular equations have solutions which are generic representations of the symmetry group.

Journal of Physics: Conference Series **330** (2011) 012014 doi:10.1088/1742-6596/330/1/012014

2. Harmonic oscillator in $D = 2$

2.1. Wavefunctions

In $D = 2$ with parameterizations (36), the Schrodinger equation becomes

$$\left[-\partial_\rho^2 - \frac{1}{\rho}\partial_\rho + \frac{1}{\rho^2}\mathbf{M}^2 + \rho^2 - \varepsilon\right]\psi = 0 \tag{41}$$

and the group generator is

$$\begin{array}{ll} M^{12} = x^1p^2 - x^2p^1 = -i\partial_\phi & \mathrm{O}(2) \\ M^{01} = x^0p^1 - x^1p^0 = -i\epsilon(\rho)\,\partial_\beta & \mathrm{O}(1{,}1) \end{array} \tag{42}$$

where

$$\epsilon(\rho) = \frac{\rho}{|\rho|} \tag{43}$$

and the Casimir operators

$$\begin{array}{ll} \mathbf{M}^2 = M^{12}M_{12} = \eta_{11}\eta_{22}\left(-i\partial_\phi\right)^2 = -\partial_\phi^2 & \mathrm{O}(2) \\ \mathbf{M}^2 = M^{01}M_{01} = \eta_{00}\eta_{11}\left(-i\partial_\phi\right)^2 = \partial_\beta^2 & \mathrm{O}(1{,}1) \end{array} \tag{44}$$

are seen to have opposite sign. Separation of variables

$$\begin{array}{ll} \psi(\rho,\phi) = R(\rho)\,\Phi(\phi) & \mathrm{O}(2) \\ \psi(\rho,\phi) = \tilde{R}(\rho)\,\tilde{\Phi}(\beta) & \mathrm{O}(1{,}1) \end{array} \tag{45}$$

leads to

$$\begin{array}{ll} \dfrac{\rho^2}{R(\rho)}\left(-\partial_\rho^2 - \dfrac{1}{\rho}\partial_\rho + \rho^2 - \varepsilon\right)R(\rho) = \dfrac{1}{\Phi(\phi)}\partial_\phi^2\Phi(\phi) = -\Lambda^2 & \mathrm{O}(2) \\ \dfrac{\rho^2}{\tilde{R}(\rho)}\left(-\partial_\rho^2 - \dfrac{1}{\rho}\partial_\rho + \rho^2 - \varepsilon\right)\tilde{R}(\rho) = -\dfrac{1}{\tilde{\Phi}(\beta)}\partial_\beta^2\tilde{\Phi}(\beta) = \tilde{\Lambda}^2 & \mathrm{O}(1{,}1) \end{array} \tag{46}$$

where the signs of the separation constant are fixed by the requirement that the hermitian generators have real eigenvalues. The equations now take the form

$$\begin{array}{lll} \left[-\partial_\rho^2 - \dfrac{1}{\rho}\partial_\rho + \dfrac{1}{\rho^2}\Lambda^2 + \rho^2 - \varepsilon\right]R(\rho) = 0 & \Phi(\phi) = e^{i\Lambda\phi} & \mathrm{O}(2) \\ \left[-\partial_\rho^2 - \dfrac{1}{\rho}\partial_\rho - \dfrac{1}{\rho^2}\tilde{\Lambda}^2 + \rho^2 - \varepsilon\right]R(\rho) = 0 & \tilde{\Phi}(\beta) = e^{i\tilde{\Lambda}\beta} & \mathrm{O}(1{,}1). \end{array} \tag{47}$$

To solve the radial equation

$$\left[-\partial_\rho^2 - \frac{1}{\rho}\partial_\rho + \frac{1}{\rho^2}Q + \rho^2 - \varepsilon\right]R(\rho) = 0 \qquad \text{where} \qquad Q = \begin{cases} \Lambda^2 & \mathrm{O}(2) \\ -\tilde{\Lambda}^2 & \mathrm{O}(1{,}1). \end{cases} \tag{48}$$

we make the substitutions

$$x = \rho^2 \qquad \Rightarrow \qquad \begin{cases} \partial_\rho = 2x^{\frac{1}{2}}\partial_x \\ \partial_\rho^2 = 4x\partial_x^2 + 2\partial_x \end{cases}, \tag{49}$$

and

$$R(\rho) = x^{A/2}e^{-x/2}L(x) \tag{50}$$

leaving the constant A to be determined. The radial equation becomes

$$4x\frac{d^2}{dx^2}L(x) + [4 + 8A - 4x]\frac{d}{dx}L(x) + \left[\frac{4A^2}{x} - \frac{Q^2}{x} - 4A - 2 + \varepsilon\right]L(x) = 0 \tag{51}$$

which by choosing

$$A = \begin{cases} \frac{1}{2}\Lambda & \text{O(2)} \\ \frac{i}{2}\tilde{\Lambda} & \text{O(1,1)} \end{cases} \tag{52}$$

takes the form of the Laguerre equation

$$\begin{aligned} & x\frac{d^2}{dx^2}L(x) + [\Lambda - x + 1]\frac{d}{dx}L(x) + \frac{1}{2}\left[\frac{1}{2}\varepsilon - \Lambda - 1\right]L(x) = 0 && \text{O(2)} \\ & x\frac{d^2}{dx^2}L(x) + \left[i\tilde{\Lambda} - x + 1\right]\frac{d}{dx}L(x) + \frac{1}{2}\left[\frac{1}{2}\varepsilon - i\tilde{\Lambda} - 1\right]L(x) = 0 && \text{O(1,1).} \end{aligned} \tag{53}$$

Now, the solution for the O(2) oscillator becomes

$$\psi(\rho, \phi) = Ae^{-\rho^2/2}\rho^\Lambda L_n^\Lambda(\rho^2)\, e^{i\Lambda\phi} \tag{54}$$

with index

$$n = \frac{1}{2}\left(\frac{1}{2}\varepsilon - \Lambda - 1\right) \tag{55}$$

so that

$$\kappa = \frac{\omega}{2}\varepsilon = \omega(2n + \Lambda + 1). \tag{56}$$

Thus far, the only restriction on the angular eigenvalue is $2n + \Lambda \geq 0$, and so we put

$$\Lambda = m + s \qquad\qquad \kappa = \omega(2n + m + s + 1) \tag{57}$$

where s is a fixed whole number, and express the O(2) wavefunction as

$$\psi_{nm}(\rho, \phi) = A_{nm}e^{-\rho^2/2}\rho^{m+s}L_n^{m+s}(\rho^2)\, e^{i(m+s)\phi}. \tag{58}$$

The normalization is found from

$$\int_0^{2\pi} d\phi \int_0^\infty \rho d\rho\ |\psi_{nm}(\rho, \phi)|^2 = |A_{nm}|^2 \pi \int_0^\infty dx\ e^{-x}x^{(m+s)}\left|L_n^{m+s}(x)\right|^2 \tag{59}$$

$$= |A_{nm}|^2 \pi \frac{\Gamma(n + m + s + 1)}{\Gamma(n + 1)}. \tag{60}$$

Using the properties $L^{\alpha}_{\beta} = 0$ for $\beta < 0$ and $L^{\alpha}_{0} = P^{0}_{0} = 1$, the ground state wavefunction is

$$\psi_0(\rho, \phi) = A_0 e^{-\rho^2/2}\left(\rho e^{i\phi}\right)^s = A_0 e^{-(x^2+y^2)/2}\left(x^2+y^2\right)^{s/2} e^{is\arctan\left(\frac{y}{x}\right)} \tag{61}$$

and as expected, is separable in Cartesian coordinates only for $s = 0$, which recovers the product of familiar one dimensional oscillators.

For the O(1,1) solution, the requirement of real mass eigenvalue imposes $\tilde{\Lambda} = 0$ so that the O(1,1) wavefunction, also obtained by Bars [12], is

$$\psi_n(\rho, \beta) = A_n e^{-\rho^2/2} L^0_n\left(\rho^2\right) \tag{62}$$

with

$$\kappa = \frac{\omega}{2}\varepsilon = \omega\left(2n+1\right). \tag{63}$$

The ground state for this solution is the restriction to spacelike support of the $D = 2$ FKR ground state (18). Despite the absence of timelike excitations, as seen from

$$\int_{-\infty}^{\infty} d\beta \int_{-\infty}^{\infty} d\rho\ \rho\ \left|\psi_n(\rho)\right|^2 = \left|A_n\right|^2 \int_{-\infty}^{\infty} d\beta \tag{64}$$

these states are non-normalizable as they stand.

2.2. *Fock space*

A number representation appropriate to the solutions (58) and (62) consists of polar creation/annihilation operators that act on eigenstates of the total mode number N and the symmetry generator, M^{12} or M^{01}, to produce new eigenstates. The resulting representation will be equivalent to the standard Cartesian Fock space if the polar eigenstates are unitarily connected to the Cartesian number states, in which case they can be found by expressing N and M^{12} or M^{01} in terms of $\bar{a}^{\mu}$ and a^{μ} and diagonalizing the resulting operators. We consider the Cartesian multiplet φ_1 of first excited states as arising from the action of the vector multiplet of creation operators on the ground state φ_0. Then, by succesive action of creation operators, we attempt to construct the full space of oscillator states.

2.2.1. *O(2) oscillator*

In the case of the O(2) oscillator, the vector operator multiplet takes φ_0 to the Cartesian multiplet φ_1 as

$$\varphi_1 = \begin{pmatrix} \varphi_{10} \\ \varphi_{01} \end{pmatrix} = \begin{pmatrix} \bar{a}^1\varphi_0 \\ \bar{a}^2\varphi_0 \end{pmatrix} = \begin{pmatrix} \bar{a}^1 \\ \bar{a}^2 \end{pmatrix}\varphi_0. \tag{65}$$

Using (7) to replace x^{μ} and p^{μ} with $\bar{a}^{\mu}$ and a^{μ}, the angular momentum operator

$$M^{12} = x^1p^2 - x^2p^1 = -i\left(\bar{a}^1a^2 - \bar{a}^2a^1\right) \tag{66}$$

is seen to act on φ_1 as

$$M^{12}\varphi_1 = -i\left(\bar{a}^1a^2 - \bar{a}^2a^1\right)\begin{pmatrix} \bar{a}^1\varphi_0 \\ \bar{a}^2\varphi_0 \end{pmatrix} = \begin{pmatrix} i\bar{a}^2\varphi_0 \\ -i\bar{a}^1\varphi_0 \end{pmatrix} = \begin{pmatrix} 0 & i \\ -i & 0 \end{pmatrix}\varphi_1 \tag{67}$$

Journal of Physics: Conference Series **330** (2011) 012014 doi:10.1088/1742-6596/330/1/012014

and so has eigenvalues ± 1 on eigenstates

$$\tilde{\varphi}_1 = \frac{1}{\sqrt{2}} \begin{pmatrix} \varphi_{10} + i\varphi_{01} \\ -\varphi_{10} + i\varphi_{01} \end{pmatrix} = \frac{1}{\sqrt{2}} \begin{pmatrix} \bar{a}_+ \\ -\bar{a}_- \end{pmatrix} \varphi_0 \tag{68}$$

where the polar creation/annihilation operators

$$a_\pm = \frac{1}{\sqrt{2}} \left(a^1 \pm i a^2 \right) \qquad \bar{a}_\pm = \frac{1}{\sqrt{2}} \left(\bar{a}^1 \pm i \bar{a}^2 \right) \tag{69}$$

commute among themselves except for

$$[a_+, \bar{a}_-] = [a_-, \bar{a}_+] = 1. \tag{70}$$

We note that the unitary transformation to polar operators

$$\begin{pmatrix} \bar{a}_+ \\ -\bar{a}_- \end{pmatrix} \varphi_0 = \frac{1}{\sqrt{2}} \begin{pmatrix} 1 & i \\ 1 & -i \end{pmatrix} \begin{pmatrix} \bar{a}^1 \\ \bar{a}^2 \end{pmatrix} \varphi_0 \tag{71}$$

takes the Cartesian vector wavefunction multiplet

$$\varphi_1 = \frac{1}{\sqrt{2}} \begin{pmatrix} x^1 - \partial_1 \\ x^2 - \partial_2 \end{pmatrix} A_0 e^{-\mathbf{x}^2/2} = \sqrt{2} \begin{pmatrix} x^1 \\ x^2 \end{pmatrix} A_0 e^{-\mathbf{x}^2/2} \tag{72}$$

to the polar multiplet

$$\tilde{\varphi}_1 = \begin{pmatrix} x^1 + i x^2 \\ -x^1 + i x^2 \end{pmatrix} A_0 e^{-\mathbf{x}^2/2} \tag{73}$$

familiar from helicity representations.

Because the Casimir invariant $\mathbf{M}^2 = \left(M^{12} \right)^2$ is automatically diagonal in the polar basis, operators (69) are sufficient to fully characterize the O(2) oscillator. From the four available products

$$\begin{aligned}
\bar{a}_+ a_+ &= \frac{1}{2} \left(\bar{a}^1 + i\bar{a}^2 \right) \left(a^1 + i a^2 \right) = \frac{1}{2} \left(N^1 - N^2 + i\bar{a}^2 a^1 + i\bar{a}^1 a^2 \right) && (74) \\
\bar{a}_- a_- &= \frac{1}{2} \left(\bar{a}^1 - i\bar{a}^2 \right) \left(a^1 - i a^2 \right) = \frac{1}{2} \left(N^1 - N^2 - i\bar{a}^2 a^1 - i\bar{a}^1 a^2 \right) && (75) \\
\bar{a}_+ a_- &= \frac{1}{2} \left(\bar{a}^1 + i\bar{a}^2 \right) \left(a^1 - i a^2 \right) = \frac{1}{2} \left(N^1 + N^2 + i\bar{a}^2 a^1 - i\bar{a}^1 a^2 \right) && (76) \\
\bar{a}_- a_+ &= \frac{1}{2} \left(\bar{a}^1 - i\bar{a}^2 \right) \left(a^1 + i a^2 \right) = \frac{1}{2} \left(N^1 + N^2 - i\bar{a}^2 a^1 + i\bar{a}^1 a^2 \right) && (77)
\end{aligned}$$

we may form the symmetric Hermitian combinations

$$N = \bar{a}_+ a_- + \bar{a}_- a_+ = N^1 + N^2 \tag{78}$$

$$\Delta = \bar{a}_+ a_+ + \bar{a}_- a_- = N^1 - N^2 = \bar{a}^1 a^1 - \bar{a}^2 a^2 \tag{79}$$

and the antisymmetric Hermitian combinations

$$M^{12} = \bar{a}_+ a_- - \bar{a}_- a_+ = -i \left(\bar{a}^1 a^2 - \bar{a}^2 a^1 \right) \tag{80}$$

Journal of Physics: Conference Series **330** (2011) 012014 doi:10.1088/1742-6596/330/1/012014

$$Q = -i\left(\bar{a}_+ a_+ - \bar{a}_- a_-\right) = \bar{a}^1 a^2 + \bar{a}^2 a^1. \tag{81}$$

In Cartesian coordinates, the maximal set of commuting operators is $\{N, N^1, N^2\}$, and from these one may construct the Hamiltonian. Using

$$\begin{aligned}
\left[M^{12}, N^1\right] &= -i\left[\bar{a}^1 a^2 - a^1 \bar{a}^2, \bar{a}^1 a^1\right] = i\left(\bar{a}^1 a^2 + \bar{a}^2 a^1\right) = iQ && (82)\\
\left[M^{12}, N^2\right] &= -i\left[\bar{a}^1 a^2 - a^1 \bar{a}^2, \bar{a}^2 a^2\right] = -i\left(\bar{a}^1 a^2 + \bar{a}^2 a^1\right) - iQ && (83)
\end{aligned}$$

we confirm that angular momentum commutes with the total mode number

$$\left[M^{12}, N\right] = \left[M^{12}, N^1\right] + \left[M^{12}, N^2\right] = 0 \tag{84}$$

but, of course, N^1 and N^2 do not commute with M^{12} and are not separately observable in the polar representation.

From (6) the Hamiltonian in $D = 2$ is $H = \omega\left(N + 1\right)$ so comparing with (57) we find

$$N\psi_{nm} = \left(2n + m + s\right)\psi_{nm} \tag{85}$$

and using the commutation relations

$$\left[N, a_\pm\right] = -a_\pm \qquad \left[N, \bar{a}_\pm\right] = \bar{a}_\pm \tag{86}$$

$$\left[M^{12}, a_\pm\right] = \pm a_\pm \qquad \left[M^{12}, \bar{a}_\pm\right] = \pm \bar{a}_\pm \tag{87}$$

we compare

$$\begin{aligned}
N a_+ \psi_{nm} &= a_+\left(N - 1\right)\psi_{nm} = \left(2n + m + s - 1\right) a_+ \psi_{nm} && (88)\\
M^{12} a_+ \psi_{nm} &= \left(m + s + 1\right) a_+ \psi_{nm} && (89)
\end{aligned}$$

with

$$\begin{aligned}
N\psi_{n-1,m+1} &= \left(2n + m + s - 1\right)\psi_{n-1,m+1} && (90)\\
M^{12}\psi_{n-1,m+1} &= \left(m + s + 1\right)\psi_{n-1,m+1} && (91)
\end{aligned}$$

and conclude that

$$a_+ \psi_{nm} = C^+_{nm}\psi_{n-1,m+1} \tag{92}$$

where C^+_{nm} is a complex coefficient, with norm found from

$$\left\|a_+\psi_{nm}\right\|^2 = \left\langle\psi_{nm}\right|\bar{a}_- a_+\left|\psi_{nm}\right\rangle = \left|C^+_{nm}\right|^2\left\langle\psi_{n-1,m+1}\middle|\psi_{n-1,m+1}\right\rangle = \left|C^+_{nm}\right|^2 \ . \tag{93}$$

Similarly comparing,

$$\begin{aligned}
N a_- \psi_{nm} &= \left(2n + m + s - 1\right) a_- \psi_{nm} && (94)\\
M^{12} a_- \psi_{nm} &= \left(m + s - 1\right) a_- \psi_{nm} && (95)
\end{aligned}$$

$$\begin{aligned}
N \bar{a}_+ \psi_{nm} &= \left(2n + m + s + 1\right) a_+ \psi_{nm} && (96)\\
M^{12} \bar{a}_+ \psi_{nm} &= \left(m + s + 1\right) \bar{a}_+ \psi_{nm} && (97)
\end{aligned}$$

$$\begin{aligned}
N \bar{a}_- \psi_{nm} &= \left(2n + m + s + 1\right) \bar{a}_- \psi_{nm} && (98)\\
M^{12} \bar{a}_- \psi_{nm} &= \left(m + s - 1\right) \bar{a}_- \psi_{nm} && (99)
\end{aligned}$$

with

$$N\psi_{n,m-1} = [2n+(m-1)]\,\psi_{n,m-1} = (2n+s+m-1)\,\psi_{n,m-1} \tag{100}$$
$$M^{12}\psi_{n,m-1} = (m-1)\,\psi_{n,m-1} \tag{101}$$

$$N\psi_{n,m+1} = [2n+s+(m+1)]\,\psi_{n,m+1} = (2n+s+m+1)\,\psi_{n,m+1} \tag{102}$$
$$M^{12}\psi_{n,m+1} = (m+s+1)\,\psi_{n,m+1} \tag{103}$$

$$\begin{aligned} N\psi_{n+1,m-1} &= [2\,(n+1)+s+(m-1)]\,\psi_{n+1,m-1} \\ &= (2n+s+m+1)\,\psi_{n+1,m-1} \end{aligned} \tag{104}$$
$$M^{12}\psi_{n+1,m-1} = (m+s-1)\,\psi_{n+1,m-1} \tag{105}$$

leads to

$$a_-\psi_{nm} = C^-_{nm}\psi_{nm-1} \qquad \bar{a}_+\psi_{nm} = \bar{C}^+_{nm}\psi_{nm+1} \qquad \bar{a}_-\psi_{nm} = \bar{C}^-_{nm}\psi_{n+1m-1}. \tag{106}$$

We eliminate two coefficients

$$\left|\bar{C}^-_{nm}\right|^2 = \left|C^+_{nm}\right|^2 + 1 \qquad \left|\bar{C}^+_{nm}\right|^2 = \left|C^-_{nm}\right|^2 + 1 \tag{107}$$

using the commutation relations (70). Solving

$$2n+m+s = \langle nm|\,N\,|nm\rangle = \langle nm|\,\bar{a}_+a_- + \bar{a}_-a_+\,|nm\rangle = \left|C^-_{nm}\right|^2 + \left|C^+_{nm}\right|^2 \geq 0, \tag{108}$$

together with

$$m+s = \langle nm|\,M^{12}\,|nm\rangle = \langle nm|\,\bar{a}_+a_- - \bar{a}_-a_+\,|nm\rangle = \left|C^-_{nm}\right|^2 - \left|C^+_{nm}\right|^2 \tag{109}$$

and taking the coefficients to be real leads to

$$C^+_{nm} = \sqrt{n} \qquad \bar{C}^-_{nm} = \sqrt{n+1} \tag{110}$$
$$C^-_{nm} = \sqrt{n+m+s} \qquad \bar{C}^+_{nm} = \sqrt{n+m+s+1} \tag{111}$$

and so we normalize the ladder operators as

$$a_+\psi_{nm} = \sqrt{n}\;\psi_{n-1,m+1} \qquad \bar{a}_-\psi_{nm} = \sqrt{n+1}\;\psi_{n+1,m-1} \tag{112}$$
$$a_-\psi_{nm} = \sqrt{n+m+s}\;\psi_{n,m-1} \qquad \bar{a}_+\psi_{nm} = \sqrt{n+m+s+1}\;\psi_{n,m+1}\,. \tag{113}$$

Special care must be taken with the ground state (61) because (112) and (113) lead to

$$a_+\psi_0 = 0 \qquad a_-\psi_0 = \sqrt{s}\;\psi_{0,-1} \tag{114}$$

or equivalently

$$a^1\psi_0 = \sqrt{\frac{s}{2}}\psi_{0,-1} \qquad a^2\psi_0 = i\sqrt{\frac{s}{2}}\psi_{0,-1} \tag{115}$$

suggesting a state with $m=-1$ and mode number $N=s-1$. By restricting $0\leq s<1$, the state $\psi_{0,-1}$ is excluded from the Fock space. The action of a_- in (114) may then be understood

Journal of Physics: Conference Series **330** (2011) 012014 doi:10.1088/1742-6596/330/1/012014

as taking the ground state to a non-observable function which must be taken into account in calculations such as

$$N\psi_0 = (\bar{a}_+a_- + \bar{a}_-a_+)\,\psi_0 = \sqrt{s}\bar{a}_+\psi_{0,-1} = \sqrt{s}\sqrt{-1+s+1}\psi_0 = s\psi_0 \tag{116}$$

but is effectively annihilated at the end of calculations. As can be seen from (58), allowing $s = 1$ merely reproduces the $s = 0$ states, shifted as $m \to m-1$, with $\psi_{0,-1}$ as the unique ground state.

Excited states may be constructed from the ground state as

$$\zeta_{\alpha\beta} = \frac{1}{N_{\alpha\beta}}\,(\bar{a}_+)^\alpha\,(\bar{a}_-)^\beta\,\psi_0 \tag{117}$$

with normalization coefficient $1/N_{\alpha\beta}$. It follows from (112) that

$$(\bar{a}_-)^\beta\,\psi_0 = \sqrt{\beta!}\;\psi_{\beta,-\beta} \tag{118}$$

and from (113) that

$$(\bar{a}_+)^\alpha\,\psi_{\beta,-\beta} = \sqrt{\frac{\Gamma(s+\alpha+1)}{\Gamma(s+1)}}\;\psi_{\beta,-\beta+\alpha} \tag{119}$$

and so we take

$$N_{\alpha\beta} = \sqrt{\beta!\frac{\Gamma(s+\alpha+1)}{\Gamma(s+1)}} \tag{120}$$

which reduces to $\sqrt{\alpha!\beta!}$ in the case $s = 0$. Operating on these states with the total mode operator (78)

$$N\zeta_{\alpha\beta} = \frac{1}{N_{\alpha\beta}}\,(\bar{a}_+a_- + \bar{a}_-a_+)\,(\bar{a}_+)^\alpha\,(\bar{a}_-)^\beta\,\psi_0 \tag{121}$$

and using the commutation relations (70) and the identity

$$[B, A] = c, \quad \text{a c-number} \qquad \Rightarrow \qquad [B, A^n] = cnA^{n-1} \tag{122}$$

we calculate

$$\bar{a}_+a_-\,(\bar{a}_+)^\alpha\,(\bar{a}_-)^\beta\,\psi_0 = \bar{a}_+\left[(\bar{a}_+)^\alpha\,a_- + \alpha\,(\bar{a}_+)^{\alpha-1}\right](\bar{a}_-)^\beta\,\psi_0 \tag{123}$$

$$= \left[(\bar{a}_+)^{\alpha+1}\,(\bar{a}_-)^\beta\,a_- + \alpha\,(\bar{a}_+)^\alpha\,(\bar{a}_-)^\beta\right]\psi_0 \tag{124}$$

$$= (\bar{a}_+)^\alpha\,(\bar{a}_-)^\beta\,(\bar{a}_+a_- + \alpha)\,\psi_0 \tag{125}$$

$$\bar{a}_-a_+\,(\bar{a}_+)^\alpha\,(\bar{a}_-)^\beta\,\psi_0 = (\bar{a}_+)^\alpha\,\bar{a}_-a_+\,(\bar{a}_-)^\beta\,\psi_0 \tag{126}$$

$$= (\bar{a}_+)^\alpha\,\bar{a}_-\left[(\bar{a}_-)^\beta\,a_+ + \beta\,(\bar{a}_-)^{\beta-1}\right]\psi_0 \tag{127}$$

$$= (\bar{a}_+)^\alpha\,(\bar{a}_-)^\beta\,(\bar{a}_-a_+ + \beta)\,\psi_0 \tag{128}$$

which combine to

$$N\zeta_{\alpha\beta} = \frac{1}{N_{\alpha\beta}}\,(\bar{a}_+)^\alpha\,(\bar{a}_-)^\beta\,(\alpha + \beta + \bar{a}_+a_- + \bar{a}_-a_+)\,\psi_0. \tag{129}$$

Using (116) we obtain

$$N\zeta_{\alpha\beta} = (\alpha + \beta + s)\,\zeta_{\alpha\beta} \tag{130}$$

so that the states $\zeta_{\alpha\beta}$ have total mode number given by

$$N = \alpha + \beta + s = \alpha + \beta + N_{\text{ground state}}\ . \tag{131}$$

A similar calculation using (80) leads to

$$M^{12}\zeta_{\alpha\beta} = (\alpha - \beta + s)\,\zeta_{\alpha\beta} \tag{132}$$

so that the states $\zeta_{\alpha\beta}$ have angular momentum

$$M^{12} = \alpha - \beta + s = m + M^{12}_{\text{ground state}} \qquad m = \alpha - \beta\ . \tag{133}$$

Comparing (130) with (85) we see that

$$\alpha + \beta = 2n + m \tag{134}$$

which together with (133) and (131) provide expressions for the principal quantum number n and the integer part of the angular momentum m

$$n = \frac{1}{2}(\alpha + \beta - m) = \beta \qquad m = \alpha - \beta = N - s - 2\beta \tag{135}$$

and fixes α as

$$\alpha = 2n + m - \beta = n + m = N - s - m. \tag{136}$$

Thus, the states $\zeta_{\alpha\beta}$ defined in (117) can be identified with explicit solutions $\psi^{(N)}_{n,m}$ through

$$\zeta_{\alpha\beta} = \psi^{(\alpha+\beta+s)}_{\beta,\alpha-\beta} \qquad \psi^{(N)}_{n,m} = \zeta_{N-s-m,n} \tag{137}$$

for which $n = \beta = 0, 1, ..., N - s$ characterizes the $(N - s + 1)$-fold multiplicity of states with mode number N. Equivalently, the multiplicity can be enumerated by the angular momentum m. For the case $s = 0$, this simple multiplicity structure is identical to the Cartesian picture in which there are $N + 1$ ways to build a state of total mode number N from a pair of one dimensional oscillators.

Bars has observed [12] that the harmonic oscillator Hamiltonian in D dimensional Fock space possesses a symmetry generated by the products $\bar{a}^\mu a^\nu$ of the ladder operators. Because such products replace one ν-mode of the oscillator with one μ-mode, the total mode number, and therefore the total mass/energy, is conserved. The traceless part of the generators

$$J^{\mu\nu} = \bar{a}^\mu a^\nu - \frac{1}{D}\eta^{\mu\nu}\eta_{\lambda\rho}\bar{a}^\lambda a^\rho \tag{138}$$

generates an SU$(D-1,1)$ or SU(D) dynamical symmetry of the Hamiltonian, while the trace

$$\eta_{\lambda\rho}\bar{a}^\lambda a^\rho = \sum_\mu \eta_{\mu\mu}N^\mu = N \tag{139}$$

is the total mode number and differs from the Hamiltonian by a c-number. The antisymmetric part of the generators

$$\frac{1}{2}(J^{\mu\nu} - J^{\nu\mu}) = M^{\mu\nu} = \bar{a}^\mu a^\nu - \bar{a}^\nu a^\mu \tag{140}$$

generates the SO$(D-1,1)$ or SO(D) symmetry of the Hamiltonian. Bars argues that harmonic oscillator states should belong to representations of the SU dynamical symmetry as well as to representations of the SO symmetry, imposing additional constraints on admissible solutions.

For the O(2) Cartesian ladder operators, the traceless operator is

$$J = \begin{bmatrix} \frac{1}{2}\left(\bar{a}^1 a^1 - \bar{a}^2 a^2\right) & \bar{a}^1 a^2 \\ \bar{a}^2 a^1 & -\frac{1}{2}\left(\bar{a}^1 a^1 - \bar{a}^2 a^2\right) \end{bmatrix} \tag{141}$$

with antisymmetric part equal to the angular momentum operator

$$iM^{12} = \bar{a}^1 a^2 - \bar{a}^2 a^1 \tag{142}$$

and symmetric part given by

$$S = \frac{1}{2}\left(J + J^\mathsf{T}\right) = \frac{1}{2}\begin{bmatrix} \bar{a}^1 a^1 - \bar{a}^2 a^2 & \bar{a}^1 a^2 + \bar{a}^2 a^1 \\ \bar{a}^1 a^2 + \bar{a}^2 a^1 & -\bar{a}^1 a^1 + \bar{a}^2 a^2 \end{bmatrix} = \frac{1}{2}\begin{bmatrix} \Delta & Q \\ Q & -\Delta \end{bmatrix} \tag{143}$$

where we use (79) and (81). Directly calculating

$$\left[\tfrac{1}{2}M^{12}, \tfrac{1}{2}\Delta\right] = \tfrac{1}{4}\left[M^{12}, N^1\right] - \tfrac{1}{4}\left[M^{12}, N^2\right] = i\tfrac{1}{2}Q \tag{144}$$

$$\left[\tfrac{1}{2}\Delta, \tfrac{1}{2}Q\right] = \tfrac{1}{4}\left[\bar{a}^1 a^1 - \bar{a}^2 a^2, \bar{a}^1 a^2 + \bar{a}^2 a^1\right] = \tfrac{1}{2}\left(\bar{a}^1 a^2 - \bar{a}^2 a^1\right) = i\tfrac{1}{2}M^{12} \tag{145}$$

$$\left[\tfrac{1}{2}Q, \tfrac{1}{2}M^{12}\right] = -i\tfrac{1}{4}\left[\bar{a}^1 a^2 + \bar{a}^2 a^1, \bar{a}^1 a^2 - \bar{a}^2 a^1\right] = \tfrac{1}{2}i\left(\bar{a}^1 a^1 - \bar{a}^2 a^2\right) = i\tfrac{1}{2}\Delta \tag{146}$$

we verify that the three independent operators $\left\{\frac{1}{2}M^{12}, \frac{1}{2}\Delta, \frac{1}{2}Q\right\}$ satisfy the SU(2) algebra. Equation (84) confirms that M commutes with total mode number N; similarly,

$$[N, \Delta] = \left[N^1 + N^2, N^1 - N^2\right] = 0 \tag{147}$$

$$[N, Q] = \left[\bar{a}^1 a^1 + \bar{a}^2 a^2, \bar{a}^1 a^2 + \bar{a}^2 a^1\right] = 0 \tag{148}$$

so this SU(2) is indeed a symmetry of the Hamiltonian. Of the three operators, Δ is observable in Cartesian coordinates, while M^{12} is observable in polar coordinates. However, since the ground state satisfies $M^{12}\psi_0 = s\psi$, the SU(2) symmetry is spontaneous broken for $s > 0$. From (58), we see that the $s > 0$ ground state is centered on a ring of radius $\rho = \sqrt{s}$, so that the operation $e^{i\theta M^{12}}$ rotates the ground state on this ring. Nevertheless, this infinite degeneracy merely contributes a phase.

Expanding the SU(2) Casimir operator in terms of the creation and annihilation operators, (79) — (81) lead to

$$\left(\tfrac{1}{2}M\right)^2 + \left(\tfrac{1}{2}\Delta\right)^2 + \left(\tfrac{1}{2}Q\right)^2 = \tfrac{1}{4}N\left(N+2\right) = \tfrac{1}{2}N\left(\tfrac{1}{2}N + 1\right). \tag{149}$$

Since the Casimir eigenvalue $\frac{1}{2}N$ of a unitary representation of SU(2) must be integral or half-integral, the $s \neq 0$ solutions must be non-unitarity representations of the SU(2) dynamical symmetry[2]. This result is unsurprising, because we have seen from the form of the wavefunctions that for $s \neq 0$ there is no unitary transformation from the unitary Cartesian solution to the polar solution. The implications for these solutions as representations of the classical O(2) symmetry

[2] I am grateful to Itzhak Bars for this observation. See also [14]

are less clear. In appendix C, we verify that the solutions (58) satisfy the creation/annihilation ladder representation in polar coordinates. We may also construct new ladder operators

$$J^{+} = \frac{1}{2}(Q - i\Delta) = -i\bar{a}_{+}a_{+} \qquad J^{-} = \frac{1}{2}(Q + i\Delta) = i\bar{a}_{-}a_{-} \tag{150}$$

which using commutation relations (144) — (145) satisfy

$$\left[\frac{1}{2}M^{12}, J^{\pm}\right] = \pm J^{\pm}. \tag{151}$$

As is verified by comparison with (112) and (113), $J^{\pm}$ raise/lower the M^{12} eigenvalue by 2, leaving the mode number $N = 2n + m + s$ unchanged. The SU(2) symmetry thus enriches the description of the O(2) oscillator, by providing an algebraic transformation among the degenerate energy eigenstates.

2.2.2. O(1,1) oscillator Unlike O(2), we cannot construct excited states for the O(1,1) oscillator by acting on the ground state with the Cartesian vector multiplet of creation operators. To see this, we recall that the eigenvalue of the Lorentz generator

$$M^{01} = x^0p^1 - x^1p^0 = -i\left(\bar{a}^0a^1 - \bar{a}^1a^0\right) \tag{152}$$

on the O(1,1) wavefunctions (62) is identically zero, and so (152) must annihilate these states. Writing the multiplet

$$\varphi_1 = \begin{pmatrix} \varphi_{10} \\ \varphi_{01} \end{pmatrix} = \begin{pmatrix} \bar{a}^0\varphi_0 \\ \bar{a}^1\varphi_0 \end{pmatrix} = \begin{pmatrix} \bar{a}^0 \\ \bar{a}^1 \end{pmatrix}\varphi_0 \tag{153}$$

we immediately see that on such states

$$M^{01}\varphi_1 = -i\left(\bar{a}^0a^1 - \bar{a}^1a^0\right)\begin{pmatrix} \bar{a}^0\varphi_0 \\ \bar{a}^1\varphi_0 \end{pmatrix} = \begin{pmatrix} -i\bar{a}^1\varphi_0 \\ -i\bar{a}^0\varphi_0 \end{pmatrix} = \begin{pmatrix} 0 & -i \\ -i & 0 \end{pmatrix}\varphi_1 \tag{154}$$

the Hermitian operator M^{01} would have eigenvalues $\pm i$. The coordinate form of (153) is

$$\varphi_1 = \frac{1}{\sqrt{2}}\,\rho\begin{bmatrix} \epsilon(\rho)\sinh\beta \\ \cosh\beta \end{bmatrix}e^{-\rho^2/2} = \frac{1}{\sqrt{2}}\,\epsilon(\rho)\begin{bmatrix} x^0 \\ x^1 \end{bmatrix}e^{-x^2/2} \tag{155}$$

which, as in the O(2) case, represents a vector wavefunction multiplet, but is not consistent with the O(1,1) wavefunctions. The generators of the dynamical symmetry are

$$J = \begin{bmatrix} \frac{1}{2}\left(-\bar{a}^0a^0 - \bar{a}^1a^1\right) & \bar{a}^0a^1 \\ \bar{a}^1a^0 & \frac{1}{2}\left(\bar{a}^0a^0 + \bar{a}^1a^1\right) \end{bmatrix} \tag{156}$$

from which we identify the operators

$$M^{01} = -i\left(\bar{a}^0a^1 - \bar{a}^1a^0\right) \qquad \Delta^{01} = -\bar{a}^0a^0 - \bar{a}^1a^1 \qquad Q^{01} = \bar{a}^0a^1 + \bar{a}^1a^0 \tag{157}$$

that are seen to annihilate the ground state. One may verify that these operators satisfy the SU(1,1) algebra

$$[K_1, K_2] = -iK_0 \qquad [K_0, K_1] = iK_2 \qquad [K_2, K_0] = iK_1 \tag{158}$$

Journal of Physics: Conference Series **330** (2011) 012014 doi:10.1088/1742-6596/330/1/012014

where

$$K_0 = \frac{1}{2}Q^{01} \qquad K_1 = \frac{1}{2}M^{01} \qquad K_2 = \frac{1}{2}\Delta^{01}. \tag{159}$$

We notice that Q^{01} rather than M^{01} acts as the rotation generator in this SU(1,1), but like M^{01}, has eigenvalues $\pm i$ on the vector multiplet, so that the implementation of the ladder operators

$$\left[K_0, K^{\pm}\right] = \pm iK^{\pm} \qquad K^{\pm} = K_1 \pm iK_2 = \frac{1}{2}M^{01} \pm i\frac{1}{2}\Delta^{01} \tag{160}$$

provides no additional insight. Bars [12] has shown that since the ground state has zero M^{01} eigenvalue, the states (62) can be obtained through repeated application of the scalar operator

$$\bar{a} \cdot \bar{a} = -\left(\bar{a}^0\right)^2 + \left(\bar{a}^1\right)^2 = \frac{1}{2}\left(\rho^2 + \partial_\rho^2 + \frac{1}{\rho}\partial_\rho - \frac{1}{\rho^2}\mathbf{M}^2 - 2\rho\partial_\rho - 2\right) \tag{161}$$

as

$$\psi_n\left(\rho, \beta\right) = A_n e^{-\rho^2/2} L_n^0\left(\rho^2\right) = \alpha_n\left(\bar{a} \cdot \bar{a}\right)^n e^{-\rho^2/2} \tag{162}$$

for appropriate α_n. He has argued that since all unitary representations of a non-compact group must be either a singlet or infinite dimensional, these states are naturally created by the singlet operator. Since

$$\left[\bar{a} \cdot \bar{a}, M^{01}\right] \psi_n\left(\rho, \beta\right) = 0 \tag{163}$$

it follows that all states have zero M^{01} eigenvalue. Nevertheless, the states $K_0\psi_n\left(\rho, \beta\right)$ are not singlet states for $n > 0$ as can be seen from

$$\left[\frac{1}{2}Q^{01}, \bar{a} \cdot \bar{a}\right] = \bar{a}^0\bar{a}^1 - \bar{a}^1\bar{a}^0 \tag{164}$$

so that, for example the state

$$K_0\psi_1 = K_0\bar{a} \cdot \bar{a}\psi_0 = \bar{a} \cdot \bar{a}K_0\psi_0 + \left(\bar{a}^0\bar{a}^1 - \bar{a}^1\bar{a}^0\right)\psi_0 = \left(\bar{a}^0\bar{a}^1 - \bar{a}^1\bar{a}^0\right)\psi_0 \tag{165}$$

is not in the solution space. Therefore, the significance of the SU(1,1) symmetry is unclear for the wavefunction solutions, as well as for the vector multiplet.

The O(1,1) oscillator thus differs from O(2) in significant ways. For O(2), the properties of the creation/annihilation operators provide a solution equivalent to the explicit coordinate wavefunctions, in addition to a ladder representation of excited modes. Moreover, the states provide a non-trivial representation of the O(2) and SU(2) symmetries, with raising/lowering operators for angular momentum. In particular, the $s = 0$ polar solution is related to the Cartesian solution through the unitary transformation that diagonalizes the group generator. For O(1,1) however, all wavefunctions belong to the singlet representation of the Lorentz group, and the ladder representation consists only of the scalar combinations of creation/annihilation operators. In particular, the SU(1,1) symmetry does not lead to a ladder representation for these solutions. Defining creation and annihilation operators

$$\bar{A} = \frac{1}{2}\bar{a} \cdot \bar{a} \qquad A = \frac{1}{2}a \cdot a \tag{166}$$

that satisfy

$$\left[A, \bar{A}\right] = 1 + N \qquad \left[N, \bar{A}\right] = 2\bar{A} \qquad \left[N, A\right] = -2A \tag{167}$$

the singlet representation is seen to be equivalent to a one dimensional oscillator. The number operator $\mathcal{N} = \bar{A}A$ satisfies ladder relations

$$[\mathcal{N}, \bar{A}] = \bar{A}\,(1+N) \qquad [\mathcal{N}, A] = -\,(1+N)\,A \tag{168}$$

and since

$$[\mathcal{N}, N] = \bar{A}\,[A, N] + \left[\bar{A}, N\right] A = 0 \tag{169}$$

it shares mode eigenstates with N. Using the general identity

$$\left[a, b^k\right] = \sum_{j=0}^{k-1} b^j\,[a,b]\,b^{k-(j+1)} \tag{170}$$

one finds

$$\mathcal{N}\psi_k = \begin{cases} 0 & , \quad k=0 \\ \left[k^2 + 2\,(k-1)\right]\psi_k & , \quad k = 1,2,3,\cdots \end{cases} \tag{171}$$

where, as in (162), the states are generated from the ground state as

$$\psi_k = \alpha_k \bar{A}^k \psi_0 \tag{172}$$

with normalizations α_k appropriate to the eigenvalues of $\mathcal{N}$. While we have obtained an essentially one dimensional solution providing a unitary singlet representation of O(1,1), we have not succeeded in identifying an infinite dimensional representation capable of providing a reasonable description of the infinite degeneracy expected in an invariant-mass system under boost. Such a representation would require either an infinitely degenerate ground state (such as we will obtain for the O(2,1) oscillator) or an infinite dimensional multiplet of creation operators to act on a singlet ground state.

3. Harmonic oscillator in $D = 3$

3.1. Wavefunctions

In $D = 3$ with parameterizations (36), the Schrodinger equation becomes

$$\left[-\partial_\rho^2 - \frac{2}{\rho}\partial_\rho + \frac{1}{\rho^2}\mathbf{M}^2 + \rho^2 - \varepsilon\right]\psi = 0 \tag{173}$$

with group generators

$$L^{\mu\nu} = x^\mu p^\nu - x^\nu p^\mu \tag{174}$$

which we notate

$$\begin{array}{llll} L^1 = x^2p^3 - x^3p^2 & L^2 = x^3p^1 - x^1p^3 & M = x^1p^2 - x^2p^1 & \mathrm{O(3)} \\ K^1 = x^0p^1 - x^1p^0 & K^2 = x^0p^2 - x^2p^0 & M = x^1p^2 - x^2p^1 & \mathrm{O(2,1)} \end{array} \tag{175}$$

so that the parameterizations (36) diagonalize the M angular momentum component

$$M = -i\partial_\phi. \tag{176}$$

Journal of Physics: Conference Series **330** (2011) 012014 doi:10.1088/1742-6596/330/1/012014

The Casimir operators in these coordinates are

$$\mathbf{M}^2 = \begin{cases} L_1^2 + L_2^2 + M^2 = -\partial_\theta^2 - \dfrac{\cos\theta}{\sin\theta}\partial_\theta - \dfrac{1}{\sin^2\theta}\partial_\phi^2 & \mathrm{O}(3) \\ M^2 - K_1^2 - K_2^2 = \partial_\beta^2 + \dfrac{\sinh\beta}{\cosh\beta}\partial_\beta - \dfrac{1}{\cosh^2\beta}\partial_\phi^2 & \mathrm{O}(2{,}1) \end{cases} \tag{177}$$

so writing a separation of variables

$$\begin{aligned} \psi(\rho,\theta,\phi) &= R(\rho)\,F(\theta)\,\Phi(\phi) \qquad \mathrm{O}(3) \\ \psi(\rho,\beta,\phi) &= R(\rho)\,G(\beta)\,\Phi(\phi) \qquad \mathrm{O}(2{,}1) \end{aligned} \tag{178}$$

leads to the common angular function

$$\Phi(\phi) = e^{i\Lambda_1\phi} \tag{179}$$

allowing the replacement of $M = -i\partial_\phi$ in (177) by its eigenvalue Λ_1. A second separation of variables, associated with the eigenvalue equation

$$\mathbf{M}^2\psi = \Lambda_2\psi \tag{180}$$

for the Casimir operators, leads to

$$\left(-\mathbf{M}^2 + \Lambda_2\right)F(\theta) = \left(\partial_\theta^2 + \frac{\cos\theta}{\sin\theta}\partial_\theta - \frac{\Lambda_1^2}{\sin^2\theta} + \Lambda_2\right)F(\theta) = 0 \tag{181}$$

$$\left(\mathbf{M}^2 - \Lambda_2\right)G(\beta) = \left(\partial_\beta^2 + \frac{\sinh\beta}{\cosh\beta}\partial_\beta + \frac{\Lambda_1^2}{\cosh^2\beta} - \Lambda_2\right)G(\beta) = 0 \tag{182}$$

which may be approached in two inequivalent ways. Method I follows the approach generally applied to the classical central force problems, noting the form of the first order derivative terms to make the substitutions

$$z = \cos\theta \qquad \zeta = \sinh\beta \qquad \Lambda_2 = l\,(l+1) \qquad \Lambda_1 = m \tag{183}$$

and

$$F(\theta) \to P_l^m(z) \qquad G(\beta) \to \hat{P}_l^m(\zeta) \tag{184}$$

so that (181) and (182) become associated Legendre equations in the respective forms

$$\left[\left(1-z^2\right)\partial_z^2 - 2z\partial_z + l\,(l+1) - \frac{m^2}{1-z^2}\right]P_l^m(z) = 0 \tag{185}$$

$$\left[\left(1+\zeta^2\right)\partial_\zeta^2 + 2\zeta\partial_\zeta - l\,(l+1) + \frac{m^2}{1+\zeta^2}\right]\hat{P}_l^m(\zeta) = 0. \tag{186}$$

Notice that (186) can be obtained from (185) by putting $z = i\zeta$. Method II produces an inequivalent set of solutions obtained through the substitution

$$\begin{aligned} z &= \frac{\cos\theta}{\sin\theta} & \zeta &= \frac{\sinh\beta}{\cosh\beta} \\ F_l^m(z) &= \left(1+z^2\right)^{\frac{1}{4}}\hat{P}_l^m(z) & G_l^m(\zeta) &= \left(1-\zeta^2\right)^{\frac{1}{4}}P_l^m(\zeta) \end{aligned} \tag{187}$$

Journal of Physics: Conference Series **330** (2011) 012014 doi:10.1088/1742-6596/330/1/012014

leading to associated Legendre equations

$$\left[\left(1+z^2\right)\partial_z^2 + 2z\partial_z - m\left(m+1\right) + \frac{l^2}{1+z^2}\right]\hat{P}_m^l\left(z\right) = 0 \tag{188}$$

$$\left[\left(1-\zeta^2\right)\partial_\zeta^2 - 2\zeta\partial_\zeta + m\left(m+1\right) - \frac{l^2}{1-\zeta^2}\right]P_m^l\left(\zeta\right) = 0. \tag{189}$$

Notice that the constants l and m have reversed roles with respect to equations (185) and (186), having been introduced to satisfy

$$m\left(m+1\right) = \Lambda_1^2 - \frac{1}{4} \quad \longrightarrow \quad \Lambda_1 = m + \frac{1}{2} \tag{190}$$

$$l^2 = \Lambda_2 + \frac{1}{4} \quad \longrightarrow \quad \Lambda_2 = l^2 - \frac{1}{4}\,. \tag{191}$$

Comparing (183) and (190), we write $\Lambda_1 = m + s$, so that (179) becomes

$$\Phi\left(\phi\right) = e^{i\Lambda_1\phi} = e^{i(m+s)\phi} \tag{192}$$

where the orbital angular momentum may be integral or half-integral as characterized by

$$s = \begin{cases} 0 & , \text{ method I} \\ 1/2 & , \text{ method II} \end{cases}. \tag{193}$$

Replacing $\mathbf{M}^2$ with its eigenvalue in (173) leads to the radial equation

$$\left[-\partial_\rho^2 - \frac{2}{\rho}\partial_\rho + \frac{1}{\rho^2}\Lambda_2 + \rho^2 - \varepsilon\right]R\left(\rho\right) = 0 \tag{194}$$

in which we again make the substitutions (49) and (50), and put

$$A = \sqrt{\frac{1}{4} + \Lambda_2} - \frac{1}{2} = l - s \tag{195}$$

to obtain the Laguerre equation,

$$\left[x\frac{d^2}{dx^2} + \left(l - s - x + \frac{3}{2}\right)\frac{d}{dx} + \frac{1}{2}\left(\frac{1}{2}\varepsilon - \left(l-s\right) - \frac{3}{2}\right)\right]L_n^\alpha\left(x\right) = 0 \tag{196}$$

with

$$\alpha = l - s + \frac{1}{2} \qquad n = \frac{1}{2}\left(\frac{1}{2}\varepsilon - \left(l-s\right) - \frac{3}{2}\right). \tag{197}$$

From $\kappa = \frac{\omega}{2}\varepsilon$ the spectra are given by

$$\kappa = \omega\left(2n + l + 3/2 - s\right) \tag{198}$$

the O(3) wavefunctions ψ_{nlm} and O(2,1) wavefunctions $\hat{\psi}_{nlm}$ up to normalization are shown in figure 1:

Figure 1. Solutions to eigenvalue equation

	O(3)	O(2,1)
$s=0$	$e^{-\frac{\rho^2}{2}}\rho^l L_n^{l+\frac{1}{2}}\left(\rho^2\right) P_l^m\left(\cos\theta\right)e^{im\phi}$	$e^{-\frac{\rho^2}{2}}\rho^l L_n^{l+\frac{1}{2}}\left(\rho^2\right) \hat{P}_l^m\left(\sinh\beta\right)e^{im\phi}$
$s=\frac{1}{2}$	$e^{-\frac{\rho^2}{2}}\rho^{l-\frac{1}{2}} L_n^{l}\left(\rho^2\right) \dfrac{\hat{P}_m^l\left(\cot\theta\right)}{\sqrt{\lvert\sin\theta\rvert}}e^{i\left(m+\frac{1}{2}\right)\phi}$	$e^{-\frac{\rho^2}{2}}\rho^{l-\frac{1}{2}} L_n^{l}\left(\rho^2\right) \dfrac{P_m^l\left(\tanh\beta\right)}{\sqrt{\cosh\beta}}e^{i\left(m+\frac{1}{2}\right)\phi}$

The ground states (taking $n=l=m=0$) are

$$\begin{aligned}\psi_0\left(\rho,\theta,\phi\right) &= A_0 e^{-\rho^2/2}\frac{e^{is\phi}}{\left(\rho\left|\sin\theta\right|\right)^s} = A_0 e^{-\left(x^2+y^2+z^2\right)/2}\frac{e^{is\arctan\left(\frac{y}{x}\right)}}{\left(x^2+y^2\right)^{s/2}} \qquad \text{O(3)}\\ \psi_0\left(\rho,\beta,\phi\right) &= A_0 e^{-\rho^2/2}\frac{e^{is\phi}}{\left(\rho\cosh\beta\right)^s} = A_0 e^{-\left(x^2+y^2-t^2\right)/2}\frac{e^{is\arctan\left(\frac{y}{x}\right)}}{\left(x^2+y^2\right)^{s/2}} \qquad \text{O(2,1)}\end{aligned} \tag{199}$$

which, as expected, are separable and recover the standard solutions in Cartesian coordinates only for $s=0$. In particular, for O(2,1), the condition $s=0$ recovers the FKR ground state restricted to spacelike support. Nevertheless, the choice $s=1/2$ in (198) leads to a lowering of the mass/energy, just as found by Horwitz and Arshansky [7].

For $s=0$, the O(3) wavefunctions ψ_{nlm} depend on θ and ϕ through the spherical harmonics

$$Y_l^m\left(\theta,\phi\right) = C_{lm}P_l^m\left(\cos\theta\right)e^{im\phi} \tag{200}$$

and thus provide the familiar $(2l+1)$-dimensional representation of O(3) through

$$\mathbf{M}^2 Y_l^m\left(\theta,\phi\right) = l\left(l+1\right)Y_l^m\left(\theta,\phi\right) \qquad MY_l^m\left(\theta,\phi\right) = mY_l^m\left(\theta,\phi\right) \tag{201}$$

$$L^{\pm}Y_l^m\left(\theta,\phi\right) = \sqrt{\left(l\mp m\right)\left(l\pm m+1\right)}Y_l^m\left(\theta,\phi\right) \tag{202}$$

with allowed values

$$l=0,1,... \qquad m=-l,-l+1,...,l-1,l. \tag{203}$$

Similarly, for $s=0$ the O(2,1) wavefunctions $\hat{\psi}_{nlm}$ depend on β and ϕ through the functions

$$\hat{Y}_l^m\left(\beta,\phi\right) = C_{lm}\hat{P}_l^m\left(\sinh\beta\right)e^{im\phi} \tag{204}$$

and it follows from (182) and (183) that

$$\mathbf{M}^2\hat{\psi}_{nlm} = l\left(l+1\right)\hat{\psi}_{nlm} \qquad M\hat{\psi}_{nlm} = m\hat{\psi}_{nlm}\;. \tag{205}$$

The remaining O(2,1) boost generators are

$$K^{\pm} = K^1 \pm iK^2 = -ie^{\pm i\phi}\left(\partial_\beta \pm i\frac{\sinh\beta}{\cosh\beta}\partial_\phi\right) \tag{206}$$

and for $s = 0$, where we set $\zeta = \sinh\beta$, they take the form

$$K^{\pm} = -i\frac{e^{\pm i\phi}}{\left(1+\zeta^2\right)^{1/2}}\left[\left(1+\zeta^2\right)\partial_\zeta \pm i\zeta\partial_\phi\right] \tag{207}$$

and

$$K^{\pm}\hat{Y}_l^m\left(\beta,\phi\right) = -iC_{lm}\frac{e^{i(m\pm1)\phi}}{\left(1+\zeta^2\right)^{1/2}}\left[\left(1+\zeta^2\right)\partial_\zeta\hat{P}_l^m\left(\zeta\right) \mp m\zeta\hat{P}_l^m\left(\zeta\right)\right]. \tag{208}$$

Using the identities [15]

$$\left(1+\zeta^2\right)\frac{d}{d\zeta}\hat{P}_\nu^\mu\left(\zeta\right) = \sqrt{1+\zeta^2}\hat{P}_\nu^{\mu+1}\left(\zeta\right) + \mu\zeta\hat{P}_\nu^\mu\left(\zeta\right) \tag{209}$$

$$\left(1+\zeta^2\right)\frac{d}{d\zeta}\hat{P}_\nu^\mu\left(\zeta\right) = \left(\nu-\mu+1\right)\left(\nu+\mu\right)\sqrt{1+\zeta^2}\hat{P}_\nu^{\mu-1}\left(\zeta\right) - \mu\zeta\hat{P}_\nu^\mu\left(\zeta\right) \tag{210}$$

we find

$$K^+\hat{Y}_l^m = -iC_{lm}\frac{e^{i(m+1)\phi}}{\left(1+\zeta^2\right)^{1/2}}\left(\sqrt{1+\zeta^2}\hat{P}_l^{m+1}\left(\zeta\right) + m\zeta\hat{P}_l^m\left(\zeta\right) - m\zeta\hat{P}_l^m\left(\zeta\right)\right) \tag{211}$$

$$= -iC_{lm}e^{i(m+1)\phi}\hat{P}_l^{m+1}\left(\zeta\right) \tag{212}$$

$$K^-\hat{Y}_l^m = -iC_{lm}\frac{e^{i(m-1)\phi}}{\left(1+\zeta^2\right)^{1/2}}\Big(\left(l-m+1\right)\left(l+m\right)\sqrt{1+z^2}\hat{P}_l^{m-1}\left(\zeta\right) - m\zeta\hat{P}_l^m\left(\zeta\right) + m\zeta\hat{P}_l^m\left(\zeta\right)\Big) \tag{213}$$

$$= -i\left(l-m+1\right)\left(l+m\right)C_{lm}e^{i(m-1)\phi}\hat{P}_l^{m-1}\left(\zeta\right) \tag{214}$$

and so that the boost operators $K^{\pm}$ raise and lower the m eigenvalue as

$$K^{\pm}\hat{Y}_l^m\left(\beta,\phi\right) = \sqrt{\left(l\mp m\right)\left(l\pm m+1\right)}\hat{Y}_l^{m\pm1}\left(\beta,\phi\right) \tag{215}$$

comparable to the action of $L^{\pm}$ in (202). It follows from (205) and (215) that the hyperangular wavefunctions $\hat{\psi}_{nlm}$ provide a $(2l+1)$-dimensional representation of O(2,1) with the same multiplicity structure found in the $s=0$ solutions for O(3). Since the unitary representations of the non-compact group O(2,1) should be infinite-dimensional, the $s=0$ solutions, which as noted previously are equivalent to the FKR solutions, must violate unitarity on algebraic grounds.

The $s = 1/2$ wavefunctions ψ_{nlm} for O(3) depend on θ and ϕ through the angular functions

$$\hat{\chi}_m^l\left(\theta,\phi\right) = F_m^l\left(z\right)e^{i(m+1/2)\phi} = C_{lm}\left(1+z^2\right)^{\frac{1}{4}}\hat{P}_m^l\left(z\right)e^{i(m+1/2)\phi} \tag{216}$$

where $z = \cot\theta$, and it follows from (181), (190) and (191) that

$$\mathbf{M}^2\hat{\chi}_m^l\left(\theta,\phi\right) = \left(l^2-1/4\right)\hat{\chi}_m^l\left(\theta,\phi\right) \qquad M\hat{\chi}_m^l\left(\theta,\phi\right) = \left(m+1/2\right)\hat{\chi}_m^l\left(\theta,\phi\right). \tag{217}$$

Similarly, the $s = 1/2$ wavefunctions $\hat{\psi}_{nlm}$ for O(2,1) depend on β and ϕ through the hyperangular functions

$$\chi_m^l\left(\beta,\phi\right) = G_m^l\left(\zeta\right)e^{i(m+1/2)\phi} = C_{lm}\left(1-\zeta^2\right)^{\frac{1}{4}}P_m^l\left(\zeta\right)e^{i(m+1/2)\phi} \tag{218}$$

where $\zeta = \tanh\beta$ and

$$\mathbf{M}^2 \chi^l_m(\beta,\phi) = \left(l^2 - 1/4\right)\chi^l_m(\beta,\phi) \qquad M\chi^l_m(\beta,\phi) = (m+1/2)\,\chi^l_m(\beta,\phi)\,. \tag{219}$$

In terms of the parameters (187) for $s = 1/2$, the non-diagonal operators for O(3) and O(2,1) take the forms

$$L^\pm = L^1 \pm iL^2 = e^{\pm i\phi}\left[\pm\left(1+z^2\right)\partial_z - iz\partial_\phi\right] \qquad \mathrm{O}(3) \tag{220}$$

$$K^\pm = K^1 \pm iK^2 = e^{\pm i\phi}\left[-i\left(1-\zeta^2\right)\partial_\zeta \pm \zeta\partial_\phi\right] \qquad \mathrm{O}(2,1). \tag{221}$$

For O(3)

$$L^\pm \hat\chi^l_m(\theta,\phi) = C_{lm}e^{\pm i\phi}\left[\pm\left(1+z^2\right)\partial_z + (m+1/2)\,z\right]\left(1+z^2\right)^{\frac14}\hat P^l_m(z)\,e^{i(m+1/2)\phi} \tag{222}$$

where

$$\left(1+z^2\right)\partial_z\left(1+z^2\right)^{\frac14}\hat P^l_m(z) = \left(1+z^2\right)^{\frac14}\left[\frac12 z\hat P^l_m(z) + \left(1+z^2\right)\frac{d}{dz}\hat P^l_m(z)\right] \tag{223}$$

so that

$$L^\pm \hat\chi^l_m(\theta,\phi) = C_{lm}e^{i(m+1/2\pm1)\phi}\left(1+z^2\right)^{\frac14}\left[\pm\left(1+z^2\right)\frac{d}{dz} + (m+1/2\pm1/2)\,z\right]\hat P^l_m(z) \tag{224}$$

and using the identities [15]

$$\left(1+z^2\right)\frac{d}{dz}\hat P^\mu_\nu = -\left(\mu-\nu-1\right)\hat P^\mu_{\nu+1} - (\nu+1)\,z\hat P^\mu_\nu = (\mu+\nu)\,\hat P^\mu_{\nu-1} + \nu z\hat P^\mu_\nu \tag{225}$$

one is led to

$$\begin{aligned} L^+\hat\chi^l_m(\theta,\phi) &= C_{lm}e^{i(m+1/2+1)\phi}\left(1+z^2\right)^{\frac14}\Big[-(l-m-1)\,\hat P^l_{m+1} - (m+1)\,z\hat P^l_m \\ &\qquad + (m+1)\,z\hat P^l_m(z)\Big] \qquad (226) \\ &= -(l-m-1)\,C_{lm}e^{i(m+1/2+1)\phi}\left(1+z^2\right)^{\frac14}\hat P^l_{m+1} \qquad (227) \end{aligned}$$

and

$$\begin{aligned} L^-\hat\chi^l_m(\theta,\phi) &= C_{lm}e^{i(m+1/2-1)\phi}\left(1+z^2\right)^{\frac14}\Big[-\Big((l+m)\,\hat P^l_{m-1} + mz\hat P^l_m\Big) \\ &\qquad + mz\hat P^l_m(z)\Big] \qquad (228) \\ &= -(l+m)\,C_{lm}e^{i(m+1/2-1)\phi}\left(1+z^2\right)^{\frac14}\hat P^l_{m-1}. \qquad (229) \end{aligned}$$

The actions of $L^\pm$ and a similar calculation for $K^\pm$ using the identities [15]

$$\left(1-\zeta^2\right)\frac{d}{d\zeta}P^\mu_\nu = (\mu-\nu-1)\,P^\mu_{\nu+1} + (\nu+1)\,\zeta P^\mu_\nu = (\mu+\nu)\,P^\mu_{\nu-1} - \nu z P^\mu_\nu \tag{230}$$

leads to

$$L^\pm\hat\chi^l_m(\theta,\phi) = \hat c(l,m)\,\hat\chi^l_{m\pm1}(\theta,\phi) \qquad K^\pm\chi^l_m(\beta,\phi) = c(l,m)\,\chi^l_{m\pm1}(\beta,\phi) \tag{231}$$

where $\hat{c}(l,m)$ and $c(l,m)$ are combinations of the eigenvalues. Since $L^{\pm}$ and $K^{\pm}$ act on the lower index (the associated Legendre functions P_ν^μ and $\hat{P}_\nu^\mu$ are nonzero for $\nu \geq 0$ and $\nu \geq |\mu|$), there is no upper bound on the action of the raising operators L^+ and K^+, but from

$$\hat{P}_n^n = \frac{(2n)!}{2^n n!}\left(1+z^2\right)^{n/2} \qquad P_n^n(\zeta) = (-1)^n \frac{(2n)!}{2^n n!}\left(1+\zeta^2\right)^{n/2} \tag{232}$$

we find the lower bounds

$$L^- \hat{\chi}_m^m(\theta,\phi) = K^- \chi_m^m(\beta,\phi) = 0. \tag{233}$$

The functions F and G therefore provide infinite-dimensional representations of O(3) and O(2,1), leading to mass/energy states of infinite degeneracy, similarly to the hyperangular functions found in [7]. Although the $s = 1/2$ wavefunctions ψ_{nlm} are thus unacceptable for O(3), the $s = 1/2$ wavefunctions $\hat{\psi}_{nlm}$ for O(2,1) provide a faithful representation of the angular momentum and boost structure of the symmetry. Because O(2,1) is both non-compact and non-abelian, the infinite degeneracy is expressed through an infinite-dimensional group representation, and not merely absorbed into the phase of the wavefunction as was seen for O(1,1).

3.2. Fock space

To obtain a number representation in $D = 3$, we must simultaneously diagonalize the operators $\mathbf{M}^2$ and M expressed in terms of creation/annihilation operators $\left(\bar{a}^1, \bar{a}^2, \bar{a}^3\right)$ for O(3) and $\left(\bar{a}^0, \bar{a}^1, \bar{a}^2\right)$ for O(2,1). Since

$$M = -i\left(\bar{a}^1 a^2 - \bar{a}^2 a^1\right) \qquad \Rightarrow \qquad \left[M, \bar{a}^3\right] = \left[M, \bar{a}^0\right] = 0 \tag{234}$$

we regard the direction

$$x^{\|} = \begin{cases} x^3, & \text{O(3)} \\ x^0, & \text{O(2,1)} \end{cases} \tag{235}$$

as longitudinal, orthogonal to the $x^1 - x^2$ subspace, chosen as the plane of observable angular momentum M. Therefore, by extending the procedure in $D = 2$, the operator M is found to have eigenvalues $\{0, \pm 1\}$ on first level excited eigenstates

$$\tilde{\varphi}_1 = \frac{1}{\sqrt{2}}\left(\bar{a}_+, \sqrt{2}\,\bar{a}^{\|}, -\bar{a}_-\right)\varphi_0 \tag{236}$$

where $\bar{a}_\pm$ are as defined in (69). However, the Casimir operators

$$\mathbf{M}^2 = \frac{1}{2}M^{\mu\nu}M_{\mu\nu} = -\frac{1}{2}\left(\bar{a}^\mu a^\nu - \bar{a}^\nu a^\mu\right)\left(\bar{a}_\mu a_\nu - \bar{a}_\nu a_\mu\right) = N^2 + N - (\bar{a}\cdot\bar{a})(a\cdot a) \tag{237}$$

with total mode number

$$N = \begin{cases} \bar{a}^1 a^1 + \bar{a}^2 a^2 + \bar{a}^3 a^3 = \bar{a}_+ a_- + \bar{a}_- a_+ + \bar{a}^3 a^3\,, & \text{O(3)} \\ \bar{a}^1 a^1 + \bar{a}^2 a^2 + \bar{a}^0 a^0 = \bar{a}_+ a_- + \bar{a}_- a_+ - \bar{a}^0 a^0\,, & \text{O(2,1)} \end{cases} \tag{238}$$

and scalar operators

$$\bar{a}\cdot\bar{a} = \begin{cases} 2\bar{a}_+\bar{a}_- + \bar{a}^3\bar{a}^3\,, & \text{O(3)} \\ 2\bar{a}_+\bar{a}_- - \bar{a}^0\bar{a}^0\,, & \text{O(2,1)} \end{cases} \tag{239}$$

$$a \cdot a = \begin{cases} 2a_+a_- + a^3a^3 \,, & \mathrm{O}(3) \\ 2a_+a_- - a^0a^0 \,, & \mathrm{O}(2,1) \end{cases} \tag{240}$$

remain non-diagonal on the states (236). Nevertheless, $\mathbf{M}^2$ must be block diagonal with respect to N and M and so studying the expected multiplicity of the mass/energy states leads to a characterization of the oscillator states. Recall that despite the sign of the term $-\bar{a}^0a^0$ in the second of (238), the total mass/energy of the O(2,1) oscillator was found in equation (198) to be positive definite; because the solutions in figure 1 are of the FKR type for which (13) requires $n^0 \leq 0$, the timelike modes contribute positive mass/energy. We therefore expect that the O(3) and O(2,1) will have similar multiplicity structure, which was seen to hold for the wavefunctions ψ_{nlm} and $\hat{\psi}_{nlm}$, and we may study their common eigenvalue content. From

$$\left[M, N^1\right] \neq 0 \qquad \left[M, N^2\right] \neq 0 \qquad \left[N, N^{\|}\right] = \left[M, N^{\|}\right] = 0, \tag{241}$$

the matrix representation of $\mathbf{M}^2$ reduces to coherent subspaces labeled by eigenvalues N and $n^{\|}$. A convenient parameterization of Cartesian states is

$$\begin{pmatrix} n^1 \\ n^2 \\ n^{\|} \end{pmatrix} = \begin{pmatrix} k \\ N - n^{\|} - k \\ n^{\|} \end{pmatrix} \tag{242}$$

with

$$n^{\|} = 0, 1, ..., N, \qquad k = 0, 1, ..., \left(N - n^{\|}\right). \tag{243}$$

The number of states for given N and $n^{\|}$ is therefore $N - n^{\|} + 1$, and the total number of states with mode number N is

$$\sum_{n^{\|}=0}^{N} \left(N + 1 - n^{\|}\right) = \left(N+1\right)\left(N+1\right) - \frac{N\left(N+1\right)}{2} = \frac{\left(N+1\right)\left(N+2\right)}{2} \,. \tag{244}$$

For $s = 0$, we extend (117) and construct excited states through

$$\zeta_{\alpha\beta\gamma} = \frac{1}{\sqrt{\alpha!\beta!\gamma!}} \left(\bar{a}_+\right)^\alpha \left(\bar{a}_-\right)^\beta \left(\bar{a}^{\|}\right)^\gamma \psi_0 \tag{245}$$

which are eigenstates of N and M with

$$N\zeta_{\alpha\beta\gamma} = \left(\alpha + \beta + \gamma\right)\zeta_{\alpha\beta\gamma} \qquad\qquad M\zeta_{\alpha\beta\gamma} = \left(\alpha - \beta\right)\zeta_{\alpha\beta\gamma} \tag{246}$$

so that the states $\zeta_{\alpha\beta\gamma}$ are precisely the states found by diagonalizing M in the Cartesian picture. Acting on (245) with (237) leads to

$$\begin{aligned} \mathbf{M}^2\zeta_{\alpha\beta\gamma} &= \left[N\left(N+1\right) - 4\alpha\beta - \gamma\left(\gamma-1\right)\right]\zeta_{\alpha\beta\gamma} \\ &\quad - 2\sqrt{\left(\alpha+1\right)\left(\beta+1\right)\gamma\left(\gamma-1\right)}\zeta_{(\alpha+1)(\beta+1)(\gamma-2)} \\ &\quad - 2\sqrt{\alpha\beta\left(\gamma+2\right)\left(\gamma+1\right)}\zeta_{(\alpha-1)(\beta-1)(\gamma+2)} \end{aligned} \tag{247}$$

so that the states $\zeta_{\alpha\beta\gamma}$ are not generally eigenstates of $\mathbf{M}^2$, but as expected are mixtures of states with $(\alpha \pm 1, \beta \pm 1, \gamma \mp 2)$ and fixed M eigenvalue

$$m = \left(\alpha \pm 1\right) - \left(\beta \pm 1\right) = \alpha - \beta. \tag{248}$$

It follows from (247) that

$$\mathbf{M}^2\zeta_{N00} = N\left(N+1\right)\zeta_{N00} \qquad M\zeta_{N00} = N\zeta_{N00} \tag{249}$$

$$\mathbf{M}^2\zeta_{0N0} = N\left(N+1\right)\zeta_{0N0} \qquad M\zeta_{0N0} = -N\zeta_{0N0} \tag{250}$$

and so the allowed eigenvalues of M

$$m = \alpha - \beta = -l, -l+1, ..., l-1, l \tag{251}$$

are consistent with the parameter range

$$\alpha, \beta = 0, 1, ..., N. \tag{252}$$

Generally, as demonstrated in appendix D by exploiting the invariance of $\mathrm{tr}\left(\mathbf{M}^2\right)$ under unitary transformations, the Casimir content of the states $\zeta_{\alpha\beta\gamma}$ is

$$l = N, N-2, ..., N - \mathrm{int}\left(N/2\right) \tag{253}$$

and since the multiplicity of l-states is $2l+1$, the multiplicity of states with total mode number N is

$$\sum_{k=0}^{\mathrm{int}(N/2)} 2\left(N-2k\right)+1 = \frac{\left(N+1\right)\left(N+2\right)}{2} \tag{254}$$

in agreement with (244). Since diagonalization of $\mathbf{M}^2$ does not mix states of different m, the mode number N of states ψ_{lm} and $\hat{\psi}_{lm}$ must depend on l, with Casimir eigenvalues given in (253), but not on m. The principal quantum number n complements the contribution of l to energy, incrementing by 2 when l is decremented by 1. Thus, the mode number is

$$N = 2n + l, \qquad n = 0, 1, 2, ..., N \tag{255}$$

and the total mass/energy becomes

$$\kappa = \omega\left(2n + l + \frac{3}{2}\right) \tag{256}$$

in agreement with the solution (198) to the Schrodinger equation for $s = 0$.

According to (245) — (247) the $N = 1$ states constitute the $l = 1$ vector multiplet

$$\zeta^{(1)} = \begin{pmatrix} \zeta_{001} \\ \zeta_{010} \\ -\zeta_{100} \end{pmatrix} = \begin{pmatrix} \bar{a}_- \\ \bar{a}^{\|} \\ -\bar{a}_+ \end{pmatrix} \psi_0 \tag{257}$$

which we order according to the eigenvalues $m = -1, 0, 1$ found by diagonalizing M on the $N = 1$ multiplet of Cartesian states

$$\varphi^{(1)} = \begin{pmatrix} \varphi_{100} \\ \varphi_{010} \\ \varphi_{001} \end{pmatrix} = \begin{pmatrix} \bar{a}_1 \\ \bar{a}_2 \\ \bar{a}^{\|} \end{pmatrix} \varphi_0. \tag{258}$$

Applying the creation/annihilation operators in polar parameterizations (36)

$$\bar{a}_\pm = \frac{1}{2}e^{\pm i\phi}\left(\rho\sin\theta - \sin\theta\partial_\rho - \frac{\cos\theta}{\rho}\partial_\theta \mp \frac{i}{\rho\sin\theta}\partial_\phi\right) \tag{259}$$

Journal of Physics: Conference Series **330** (2011) 012014 doi:10.1088/1742-6596/330/1/012014

$$\bar{a}^3 = \frac{1}{\sqrt{2}}\left(\rho\cos\theta - \cos\theta\partial_\rho + \frac{\sin\theta}{\rho}\partial_\beta\right) \tag{260}$$

for O(3), and

$$\bar{a}_\pm = \frac{1}{2}e^{\pm i\phi}\left(\rho\cosh\beta - \cosh\beta\partial_\rho + \frac{\sinh\beta}{\rho}\partial_\beta \mp \frac{i}{\rho\cosh\beta}\partial_\phi\right) \tag{261}$$

$$\bar{a}^0 = \frac{1}{\sqrt{2}}\left(\rho\sinh\beta - \sinh\beta\partial_\rho + \frac{\cosh\beta}{\rho}\partial_\beta\right) \tag{262}$$

for O(2,1), to the ground states

$$\psi_0 = \hat{\psi}_0 = A_0 e^{-\rho^2/2} \tag{263}$$

we obtain

$$\zeta^{(1)} = A_0\rho\begin{pmatrix}\sin\theta e^{-i\phi}\\ \sqrt{2}\cos\theta\\ -\sin\theta e^{i\phi}\end{pmatrix}e^{-\rho^2/2} \qquad \text{O(3)} \tag{264}$$

$$\zeta^{(1)} = A_0\rho\begin{pmatrix}\cosh\beta e^{-i\phi}\\ \sqrt{2}\sinh\beta\\ -\cosh\beta e^{i\phi}\end{pmatrix}e^{-\rho^2/2} \qquad \text{O(2,1)}. \tag{265}$$

Using

$$P_1^1(z) = -\sqrt{1-z^2} \qquad P_1^0(z) = z \qquad P_1^{-1}(z) = \sqrt{1-z^2} \tag{266}$$

$$\hat{P}_1^1(\zeta) = -\sqrt{1+\zeta^2} \qquad \hat{P}_1^0(\zeta) = \zeta \qquad \hat{P}_1^{-1}(\zeta) = \sqrt{1+\zeta^2} \tag{267}$$

wavefunctions (264) and (265) are seen to agree with the $l = 1$ vector multiplet found for $s = 0$ in figure 1, corresponding to the well-known property of the spherical harmonics that the three components of $Y_{l=1}^m(\theta,\phi)$ and $\hat{Y}_{l=1}^m(\beta,\phi)$ form a unit vector.

The first level of excited states was found by acting on the ground state with the operator multiplet $(\bar{a}_-, \bar{a}^\parallel, -\bar{a}_+)$ which we regard as the fundamental representation of a set of irreducible tensor operators constructed successively by taking irreducible tensor products

$$\bar{a}_m^{(j\pm1)} = \sum_{m_2=-1,0,1}\langle j\ m-m_2\ 1\ m_2 \mid j\ 1\ j\pm1\ m\rangle\, \bar{a}_{m-m_2}^{(j)}\bar{a}_{m_2}^{(1)} \tag{268}$$

where $\bar{a}_m^{(j\pm1)}$ is an irreducible tensor operator of rank $j\pm1$, $\bar{a}_{m_2}^{(1)}$ is the vector operator, and $\langle j\ m-m_2\ 1\ m_2 \mid j\ 1\ j\pm1\ m\rangle$ is the appropriate Clebsch-Gordan coefficient. Thus, according to (244), the $N = 2$ states have total multiplicity of 6, which by (253) must include the five $l = 2$ states and the $l = 0$ singlet state. The two irreducible tensor operators that can be constructed from the vector operator are the singlet ($l = m = 0$)

$$\bar{a}_0^{(0)} = -\frac{1}{\sqrt{3}}(2\bar{a}_+\bar{a}_- + \bar{a}_3\bar{a}_3) = -\frac{1}{\sqrt{3}}\bar{a}\cdot\bar{a} \tag{269}$$

Journal of Physics: Conference Series **330** (2011) 012014 doi:10.1088/1742-6596/330/1/012014

and the $l = 2$ operators

$$\bar{a}^{(2)}_{-2} = \bar{a}^{(1)}_{-1}\bar{a}^{(1)}_{-1} = (\bar{a}_-)^2 \tag{270}$$

$$\bar{a}^{(2)}_{-1} = \frac{1}{\sqrt{2}}\left(\bar{a}^{(1)}_{0}\bar{a}^{(1)}_{-1} + \bar{a}^{(1)}_{-1}\bar{a}^{(1)}_{0}\right) = -\sqrt{2}\bar{a}_-\bar{a}_3 \tag{271}$$

$$\bar{a}^{(2)}_{0} = \frac{1}{\sqrt{6}}\left(\bar{a}^{(1)}_{1}\bar{a}^{(1)}_{-1} + 2\bar{a}^{(1)}_{0}\bar{a}^{(1)}_{0} + \bar{a}^{(1)}_{-1}\bar{a}^{(1)}_{1}\right) = -\frac{2}{\sqrt{6}}\left(\bar{a}_+\bar{a}_- - \bar{a}_3\bar{a}_3\right) \tag{272}$$

$$\bar{a}^{(2)}_{1} = \frac{1}{\sqrt{2}}\left(\bar{a}^{(1)}_{1}\bar{a}^{(1)}_{0} + \bar{a}^{(1)}_{1-1}\bar{a}^{(1)}_{1}\right) = \sqrt{2}\bar{a}_+\bar{a}_3 \tag{273}$$

$$\bar{a}^{(2)}_{2} = \bar{a}^{(1)}_{1}\bar{a}^{(1)}_{1} = (\bar{a}_+)^2 \tag{274}$$

which are precisely the operators found by diagonalizing the matrix representation of $\mathbf{M}^2$. In this way, the complete set of spherical polar harmonic oscillators in 3 dimensions can be constructed from the ground state.

For the $s = 1/2$ wave functions, the action of the operator multiplet $(\bar{a}_-, \bar{a}^{\|}, -\bar{a}_+)$ on the ground state does not lead to excited wavefunctions. In particular, acting with (261) and (262) on the O(2,1) ground state in (199) one finds

$$\begin{pmatrix} \bar{a}_- \\ \bar{a}^0 \\ \bar{a}_+ \end{pmatrix} \psi_0 = \begin{pmatrix} e^{-i\phi}\rho\cosh\beta\hat{\psi}_0 \\ \sqrt{2}\rho\sinh\beta \\ e^{i\phi}\left(\rho\cosh\beta + \dfrac{1}{2\rho\cosh\beta}\right) \end{pmatrix} \psi_0 \tag{275}$$

which is not an excited state and not even a vector in the polar representation. Expressing the singlet operator $\bar{a}\cdot\bar{a}$ in the O(2,1) parameterization,

$$\bar{a}\cdot\bar{a} = \frac{1}{2}\left(\rho^2 + \partial^2_\rho + \frac{2}{\rho}\partial_\rho - \frac{1}{\rho^2}\mathbf{M}^2 - 2\partial_\rho - 3\right) \tag{276}$$

and comparing with the Schrodinger equation (173) we may write

$$(\bar{a}\cdot\bar{a})\,\psi_0 = \left(2\rho^2 - 2\partial_\rho - \varepsilon_0 - 3\right)\psi_0 = \left(2\rho^2 + \frac{1+2\rho^2}{\rho} - 5\right)\psi_0 \tag{277}$$

which acts only on the radial coordinate. Examining the $s = 1/2$ wavefunctions for O(2,1) in figure 1, we see that (277) can only correspond to a state if the ρ-dependent factor on the RHS is equivalent to raising the value of l. However the Legendre function $P^l_m(\tanh\beta)$ must vanish for $l > m$, and since the singlet operator does not contain β or ϕ, it cannot affect m and so cannot produce a wavefunction in the Hilbert space by operating on ψ_0. Apparently, the vector operator multiplet belongs only to the $s = 0$ vector representations of the symmetry groups, and not to the infinite-dimensional $s = 1/2$ representations. It may be possible to construct an appropriate ladder representation of creation/annihilation operators though a multipole expansion of the Hamiltonian, corresponding to an infinite summation of the associated Legendre functions. This will be discussed in a subsequent paper.

4. Conclusion

The goal of this study has been to explore polar solutions for the harmonic oscillator in two and three Euclidean and Lorentzian dimensions, in order to construct Fock space descriptions

Journal of Physics: Conference Series **330** (2011) 012014 doi:10.1088/1742-6596/330/1/012014

equivalent to the Hilbert space wavefunctions found by solution of the Schrodinger equation in (hyper)spherical coordinates. Although this work was initially motivated by particular interest in the infinite dimensional representation of O(2,1) symmetric oscillator, we have seen that the methods which were successful for O(2), O(1,1), and the finite dimensional representation of O(3), fail for the case of primary interest. Similarly, no infinite dimensional representation was found for O(1,1) in either Hilbert space or Fock space.

The familiar treatment of the oscillator by separation into Cartesian modes, whether in Hilbert space or Fock space, does not not directly address algebraic symmetries, but implicitly chooses as a maximal commuting set of operators the Hamiltonian and its orthogonal components $\{K, K_\mu\}$ where

$$\psi(x) = \prod_\mu \psi(x^\mu) \qquad\qquad K = \sum_\mu K^\mu. \tag{278}$$

As seen explicitly, for example in (269) — (274), the resulting solutions mix eigenstates of the Casimir invariant associated with the orthogonal symmetry, as well as a group generator that may be diagonalized in a polar representation. Although the degeneracy of the mass/energy is evident from the counting of total mode number, there is little motivation in the Cartesian approach to treat excited states as members of a multiplet.

In the polar approach to the Schrodinger equation, on the other hand, the orthogonal symmetry guides the separation of variables, and so the wavefunctions emerge naturally as irreducible representations of the group. Because the polar operators do not admit an obvious Dirac factorization, we built Fock space descriptions by decomposing the fundamental vector multiplet of Cartesian modes into eigenstates of the diagonal group generator and then constructing higher-level excited states as irreducible tensor products. The choice of the Cartesian vector multiplet as fundamental representation was thus made for reasons of convenience, and in retrospect, it is not be entirely surprising that this finite-dimensional multiplet structure does not yield any of the desired infinite-dimensional representations. It it hoped that a different factorization of the Hamiltonian, such as a multipole expansion that naturally produces an infinite sum of terms, may lead to a faithful representation of the O(2,1) symmetry equivalent to the wavefunctions found in section 3.

Aside from possible applications to relativistic quantum field theory and string theory, a principal motivation for developing a consistent Fock space approach to the O(2,1) oscillator is to obtain a deeper understanding of the Horwitz-Arshansky bound states, and in particular, the question of ghost states. For O(1,1), we demonstrated that the oscillator is equivalent to a one-dimensional system with commutation relations between the ladder operators A and $\bar{A}$ that cannot produce ghosts. However, as was seen in section 1.2 for the FKR oscillator, the identification of ghost states is not straightforward even when the creation/annihilation operators are known. Except for the KN oscillator, whose states have positive definite norm by construction, the integration by parts implicit in the use of commutation relations to evaluate the norm of timelike modes presents difficulties that leave interpretation questionable. A detailed answer to these questions will require a Fock space description that has yet to be found.

Acknowledgments

The author would like like to thank Larry Horwitz and Itzhak Bars for valuable discussions and thoughtful comments.

Appendix A. Virial theorem

As in nonrelativistic quantum mechanics, the Ehrenfest theorem for an observable $\hat{O}$ and Hamiltonian

$$K = \frac{1}{2}p^2 + V(x) \tag{A.1}$$

is

$$\frac{d}{d\tau}\left\langle \psi \left| \hat{O} \right| \psi \right\rangle = i\left\langle \psi \left| \left[K, \hat{O}\right] \right| \psi \right\rangle + \left\langle \psi \left| \partial_\tau \hat{O} \right| \psi \right\rangle . \tag{A.2}$$

Generally then, we have

$$\frac{d}{d\tau}\langle x^\mu p_\mu \rangle = i\left\langle \left[\frac{1}{2}p^2, x^\mu\right] p_\mu \right\rangle + i\left\langle x^\mu \left[V(x), p_\mu\right]\right\rangle = \langle p^2 \rangle - \langle x^\mu \partial_\mu V(x) \rangle \tag{A.3}$$

and on eigenstates of the Hamiltonian,

$$\frac{d}{d\tau}\langle x^\mu p_\mu \rangle = \frac{d}{d\tau}\left\langle \psi(0) \left| e^{-i\kappa\tau}(x^\mu p_\mu) e^{i\kappa\tau} \right| \psi(0) \right\rangle = \frac{d}{d\tau}\langle \psi(0) | x^\mu p_\mu | \psi(0) \rangle = 0 \tag{A.4}$$

so that

$$\langle p^2 \rangle = \langle x^\mu \partial_\mu V(x) \rangle . \tag{A.5}$$

For the harmonic oscillator

$$V(x) = \frac{1}{2}\omega^2 x^2 \longrightarrow x^\mu \partial_\mu V(x) = \omega^2 x^2 \tag{A.6}$$

so that on states of spacelike support

$$\langle p^2 \rangle = \langle \omega^2 x^2 \rangle > 0. \tag{A.7}$$

Appendix B. Differential operators in polar coordinates

$$\begin{bmatrix} \partial_1 \\ \partial_2 \end{bmatrix} = \begin{bmatrix} \cos\theta\partial_r - \frac{1}{r}\sin\theta\partial_\theta \\ \sin\theta\partial_r + \frac{1}{r}\cos\theta\partial_\theta \end{bmatrix} \qquad \mathrm{O(2)} \tag{B.1}$$

$$\begin{bmatrix} \partial_0 \\ \partial_1 \end{bmatrix} = \begin{bmatrix} -\frac{|\rho|}{\rho}\sinh\beta\partial_\rho + \frac{1}{|\rho|}\cosh\beta\partial_\beta \\ \cosh\beta\partial_\rho - \frac{1}{\rho}\sinh\beta\partial_\beta \end{bmatrix} \qquad \mathrm{O(1,1)} \tag{B.2}$$

$$\begin{bmatrix} \partial_1 \\ \partial_2 \\ \partial_3 \end{bmatrix} = \begin{bmatrix} \cos\phi\sin\theta\partial_r + \frac{\cos\theta\cos\phi}{r}\partial_\theta - \frac{\sin\phi}{r\sin\theta}\partial_\phi \\ \sin\theta\sin\phi\partial_r + \frac{\cos\theta\sin\phi}{r}\partial_\theta + \frac{\cos\phi}{r\sin\theta}\partial_\phi \\ \cos\theta\partial_r - \frac{\sin\theta}{r}\partial_\theta \end{bmatrix} \qquad \mathrm{O(3)} \tag{B.3}$$

$$\begin{bmatrix} \partial_0 \\ \partial_1 \\ \partial_2 \end{bmatrix} = \begin{bmatrix} -\sinh\beta\partial_\rho + \dfrac{\cosh\beta}{\rho}\partial_\beta \\ \cosh\beta\cos\phi\partial_\rho - \dfrac{\sinh\beta\cos\phi}{\rho}\partial_\beta - \dfrac{\sin\phi}{\rho\cosh\beta}\partial_\phi \\ \cosh\beta\sin\phi\partial_\rho - \dfrac{\sinh\beta\sin\phi}{\rho}\partial_\beta + \dfrac{\cos\phi}{\rho\cosh\beta}\partial_\phi \end{bmatrix} \qquad \mathrm{O}(2,1) \tag{B.4}$$

Appendix C. Coordinate ladder representation for O(2)

To verify that the solutions (58) form the basis for a representation of the operator algebra, we express the creation/annihilation operators in polar coordinates. Combining (7) and (69) as

$$a_\pm = \frac{1}{2}\left[(x+\partial_x) \pm i(y+\partial_y)\right] = \frac{1}{2}\left[x \pm iy + (\partial_x \pm i\partial_y)\right] \tag{C.1}$$

$$\bar{a}_\pm = \frac{1}{2}\left[(x-\partial_x) \pm i(y-\partial_y)\right] = \frac{1}{2}\left[x \pm iy - (\partial_x \pm i\partial_y)\right] \tag{C.2}$$

we obtain the polar expressions

$$a_\pm = \frac{1}{2}e^{\pm i\phi}\left(\rho + \frac{\partial}{\partial\rho} \pm \frac{i}{\rho}\frac{\partial}{\partial\phi}\right) \qquad \bar{a}_\pm = \frac{1}{2}e^{\pm i\phi}\left[\rho - \left(\frac{\partial}{\partial\rho} \pm \frac{i}{\rho}\frac{\partial}{\partial\phi}\right)\right]. \tag{C.3}$$

Applying the annihilation operators to the ground state (61), we recover (114) in the explicit form

$$a_+\psi_0 = \frac{1}{2}\sqrt{\frac{1}{\pi\Gamma(s+1)}}\left(\rho + (s-\rho^2)\frac{1}{\rho} - \frac{1}{\rho}s\right)e^{-\rho^2/2}\rho^s e^{i\phi(s+1)} = 0 \tag{C.4}$$

$$a_-\psi_0 = \sqrt{s}\sqrt{\frac{1}{\pi\Gamma(s)}}e^{-\rho^2/2}\left(\rho e^{i\phi}\right)^{s-1}, \tag{C.5}$$

where the result in (C.5) is formally equivalent to $\sqrt{s}\psi_{0,-1}$ but as discussed above, does not correspond to any state in the Fock space, and we treat as annihilation. For the general state (58), using the notation of (49) $x = \rho^2$, the ρ derivative is

$$\frac{\partial}{\partial\rho}\left[e^{-\rho^2/2}L_n^{m+s}(\rho^2)\left(\rho e^{i\phi}\right)^{m+s}\right] = \frac{e^{-\rho^2/2}\left(\rho e^{i\phi}\right)^{m+s}}{\rho}\left[-\rho^2 + m + s + 2x\frac{d}{dx}\right]L_n^{m+s}(x) \tag{C.6}$$

providing

$$\frac{\partial}{\partial\rho}\psi_{nm} = \left(-\rho + \frac{m+s}{\rho}\right)\psi_{nm} + \frac{2}{\rho}A_{nm}e^{-\rho^2/2}\left(\rho e^{i\phi}\right)^{m+s}x\frac{d}{dx}L_n^{m+s}(x) \tag{C.7}$$

and the ϕ derivative is

$$\frac{i}{\rho}\frac{\partial}{\partial\phi}\psi_{nm} = -\frac{m+s}{\rho}\psi_{nm} \tag{C.8}$$

so that

$$\left(\frac{\partial}{\partial\rho} + \frac{i}{\rho}\frac{\partial}{\partial\phi}\right)\psi_{nm} = -\rho\psi_{nm} + \frac{2}{\rho}A_{nm}e^{-\rho^2/2}\left(\rho e^{i\phi}\right)^{m+s}x\frac{d}{dx}L_n^{m+s}(x) \tag{C.9}$$

$$\left(\frac{\partial}{\partial\rho} - \frac{i}{\rho}\frac{\partial}{\partial\phi}\right)\psi_{nm} = \left(-\rho + 2\frac{m+s}{\rho}\right)\psi_{nm} + \frac{2}{\rho}A_{nm}e^{-\rho^2/2}\left(\rho e^{i\phi}\right)^{m+s}x\frac{d}{dx}L_n^{m+s}(x). \tag{C.10}$$

Journal of Physics: Conference Series **330** (2011) 012014 doi:10.1088/1742-6596/330/1/012014

Then, using the identity [15]

$$\frac{d}{dx}L_a^b(x) = -L_{a-1}^{b+1}(x) \tag{C.11}$$

for the Laguerre functions, we calculate

$$\begin{aligned} a_+\psi_{nm}(\rho,\phi) &= \frac{1}{2}e^{i\phi}\left(\rho + \frac{\partial}{\partial\rho} + \frac{i}{\rho}\frac{\partial}{\partial\phi}\right)\psi_{nm} & \text{(C.12)}\\ &= e^{i\phi}\frac{1}{\rho}A_{nm}e^{-\rho^2/2}\left(\rho e^{i\phi}\right)^{m+s} x\frac{d}{dx}L_n^{m+s}(x) & \text{(C.13)}\\ &= A_{nm}e^{-\rho^2/2}\left(\rho e^{i\phi}\right)^{m+s} e^{i\phi}\frac{\rho^2}{\rho}\left(-L_{n-1}^{m+s+1}\right) \ ,\ n>0 & \text{(C.14)}\\ &= -A_{nm}e^{-\rho^2/2}\left(\rho e^{i\phi}\right)^{m+s+1} L_{n-1}^{m+s+1} \ ,\ n>0. & \text{(C.15)} \end{aligned}$$

Using (60) for A_{nm} we obtain

$$a_+\psi_{nm} = (-1)^{n+1}\sqrt{\frac{\Gamma(n+1)}{\pi\Gamma(n+m+s+1)}}L_{n-1}^{m+s+1}(x)\left(\rho e^{i\phi}\right)^{m+s+1}e^{-\frac{\rho^2}{2}} = \sqrt{n}\psi_{n-1,m+1} \tag{C.16}$$

as required by the first of (112). The second lowering operator acts as

$$\begin{aligned} a_-\psi_{nm}(\rho,\phi) &= \frac{1}{2}e^{-i\phi}\left(\rho + \frac{\partial}{\partial\rho} - \frac{i}{\rho}\frac{\partial}{\partial\phi}\right)\psi_{nm} & \text{(C.17)}\\ &= \frac{m+s}{\rho}e^{-i\phi}\psi_{nm} + e^{-i\phi}\frac{1}{\rho}A_{nm}e^{-\rho^2/2}\left(\rho e^{i\phi}\right)^{m+s} x\frac{d}{dx}L_n^{m+s}(x) & \text{(C.18)} \end{aligned}$$

so the identities [15]

$$\begin{aligned} x\frac{d}{dx}L_a^b(x) &= aL_a^b(x) - (a+b)L_{a-1}^b(x) & \text{(C.19)}\\ L_a^{b-1}(x) &= L_a^b(x) - L_{a-1}^b(x) & \text{(C.20)} \end{aligned}$$

lead to

$$\begin{aligned} a_-\psi_{nm}(\rho,\phi) &= A_{nm}e^{-\rho^2/2}\left(\rho e^{i\phi}\right)^{m+s-1}\Big[(m+s)L_n^{m+s}(x) \\ &\qquad + nL_n^{m+s}(x) - (n+m+s)L_{n-1}^{m+s}(x)\Big] & \text{(C.21)}\\ &= (n+m+s)A_{nm}e^{-\rho^2/2}\left(\rho e^{i\phi}\right)^{m+s-1}\left(L_n^{m+s}(x) - L_{n-1}^{m+s}(x)\right) & \text{(C.22)}\\ &= (n+m+s)A_{nm}e^{-\rho^2/2}\left(\rho e^{i\phi}\right)^{m+s-1}L_n^{m+s-1}(x) & \text{(C.23)} \end{aligned}$$

providing

$$\begin{aligned} a_-\psi_{nm} &= (-1)^n\sqrt{\frac{\Gamma(n+1)(m+s+n)^2}{\pi\Gamma(n+m+s+1)}}e^{-\frac{\rho^2}{2}}L_n^{m+s-1}(x)\left(\rho e^{i\phi}\right)^{m+s-1} \\ &= \sqrt{n+m+s}\,\psi_{n,m-1} & \text{(C.24)} \end{aligned}$$

as required by the first of (113). The raising operator $\bar{a}_+$ acts as

$$\bar{a}_+\psi_{nm}(\rho,\phi) = \frac{1}{2}e^{i\phi}\left[\rho - \left(\frac{\partial}{\partial\rho} + \frac{i}{\rho}\frac{\partial}{\partial\phi}\right)\right]\psi_{nm}(\rho,\phi) \quad \text{(C.25)}$$

$$= \rho e^{i\phi}\psi_{nm} - e^{i\phi}A_{nm}\frac{e^{-\rho^2/2}\left(\rho e^{i\phi}\right)^{m+s}}{\rho}x\frac{d}{dx}L_n^{m+s}(x) \quad \text{(C.26)}$$

so applying identity (C.11) and the identities [15]

$$L_a^b(x) = L_{a-1}^b(x) + L_a^{b-1}(x) \quad \text{(C.27)}$$

we calculate

$$\bar{a}_+\psi_{nm}(\rho,\phi) = \rho e^{i\phi}A_{nm}e^{-\rho^2/2}\left(\rho e^{i\phi}\right)^{m+s}\left(L_n^{m+s}(x) - \frac{d}{dx}L_n^{m+s}(x)\right) \quad \text{(C.28)}$$

$$= A_{nm}e^{-\rho^2/2}\left(\rho e^{i\phi}\right)^{m+s+1}\left(L_n^{m+s}(x) + L_{n-1}^{m+s+1}(x)\right) \quad \text{(C.29)}$$

$$= A_{nm}e^{-\rho^2/2}\left(\rho e^{i\phi}\right)^{m+s+1}L_n^{m+s+1}(x) \quad \text{(C.30)}$$

providing

$$\bar{a}_+\psi_{nm} = (-1)^n\sqrt{\frac{\Gamma(n+1)}{\pi\Gamma(n+m+s+1)}}e^{-\rho^2/2}\left(\rho e^{i\phi}\right)^{m+s+1}L_n^{m+s+1}(\rho^2) \quad \text{(C.31)}$$

$$= \sqrt{n+m+s+1}\psi_{n,m+1} \quad \text{(C.32)}$$

confirming the second of (112). Finally, the action of the second raising operator $\bar{a}_-$ is

$$\bar{a}_-\psi_{nm}(\rho,\phi) = \frac{1}{2}e^{-i\phi}\left[\rho - \left(\frac{\partial}{\partial\rho} - \frac{i}{\rho}\frac{\partial}{\partial\phi}\right)\right]\psi_{nm}(\rho,\phi) \quad \text{(C.33)}$$

$$= e^{-i\phi}\left[\left(\rho - \frac{m+s}{\rho}\right)\psi_{nm} - \frac{A_{nm}}{\rho}e^{-\frac{\rho^2}{2}}\left(\rho e^{i\phi}\right)^{m+s}x\frac{d}{dx}L_n^{m+s}(x)\right] \quad \text{(C.34)}$$

so using the identity [15]

$$x\frac{d}{dx}L_a^b(x) = (a+1)L_{a+1}^b(x) - (a+b+1-x)L_a^b(x) \quad \text{(C.35)}$$

and (C.11) leads to

$$\bar{a}_-\psi_{nm}(\rho,\phi) = -A_{nm}e^{-\frac{\rho^2}{2}}\left(\rho e^{i\phi}\right)^{m+s-1}\left[(m+s-x)L_n^{m+s}(x) + x\frac{d}{dx}L_n^{m+s}(x)\right] \quad \text{(C.36)}$$

$$= -A_{nm}e^{-\frac{\rho^2}{2}}\left(\rho e^{i\phi}\right)^{m+s-1}\left[(n+1)L_{n+1}^{m+s-1}(x)\right] \quad \text{(C.37)}$$

we confirm

$$\bar{a}_-\psi_{nm} = -(-1)^n\sqrt{\frac{(n+1)^2\Gamma(n+1)}{\pi\Gamma(n+m+s+1)}}e^{-\frac{\rho^2}{2}}L_{n+1}^{m+s-1}(\rho^2)\left(\rho e^{i\phi}\right)^{m+s-1} \quad \text{(C.38)}$$

$$= \sqrt{n+1}\psi_{n+1,m-1} \quad \text{(C.39)}$$

so that the solutions (58) belong to the ladder representation for any value of s.

Unlike the angular momentum $M = -i\partial_\phi$, which is diagonal in polar coordinates, the remaining SU(2) generators are most conveniently expressed in Cartesian coordinates

$$\Delta = \bar{a}_+ a_+ + \bar{a}_- a_- = N^1 - N^2 = \frac{1}{2}\left(x^2 - y^2 - \frac{\partial^2}{\partial x^2} + \frac{\partial^2}{\partial y^2}\right) \tag{C.40}$$

$$Q = -i\left(\bar{a}_+ a_+ - \bar{a}_- a_-\right) = xy - \frac{\partial}{\partial x}\frac{\partial}{\partial y} \tag{C.41}$$

from which it follows that

$$M\psi_0 = s\,\psi_0 \tag{C.42}$$

$$\Delta\psi_0 = s\left[\frac{x^2 + y^2 - s + 1}{(x + iy)^2}\right]\psi_0 \tag{C.43}$$

$$Q\psi_0 = is\left[\frac{x^2 + y^2 - s + 1}{(x + iy)^2}\right]\psi_0 \tag{C.44}$$

and again we see that the SU(2) symmetry of the Hamiltonian is spontaneously broken for $s \neq 0$.

Since the operator Δ is diagonal in Cartesian coordinates, its action on the states is of special interest. From (117) and (120) and the commutator

$$\left[\left(\bar{a}_+ a_+ + \bar{a}_- a_-\right), \left(\bar{a}_+\right)^\alpha \left(\bar{a}_-\right)^\beta\right] = \alpha \bar{a}_+^{\alpha-1}\left(\bar{a}_-\right)^{\beta+1} + \beta\left(\bar{a}_+\right)^{\alpha+1}\bar{a}_-^{\beta-1} \tag{C.45}$$

we obtain

$$\Delta\zeta_{\alpha\beta} = \alpha\sqrt{\frac{\beta+1}{s+\alpha}}\zeta_{\alpha-1,\beta+1} + \beta\sqrt{\frac{s+\alpha+1}{\beta}}\zeta_{\alpha+1,\beta-1} + \left(\bar{a}_+\right)^\alpha\left(\bar{a}_-\right)^\beta \Delta\psi_0 \tag{C.46}$$

which cannot be diagonalized unless $s = 0$ because (C.43) shows that the ground state is not an eigenstate of Δ. Using (137) to construct the $s = 0$ multiplets of for given N, it is easily shown, case by case, that diagonalization of Δ recovers the standard Cartesian description of the oscillator. Since Δ cannot be diagonalized on states with $s \neq 0$, there is no unitary combination of the spherical states $\psi_{nm}^{s\neq 0}$ equivalent to the familiar Cartesian states of the harmonic oscillator.

Appendix D. Multiplicity of $D = 3$ Casimir

States $\zeta_{\alpha\beta\gamma}$ have $n = \alpha + \beta + \gamma$and $m = \alpha - \beta$ but are mixtures of l-states. The cases $(\alpha,\beta,\gamma) = (n,0,0)$ and $(\alpha,\beta,\gamma) = (0,n,0)$ have eigenvalues

$$\mathbf{M}^2\zeta_{n00} = n(n+1)\zeta_{n00} \qquad M\zeta_{n00} = n\zeta_{n00} \tag{D.1}$$

$$\mathbf{M}^2\zeta_{0n0} = n(n+1)\zeta_{0n0} \qquad M\zeta_{0n0} = -n\zeta_{n00} \tag{D.2}$$

The allowed eigenvalues of M

$$m = \alpha - \beta = -l, -l+1, ..., l-1, l \tag{D.3}$$

are consistent with

$$\alpha, \beta = 0, 1, ..., l_{\max} = n. \tag{D.4}$$

Journal of Physics: Conference Series **330** (2011) 012014 doi:10.1088/1742-6596/330/1/012014

The Casimir $\mathbf{M}^2$ mixes (α, β, γ)-states with $(\alpha \pm 1, \beta \pm 1, \gamma \mp 2)$-states, but the mixed states have the same eigenvalues of M

$$m = \alpha - \beta = (\alpha \pm 1) - (\beta \pm 1)\,. \tag{D.5}$$

The angular momentum content of the states $\zeta_{\alpha\beta\gamma}$ is

$$l = n, n-2, n-4, ..., n - 2 \cdot \text{int}\left(\frac{n}{2}\right) = \begin{cases} l_{\max}, l_{\max} - 2, ..., 0 & l_{\max} \text{ even} \\ l_{\max}, l_{\max} - 2, ..., 1 & l_{\max} \text{ odd} \end{cases}. \tag{D.6}$$

Since the multiplicity of l-states is $2l + 1$ the total multiplicity of n-states is

$$\sum_{k=0}^{\text{int}(n/2)} [2\,(2k) + 1] = \frac{(n+1)\,(n+2)}{2} \tag{D.7}$$

as required. For given values of $n = \alpha + \beta + \gamma$ and $m = \alpha - \beta$, the possible (α, β, γ)-states are thus

$$\alpha = m + k, \qquad \beta = k, \qquad \gamma = n - m - 2k, \qquad k = 0, 1, 2, ..., \text{int}\left(\frac{n-m}{2}\right) \tag{D.8}$$

and the on-diagonal elements of $\mathbf{M}^2$ for given n and m are

$$\mathbf{M}^2{}_{\text{on-diagonal}} = (2n - m)\,(m+1) + 2k\,(2n - 4m - 1) - 8k^2. \tag{D.9}$$

Since $\text{tr}\left(\mathbf{M}^2\right)$ is invariant under diagonalization, the sum of eigenvalues cacluated from the diagonal elements is

$$\begin{aligned} \text{tr}\left(\mathbf{M}^2\right) &= \sum_{k=0}^{\text{int}\left(\frac{n-m}{2}\right)} \mathbf{M}^2{}_{\text{on-diagonal}} \\ &= \begin{cases} \frac{1}{6}n^3 + \frac{3}{4}n^2 + \frac{5}{6}n - \frac{1}{6}m^3 + \frac{1}{4}m^2 + \frac{1}{6}m & ,\ n - m \text{ even} \\ \frac{1}{6}n^3 + \frac{3}{4}n^2 + \frac{5}{6}n - \frac{1}{6}m^3 - \frac{1}{4}m^2 + \frac{1}{6}m + \frac{1}{4} & ,\ n - m \text{ odd} \end{cases} \end{aligned} \tag{D.10}$$

For given values of $n = \alpha + \beta + \gamma$ and $m = \alpha - \beta$, the possible l-states are

$$l = n - 2k \qquad k = 0, 1, ..., \text{int}\left(\frac{n-m}{2}\right). \tag{D.11}$$

So, on the other hand, for given n and m the sum of eigenvalues is must be the sum of $l\,(l+1)$

$$\begin{aligned} \text{tr}\left(\mathbf{M}^2\right) &= \sum_{k=0}^{\text{int}\left(\frac{n-m}{2}\right)} (n - 2k)\,(n - 2k + 1) \\ &= \begin{cases} \frac{1}{6}n^3 + \frac{3}{4}n^2 + \frac{5}{6}n - \frac{1}{6}m^3 + \frac{1}{4}m^2 + \frac{1}{6}m & ,\ n - m \text{ even} \\ \frac{1}{6}n^3 + \frac{3}{4}n^2 + \frac{5}{6}n - \frac{1}{6}m^3 - \frac{1}{4}m^2 + \frac{1}{6}m + \frac{1}{4} & ,\ n - m \text{ odd} \end{cases} \end{aligned} \tag{D.12}$$

as required.

Journal of Physics: Conference Series **330** (2011) 012014 doi:10.1088/1742-6596/330/1/012014

References

[1] Feynman R P, Kislinger M, and Ravndal F 1971 *Phys. Rev.* D **3** 2706
[2] Stueckelberg E C G 1941 *Helv. Phys. Acta* **14** 322
Stueckelberg E C G 1941 *Helv. Phys. Acta* **14** 588
[3] Feynman R P 1950 *Phys. Rev.* **80** 440
Feynman R 1948 *Rev P. Mod. Phys.* **20** 367
[4] Land M C, Shnerb N, Horwitz L P 1995 *J. Math. Phys.* **36** 3263
[5] Land M C and Horwitz L P 1991 *Found. of Phys. Lett.* **4** 61
[6] Kim Y S and Marilyn Noz E 1973 *Phys. Rev.* D **8** 3521
[7] Horwitz L P and Arshansky R 1989 *J. Math. Phys.* **30** 66
Horwitz L P and Arshansky R 1989 *J. Math. Phys.* **30** 380
[8] Horwitz L P and Piron C 1973 *Helv. Phys. Acta* **48** 316
[9] Land M C and Horwitz L P 1995 *J. Phys. A: Math. and Gen.* **28** 3289
[10] Land M C 2001 *Found of Phys.* **31** 967
[11] Kastrup H 2007 *Annalen Phys.* **16** 439
[12] Bars I 2009 *Phys. Rev.* D **79**, 045009
[13] Itzykson C and Zuber J-B 1980 *Quantum Field Theory*, (New York: McGraw-Hill)
[14] Schwinger J 1952 *On angular momentum* in 1965 *Quantum Theory of Angular Momentum* ed L C Biedenharn and H Van Dam (New York: Academic Press) p 229 and available at http://www.osti.gov/accomplishments/documents/fullText/ACC0111.pdf
Mejía F M and Pleitez V 2002 *Rev. Bras. Ens. Fis.* **24** 41
[15] Gradshteyn I S and Ryzhik I M 1994 *Table of Integrals, Series, and Products* (London: Academic Press)

Journal of Physics: Conference Series **330** (2011) 012015 doi:10.1088/1742-6596/330/1/012015

Abraham-Lorentz-Dirac equation in 5D Stuekelberg electrodynamics

Martin Land

Department of Computer Science, Hadassah College, 37 HaNeviim Street, Jerusalem

E-mail: `martin@hadassah.ac.il`

Abstract. We derive the Abraham-Lorentz-Dirac (ALD) equation in the framework of the electrodynamic theory associated with Stueckelberg manifestly covariant canonical mechanics. In this framework, a particle worldline is traced out through the evolution of an event $x^\mu(\tau)$. By admitting unconstrained commutation relations between the positions and velocities, the associated electromagnetic gauge fields are in general dependent on the parameter τ, which plays the role of time in Newtonian mechanics. Standard Maxwell theory emerges from this system as a τ-independent equilibrium limit. In this paper, we calculate the τ-dependent field induced by an arbitrarily evolving event, and study the long-range radiation part, which is seen to be an on-shell plane wave of the Maxwell type. Following Dirac's method, we obtain an expression for the finite part of the self-interaction, which leads to the ALD equation that generalizes the Lorentz force. This third-order differential equation is then converted to an integro-differential equation, identical to the standard Maxwell expression, except for the τ-dependence of the field. By studying this τ-dependence in detail, we show that field can be removed from the integration, so that the Lorentz force depends only on the instantaneous external field and an integral over dynamical variables of the event evolution. In this form, pre-acceleration of the event by future values of the field is not present.

1. Introduction

In classical electrodynamics, fields accelerate charges through the Lorentz force

$$m\ddot{z}^\mu = eF^{\mu\nu}(z)\,\dot{z}_\nu \tag{1}$$

and current densities defined on charge worldlines

$$J^\mu(x) = \int d\tau\ \dot{z}^\mu(\tau)\,\delta^4\left[x - z(\tau)\right] \tag{2}$$

induce fields through Maxwell's equations

$$\partial_\nu F^{\mu\nu}(x) = eJ^\mu(x) \qquad\qquad \partial_{[\lambda}F_{\mu\nu]}(x) = 0\ . \tag{3}$$

In general configurations, multiple charges interact through a complex system of mutual interactions, but as a first approximation to the motion of a single charge, the field may be treated as fixed, neglecting both its evolution under recoil of the sources producing it and the

Journal of Physics: Conference Series **330** (2011) 012015 doi:10.1088/1742-6596/330/1/012015

reaction field induced by the accelerating charge itself. The manifestly covariant system (1) – (3) employs an invariant evolution parameter τ often associated with proper time; this association is at best inexact, ignoring the diversity of proper times among the many particles as well as the presumed τ-reparameterization invariance of the system.

The Abraham-Lorentz-Dirac (ALD) equation [1, 2]

$$m\ddot{z}^\mu = eF^{\mu\nu}_{ext}(z)\,\dot{z}_\nu + \frac{2}{3}e^2\left(\dddot{z}^\mu - \dot{z}^\mu\ddot{z}^2\right) \tag{4}$$

improves the fixed field approximation, still treating the field $F^{\mu\nu}_{ext}$ as external to the dynamics, but incorporating a generic Green's function solution

$$F^{\mu\nu}_{rad}(x) = \partial^\mu A^\nu - \partial^\nu A^\mu \qquad\qquad A^\nu(x) = -e\int d^4x'\, D_{rad}(x-x')\,J^\nu(x') \tag{5}$$

to the Maxwell equations to account for the transfer of energy-momentum from the charge to the field. Thus, ALD provides an effective one-particle theory for a charge in an external electromagnetic field, but being a third order differential equation, admits run-away solutions of the type

$$\dot{z}^0 = \cosh\left[\tau\left(e^{\tau/\tau_0}-1\right)\right] \qquad\qquad \dot{z}^1 = \sinh\left[\tau\left(e^{\tau/\tau_0}-1\right)\right] \tag{6}$$

for the free particle case $F^{\mu\nu}_{ext} = 0$, where

$$\tau_0 = \frac{2}{3}\frac{e^2}{4\pi m} \sim 10^{-24} \text{ sec} \tag{7}$$

is the time scale associated with a photon crossing a classical electron radius. This unphysical solution may be eliminated under the assumption of bounded acceleration, by converting (4) to the integro-differential equation

$$m\ddot{x}^\mu(\tau) = \int_\tau^\infty ds\, \frac{1}{\tau_0}e^{-(s-\tau)/\tau_0}\left[eF^{\mu\nu}_{ext}\left(x(s)\right)\dot{x}_\nu(s) - \tau_0 m\ddot{x}^2(s)\,\dot{x}^\mu(s)\right]. \tag{8}$$

Although (8) suppresses run-away solutions, its causal behavior presents a new difficulty in the form of pre-acceleration. If the external field is a constant switched on at $\tau = 0$, so that

$$\frac{e}{\tau_0}F^{\mu\nu}_{ext}\left(x(\tau)\right)\dot{x}_\nu(\tau) \to \frac{1}{\tau_0}F^\mu(\tau) = \frac{1}{\tau_0}F^\mu_0\,\theta(\tau), \tag{9}$$

then one obtains a non-zero pre-acceleration

$$m\ddot{x}^\mu(\tau) \simeq \frac{1}{\tau_0}F^\mu_0\int_\tau^\infty ds\; e^{-(s-\tau)/\tau_0}\theta(s) = \begin{cases} F^\mu_0 e^{-|\tau|/\tau_0} & , \quad \tau < 0 \\ F^\mu_0 & , \quad \tau > 0 \end{cases} \tag{10}$$

effective over a period on the order of τ_0 *before* the field is applied. It may be argued that on this time scale, problems of classical microcausality are overtaken by quantum effects, empirically and theoretically. In particular, Johnson and Hu [3] have derived a modified ALD equation as a semiclassical limit in QED and shown that causality is enforced by the stochastic nature of the particle-field system. Nevertheless, a violation of classical causality posed in classical electrodynamics seems to raise questions about the interpretation of the underlying framework, beyond issues of approximation and measurement.

Journal of Physics: Conference Series **330** (2011) 012015 doi:10.1088/1742-6596/330/1/012015

In this paper, we develop the Abraham-Lorentz-Dirac equation in a formulation of classical electrodynamics associated with a manifestly covariant canonical mechanics of interacting spacetime events. This approach, first suggested by Stueckelberg [4] in modeling pair creation/annihilation, regards the charged event $x^\mu(\tau)$ as the source of an electromagnetic field, so that event currents and fields are defined locally at a given spacetime point x^μ and given moment τ. Microscopic details of event dynamics may differ from the usual approach, but standard Maxwell theory emerges as the τ-static equilibrium limit. Because the relationships among events, particles, currents, and fields are determined at a finer resolution, this formalism leads to a generalization of (8) for which the causal structure can be given a natural and consistent interpretation. In the next section, we review the essential features of Stueckelberg electrodynamics, then characterize the induced radiation field in section 3, and derive the ALD equation for the Stueckelberg framework in section 4.

2. Stueckelberg off-shell electrodynamics

In seeking a classical description of pair creation/annihilation as a single worldline generated dynamically by the evolution of an event $x^\mu(\tau)$, Stueckelberg proposed [4] a generalized Lorentz force of the form

$$m\ddot{x}^\mu(\tau) = e_0 f^{\mu\nu}(x,\tau)\,\dot{x}_\nu(\tau) + e_0\varepsilon^\mu(x,\tau) \tag{11}$$

where $f^{\mu\nu}(x,\tau)$ is the electromagnetic field tensor (made τ-dependent), and $\varepsilon^\mu(x,\tau)$ is a field strength vector introduced to permit continuous reversal of the time direction $\dot{x}^0(\tau)$. Without the additional vector field, the value of $\dot{x}^2$ is conserved through

$$\frac{d}{d\tau}\left(\dot{x}^2\right) = 2\dot{x}_\mu\ddot{x}^\mu = \frac{2e}{m}\dot{x}_\mu f^{\mu\nu}\dot{x}_\nu = 0 \tag{12}$$

and so the particle cannot evolve smoothly from $\dot{x}^0 > 0$ to $\dot{x}^0 < 0$ through spacelike motion with $\left(\dot{x}^0\right)^2 < \dot{\mathbf{x}}^2$. But Stueckelberg found no physical justification for the vector field and proceeded to develop the τ-parameterized covariant canonical quantum mechanics

$$i\partial_\tau\psi(x,\tau) = \frac{1}{2m}\left[p^\mu - eA^\mu(x)\right]\left[p_\mu - eA_\mu(x)\right]\psi(x,\tau) \tag{13}$$

later used by Feynman [5], Schwinger [6], DeWitt [7] and others in developing QED. Stueckelberg required an independent parameter because, by construction, Einstein coordinate time x^0 does not increase monotonically, and he regarded τ as a physical time of system evolution, playing the role of Newtonian time in nonrelativistic mechanics. The Hamiltonian system associated with (13) is then a symplectic mechanics for which manifest Poincaré covariance plays the role of Galilean covariance in Newtonian mechanics (see also [8, 9]).

Stueckelberg's generalized Lorentz force was later obtained from first principles in the framework of a fundamental gauge theory [10, 11], called off-shell electrodynamics because the non-conservation of $\dot{x}^2(\tau)$ found from (11) is associated with mass exchange between particle and field. Quite generally [12], this framework is implicitly incorporated as the underlying electrodynamics whenever one writes equations of motion

$$m\ddot{x}^\mu = F^\mu(x,\dot{x},\tau) \tag{14}$$

on the unconstrained phase space defined by

$$[x^\mu, x^\nu] = 0 \qquad\qquad m[x^\mu, \dot{x}^\nu] = -i\eta^{\mu\nu}, \tag{15}$$

with metric

$$\eta^{\mu\nu} = \text{diag}(-1,1,1,1). \tag{16}$$

and $\mu,\nu = 0,\cdots,3$. The apparently naive commutation relations (15) are actually sufficient [13] to establish the self-adjointness of (14) from which one is necessarily led to the Lorentz force (13) along with homogeneous field equations

$$\partial_\mu f_{\nu\rho} + \partial_\nu f_{\rho\mu} + \partial_\rho f_{\mu\nu} = 0 \qquad \partial_\mu \varepsilon_\nu - \partial_\nu \varepsilon_\mu + \partial_\tau f_{\mu\nu} = 0. \tag{17}$$

The fields $f_{\mu\nu}(x,\tau)$ and $\varepsilon_\mu(x,\tau)$ have been called pre-Maxwell fields because of their explicit τ-dependence. Adopting formal designations

$$x^5 = \tau \qquad \partial_5 = \partial_\tau \qquad \varepsilon_\mu = f_{5\mu} \tag{18}$$

and

$$\mu,\nu = 0,\cdots,3 \qquad \alpha,\beta = 0,1,2,3,5 \tag{19}$$

we summarize the Lorentz force (11) and field equations (17) as

$$m\ddot{x}_\mu = e_0 f_{\mu\alpha}(x,\tau)\,\dot{x}^\alpha \quad \Rightarrow \quad \frac{d}{d\tau}(-\tfrac{1}{2}m\dot{x}^2) = e_0 f_{5\alpha}(x,\tau)\,\dot{x}^\alpha \tag{20}$$

$$\partial_{[\alpha} f_{\beta\gamma]} = 0 \quad \Rightarrow \quad f_{\alpha\beta}(x,\tau) = \partial_\alpha a_\beta(x,\tau) - \partial_\beta a_\alpha(x,\tau) \tag{21}$$

from which the Lagrangian is uniquely determined [14] as

$$L = \tfrac{1}{2}m\dot{x}^\mu \dot{x}_\mu + e_0\,\dot{x}^\alpha\, a_\alpha(x,\tau). \tag{22}$$

Writing the canonical momentum

$$p_\mu = \frac{\partial L}{\partial \dot{x}^\mu} = m\dot{x}_\mu + e_0 a_\mu(x,\tau) \tag{23}$$

and transforming to the Hamiltonian

$$K = p_\mu \dot{x}^\mu - L \tag{24}$$

the Stueckelberg-Schrodinger equation

$$[i\partial_\tau + e_0 a_5(x,\tau)]\,\psi(x,\tau) = \frac{1}{2m}\left[p^\mu - e_0 a^\mu(x,\tau)\right]\left[p_\mu - e_0 a_\mu(x,\tau)\right]\psi(x,\tau) \tag{25}$$

is seen to be locally gauge invariant [11] under τ-dependent gauge transformations of the type

$$\psi \to e^{ie_0\Lambda(x,\tau)}\psi \qquad a_\mu \to a_\mu + \partial_\mu\Lambda(x,\tau) \qquad a_5 \to a_5 + \partial_\tau\Lambda(x,\tau). \tag{26}$$

The electromagnetic field $f_{\alpha\beta}(x,\tau)$ is made a dynamical quantity by adding a kinetic term to the action. In analogy to standard Maxwell theory, one may adopt the formal designation $f^{\mu 5} = \eta^{55} f^\mu{}_5 = -f^\mu{}_5$ and chose the form

$$-\frac{\lambda}{4} f^{\alpha\beta}(x,\tau)\, f_{\alpha\beta}(x,\tau) \tag{27}$$

Journal of Physics: Conference Series **330** (2011) 012015 doi:10.1088/1742-6596/330/1/012015

which is a gauge invariant Lorentz scalar and contains only first order derivatives of the fields. The electromagnetic part of the action is now

$$S_{field} = \int d^4x d\tau \left\{ e_0 \dot{z}^\alpha a_\alpha (x,\tau) \delta^4 [x - z(\tau)] - \frac{\lambda}{4} f^{\alpha\beta} (x,\tau) f_{\alpha\beta} (x,\tau) \right\} \tag{28}$$

leading to inhomogeneous field equations

$$\partial_\beta f^{\alpha\beta} (x,\tau) = \frac{e_0}{\lambda} j^\alpha (x,\tau) = e j^\alpha (x,\tau) = e \dot{z}^\alpha (\tau) \delta^4 [x - z(\tau)] \tag{29}$$

and a conserved 5-current

$$\partial_\mu j^\mu (x,\tau) + \partial_\tau j^5 (x,\tau) = 0. \tag{30}$$

As in nonrelativistic mechanics, equation (30) suggests the interpretation of $j^5 (x,\tau)$ as the probability density at τ of finding the event at x. Since $\partial_\mu j^\mu \neq 0$, $j^\mu (x,\tau)$ cannot be identified as the source current in Maxwell's equations. However, under the boundary conditions $j^5 \to 0$, pointwise, as $\tau \to \pm\infty$, integration of (30) over τ, leads to $\partial_\mu J^\mu = 0$, where

$$J^\mu(x) = \int_{-\infty}^{\infty} d\tau \; j^\mu (x,\tau) \; . \tag{31}$$

This integration has been called concatenation [15] and links the event current $j^\mu (x,\tau)$ with the particle current $J^\mu(x)$ defined on the entire particle worldline. Similarly,

$$\left. \begin{array}{c} \partial_\beta f^{\alpha\beta} (x,\tau) = e j^\alpha (x,\tau) \\ \\ \partial_{[\alpha} f_{\beta\gamma]} = 0 \end{array} \right\} \xrightarrow[\int d\tau]{} \left\{ \begin{array}{c} \partial_\nu F^{\mu\nu} (x) = e J^\mu (x) \\ \\ \partial_{[\mu} F_{\nu\rho]} = 0 \end{array} \right. \tag{32}$$

where

$$F^{\mu\nu}(x) = \int_{-\infty}^{\infty} d\tau f^{\mu\nu} (x,\tau) \qquad A^\mu(x) = \int_{-\infty}^{\infty} d\tau a^\mu (x,\tau) \tag{33}$$

extracting standard Maxwell theory as the equilibrium limit of event dynamics. It is seen from (25) and (33) that e_0 and λ must have dimensions of time, so that the dimensionless ratio $e = e_0/\lambda$ can be identified as the Maxwell charge.

As in the non-relativistic case, the two-body action-at-a-distance potential in the Horwitz-Piron theory [10] may be understood as the approximation $-e_0 a_5 (x,\tau) \longrightarrow V(x)$. Within this framework, solutions have been found for the generalizations of the standard central force problem, including potential scattering [16] and bound states [17, 18]. Examination of radiative transitions [19], in particular the Zeeman [20] and Stark effects [21], indicate that all five components of the gauge potential are necessary for an adequate explanation of observed phenomenology.

The wave equation derived from (29) is

$$\partial_\alpha \partial^\alpha a^\beta (x,\tau) = \left(\partial_\mu \partial^\mu - \partial_\tau^2\right) a^\beta (x,\tau) = -e j^\beta (x,\tau) \tag{34}$$

for which the Greens function [22] found from

$$\left(\partial_\mu \partial^\mu - \partial_\tau^2\right) G(x,\tau) = -\delta^4 (x,\tau) \tag{35}$$

is

$$G(x,\tau) = -\frac{1}{2\pi} \delta(x^2) \delta(\tau) - \frac{1}{2\pi^2} \frac{\partial}{\partial x^2} \frac{\theta(x^2 - \tau^2)}{\sqrt{x^2 - \tau^2}} = D(x) \delta(\tau) - G_{correlation} (x,\tau) . \tag{36}$$

The first term has support on the lightcone at instantaneous τ, and recovers the standard Maxwell Greens function under concatenation. The second term has spacelike support ($x^2 > \tau^2 \geq 0$) and vanishes under concatenation, so it may contribute to correlations but not to Maxwell potentials. Potentials obtained from (36) as

$$a^\beta(x,\tau) = -e\int d^4x'd\tau'\ G\left(x-x',\tau-\tau'\right)\ j^\beta(x',\tau') \tag{37}$$

recover Maxwell potentials under concatenation

$$\begin{aligned} A^\mu(x) &= \int d\tau\ \left\{-e\int d^4x'd\tau'\ G\left(x-x',\tau-\tau'\right)\ j^\beta(x',\tau')\right\} \\ &= -e\int d^4x'\ D\left(x-x'\right)\int d\tau'\ j^\mu(x',\tau') \\ &= -e\int d^4x'\ D\left(x-x'\right) J^\mu(x'). \end{aligned} \tag{38}$$

To study the low energy Coulomb problem, the source is taken to be the "static" event

$$z(\tau) = (\tau,0,0,0) \tag{39}$$

from which one calculates

$$\begin{aligned} a^0(x,\tau) &= -e\int d^4x'd\tau' D\left(x-x'\right)\delta\left(\tau-\tau'\right)\delta\left(x^{0\prime}-\tau'\right)\delta^3(\mathbf{x}) \\ &= -\frac{e}{2\pi}\,\delta\left[(x-\tau)^2\right]\theta\left(x^0\right) \\ &= -\frac{e}{4\pi\left|\mathbf{x}\right|}\,\delta\left(x^0-\tau-\left|\mathbf{x}\right|\right) \end{aligned} \tag{40}$$

Although concatenation of (40) recovers the correct Coulomb potential

$$A^0(x) = -\frac{e}{4\pi\left|\mathbf{x}\right|}\int d\tau\ \delta\left(x^0-\tau-\left|\mathbf{x}\right|\right) = -\frac{e}{4\pi\left|\mathbf{x}\right|} \tag{41}$$

the δ-function in $a^0(x,\tau)$ results in a microscopic dynamics that cannot reproduce expected low-energy interactions between a pair of charges. In particular, a second "static" event given by

$$\zeta(\tau) = \left(u^0\tau+\alpha,\beta,0,0\right) \tag{42}$$

will generally not experience any interaction with $z(\tau)$, except for appropriately tuned values of $\dot{\zeta}^0 = u^0$ and the offset $\zeta^0(0) = \alpha$ that determines the τ-synchronization of the two events.

Since off-shell quantum theory does provide a correct description of Coulomb scattering for sharp asymptotic mass states, which contain no information about τ-synchronization, it was suggested in [23] that the electromagnetic interaction between events should be modified to relax the deterministic synchronization expressed in (20) and (29). Under this modification, the source $j^\beta_\varphi(x,\tau)$ for the pre-Maxwell fields is taken to be a smoothed current density induced by an ensemble of events $z^\beta(\tau+\delta\tau)$ along a particle worldline, where $\delta\tau$ is given by a normalized distribution $\varphi(\tau)$. From (29) the smoothed source current is given as the ensemble average

$$j^\alpha_\varphi(x,\tau) = \int_{-\infty}^{\infty} ds\ \varphi(\tau-s)\,\dot{z}^\alpha(s)\,\delta^4\left[x-z(s)\right] \tag{43}$$

or equivalently as an ensemble of sharp currents $j^\beta(x,\tau)$ induced by a single event

$$j^\alpha(x,\tau) \longrightarrow j^\alpha_\varphi(x,\tau) = \int_{-\infty}^{\infty} ds\ \varphi(\tau - s)\ j^\alpha(x,s)\,. \tag{44}$$

The fields $f^{\alpha\beta}(x,\tau)$ and $a^\beta(x,\tau)$ are induced by the smoothed current density, so the electrodynamic system that mediates between sharp events $z^\beta(\tau)$ and $\zeta^\beta(\tau)$ explicitly introduces a statistical structure to the event-event interaction. Because Maxwell currents are defined along the entire worldline, this procedure preserves the concatenated current

$$J^\mu(x) = \int_{-\infty}^{\infty} d\tau\ j^\mu_\varphi(x,\tau) = \int_{-\infty}^{\infty} ds \left[\int_{-\infty}^{\infty} d\tau \varphi(\tau - s)\, j^\mu(x,s)\right] = \int_{-\infty}^{\infty} ds\ j^\mu(x,s)\,. \tag{45}$$

Taking the distribution to be

$$\varphi(\tau) = \frac{1}{2\lambda} e^{-|\tau|/\lambda} \qquad\qquad \int_{-\infty}^{\infty} d\tau\ \varphi(\tau) = 1 \tag{46}$$

the low energy Coulomb field becomes a Yukawa-type potential with the correct non-relativistic limit for large λ

$$m\ddot{\mathbf{x}} = -e_0 \nabla\left[a^0(x,\tau) + a^5(x,\tau)\right] \rightarrow m\ddot{\mathbf{x}} = -2e_0\nabla a^0_\varphi(x,\tau) = e^2\nabla\left[\frac{e^{-|\mathbf{x}|/\lambda}}{4\pi|\mathbf{x}|}\right]. \tag{47}$$

The distribution $\varphi(\tau)$ provides a cutoff for the photon mass spectrum, which we take to be the conventional experimental uncertainty in photon mass ($\Delta m_\gamma \simeq 10^{-17}\,\mathrm{eV}$ [24]), leading to a value of about 400 seconds for λ. The limit $\lambda \to 0$ restores $\varphi(\tau) \to \delta(\tau)$ and the limit $\lambda \to \infty$ restores standard Maxwell theory. Since the form of $\varphi(\tau)$ given in (44) represents the distribution of interarrival times of events in a Poisson-distributed stochastic process, this choice suggests an information-theoretic interpretation for the underlying the relationship between the current density and the ensemble of events from which it is induced.

The smoothed current can be introduced through the action [25], by adding a higher τ-derivative term to the electromagnetic part. The substitution

$$S_{em} \rightarrow \int d^4x d\tau \left[e_0 j^\alpha a_\alpha - \frac{\lambda}{4} f^{\alpha\beta}(x,\tau) f_{\alpha\beta}(x,\tau) - \frac{\lambda^3}{4}\left[\partial_\tau f^{\alpha\beta}(x,\tau)\right]\left[\partial_\tau f_{\alpha\beta}(x,\tau)\right]\right] \tag{48}$$

preserves Lorentz and gauge invariance, and leaves the action first order in spacetime derivatives. Defining a field interaction kernel

$$\Phi(\tau) = \delta(\tau) - \lambda^2\delta''(\tau) = \frac{1}{2\pi}\int d\kappa\left[1 + (\lambda\kappa)^2\right] e^{-i\kappa\tau} \tag{49}$$

which is seen from

$$\int_{-\infty}^{\infty} ds\ \Phi(\tau - s)\varphi(s) = \delta(\tau) \rightarrow \varphi(\tau) = \int \frac{d\kappa}{2\pi}\frac{e^{-i\kappa\tau}}{1 + (\lambda\kappa)^2} = \frac{1}{2\lambda} e^{-|\tau|/\lambda} \tag{50}$$

to be the inverse function to $\varphi(\tau)$, the action becomes

$$S_{em} = \int d^4x d\tau\ e_0 j^\alpha a_\alpha - \frac{\lambda}{4}\int d^4x\, d\tau\, ds\ f^{\alpha\beta}(x,\tau)\,\Phi(\tau - s) f_{\alpha\beta}(x,s)\,. \tag{51}$$

Journal of Physics: Conference Series **330** (2011) 012015 doi:10.1088/1742-6596/330/1/012015

The Euler-Lagrange equations

$$\partial_\beta f_\Phi^{\alpha\beta}(x,\tau) = \partial_\beta \int ds\, \Phi(\tau - s) f^{\alpha\beta}(x,s) = e j^\alpha(x,\tau) \tag{52}$$

describe a sharp field induced by a sharp event current, and using (50) can be inverted to recover

$$\partial_\beta f^{\alpha\beta}(x,\tau) = e j_\varphi^\alpha(x,\tau) = e \int ds\, \varphi(\tau - s)\, j^\alpha(x,s)\,. \tag{53}$$

The action (51), in which the statistical synchronization performed by $\Phi(\tau - s)$ is made explicit, has the advantage of permitting the usual study of symmetries and being amenable to second quantization, where the factor $\left[1 + (\lambda\kappa)^2\right]^{-1}$ provides a natural mass cutoff for the off-shell photon that renders off-shell quantum field theory super-renormalizable at two-loop order [25].

3. Radiation fields

In this section we calculate the electromagnetic field induced by an arbitrarily evolving spacetime event, in order to identify and characterize the radiation part. In particular, we obtain the five Liénard-Wiechert potentials and the field strength tensor, which is conveniently expressed as a Clifford product of a pair orthogonal vectors. This bivector formulation simplifies calculation of field invariants, the mass-energy-momentum tensor, and the plane wave expansion. The radiation field is shown to be an equilibrium field of the standard Maxwell type.

3.1. Liénard-Wiechert potential

Beginning with a generic spacetime event $X^\mu(\tau)$ and using (29) and (44), we write the smoothed current

$$j_\varphi^\alpha(x,\tau) = \int ds\, \varphi(\tau - s)\, j^\alpha(x,s) = \int ds\, \varphi(\tau - s)\, \dot{X}^\alpha(s)\, \delta^4\left[x - X(s)\right] \tag{54}$$

where

$$\dot{X}^5(\tau) = \dot{\tau} = 1. \tag{55}$$

We choose the τ-instantaneous part of the Greens function (36), which recovers the Maxwell propagator under concatenation (the correlation Greens function has been studied in [26] and applied to radiation reaction in [27]). The Liénard-Wiechert potential $a^\alpha(x,\tau)$ induced by the current (54) is

$$a^\alpha(x,\tau) = -e \int d^4x' d\tau' G\left(x - x', \tau - \tau'\right) j_\varphi^\alpha\left(x', \tau'\right) \tag{56}$$

$$= -\frac{e}{2\pi} \int ds\, \varphi(\tau - s)\, \dot{X}^\alpha(s)\, \delta\left[\left(x - X(s)\right)^2\right] \theta^{ret} \tag{57}$$

$$= -\frac{e}{2\pi} \varphi(\tau - s) \frac{\dot{X}^\alpha(s)}{2\left(x^\mu - X^\mu(s)\right)\dot{X}_\mu(s)} \tag{58}$$

where we used the identity

$$\int d\tau f(\tau)\, \delta\left[g(\tau)\right] = \left.\frac{f(s)}{|g'(s)|}\right|_{s=g^{-1}(0)} \tag{59}$$

Journal of Physics: Conference Series **330** (2011) 012015 doi:10.1088/1742-6596/330/1/012015

and the retarded time s satisfies

$$[x - X(s)]^2 = 0 \qquad \theta^{ret} = \theta\left(x^0 - X^0(s)\right) = 1. \tag{60}$$

Introducing the timelike velocity

$$u^\alpha = \dot{X}^\alpha(s), \tag{61}$$

the vector from event $X(s)$ to observation point x

$$z^\mu = x^\mu - X^\mu(s) \Rightarrow \dot{z}^\mu = -u^\mu, \tag{62}$$

and the scalar function

$$R = \frac{1}{2}\frac{d}{ds}(x - X(s))^2 = -z^\mu u_\mu = -z \cdot u \geq 0, \tag{63}$$

we find

$$a^\alpha(x,\tau) = -e\frac{1}{2\lambda}\frac{u^\alpha}{4\pi R}e^{-|\tau - s|/\lambda} \tag{64}$$

where the nonnegativity of R follows from (60) and (61).

To calculate the field strengths, we need derivatives of the Liénard-Wiechert potential. The spacetime derivative is most conveniently found by applying the identity (59) to expression (57)

$$\partial^\mu a^\beta(x,\tau) = -\frac{e}{2\pi}\int ds\; \varphi(\tau - s)\dot{X}^\alpha(s)\theta^{ret}\partial^\mu\delta\left((x - X(s))^2\right) \tag{65}$$

$$= \frac{e}{2\pi}\int ds\; \varphi(\tau - s)\dot{X}^\beta(s)\theta^{ret}\delta'\left[(x - X(s))^2\right]\left[-2\left(x^\mu - X^\mu(s)\right)\right] \tag{66}$$

$$= \frac{e}{2\pi}\int ds\; \varphi(\tau - s)\frac{\dot{X}^\beta(s)\left[x^\mu - X^\mu(s)\right]}{\dot{X}(s)\cdot(x - X(s))}\theta^{ret}\frac{d}{ds}\delta\left[(x - X(s))^2\right] \tag{67}$$

and integrating by parts to obtain

$$\partial^\mu a^\beta(x,\tau) = -\frac{e}{2\pi}\int ds\frac{d}{ds}\left[\varphi(\tau - s)\frac{\dot{X}^\beta(s)\left[x^\mu - X^\mu(s)\right]}{\dot{X}(s)\cdot(x - X(s))}\right]\theta^{ret}\delta\left[(x - X(s))^2\right] \tag{68}$$

$$= -\frac{e}{4\pi}\frac{1}{R}\frac{d}{ds}\left[\varphi(\tau - s)\frac{z^\mu u^\beta}{R}\right]. \tag{69}$$

Since

$$\frac{d}{d\tau}\varphi(\tau) = \frac{1}{2\lambda}\frac{d}{d\tau}e^{-|\tau|/\lambda} = -\frac{1}{2\lambda^2}\epsilon(\tau)e^{-|\tau|/\lambda} = -\frac{1}{\lambda}\epsilon(\tau)\varphi(\tau) \tag{70}$$

we obtain the τ-derivative directly from (64) as

$$\partial_\tau a_\mu(x,\tau) = -e\dot{\varphi}(\tau - s)\frac{u_\mu}{4\pi R} = e\epsilon(\tau - s)\varphi(\tau - s)\frac{u_\mu}{4\pi\lambda R}\quad. \tag{71}$$

3.2. Field strengths

From (69) the spacetime components of the field strength tensor takes the anti-symmetric form

$$f^{\mu\nu} = -\frac{e}{4\pi}\frac{1}{R}\frac{d}{ds}\left[\varphi(\tau - s)\frac{z^\mu u^\nu - z^\nu u^\mu}{R}\right]. \tag{72}$$

Using (62), (63), and

$$\dot{R} = -\frac{d}{d\tau}(z \cdot u) = u^2 - z \cdot \dot{u} \tag{73}$$

the derivatives split into a retarded field exhibiting $|\mathbf{z}|^{-2}$ far-field behavior

$$f^{\mu\nu}_{ret} = e\varphi(\tau - s)\left[\frac{(z^\mu u^\nu - z^\nu u^\mu)\, u^2}{4\pi\,(u \cdot z)^3} - \epsilon(\tau - s)\frac{(z^\mu u^\nu - z^\nu u^\mu)}{4\pi\lambda\,(u \cdot z)^2}\right] , \tag{74}$$

and a long-range radiation field exhibiting $|\mathbf{z}|^{-1}$ far-field behavior

$$f^{\mu\nu}_{rad} = e\varphi(\tau - s)\frac{(z^\mu \dot{u}^\nu - z^\nu \dot{u}^\mu)(u \cdot z) - (z^\mu u^\nu - z^\nu u^\mu)(\dot{u} \cdot z)}{4\pi\,(u \cdot z)^3} . \tag{75}$$

Because we take $\lambda \simeq 400$ seconds, we include the λ^{-1} term in the retarded field. Using (69) to find $\partial_\mu a_5$ and (71) to find $\partial_\tau a_\mu$ we calculate the fifth component fields as

$$f^{ret}_{\mu 5} = -e\varphi(\tau - s)\left[\frac{z_\mu u^2 - u_\mu (u \cdot z)}{4\pi\,(u \cdot z)^3} - \epsilon(\tau - s)\frac{z_\mu - u_\mu (u \cdot z)}{4\pi\lambda\,(u \cdot z)^2}\right] \tag{76}$$

$$f^{rad}_{\mu 5} = -e\varphi(\tau - s)\frac{(\dot{u} \cdot z)\, z_\mu}{4\pi\,(u \cdot z)^3} . \tag{77}$$

3.3. Electromagnetic two form

For notational simplicity, we will represent the field as a bivector

$$f = \frac{1}{2} f^{\alpha\beta}(\mathbf{e}_\alpha \wedge \mathbf{e}_\beta) = \frac{1}{2} f^{\alpha\beta}(\mathbf{e}_\alpha \otimes \mathbf{e}_\beta - \mathbf{e}_\beta \otimes \mathbf{e}_\alpha) \tag{78}$$

in a Clifford algebra (see [28] and references contained therein) over the formal 5D space with basis vectors

$$\mathbf{e}_\alpha \cdot \mathbf{e}_\beta = \eta_{\alpha\beta} \qquad \alpha, \beta = 0, 1, 2, 3, 5. \tag{79}$$

In this notation, expression (72) for the spacetime component of the electromagnetic field becomes

$$f_{spacetime} = \frac{1}{2} f^{\mu\nu}(\mathbf{e}_\mu \wedge \mathbf{e}_\nu) = -\frac{e}{4\pi}\frac{1}{R}\frac{d}{ds}\left[\varphi(\tau - s)\frac{z \wedge u}{R}\right] \tag{80}$$

and the radiation part (75) is

$$f_{spacetime} = e\varphi(\tau - s)\frac{(z \wedge \dot{u})(u \cdot z) - (z \wedge u)(\dot{u} \cdot z)}{4\pi\,(u \cdot z)^3} \tag{81}$$

$$= e\varphi(\tau - s)\ z \wedge \frac{\dot{u}\,(u \cdot z) - u\,(\dot{u} \cdot z)}{4\pi\,(u \cdot z)^3} . \tag{82}$$

Similarly, the fifth component radiation field (77) takes the form

$$f_5 = \frac{1}{2} f^{\mu 5}(\mathbf{e}_\mu \wedge \mathbf{e}_5) = -e\varphi(\tau - s)\frac{(\dot{u} \cdot z)\ z \wedge \mathbf{e}_5}{4\pi\,(u \cdot z)^3} . \tag{83}$$

Introducing the scalar quantity

$$Q = -\dot{u} \cdot z \tag{84}$$

Journal of Physics: Conference Series **330** (2011) 012015 doi:10.1088/1742-6596/330/1/012015

and the vector

$$w = \dot{u}R - uQ = -\left[\dot{u}\,(u \cdot z) - u\,(\dot{u} \cdot z)\right] \tag{85}$$

the radiation fields assume the form

$$f_{spacetime} = e\varphi(\tau - s)\,\frac{z \wedge w}{4\pi R^3} \tag{86}$$

$$f_5^{rad} = -e\varphi\,(\tau - s)\,\frac{z \wedge (Q\mathbf{e}_5)}{4\pi R^3}\,. \tag{87}$$

Designating formal 5-velocity and 5-acceleration as

$$U = \frac{d}{d\tau}\,(X\,(\tau)\,,\tau) = \left(\dot{X}\,(\tau)\,,1\right) = u + \mathbf{e}_5 \qquad \dot{U} = \frac{d}{d\tau}\left(\dot{X}\,(\tau)\,,1\right) = (\dot{u}, 0) = \dot{u} \tag{88}$$

so that

$$R = -z \cdot U \qquad Q = -z \cdot \dot{U} \tag{89}$$

the radiation fields may be combined in the expression

$$f = f_{spacetime} + f_5 = e\varphi(\tau - s)\,\frac{z \wedge W}{4\pi R^3}. \tag{90}$$

where the 5D vector

$$W = \dot{U}R - UQ = w - Q\mathbf{e}_5 \tag{91}$$

depends on the velocity and acceleration of the event, characterizing the directionality of the motion inducing the radiation.

3.4. Field invariants

From (89) and the Clifford identity

$$a \cdot (b \wedge c) = (a \cdot b)\,c - (a \cdot c)\,b \tag{92}$$

the vector W can be expressed as

$$W = U\left(z \cdot \dot{U}\right) - \dot{U}\,(z \cdot U) = z \cdot \left(U \wedge \dot{U}\right) \tag{93}$$

describing the projection of z into the plane spanned by the vectors U and $\dot{U}$, and reflected through the direction of U. The vector W is orthogonal to z, which follows from the Clifford identity

$$(a \wedge b) \cdot (c \wedge d) = [(a \wedge b) \cdot c] \cdot d = (a \cdot c)\,(b \cdot d) - (b \cdot c)\,(a \cdot d) \tag{94}$$

as

$$z \cdot W = z \cdot \left[z \cdot \left(U \wedge \dot{U}\right)\right] = (z \wedge z) \cdot \left(U \wedge \dot{U}\right) \tag{95}$$

or directly as

$$z \cdot W = z \cdot \left(\dot{U}R - UQ\right) = \left(z \cdot \dot{U}\right)R - (z \cdot U)\,Q = -QR + RQ = 0. \tag{96}$$

The radiation field thus lies in the plane spanned by the orthogonal vectors z and W, where W is normal to z in the plane spanned by U and $\dot{U}$. It also follows from (96), (60) and the Clifford identity (92) that

$$z \cdot (z \wedge W) = z^2 W - z\,(z \cdot W) = 0 \tag{97}$$

so that the plane $z \wedge W$ is null with respect to the lightlike observation vector z. Writing the components of the field bivector (90) using

$$(z \wedge W)_{\beta\gamma} = z_\beta W_\gamma - z_\gamma W_\beta \tag{98}$$

the field invariant $\epsilon^{\alpha\beta\gamma\delta\varepsilon} f_{\beta\gamma} f_{\delta\varepsilon}$ is seen to vanish identically because the product of vectors satisfies

$$\epsilon^{\alpha\beta\gamma\delta\varepsilon} \left(z_\beta W_\gamma - z_\gamma W_\beta\right)\left(z_\delta W_\varepsilon - z_\varepsilon W_\delta\right) = 4\epsilon^{\alpha\beta\gamma\delta\varepsilon} z_\beta W_\gamma z_\delta W_\varepsilon = 0. \tag{99}$$

Similarly, since $f^{\alpha\beta} f_{\alpha\beta}$ includes $(z \wedge W)^{\alpha\beta} (z \wedge W)_{\alpha\beta}$ it must vanish because using (94) we find

$$(z \wedge W) \cdot (z \wedge W) = (z \cdot W)^2 - z^2 W^2 = 0. \tag{100}$$

Thus, the radiation field (90) is seen to be a null field, satisfying

$$z \cdot f = 0 \longrightarrow z_\mu f^{\mu\alpha} = 0 \tag{101}$$

$$f \cdot f = 0 \longrightarrow f^{\alpha\beta} f_{\alpha\beta} = f^{\mu\nu} f_{\mu\nu} = f^{5\nu} f_{5\nu} = 0 \tag{102}$$

$$\epsilon^{\alpha\beta\gamma\delta\varepsilon} f_{\beta\gamma} f_{\delta\varepsilon} = 0 \longrightarrow \epsilon^{\mu\nu\rho\sigma} f_{\mu\nu} f_{\rho\sigma} = \epsilon^{\mu\nu\rho\sigma} f_{\sigma 5} f_{\rho\nu} = 0. \tag{103}$$

3.5. Mass-energy-momentum tensor

In [11] the mass-energy-momentum tensor was derived as the Noether current associated with the translation invariance of electromagnetic action in the form (28). Studying the translation invariance of the modified action (51), one finds the conservation law

$$\partial_\alpha T^{\alpha\beta} = e_0 f^{\alpha\beta} j_\alpha \tag{104}$$

for conserved tensor

$$T^{\alpha\beta} = -\lambda\left(g^{\alpha\beta} f^{\delta\gamma} f^{\Phi}_{\delta\gamma} - f^{\alpha}_{\ \gamma} f^{\beta\gamma}_{\Phi}\right), \tag{105}$$

where the modified field f_Φ is given in (52) and the sharp current j_α is defined in (29). Integrating (104) over spacetime, leads to

$$\frac{d}{d\tau} \int d^4x \; T^{5\alpha} = e_0 \int d^4x \; f^{\alpha\beta}(x,\tau)\, \dot{X}_\alpha \delta^4(x - X) = e_0 f^{\alpha\beta}(X,\tau)\, \dot{X}_\alpha \tag{106}$$

which by comparison with the Lorentz force (20) provides

$$\frac{d}{d\tau} \int d^4x \; \left(T^{5\mu} + m\dot{X}^\mu\right) = 0 \qquad\qquad \frac{d}{d\tau} \int d^4x \left(T^{55} - \tfrac{1}{2} m\dot{X}^2\right) = 0, \tag{107}$$

expressing the instantaneous conservation of total energy-momentum-mass for the combined field and event over all spacetime [29]. Since the τ-dependence of (90) resides in $\varphi(\tau - s)$, the τ-integration in (52) merely inverts $\varphi(\tau - s)$ to $\delta(\tau - s)$, and since

$$\delta(\tau - s)\, \varphi(\tau - s) = \delta(\tau - s)\, \varphi(0) = \frac{1}{2\lambda} \delta(\tau - s) \tag{108}$$

terms of the type $f^{\alpha\beta} f^{\Phi}_{\delta\gamma}$ become

$$f^{\alpha\beta} f^{\Phi}_{\delta\gamma} = \frac{1}{2\lambda} \delta(\tau - s) \left(\frac{e}{4\pi R^3}\right)^2 (z \wedge W)^{\alpha\beta} (z \wedge W)_{\delta\gamma}. \tag{109}$$

Journal of Physics: Conference Series **330** (2011) 012015 doi:10.1088/1742-6596/330/1/012015

In light of (102), tensor (105) reduces to

$$T^{\alpha\beta} = \lambda f^{\alpha}_{\ \gamma} \, f^{\beta\gamma}_{\Phi} \tag{110}$$

containing the products

$$\begin{aligned}(z^{\alpha} W_{\gamma} - z_{\gamma} W^{\alpha}) \left(z^{\beta} W^{\gamma} - z^{\gamma} W^{\beta}\right) &= z^2 W^{\alpha} W^{\beta} + W^2 z^{\alpha} z^{\beta} - z \cdot W \left(z^{\alpha} W^{\beta} + z^{\alpha} W^{\beta}\right) \\ &= W^2 z^{\alpha} z^{\beta}\end{aligned} \tag{111}$$

and providing the simple expression

$$T^{\alpha\beta} = \frac{1}{2} \delta(\tau - s) \left(\frac{e}{4\pi R^3}\right)^2 W^2 z^{\alpha} z^{\beta}. \tag{112}$$

Because the observation vector z has no 5-component, this becomes

$$T^{\alpha 5} = 0 \qquad T^{\mu\nu} = \frac{1}{2} \delta(\tau - s) \left(\frac{e}{4\pi R^3}\right)^2 W^2 z^{\mu} z^{\nu} \tag{113}$$

so that the total (over all spacetime) mass and energy-momentum carried by the radiation field vanishes.

Integrating (104) over space and τ — equivalent to the space integration of the concatenated dynamics — we recover an expression for the Maxwell Poynting vector

$$\frac{d}{dt} \int d^3x \; d\tau \; T^{0\mu} = e_0 \int d^3x \; d\tau \; f^{\alpha\mu}(x, \tau) \, \dot{X}_{\alpha}(\tau) \, \delta^4 [x - X(\tau)] \tag{114}$$

where using (113), the energy-momentum

$$P^{\mu} = \int d^3x \; d\tau \; T^{0\mu} = \frac{1}{2} \left(\frac{e}{4\pi R^3}\right)^2 W^2 z^0 z^{\mu} \tag{115}$$

is found to be oriented along the observation vector z^{μ}. We notice that the RHS of (114) differs from the Maxwell formulation through the τ-dependence of $f^{\alpha\beta}(x, \tau)$. In the equilibrium limit $f^{\mu\nu}(x, \tau) \to \frac{1}{\lambda} F^{\mu\nu}(x)$, (114) is seen to recover the standard form of energy-momentum conservation

$$\frac{d}{dt} \int d\tau \; T^{0\mu} = e F^{\mu\nu}(x) \int d\tau \; \dot{X}_{\nu}(\tau) \, \delta^4 [x - X(\tau)] = e F^{\nu\mu}(x) \, J_{\nu}(x) \, . \tag{116}$$

3.6. Vector field picture

To compare the radiation field (90) with the standard Maxwell field, we write $f^{\alpha\beta}$ in vector components [11] as

$$e^i = f^{0i} \qquad b^i = \epsilon^{ijk} f_{jk} \qquad \varepsilon^{\mu} = f^{5\mu} = \left(\varepsilon^0, \varepsilon\right) \tag{117}$$

for which the field equations are

$$\begin{array}{ll} \nabla \cdot \mathbf{e} - \partial_{\tau} \varepsilon^0 = e j^0 & \nabla \times \mathbf{e} + \partial_0 \mathbf{b} = 0 \\ \nabla \times \mathbf{b} - \partial_0 \mathbf{e} - \partial_{\tau} \varepsilon = e \mathbf{j} & \nabla \cdot \mathbf{b} = 0 \\ \nabla \cdot \varepsilon + \partial_0 \varepsilon^0 = e j^5 & \nabla \times \varepsilon + \partial_{\tau} \mathbf{b} = 0 \\ \multicolumn{2}{c}{\nabla \varepsilon^0 - \partial_{\tau} \mathbf{e} + \partial_0 \varepsilon = 0} \end{array} \tag{118}$$

and the Poynting vectors are

$$T^{00} = \frac{\lambda}{2}\left[\mathbf{e}\cdot\mathbf{e}^{\Phi}+\mathbf{b}\cdot\mathbf{b}^{\Phi}-\varepsilon\cdot\varepsilon^{\mathbf{\Phi}}-\varepsilon^{0}\varepsilon_{\Phi}^{0}\right] \qquad T^{0i}=\lambda\left[\mathbf{e}\times\mathbf{b}^{\Phi}-\varepsilon^{0}\varepsilon^{\mathbf{\Phi}}\right]^{i} \tag{119}$$

$$T^{55} = \frac{\lambda}{2}\left[-\mathbf{e}\cdot\mathbf{e}^{\Phi}+\mathbf{b}\cdot\mathbf{b}^{\Phi}+\varepsilon\cdot\varepsilon^{\mathbf{\Phi}}-\varepsilon^{0}\varepsilon_{\Phi}^{0}\right] \qquad T^{5i}=\lambda\left[\varepsilon^{0}\mathbf{e}^{\Phi}+\varepsilon\times\mathbf{b}^{\Phi}\right]^{i}. \tag{120}$$

From the invariants (102) and (103) it follows that

$$-\mathbf{e}\cdot\mathbf{e}+\mathbf{b}\cdot\mathbf{b}=\varepsilon\cdot\varepsilon-\varepsilon^{0}\varepsilon^{0}=0 \quad\Rightarrow\quad \begin{cases} T^{00}=\lambda\left[\mathbf{e}\cdot\mathbf{e}^{\Phi}-\varepsilon\cdot\varepsilon^{\mathbf{\Phi}}\right] \\ T^{55}=0 \end{cases} \tag{121}$$

$$\mathbf{e}\times\mathbf{b}=(\mathbf{e}\cdot\mathbf{e})\,\hat{\varepsilon} \qquad \varepsilon^{0}\mathbf{e}=-\varepsilon\times\mathbf{b} \quad\Rightarrow\quad \begin{cases} T^{0i}=\lambda\left[\mathbf{e}\cdot\mathbf{e}^{\Phi}-\varepsilon\cdot\varepsilon^{\mathbf{\Phi}}\right]\hat{\varepsilon}^{i} \\ T^{5i}=0 \end{cases} \tag{122}$$

where from (77)

$$\hat{\varepsilon}=\frac{\varepsilon}{|\varepsilon|}=\hat{\mathbf{z}} \tag{123}$$

again indicating that the Maxwell Poynting vector is oriented along the observation vector.

A general plane wave expansion is taken by writing the Fourier transform

$$f^{\alpha\beta}(x,\tau)=\frac{1}{(2\pi)^{5}}\int d^{4}k\,d\kappa\; e^{i(k\cdot x-\kappa\tau)}f^{\alpha\beta}(k,\kappa) \tag{124}$$

for which the wave equation (34) in vacuum gives

$$k^{\alpha}k_{\alpha}=k^{\mu}k_{\mu}-\kappa^{2}=\mathbf{k}^{2}-(k^{0})^{2}-\kappa^{2}=0 \tag{125}$$

so that concatenation of $f^{\alpha\beta}(x,\tau)$ forces the photon onto the $\kappa\to 0$ zero-mass shell. The general plane wave solution to the 3-vector field equations (118) in terms of the transverse component of $\mathbf{e}$ and the longitudinal component of ε was given in [29] as

$$\mathbf{e}=\mathbf{e}_{\perp}-\frac{\kappa}{k^{0}}\,\varepsilon_{\parallel} \qquad \mathbf{h}=\frac{1}{k^{0}}\mathbf{k}\times\mathbf{e}_{\perp} \qquad \varepsilon=\varepsilon_{\parallel}+\frac{\kappa}{k^{0}}\,\mathbf{e}_{\perp} \qquad \varepsilon^{0}=\frac{1}{k^{0}}\mathbf{k}\cdot\varepsilon_{\parallel}. \tag{126}$$

On the zero-mass shell, the $\mathbf{e}$ and $\mathbf{h}$ fields take the Maxwell form — mutually orthogonal with equal amplitude, and both orthogonal to the propagation vector $\mathbf{k}$ — and the ε field decouples from $\mathbf{e}$ and $\mathbf{h}$, becoming purely longitudinal. In terms of solution (126) the components of the energy-momentum tensor are

$$\begin{aligned} T^{00}&=\lambda\left(\mathbf{e}_{\perp}\cdot\mathbf{e}_{\perp}^{\Phi}-\varepsilon_{\parallel}\cdot\varepsilon_{\parallel}^{\mathbf{\Phi}}\right) & T^{0i}&=\frac{k^{i}}{k^{0}}T^{00} \\ T^{55}&=\left(\frac{\kappa}{k^{0}}\right)^{2}T^{00} & T^{5\mu}&=\frac{\kappa}{k^{0}}\frac{k^{\mu}}{k^{0}}T^{00} \end{aligned} \tag{127}$$

and we see that $T^{5\alpha}$ vanishes on the zero-mass shell. Comparison of (113) with (127) for the $\kappa=0$ case shows that the radiation field from the arbitrary event has the character of an equilibrium plane wave solution.

Journal of Physics: Conference Series **330** (2011) 012015 doi:10.1088/1742-6596/330/1/012015

4. Abraham-Lorentz-Dirac equation

In the previous section we examined the field produced by an arbitrarily evolving event $X^\mu(\tau)$, and in particular studied the characteristics of its radiation field, identified by the far-field behavior. In this section, we sharpen the characterization of the radiation field, following Dirac's analysis [1] of the Greens functions, and obtain the ALD equation by specifically accounting for the field produced by the event's own evolution.

4.1. Radiation reaction

The Lorentz force (20) describes the motion of an event under the influence of a general electromagnetic field $f^{\alpha\beta}(x,\tau)$. Depending on the known boundary conditions, $f^{\alpha\beta}(x,\tau)$ may be described as an initial incoming field and the retarded reaction, or as a final outgoing field and the advanced reaction, according to

$$f^{\alpha\beta}(x,\tau) = f^{\alpha\beta}_{in}(x,\tau) + f^{\alpha\beta}_{ret}(x,\tau) = f^{\alpha\beta}_{out}(x,\tau) + f^{\alpha\beta}_{adv}(x,\tau) \tag{128}$$

where

$$f^{\alpha\beta}_{ret}(x,\tau) \xrightarrow[\tau\to-\infty]{} 0 \qquad\qquad f^{\alpha\beta}_{adv}(x,\tau) \xrightarrow[\tau\to\infty]{} 0. \tag{129}$$

Thus, when an incoming field $f^{\alpha\beta}_{in}(x,\tau)$ impinges on the event, the Lorentz force can be expressed as

$$m\ddot{X}^\mu(\tau) = \lambda e\left[f^{\mu\beta}_{in}(X,\tau) + f^{\mu\beta}_{ret}(X,\tau)\right]\dot{X}_\beta(\tau) \tag{130}$$

where $f^{\alpha\beta}_{ret}(x,\tau)$ is produced by the event $X^\mu(\tau)$ itself through the field equation (53). However, the interaction of the event with its own field includes an infinite part, and to identify the finite part, Dirac argued that the outgoing field can be expressed as the sum of the incoming field and a radiation reaction produced when the incoming field accelerates the event. Thus,

$$f^{\alpha\beta}_{out} = f^{\alpha\beta}_{in} + f^{\alpha\beta}_{rad} \tag{131}$$

which combines with (128) to provide

$$f^{\alpha\beta}_{rad} = f^{\alpha\beta}_{out} - f^{\alpha\beta}_{in} = \left(f^{\alpha\beta} - f^{\alpha\beta}_{adv}\right) - \left(f^{\alpha\beta} - f^{\alpha\beta}_{ret}\right) = f^{\alpha\beta}_{ret} - f^{\alpha\beta}_{adv}\,. \tag{132}$$

Considering the combinations

$$f^{\alpha\beta}_{ret} = \frac{1}{2}\left(f^{\alpha\beta}_{ret} - f^{\alpha\beta}_{adv}\right) + \frac{1}{2}\left(f^{\alpha\beta}_{ret} + f^{\alpha\beta}_{adv}\right) = \frac{1}{2}f^{\alpha\beta}_{rad} + \frac{1}{2}\left(f^{\alpha\beta}_{ret} + f^{\alpha\beta}_{adv}\right) \tag{133}$$

the radiation reaction $f^{\alpha\beta}_{rad}$ is found to be finite, while the the infinite part of the self-interaction $f^{\alpha\beta}_{ret} + f^{\alpha\beta}_{adv}$ is treated by mass renormalization. The finite part of the Lorentz force is then

$$m\ddot{X}^\mu = \lambda e\left[\frac{1}{2}f^{\mu\beta}_{rad} + f^{\mu\beta}_{in}\right]\dot{X}_\beta\,. \tag{134}$$

From the prescription (132) the potential $a^\beta_{rad}(x,\tau)$ is found by making the substitution

$$\theta^{ret} \longrightarrow \theta^{rad} = \theta^{ret} - \theta^{adv} = \epsilon\left(x^0 - X^0\right) \tag{135}$$

Journal of Physics: Conference Series **330** (2011) 012015 doi:10.1088/1742-6596/330/1/012015

in (57), so that the self-interaction term in (134) is the force

$$F^{\mu}_{self}(\tau) = \frac{1}{2}\lambda e f^{\mu\beta}_{rad}(X(\tau), \tau)\, \dot{X}_{\beta}(\tau) \tag{136}$$

evaluated at the event $X^{\mu}(\tau)$. The field must therefore be calculated from

$$\partial^{\mu} a^{\beta}_{rad}(x, \tau) = -\frac{e}{2\pi}\int ds \frac{d}{ds}\left[\varphi(\tau - s)\frac{\dot{X}^{\beta}(s)\left[x^{\mu} - X^{\mu}(s)\right]}{\dot{X}(s)\cdot\left[x - X(s)\right]}\right]\theta^{rad}\delta\left[(x - X(s))^2\right] \tag{137}$$

in the limit

$$x^{\mu} - X^{\mu}(s) \to X^{\mu}(\tau) - X^{\mu}(s) \tag{138}$$

which by conditions (62) and (60) require

$$\left[X(\tau) - X(s)\right]^2 = 0 \qquad \Rightarrow \qquad s \to \tau. \tag{139}$$

Dirac showed [1] that expressions of the type (137) have a finite limit as $X^{\mu}(s) \to X^{\mu}(\tau)$, found by performing a Taylor expansion along worldline $X(\tau)$, that is, evaluating

$$\partial^{\mu} a^{\beta}_{rad}(X(\tau'), \tau) \tag{140}$$

where

$$\tau' = \tau + h \tag{141}$$

$$X(\tau') - X(\tau) = h\dot{X}(\tau) + \frac{h^2}{2}\ddot{X}(\tau) + \frac{h^3}{3!}\dddot{X}(\tau) + \cdots \tag{142}$$

$$\dot{X}(\tau') = \dot{X}(\tau) + h\ddot{X}(\tau) + \frac{h^2}{2}\dddot{X}(\tau) + \cdots \tag{143}$$

By comparison of (137) with the expression for the field in standard Maxwell theory [1, 2]

$$\partial^{\mu} A^{\nu}_{rad}(x) = -\frac{e}{2\pi}\int ds \frac{d}{ds}\left[\frac{\dot{X}^{\nu}(s)\left[x^{\mu} - X^{\mu}(s)\right]}{\dot{X}(s)\cdot\left[x - X(s)\right]}\right]\theta^{rad}\delta\left[(x - X(s))^2\right], \tag{144}$$

the field expansion in off-shell electrodynamics is seen to take the standard form

$$f^{\mu\nu}_{rad}(X(\tau), \tau) = \frac{2}{3}\frac{e}{2\pi}\left[\dot{X}^{\mu}(\tau)\dddot{X}^{\nu}(\tau) - \dot{X}^{\nu}(\tau)\dddot{X}^{\mu}(\tau)\right] \tag{145}$$

where we use $\varphi(0) = \frac{1}{2\lambda}$. Now, the self-interaction term in (136) is

$$F^{\mu}_{self}(\tau) = \lambda\frac{2}{3}\frac{e^2}{2\pi}\left[\dot{X}^{\mu}\dddot{X}^{\nu} - \dot{X}^{\nu}\dddot{X}^{\mu}\right]\dot{X}_{\nu} = \frac{2}{3}\frac{e^2}{4\pi}\left[\dot{X}^{\mu}\dddot{X}^{\nu}\dot{X}_{\nu} - \dot{X}^2\dddot{X}^{\mu}\right]. \tag{146}$$

Since the radiation field was seen to carry no mass, the event may be considered on-shell, so that

$$\dot{X}^2 = -1 \qquad \dot{X}\cdot\ddot{X} = 0 \qquad \dot{X}\cdot\dddot{X} + \ddot{X}^2 = 0, \tag{147}$$

and the Lorentz force (134) takes the form of the usual ALD equation

$$m\ddot{x}^{\mu} = e_0 f^{\mu\beta}_{in}(x, \tau)\,\dot{x}_{\beta} + m\tau_0\left[\dddot{x}^{\mu} - \ddot{x}^2\,\dot{x}^{\mu}\right] \tag{148}$$

where we have written τ_0 as in (7), and reverted to lower case for the event trajectory

$$X^{\mu}(\tau) \to x^{\mu}(\tau) \tag{149}$$

since the field point is understood to be the event location

$$f^{\mu\beta}_{in}(x, \tau) = f^{\mu\beta}_{in}\left(x(\tau), \tau\right). \tag{150}$$

Journal of Physics: Conference Series **330** (2011) 012015 doi:10.1088/1742-6596/330/1/012015

4.2. Integro-differential equation

In order to eliminate the run-away solutions admitted by the ALD equation in the form (148), we convert it to an integro-differential equation

$$(\ddot{x}^\mu - \tau_0 \dddot{x}^\mu)\, e^{-\tau/\tau_0} = e^{-\tau/\tau_0} \left[\frac{e_0}{m} f_{in}^{\mu\beta}(x,\tau)\, \dot{x}_\beta - \tau_0 \ddot{x}^2\, \dot{x}^\mu \right] \tag{151}$$

$$\ddot{x}^\mu e^{-\tau/\tau_0} = -\int_0^\tau ds\; e^{-s/\tau_0} \left[\frac{e_0}{\tau_0 m} f_{in}^{\mu\beta}(x,s)\, \dot{x}_\beta - \ddot{x}^2\, \dot{x}^\mu \right] + \ddot{x}^\mu(0)\,. \tag{152}$$

The run-away solutions are suppressed by imposing the requirement that velocity grow less than exponentially for long times

$$\ddot{x}^\mu(\tau)\, e^{-\tau/\tau_0} \xrightarrow[\tau\to\infty]{} 0 \tag{153}$$

so that as $\tau \to \infty$ (152) becomes

$$0 = -\int_0^\infty ds\; e^{-s/\tau_0} \left[\frac{e_0}{\tau_0 m} f_{in}^{\mu\beta}(x,s)\, \dot{x}_\beta(s) - \ddot{x}^2(s)\, \dot{x}^\mu(s) \right] + \ddot{x}^\mu(0) \tag{154}$$

which is incorporated back into (152) to provide the ALD equation as

$$m\ddot{x}^\mu(\tau) = \int_\tau^\infty ds\; \frac{1}{\tau_0} e^{-(s-\tau)/\tau_0} \left[e_0 f_{in}^{\mu\beta}\left(x(s), s\right) \dot{x}_\beta(s) - m\tau_0 \ddot{x}^2(s)\, \dot{x}^\mu(s) \right]. \tag{155}$$

This expression describes an acceleration at τ that depends on values of the interaction at later times, as was seen for the standard ALD equation (8), which is recovered from (155) in the Maxwell limit

$$e_0 f_{in}^{\mu\beta}\left(x(\tau), \tau\right) = e\lambda f_{in}^{\mu\beta}\left(x(\tau), \tau\right) \to eF_{in}^{\mu\beta}\left(x(\tau)\right). \tag{156}$$

In the context of Maxwell theory, the field depends only on the spacetime location of the particle on which it acts, so that the synchronization of the particle-field interaction is sharply determined by the proper time τ of the particle motion. In the off-shell formalism, however, the τ-dependence of the electromagnetic field typically introduces a statistical synchronization between interacting events, on a time scale λ much larger than τ_0. The integration in (155) may thus be understood as analogous to this smoothed synchronization, describing a modified self-interaction between the event and a radiation field induced by an ensemble associated with the event evolution.

To clarify this interpretation, we use (52) to express $f_{in}^{\mu\beta}(x,\tau)$ in terms of $f_\Phi^{\mu\beta}(x,\tau)$, a sharp field induced by a sharp event density $j(x,\tau)$ of the type (29). From (50) and (53) the external field can be written as an ensemble average of such sharp fields

$$f_{in}^{\mu\beta}(x,\tau) = \int_{-\infty}^\infty ds\; f_\Phi^{\mu\beta}\left(x(\tau), s\right) \varphi(s-\tau) \tag{157}$$

leading to an integral expression for the Lorentz force (20)

$$m\ddot{x}^\mu(\tau) = e_0 f^{\mu\beta}\left(x(\tau), \tau\right) \dot{x}_\beta(\tau) = \int_{-\infty}^\infty ds\; e_0 f_\Phi^{\mu\beta}\left(x(\tau), s\right) \varphi(s-\tau)\, \dot{x}_\beta(\tau). \tag{158}$$

By comparison, the ALD equation (155) can be written as

$$m\ddot{x}^\mu(\tau) = \int_{-\infty}^\infty ds\; \left[e_0 f_{in}^{\mu\beta}\left(x(s), s\right) \dot{x}_\beta(s) - m\tau_0 \ddot{x}^2(s)\, \dot{x}^\mu(s) \right] \phi(s-\tau) \tag{159}$$

Journal of Physics: Conference Series **330** (2011) 012015 doi:10.1088/1742-6596/330/1/012015

where we interpret the integrating factor as a normalized distribution

$$\phi(\tau) = \frac{1}{\tau_0} e^{-\tau/\tau_0} \theta(\tau) \qquad \int_{-\infty}^{\infty} d\tau \; \phi(\tau) = 1. \tag{160}$$

Aside from the additional interaction term $\ddot{x}^2 \dot{x}^\mu$, equation (159) differs from (158) in that the statistical synchronization affected by the distribution $\phi(\tau)$ extends to both the field $f_{in}^{\mu\beta}(x, s)$ and the velocity $\dot{x}_\beta(s)$.

In order to examine the causal structure of (155) more closely, we shift the integration variable as $s \to s - \tau$

$$\begin{aligned} m\ddot{x}^\mu(\tau) &= e_0 \int_0^\infty ds \; \frac{1}{\tau_0} e^{-s/\tau_0} \; f_{in}^{\mu\beta}(x, s+\tau) \, \dot{x}_\beta(s+\tau) \\ &\quad - m\tau_0 \int_0^\infty ds \; \frac{1}{\tau_0} e^{-s/\tau_0} \; \ddot{x}^2(s+\tau) \, \dot{x}^\mu(s+\tau). \end{aligned} \tag{161}$$

and consider the genealogy of the field. Although the incoming field $f_{in}^{\mu\beta}(x, \tau)$ is taken to be external, it must have been induced by the smoothed current $j_\varphi(x, \tau)$ associated with the sharp event densities $j(x, \tau)$ distributed along the worldlines of some configuration of evolving events. Using (157) for the field induced directly by the sharp event density $j(x, \tau)$, we rewrite (155) as

$$\begin{aligned} m\ddot{x}^\mu(\tau) &= e_0 \int_0^\infty ds \; \frac{1}{\tau_0} e^{-s/\tau_0} \int_{-\infty}^{\infty} ds' \; \frac{1}{2\lambda} e^{-|s+\tau-s'|/\lambda} \; f_\Phi^{\mu\beta}(x, s') \, \dot{x}_\beta(s+\tau) \\ &\quad - m\tau_0 \int_0^\infty ds \; \frac{1}{\tau_0} e^{-s/\tau_0} \; \ddot{x}^2(s+\tau) \, \dot{x}^\mu(s+\tau) \end{aligned} \tag{162}$$

and rearrange the first line by combining the exponential functions

$$e^{-s/\tau_0} e^{-|s+\tau-s'|/\lambda} = \begin{cases} \exp\left[-s\left(\frac{1}{\tau_0} + \frac{1}{\lambda}\right) - (\tau - s')\frac{1}{\lambda}\right] & , \quad -\infty < s' \le s + \tau \\ \exp\left[-s\left(\frac{1}{\tau_0} - \frac{1}{\lambda}\right) + (\tau - s')\frac{1}{\lambda}\right] & , \quad s + \tau \le s' < \infty \end{cases} \tag{163}$$

and partitioning the s'-integration according to

$$\int_0^\infty ds \int_{-\infty}^{\infty} ds' \; \{\cdots\} = \int_0^\infty ds \left[\int_{-\infty}^{s+\tau} ds' \{\cdots\} + \int_{s+\tau}^{\infty} ds' \{\cdots\} \right]. \tag{164}$$

The relationship of time scales $\lambda \gg \tau_0$ permits the approximation

$$\exp\left[-s\left(\frac{1}{\tau_0} \pm \frac{1}{\lambda}\right)\right] = e^{-s/\tau_0} + o\left(\frac{\tau_0}{\lambda}\right) \tag{165}$$

making no assumptions about the functional forms of events or fields, and so the integrals in the first line of (155) becomes

$$\begin{aligned} &\int_0^\infty ds \frac{1}{\tau_0} e^{-s/\tau_0} \; \dot{x}_\beta(s+\tau) \times \\ &\left\{ \int_{-\infty}^{s+\tau} ds' \; \frac{1}{2\lambda} e^{-(\tau-s')/\lambda} \; f_\Phi^{\mu\beta}(x, s') + \int_{s+\tau}^{\infty} ds' \; \frac{1}{2\lambda} e^{(\tau-s')/\lambda} \; f_\Phi^{\mu\beta}(x, s') \right\}. \end{aligned} \tag{166}$$

Further partitioning the s'-integrations as

$$\int_{-\infty}^{s+\tau} ds' \frac{1}{2\lambda} e^{-(\tau-s')/\lambda} f_\Phi^{\mu\beta}(x,s') = \int_{-\infty}^{\tau} ds' \frac{1}{2\lambda} e^{-|\tau-s'|/\lambda} f_\Phi^{\mu\beta}(x,s') + \int_{\tau}^{s+\tau} ds' \frac{1}{2\lambda} e^{-(\tau-s')/\lambda} f_\Phi^{\mu\beta}(x,s') \tag{167}$$

and

$$\int_{s+\tau}^{\infty} ds' \frac{1}{2\lambda} e^{(\tau-s')/\lambda} f_\Phi^{\mu\beta}(x,s') = \int_{\tau}^{\infty} ds' \frac{1}{2\lambda} e^{-|\tau-s'|/\lambda} f_\Phi^{\mu\beta}(x,s') - \int_{\tau}^{s+\tau} ds' \frac{1}{2\lambda} e^{(\tau-s')/\lambda} f_\Phi^{\mu\beta}(x,s') \tag{168}$$

the integrals in the first line of (155) are now

$$\int_0^\infty ds \frac{1}{\tau_0} e^{-s/\tau_0}\, \dot{x}_\beta(s+\tau) \times \left\{ \int_{-\infty}^{\infty} ds' \frac{1}{2\lambda} e^{-|\tau-s'|/\lambda} f_\Phi^{\mu\beta}(x,s') + \int_{\tau}^{s+\tau} ds' \frac{1}{2\lambda} \left[e^{-(\tau-s')/\lambda} - e^{(\tau-s')/\lambda} \right] f_\Phi^{\mu\beta}(x,s') \right\} + o\left(\frac{\tau_0}{\lambda}\right) \tag{169}$$

in which, from (157), we recognize the second line as $f_{in}^{\mu\beta}(x,\tau)$. Shifting $s'-\tau \to s'/\lambda$ in the third line of (169) leads to

$$\int_{\tau}^{s+\tau} ds' \frac{1}{\lambda} \sinh\left(\frac{s'-\tau}{\lambda}\right) f_\Phi^{\mu\beta}(x,s') = \int_0^{s/\lambda} ds' \sinh(s')\; f_\Phi^{\mu\beta}(x,\lambda s'+\tau) \tag{170}$$

which for smooth f_Φ may be estimated as being of order $o\left(\tau_0^2/\lambda^2\right)$:

$$\int_0^{s/\lambda} ds' \sinh(s') f_\Phi^{\mu\beta}(x,\lambda s'+\tau) \simeq \int_0^{\tau_0/\lambda} ds' \sinh(s')\; f_\Phi^{\mu\beta}(x,\lambda s'+\tau) \tag{171}$$

$$\simeq \frac{\tau_0^2}{\lambda^2} f_\Phi^{\mu\beta}(x,\tau) - \frac{\tau_0^3}{3\lambda^2} \frac{d}{d\tau} f_\Phi^{\mu\beta}(x,\tau)\,. \tag{172}$$

Shifting the s-integral back to $s \to s-\tau$, and using the distribution $\phi(\tau)$ defined in (160) to extend the limits of integration, the ALD equation (155) now takes the form

$$m\ddot{x}^\mu(\tau) = e_0 \int_{-\infty}^{\infty} ds\; f_{in}^{\mu\beta}\Big(x(s),\tau\Big)\, \phi(s-\tau)\; \dot{x}_\beta(s) + o\left(\frac{\tau_0}{\lambda}\right) + e_0 \int_{-\infty}^{\infty} ds\; \phi(s-\tau)\; \dot{x}_\beta(s) \int_0^{(s-\tau)/\lambda} ds' \sinh(s')\, f_\Phi^{\mu\beta}\Big(x(s),\lambda s'+\tau\Big) - m\tau_0 \int_{-\infty}^{\infty} ds\; \phi(s-\tau)\; \ddot{x}^2(s)\, \dot{x}^\mu(s) \tag{173}$$

where we have made no assumptions about the τ-dependence of events or fields. In this expression, the acceleration at time τ is given by the interaction of the field at time τ with a narrow ensemble of events described by the distribution $\phi(\tau)$.

Journal of Physics: Conference Series **330** (2011) 012015 doi:10.1088/1742-6596/330/1/012015

If the field $f_{\Phi}^{\mu\beta}(x,\tau)$ varies smoothly, we see from (172) that the second line in (173) can be neglected. However, this term may be significant for external fields of the type (72), which are of the form

$$f_{in}^{\mu\beta}(x,\tau) = \varphi(\tau-\tau_{ret})\, F_{Maxwell}^{\mu\beta}(x)\Big|_{\tau_{ret}} \tag{174}$$

and associated with the sharp form

$$f_{\Phi}^{\mu\beta}(x,\tau) = \delta(\tau-\tau_{ret})\, F_{Maxwell}^{\mu\beta}(x)\Big|_{\tau_{ret}} . \tag{175}$$

Then, the integral s' integration is

$$\int_0^{(s-\tau)/\lambda} ds' \sinh(s')\, f_{\Phi}^{\mu\beta}(x,\lambda s'+\tau) = F_{Maxwell}^{\mu\beta}(x) \times$$

$$\int_0^{(s-\tau)/\lambda} ds' \; \sinh(s')\, \delta(\lambda s'+\tau-\tau_{ret}) \tag{176}$$

$$= F_{Maxwell}^{\mu\beta}(x) \sinh\left(\frac{\tau-\tau_{ret}}{\lambda}\right) \theta(s-\tau_{ret}) \tag{177}$$

and the s integration becomes

$$e_0 \int_{\tau_{ret}}^{\infty} ds\; \frac{1}{\tau_0} e^{-(s-\tau)/\tau_0}\, F_{Maxwell}^{\mu\beta}\Big(x(s)\Big)\, \dot{x}_\beta(s) \sinh\left(\frac{\tau-\tau_{ret}}{\lambda}\right) \tag{178}$$

which can be put into the form

$$e_0 e^{-(\tau-\tau_{ret})/\tau_0} \sinh\left(\frac{\tau-\tau_{ret}}{\lambda}\right) \theta(\tau-\tau_{ret}) \times \tag{179}$$

$$\int_{\tau_{ret}}^{\infty} ds\; \frac{1}{\tau_0}\; F_{Maxwell}^{\mu\beta}\Big(x(s)\Big)\, \dot{x}_\beta(s)\, e^{-(s-\tau_{ret})/\tau_0} \tag{180}$$

so that the function

$$\psi(\tau-\tau_{ret}) = e^{-(\tau-\tau_{ret})/\tau_0} \sinh\left(\frac{\tau-\tau_{ret}}{\lambda}\right) \theta(\tau-\tau_{ret}) \sim o\left(\frac{\tau_0}{\lambda}\right) \tag{181}$$

limits the significance of this term. For fields of this type, ALD takes the form

$$\begin{aligned} m\ddot{x}^{\mu}(\tau) &= e_0 \int_{-\infty}^{\infty} ds\; f_{in}^{\mu\beta}\Big(x(s),\tau\Big)\, \phi(s-\tau)\; \dot{x}_\beta(s) \\ &+ e_0 \psi(\tau-\tau_{ret}) \int_{-\infty}^{\infty} ds\; \phi(s-\tau_{ret})\; F_{Maxwell}^{\mu\beta}\Big(x(s)\Big)\, \dot{x}_\beta(s) \\ &- m\tau_0 \int_{-\infty}^{\infty} ds\; \phi(s-\tau)\; \ddot{x}^2(s)\, \dot{x}^{\mu}(s) + o\left(\frac{\tau_0}{\lambda}\right). \end{aligned} \tag{182}$$

The ALD equations, in the form (173) or (182), now appear as an interaction between the instantaneous external field and short-range ensemble averages over the specific combinations of event velocity $\dot{x}^{\mu}(\tau)$ and acceleration $\ddot{x}^{\mu}(\tau)$ that produce the radiation field (145). These averages smooth the τ-synchronization between the event and its radiation reaction, an effect that is qualitatively analogous to the statistical synchronization expressed in (53) between the event and its induced current.

Journal of Physics: Conference Series **330** (2011) 012015 doi:10.1088/1742-6596/330/1/012015

5. Conclusion

Historically, the Abraham-Lorentz equation, a nonrelativistic approximation to (4), was first obtained [30, 31] by adding an effective term to the Lorentz force to account for the energy lost by an accelerating particle to Larmor radiation. In the $|\dot{\mathbf{x}}| \ll 1$ approximation (in which case $\tau \to t$), ALD reduces to

$$\ddot{\mathbf{x}}(t) - \tau_0 \dddot{\mathbf{x}}(t) = \frac{e}{m} \mathbf{E}_{ext}(t, \mathbf{x}) \tag{183}$$

which admits a runaway solution for $\mathbf{E}_{ext} = 0$ given by

$$\dot{\mathbf{x}} = \dot{\mathbf{x}}(0)\, e^{t/\tau_0}. \tag{184}$$

Imposing once again the boundary condition (153), which merely requires that velocity grow less than exponentially over long times, equation (183) may be converted to the integro-differential equation

$$\ddot{\mathbf{x}}(t) = \frac{e}{m} \int_{-\infty}^{\infty} dt' \left[\frac{1}{\tau_0} e^{-(t'-t)/\tau_0} \theta(t'-t) \right] \mathbf{E}_{ext}\left(t', \mathbf{x}(t')\right), \tag{185}$$

suppressing the spontaneous acceleration. However, the seemingly innocent boundary condition, without which the integration over t' could not be made sensible, provides just the pretext under which the pre-acceleration evident in (185) is granted admissibility. Conversely, to insist that one cannot give meaning to the apparent violation of classical causality under the integral is equivalent to rejecting the reasonableness of the boundary condition. Jackson [32] summarizes the conventional interpretation of this situation by accepting a possible violation of microscopic causality over time scale τ_0, because it cannot disagree with experiment. On this view, since the external field cannot undergo a macroscopic binary transition in an interval smaller than τ_0, and since measurements at these time scales will be dominated by quantum effects, the violation of classical causality in classical mechanics leads to no experimental contradiction. Although the relativistic ALD equation (8) may be accepted with the same qualification, an interpretation that preserves some sense of retarded causality in the classical context would be more satisfying.

As derived in the off-shell electrodynamics associated with Stueckelberg canonical covariant mechanics, the ALD equation in the form (155) was seen to differ from the standard expression in Maxwell theory (8) only in the dependence of the external field on the invariant time parameter τ. It was shown in section 2 that the general τ-dependence of off-shell fields can be understood as an expansion of the U(1) local gauge group to include the invariant evolution parameter, but in any case, follows directly from the unconstrained commutation relations (15) among coordinates and velocities of spacetime events. Thus, off-shell electrodynamics describes a microscopic interaction between spacetime events $x^\mu(\tau)$ mediated by instantaneous τ-dependent fields $f_{\alpha\beta}(x, \tau)$. However, it was seen in (53) that while the off-shell fields are τ-dependent, they are induced by an event current associated with an ensemble of events distributed in τ along the particle worldline. This distribution introduces an underlying statistical structure to the classical event-event interaction, equivalent to relaxing the sharp τ-synchronization found in expressions (8) and (185) between an event and a field with which it interacts. Detailed consideration of the resulting τ-dependence of the external field enables the transformation of equation (155) to the form (173), in which the Lorentz force depends on the instantaneous value of the external field and short-range ensemble averages over combinations of event velocity $\dot{x}^\mu(\tau)$ and acceleration $\ddot{x}^\mu(\tau)$. These remaining averages, which involve only the dynamical variables of the event evolution, can be understood as artifacts of the suppression of the runaway solutions. On the other hand, the combinations of dynamical variables in these expressions derive from the form of the radiation field (145) emitted by the event, and therefore the integrations may be seen as qualitatively analogous to the statistical synchronization expressed in (53) between the event and the field it induces.

Journal of Physics: Conference Series **330** (2011) 012015 doi:10.1088/1742-6596/330/1/012015

References

[1] Dirac P A M 1938 *Proc. R. Soc. London* A **167** 148
[2] Barut A O 1980 *Electrodynamics and Classical Theory of Fields and Particles* (New York: Dover)
[3] Johnson P R and Hu B L 2002 *Phys. Rev.* D **65** 065015
[4] Stueckelberg E C G 1941 *Helv. Phys. Acta* **14** 322
Stueckelberg E C G 1941 *Helv. Phys. Acta* **14** 588
[5] Feynman R P 1950 *Phys. Rev.* **80** 440
Feynman R P 1948 Rev. Mod. Phys. **20** 367
[6] Schwinger J 1951 *Phys. Rev.* **82** 664
[7] DeWitt B S 1965 *Dynamical Theory of Groups and Fields* (New York: Gordon and Breach)
[8] Fock V A 1937 *Phys. Z. Sowjetunion* **12** 404
[9] Nambu Y 1950 *Prog. Theor. Phys.* **5** 82
[10] Horwitz L P and Piron C 1973 *Helv. Phys. Acta* **48** 316
[11] Saad D, Horwitz L P and Arshansky R I 1989 *Found. of Phys.* **19** 1126
[12] Land M C, Shnerb N and Horwitz L P 1995 *J. Math. Phys.* **36** 3263
[13] Hojman S A and Shepley L C 1991 *J. Math. Phys.* **32** 142
[14] Santilli R M 1990 *Foundations of Theoretical Mechanics I* (Berlin: Springer-Verlag)
[15] Arshansky R L, Horwitz P and Lavie Y 1983 *Found. of Phys.* **13** 1167
[16] Horwitz L P and Lavie Y 1982 *Phys. Rev.* D **26** 819
Arshansky R I and Horwitz L P 1989 *J. Math. Phys.* **30** 213
Arshansky R I and Horwitz L P 1988 *Phys. Lett* A **131** 222
[17] Arshansky R I and Horwitz L P 1989 *J. Math. Phys.* **30** 66
[18] Arshansky R I and Horwitz L P 1989 *J. Math. Phys.* **30** 380
[19] Land M C, Arshansky R and Horwitz L P 1994 *Found. of Phys.* **24** 563
[20] Land M C and Horwitz L P 1995 *J. Phys.* A **28** 3289
[21] Land M C 2001 *Found. of Phys.* **31** 967
[22] Land M C and Horwitz L P 1991 *Found. of Phys.* **21** 299
[23] Land M C 1996 *Found. of Phys.* **27** 19
[24] Yao W-M et. al. 2006 (Particle Data Group) *J. Phys.* G **33** 1
[25] Land M C 2003 *Found. of Phys.* **33** 1157
[26] Aharonovich I and Horwitz L P On the Green-Functions of the classical offshell electrodynamics under the manifestly covariant relativistic dynamics of Stueckelberg *Preprint* arXiv:1102.4840v2
[27] Oron O and Horwitz L P 2001 *Phys. Lett.* A **280** 265-270
[28] Land M C 2007 *Found. of Phys.* **37** 597
Land M C 2005 *Found. of Phys.* **35** 1245
[29] Land M C and Horwitz L P 1991 *Found. of Phys. Lett.* **4** 61
[30] M Abraham 1905 *Theorie der Elektrizitat* Vol. II (Leipzig: Teubner)
[31] Lorentz H A 1909 *The Theory of Electrons* (Leipzig: Teubner)
[32] Jackson J D 1975 *Classical Electrodynamics* (New York: Wiley)

Journal of Physics: Conference Series **330** (2011) 012016 doi:10.1088/1742-6596/330/1/012016

From binary black hole simulation to triple black hole simulation

Shan Bai[1], Zhoujian Cao[2], Wen-Biao Han[3], Chun-Yu Lin[4], Hwei-Jang Yo[5] and Jui-Ping Yu[6]

[1,2] Institute of Applied Mathematics, Academy of Mathematics and Systems Science, Chinese Academy of Sciences, Beijing 100190, China

[3] ICRANet and Physics Department, University of Rome "La Sapienza", Piazzale della Repubblica 10, 65122, Pescara, Italy

[4] Department of Physics, National Cheng-Kung University, Tainan 701, Taiwan

[5,6] Department of Physics, National Cheng-Kung University, Tainan 701, Taiwan

E-mail: [2] `zjcao@amt.ac.cn`, [5] `hjyo@phys.ncku.edu.tw`

Abstract. Black hole systems are among the most promising sources for a gravitational wave detection project. Now, China is planning to construct a space-based laser interferometric detector as a follow-on mission of LISA in the near future. Aiming to provide some theoretical support to this detection project on the numerical relativity side, we focus on black hole systems simulation in this work. Considering the globular galaxy, multiple black hole systems also likely to exist in our universe and play a role as a source for the gravitational wave detector we are considering. We will give a progress report in this paper on our black hole system simulation. More specifically, we will present triple black hole simulation together with binary black hole simulation. On triple black hole simulations, one novel perturbational method is proposed.

1. Introduction

Pioneered by Weber, gravitational wave detection is a tough but also a very important task for natural science development. More and more gravitational wave detectors are constructed or planned to construct including resonant bars such as EXPLORER [1], ALLEGRO [2], NIOBE [3], and AURIGA [4]; and laser interferometers such as LIGO [5], VIRGO [6], GEO600 [7], TAMA [8]) and the planned LISA [9]. Excitingly, China is also planning to construct a space-based laser interferometer as a follow-on mission of LISA. At present this project is considering shorter arm length than LISA to fill the frequency gap between LISA and LIGO detectors [10, 11]. This project is temporally named Advanced LISA mission (ALIA). Considering this project together with the above ones, the requirement for the theoretical prediction of gravitational wave signal becomes more and more urgent. As to the most important gravitational source–binary black hole(BBH) coalescence, numerical relativity(NR) plays a key role for both aiding gravitational wave data analysis [12] and helping to construct a phenomenological wave form [13] for a template bank. Aiming to give a theoretical support for the ALIA project on the NR side, we have spent much energy on black hole simulations. Here, in this paper, we will give a progress report on our BBH simulations. In previous theoretical predictions on BBH systems, BBHs are treated as an isolated system although they always locate in some galaxy with a supermassive black hole in the center. This is based on the idea that the gravitational potential induced by the supermassive black hole is ignorable compared with the strong gravitational field introduced by BBH itself. Although the curved background of the supermassive black hole is much weaker compared with the BBH's gravitational field, the nonlinearity of general relativity(GR) may make some difference due to the configuration change [14]. Moreover, the special environment of a globular galaxy makes multiple black hole systems also possibly exist. So multiple black hole systems are also interesting theoretically and realistically for NR.

NR corresponds to solving Einstein's equation without any approximation or symmetry assumption but with the aid of a supercomputer. Besides the massive computational cost involved in NR, how to make the numerical calculation stable is not a trivial problem for Einstein's equation. After the breakthroughs in 2005 and 2006 [15, 16, 17], NR developed fantastically fast. Now almost every NR group in the world can evolve black hole systems without any stability problem. Moreover, long term gravitational waves [18, 19, 20, 21, 22, 23, 24, 25, 26, 27, 28], gravitational radiation induced black hole recoil [29, 30, 31, 32, 33, 34, 35, 36, 37, 38, 39, 40, 41, 42, 43, 44], estimates of final black hole mass and spin of BBH systems [45, 46, 47, 48, 49], and other interesting topics [50, 51, 52, 53, 54, 55] have been extensively studied in the last years. So as people hoped a long time ago, NR has become an efficient tool for research in GR and other kinds of gravity theory. But in any specific application, NR may meet numerical accuracy and efficiency problems.

Based on the requirement of ALIA project, we will present binary black hole simulation and make the simulation result meaningful. As to the multiple black hole systems, we try to investigate the effect of these gravitational potentials on the evolution of BBH and especially the gravitational wave form which is interesting for detection. But the large difference of the distances among the triple black holes makes full GR simulation too computationally costly. In order to attack this problem we propose a perturbational method for triple black hole systems. Besides triple black hole systems, our perturbational method is also possibly useful to other systems, say black hole systems in the asymptotically de Sitter space which represents a possible universe background.

This paper is organized as following: In Section 2, we give a brief description of our NR code and the numerical techniques involved. Then we present our simulation results about BBH in Section 3. There both head-on collision and inspiraling cases are included. Our code is also calibrated there. In Section 4 we give an interesting result about three black hole simulation with full GR but a new perturbational method is used. Here we only present the main result about our three black hole simulation. We refer the reader to [56] for details. At last we summarize and discuss our code and simulation result. For completeness and future reference, two appendixes about gravitational wave form Ψ_4 calculation are included.

2. Numerical techniques

2.1. Code description

We updated our code for solving Einstein's equation based on the BSSN formalism from [57]. Besides the ICN method we have constructed routines to implement 4th order Runge-Kutta method in our new code. On the mesh refinement, we have developed our own infrastructure other than GrACE[58, 59] to implement these numerical techniques. Our infrastructure archives coarse-grained and fine-grained parallelism in a loose sense. We use MPI[60] to realize parallelization among different nodes and OpenMP[61] among the processors within the same node. This kind of combination of MPI and OpenMP has been applied extensively in the numerical relativity, for example the Cactus infrastructure[62, 63]. Our implementation follows very closely the ideas of Cactus about this combination. As to the mesh refinement boundary issue, we follow [64, 65] closely to deal with the interface of mesh levels. In order to make the initial data consistent with the 3rd order accuracy implemented in the prolongation and restriction during the numerical evolution, we need three initial time levels to do interpolation.

Journal of Physics: Conference Series **330** (2011) 012016 doi:10.1088/1742-6596/330/1/012016

For this end, we adopt the scheme described in [66] to prepare the initial data. With GrACE, the time step for every mesh level is set to be the Courant factor times the finest spatial resolution; while with our new infrastructure, we set the time step following the recipe of [64, 65], i.e., the Berger-Oliger algorithm is implemented for the time updating. Typically, the Courant factor is taken from $\{0.25, 0.5\}$. The results of fixed mesh refinement with GrACE and with our own infrastructure are consistent to each other. In addition, our new infrastructure provides us a "moving box" style mesh refinement. In our code, we assume [16]

$$\frac{dx^i_{BH}}{dt} = -\beta^i(x^i_{BH}), \tag{1}$$

to determine the position of black holes, and then move the corresponding boxes.

As to the spatial derivatives, we have developed 2nd, 4th, 6th and 8th order finite difference routines which include corresponding centered finite difference and lopsided finite difference. Together with different finite difference order, we also adopt Kreiss-Oliger type numerical dissipation as [64, 65]

$$\partial_t u \to \partial_t u + (-1)^{n/2}\sigma \sum_{i=1}^{3} h^{n+1} \frac{D_+^{n/2+1} D_-^{n/2+1}}{2^{n+2}} u, \tag{2}$$

where n is the finite difference order, h is the spatial spacing, $D_+u = (u_{i+1} - u_i)/h$, $D_-u = (u_i - u_{i-1})/h$, and the summation is refer to three spatial dimension. σ is the parameter to control the strength of numerical dissipation. As an example, $D_+^3 D_-^3 u$ can be explicitly written out as

$$\frac{1}{h^6}(u_{i-3} - 6u_{i-2} + 15u_{i-1} - 20u_i + 15u_{i+1} - 6u_{i+2} + u_{i+3}). \tag{3}$$

We choose

$$\begin{aligned} \partial_t \alpha &= -2\alpha K + \beta^i \alpha_{,i}, & (4)\\ \partial_t \beta^i &= \frac{3}{4} B^i + \beta^j \beta^i{}_{,j}, & (5)\\ \partial_t B^i &= \partial_t \tilde{\Gamma}^i - \eta B^i + \beta^j B^i{}_{,j} - \beta^j \tilde{\Gamma}^i{}_{,j}. & (6) \end{aligned}$$

with fixed $\eta = 2$ as our gauge condition in all of this work.

We have also constructed a multi-puncture initial data solver with the spectral method based on the LORENE library[67, 68]. The spatial infinity is included in the computational domain through compactification. Our solver is quite efficient so that we can always solve the full constraint equations for black hole systems of interest, especially the triple black hole systems in this work.

2.2. Perturbational method

As to the evolution part, the stability problem forces us to use traditional outgoing Sommerfeld boundary condition. So the large differences among the distances between black holes sets a nontrivial obstacle for multi black hole system simulation. As to the BBH system located in the supermassive black hole's potential, we take the advantage of the specific configuration and propose a perturbational method to deal with it. As to the effect of the background gravitational potential, we deal with it as a perturbation considering the gravitational field induced by the potential as weak compared with the BBH system. Specifically, we let our computational domain cover only the part near the binary. So the boundary of our computational domain is definitely non-flat. In order to take the effects of the third black hole into account, we must modify the usual Sommerfeld boundary condition. Our recipe is first to take away of the potential part introduced by the third black hole, then implement the Sommerfeld boundary condition, and finally add back the background potential part. It seems that our operation is only a linear operation although, the modification is essentially due to the nonlinear property of Einstein's equation. Of course when the third black hole becomes father and farther away, we expect the behavior of the perturbation term to approach linearity when the potential background becomes weak enough.

From the construction of our perturbational method, it can be seen that it is also possibly useful to other systems, for example, black hole systems in the asymptotically de Sitter space which represents a possible universe background. Taking the effect of cosmology on the BBH systems into consideration, it is interesting to investigate more on this problem. This will be treated in our future work.

3. Binary black hole simulations

3.1. Head-on collision simulation

We have compared the results of head-on collision with several short separation as in Fig.6 of [57], by using GrACE and our new infrastructure respectively to implement fixed mesh refinement. The result is almost the same. To test our moving box part code, we enlarge the separation from 2.303 to 8.

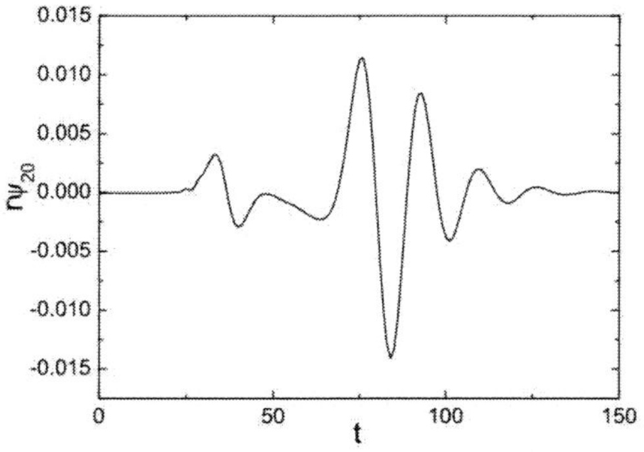

Figure 1. Simulation of two spinless black holes head on to each other along z axis and the real part of $r\Psi_{4,20}$ is plotted respect to time. The consistence with Fig.1 of [69] can be seen but note the different sign in the calculation of Ψ_4.

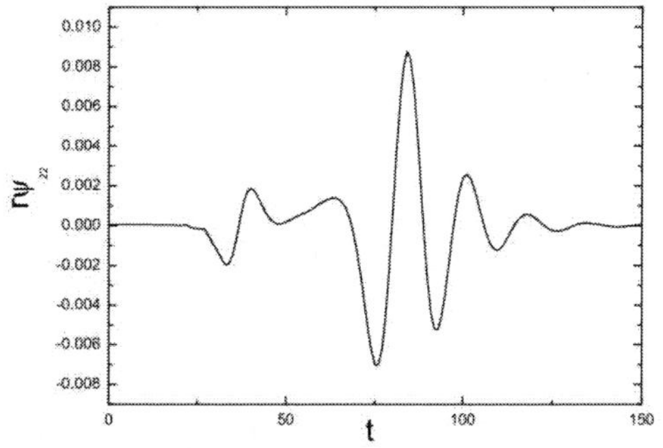

Figure 2. Simulation of two spinless black holes head on to each other along y axis and the real part of $r\Psi_{4,22}$ is plotted respect to time. The consistence with fig.4 of [38] can be seen but note the different factor of 2 in the calculation of Ψ_4.

Firstly we put two black holes at $(0, 0, \pm 4)$. This setting is same with the one in [69]. The octant symmetry is adopted for saving computation time. The outer boundary is at 128. We use 7 levels to do the simulation, and domains for each levels are initially set as

$$\begin{aligned}
0: &\ (0:128, 0:128, 0:128) \\
1: &\ (0:64, 0:64, 0:64) \\
2: &\ (0:32, 0:32, 0:32) \\
3: &\ (0:16, 0:16, 0:16)
\end{aligned}$$

$$4: \ (0:8, 0:8, 0:8)$$
$$5: \ (0:4, 0:4, 2:6)$$
$$6: \ (0:2, 0:2, 3:5)$$

where levels #5 and #6 are movable. The finest resolution is $\frac{1}{32}$. Here we set the Courant factor 0.5 and σ the numerical dissipation parameter 0.1 for moving levels while 0.5 for static levels. Under this setting, the gravitational wave Ψ_4 is dominated by $\ell = 2$, $m = 0$ mode. The result is plotted in Fig.1.

Secondly, we simulate the same system but put the two black holes at $(0, \pm 4, 0)$ initially. The grids setting is similar to the above case while putting the moving box along y axis. This simulation is same to one simulation of [38]. The resulted waveform for gravitational radiation is presented in Fig.2.

Theoretically, the results of above two simulations has a specific relation as deduced in Appendix B. In other words, we should have relation

$$C_{20z} = -2C_{20y} = -\frac{4}{\sqrt{6}}C_{2-2y} = -\frac{4}{\sqrt{6}}C_{22y}, \tag{7}$$

where $C_{\ell mi}$ denotes the corresponding ℓ, m components of Ψ_4 of simulation with black holes collision along i axis. The consistency of our result with the theoretical prediction is shown in Fig.3.

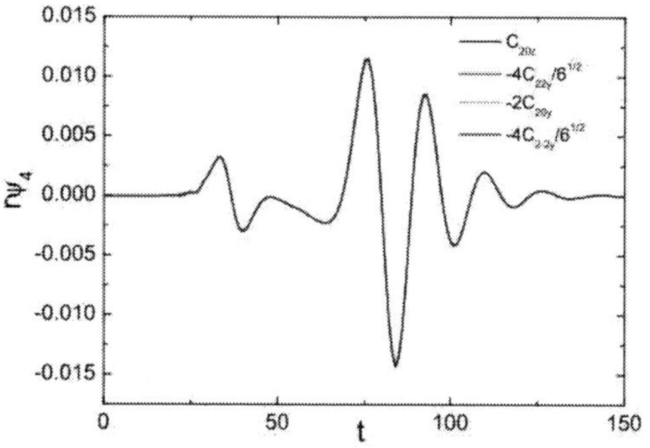

Figure 3. Waveform of head-on collision along z and y axis respectively. They are consistently related by the theoretical prediction (7).

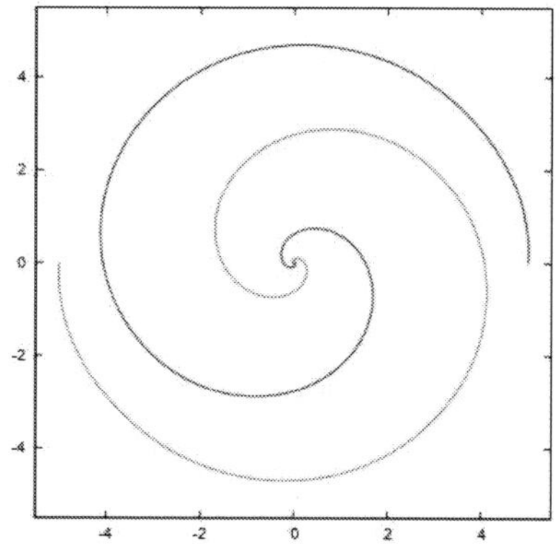

Figure 4. The evolution with Cook & Pfeiffer's initial data. The orbit of the two black holes in x-y plane behaves a well quasi circular shape.

3.2. Inspiraling binary black hole along quasi-circular orbit

Our first simulation of a quasi-circular orbit BBH is based on FMR. Due to the restriction of FMR, we choose the BBH system with some short separation. Specifically, we use the initial data with separation 10 which is available online released by Cook and Pfeiffer [70]. This initial data was also used in the UIUC group[71]. As to the excised region in the original initial data, we use global fitting method with spherical harmonic function to fill this region. Instead of caring about the accuracy, we are only aiming to test our code, therefore, the outer boundary is set at 64 and only 4 levels are used for the simulation. In the equatorial symmetry, the domains

Journal of Physics: Conference Series **330** (2011) 012016 doi:10.1088/1742-6596/330/1/012016

for these levels are set as

$$\begin{aligned} 0: &\ (-64:64,-64:64,0:64)\\ 1: &\ (-32:32,-32:32,0:32)\\ 2: &\ (-16:16,-16:16,0:16)\\ 3: &\ (-8:8,-8:8,0:8). \end{aligned}$$

The finest resolution is a coarse $\frac{1}{8}$, sufficient to test our code. The quasi-circular shape of black holes' orbit can be seen very clearly in fig.4. The resulted gravitational wave is dominated by $l=2$, $m=\pm 2$ modes. Since these two modes are only up to a complex conjugate, we plot only $\ell=m=2$ mode in Fig.5 which is consistent with [71].

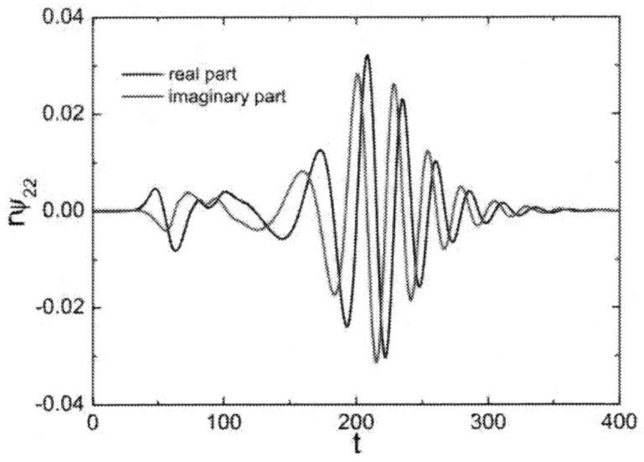

Figure 5. The $\ell=m=2$ mode of GWs detected at $r=40$ with Cook & Pfeiffer's initial data. This result is consistent with [71].

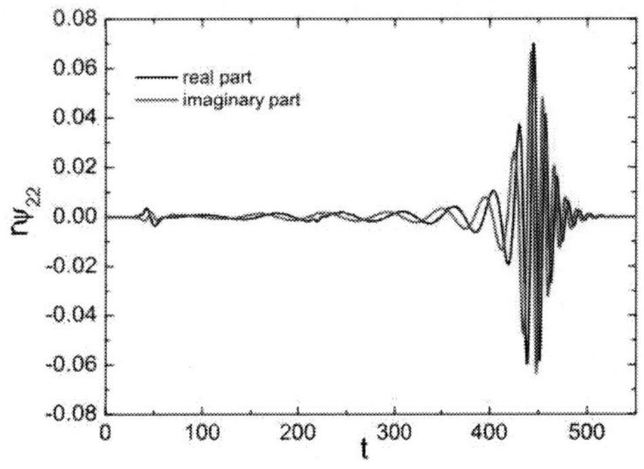

Figure 6. The $\ell=m=2$ mode of GWs detected at $r=40$ with approximate puncture type initial data. This result is consistent with [51].

We also tested our code with approximate initial data deduced in [57, 72, 51] based on puncture type initial data for BBH [73]. We choose the same parameters as in Fig.1 of [51] for the BBH system. Specifically, the punctures have mass parameters $m_1=m_2=0.4891$ and are placed at $x=\pm 4.27629813$ with momenta $p_y=\mp 0.1091832072$. In this simulation, we implement moving box techniques with our new infrastructure. We use 7 levels here and their domains are initially set as

$$\begin{aligned} 0: &\ (-64:64,-64:64,0:64)\\ 1: &\ (-36.2763:36.2763,-32.2763:32.2763,0:32)\\ 2: &\ (-20.2763:20.2763,-20.2763:20.2763,0:16)\\ 3: &\ (-12.2763:12.2763,-12.2763:12.2763,0:8)\\ 4: &\ (-8.2763:-8.2763,-8.2763:8.2763,0:5)\\ 5: &\ (-6.2763:-2.2763,-2:2,0:2)\\ &\ (2.2763:6.2763,-2:2,0:2)\\ 6: &\ (-5.2763:-3.2763,-1:1,0:1)\\ &\ (3.2763:5.2763,-1:1,0:1) \end{aligned}$$

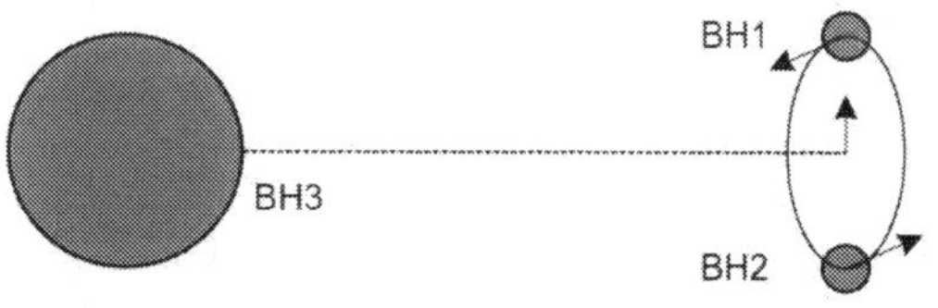

Figure 7. A schematic illustration of the BBH system in the gravitational potential of a third (massive) BH as considered in this work.

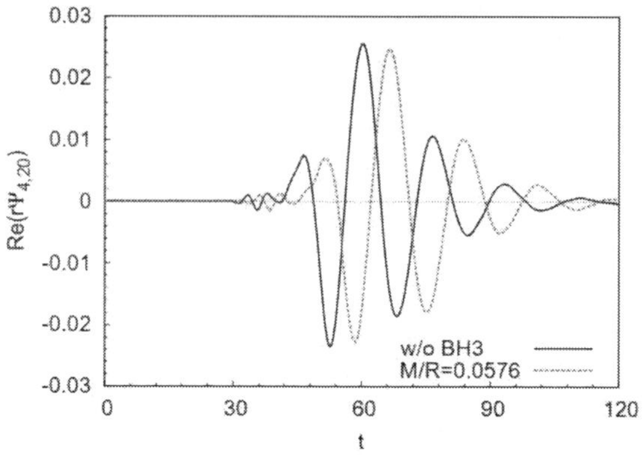

Figure 8. Comparison of the typical gravitational waveforms from the head-on collisions of a small BBH system for the head-on-freefall case. The $\ell = 2$, $m = 0$ mode of the gravitational wave detected at $r = 40$ is plotted. The red-dashed (black-solid) line represents the case with (without) the third BH. From this figure we can see that the gravitational potential of the third BH causes mainly shifts in time, decreases in amplitude, and prolongation of the wavelength in the gravitational waveform.

where levels #5 and #6 have two boxes respectively and are movable. The finest resolution is $\frac{1}{32}$. Similar to above case, we plot $\ell = m = 2$ mode of gravitational wave in Fig.6. Our result is consistent with that presented in Fig.1 of [51].

4. Triple black hole simulations

The considered triple black hole system has an obvious hierarchical structure as shown in Fig.7. The two smaller black holes form a small BBH system (denoted as SB), while this binary system and the third, massive, black hole form a bigger binary system (denoted as BB). From the point view of eccentricity e, the head-on collision ($e = 1$) and the (quasi-)circular-orbiting ($e = 0$) stand for the two limits of the astrophysical configurations existing in the universe. Here we would like to investigate the possible combinations of these limiting cases in order to offer some clues for understanding this kind of astrophysical systems. The aforementioned hierarchical structure indicates four possible limiting cases. They are the head-on-freefall case (both SB and BB are in the process of a head-on collision), the head-on-orbiting case (SB is under a head-on collision while BB is in a (quasi-)circular orbit), the inspiraling-freefall case (SB is in a (quasi-)circular orbit while BB is under a head-on collision), and the inspiraling-orbiting (both SB and BB are in a (quasi-)circular orbit). We will present the results for these four scenarios in the following subsections.

4.1. Head-on-freefall case

The results of a small BBH's head-on collisions under the background of a third (massive) BH are compared with our earlier results (see [57] for details). Consider the small BBH's head-on collision starting from the ISCO. The mass parameters of the small binary system are chosen

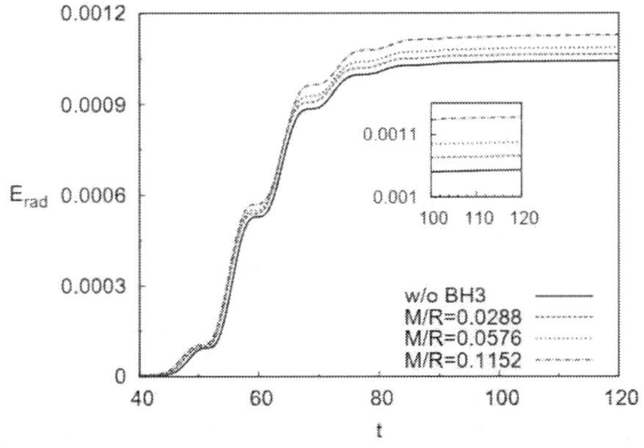

Figure 9. Accumulated radiated energy with respect to time for different gravitational potential strengths. A stronger gravitational potential results in more energy radiation.

to be $m_1 = m_2 = 0.5$, and their positions are $(0, 0, \pm1.1515)$. The settings are the same as in [57]. The third BH with mass M is placed at $(-R, 0, 0)$. In this and the next subsections we will vary M while fixing the coordinate distance at $R = 1000$.

Fig.8 shows a typical example, comparing the gravitational waveform from the small BBH's head-on collision under the influence of a third BH with the case of an isolated BBH head-on collision. From this figure we can see that the gravitational potential of the third BH causes a time delay, a decrease of the amplitude, and a prolongation of the wavelength in the gravitational waveform. It is easy to understand that the decrease in the amplitude mainly comes from the gravitational redshift caused by the potential of the third BH. However, the time delay comes from three factors: (a) the delayed merger process in which the gravitational wave is produced near the source; (b) the prolonged proper distance for propagation of the gravitational wave from the source to the detector; (c) the change in the coordinate time. It can be understood that the last one is simply caused by the change of gauge due to the third BH.

As mentioned earlier, the redshift effect due to the existence of the third BH will decrease the amplitude as well as prolong the wavelength of the gravitational waveform. The amplitude decreases as the gravitational potential of the third BH increases. And the wavelength increases as the gravitational potential of the third BH increases.

We investigate the energy radiation carried by the gravitational wave and show the result in Fig.9. The plot shows that a stronger gravitational potential results in more energy radiation. This can be understood as follows: when the small BBH system is located in the gravitational potential of a third BH, the energy radiated during the head-on collision is enhanced by the potential. In our perturbational treatment, the curved background is fixed and thus there is no back-reaction between the small BBH and the third BH. In a more realistic case, the back-reaction will speed up the head-on collision of the big binary system (which includes the small BBH and the third BH.)

The behavior of a three-body system has been analyzed in a Newtonian framework in [75]. It was found that the system could introduce higher-order modes of the gravitational wave. We found a similar phenomenon in this work. Besides distorting the lower-order modes, the existence of the third BH will also excite higher-order modes in the small BBH's gravitational radiation, especially, the $\ell = 3$ modes. The left panel of Fig.10 shows the waveform of the $\ell = 3$, $m = 1$ mode of the gravitational radiation excited in the current case with respect to the different strength of the gravitational potential. The right panel of Fig.10 shows the amplitudes of the $\ell = 3$, $m = 1$ mode of the gravitational wave, which demonstrates nonlinear growth with

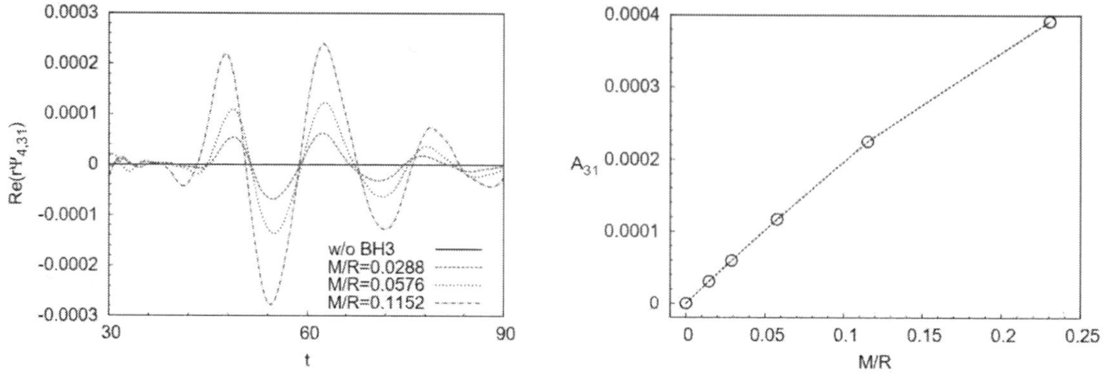

Figure 10. Left panel: Comparison of the gravitational waveforms of the $\ell = 3$, $m = 1$ mode for different potential strengths of the third BH. Right panel: Amplitude of $\Psi_{4,31}$ with respect to different potential strengths, showing a nonlinear growth pattern.

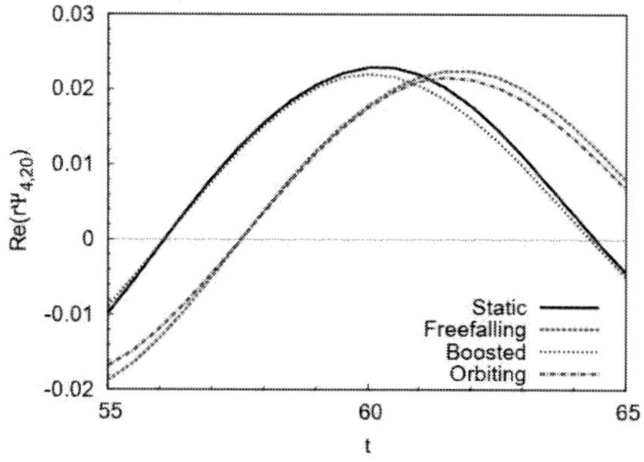

Figure 11. Comparison of the $\mathrm{Re}(r\Psi_{4,20})$ in four different situations. The solid line indicates the waveform for the head-on collision of an isolated BBH; the (red) dashed line indicates the one for the head-on collision of a BBH freefalling towards a third large BH; the (blue) dotted line indicates the one for the head-on collision of a boosted isolated BBH; the (green) dot-dashed line indicates the one for the head-on collision of a BBH orbiting around a third large BH. The results indicate that the relativistic Doppler effect on the behavior of the gravitational waveform is much less important than the gravitational redshift effect in the head-on-orbiting case.

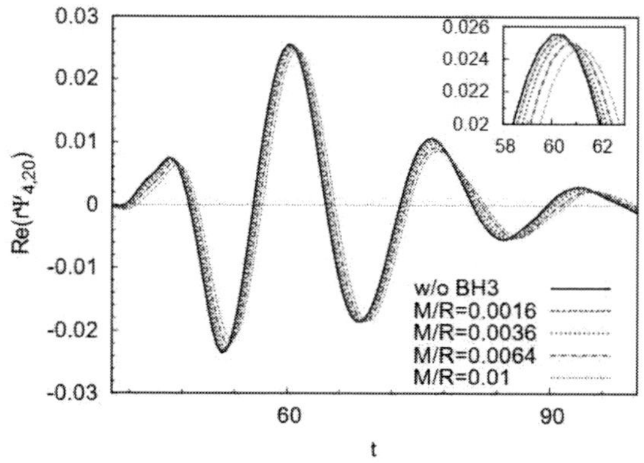

Figure 12. Comparison of the gravitational waveforms for different potential strengths of the third BH.

respect to the strength of the third BH's gravitational potential. In principle it is possible for the nonlinear growth of the $\ell = 3$, $m = 1$ mode to be used as a signature to distinguish a three-BH system from a BBH system (to some extent). However, the phenomenon could also complicate the identification of the source of the gravitational wave. Without a further detailed understanding on the waveform pattern, the gravitational wave emitted from a three-BH system as in this work, could be misidentified as one from an unequal-mass BBH system in which the higher-order modes of gravitational wave also exist.

4.2. Head-on-orbiting case

In this subsection, we study the head-on collision of a BBH while the BBH system moves along a quasi-circular orbit around a third large BH. Initially, the orbiting velocity of the center of mass (COM) of the BBH is set according to the Newtonian gravity, i.e., $v^2 \approx M/R$, in order to get a circular orbit. For the third large BH, we fix its coordinate distance at $R = 1000$ as in the previous subsection, and change the gravitational potential M/R by varying the mass parameter M. Given the linear momentum, mass parameter and the position of these BHs, the puncture initial data could be constructed with the solver described in Section 2.1.

There are at least two effects on the gravitational waveform of the BBH orbiting in the gravitational potential well of a third large BH: the relativistic Doppler effect and the gravitational redshift. In order to identify the importance of these two effects on the time delay in the waveform in the current case, we compare the waveforms obtained from the following four scenarios: (i) the head-on collision of an isolated BBH; (ii) the head-on collision of a BBH freefalling towards a third large BH; (iii) the head-on collision of a boosted isolated BBH; (iv) the head-on collision of a BBH orbiting around a third large BH. Here the mass parameters of the third BH in cases (ii) and (iv) are both $M = 14.4$, which is larger than all other mass parameters used in this subsection, and the boosting velocity in (iii) is set to be the same as the orbiting velocity in (iv) the magnitude of which is 0.12 according to $v^2 \approx M/R$ and $R = 1000$. The result is shown in Fig.11. In the plot we can see that the $\mathrm{Re}[r\Psi_{4,20}]$ waveforms of (i) and (iii) are close to each other, and the waveforms of (ii) and (iv) are close to each other. This indicates that the gravitational redshift of the third large BH should have the major effect on the time delay of the gravitational wave in the merging process of a BBH. Furthermore, the relativistic Doppler effect is shown to have little influence on the time delay of the waveform. The result allows us to neglect the relativistic Doppler effect from the orbiting velocity of the COM of a BBH around a third large BH, and to focus mainly on its gravitational redshift effect.

In this case, in order not to let the small BBH move too fast, to stay inside the computational domain and to maintain the necessary accuracy during the generation of gravitational wave, the gravitational potential strength of the third BH is set to be smaller than that in the previous subsection. The numerical results for this case indicate that all the phenomena including the time delay of the waveform, the decrease of the wave amplitude, the prolongation of the wavelength, the enhancement of the energy radiated, and the excitation of the higher-order modes are qualitatively similar to the ones in the previous case. Therefore we only show here the decrease in amplitude of the gravitational waveform and the coordinate time delay of the highest peak of the $\Psi_{4,20}$ mode with respect to different potential strengths; see Fig.12. It can be seen that the decrease in amplitude of the waveform is less than that obtained in the previous subsection due to the smaller gravitational potential strength of the third BH. Just as in the previous subsection, the orientation of the small BBH barely affects the result because of the weakness of the tidal force from the third BH.

4.3. Inspiraling-freefall case

In this and the next subsections, we will consider the effect on an inspiraling BBH of a third large BH. The two BHs initially have the irreducible masses $m_1 = m_2 \approx 0.5$ at $(0, 0, \pm\frac{D}{2})$. The third BH with a mass parameter of $M = 5 \times 10^4$ is located at $(R, 0, 0)$.

In this subsection, we study the inspiral-to-merger evolution of a BBH while the COM of the BBH freely falls towards the third large BH. The left panel of Fig.13 shows the evolution of a typical inspiraling trajectory of the BBH, with the initial coordinate separation $D = 4.5$,

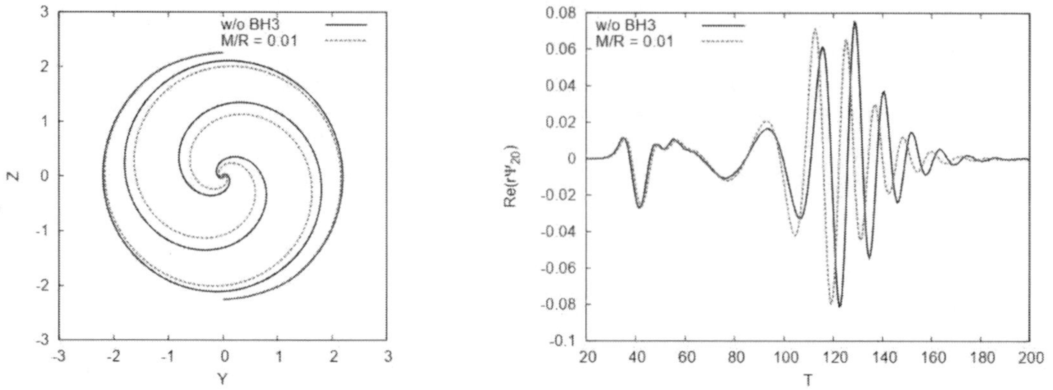

Figure 13. Left panel: Comparison of the typical inspiral trajectory of a BBH under the influence of a third BH with the one without the influence in the inspiraling-freefall case. The solid curve indicates the trajectory of an isolated BBH and shows a quasi-circular inspiral shape. The (red) dashed curve indicates the trajectory of the BBH in the gravitational potential well of the third BH with $M = 5 \times 10^4$ which is located at $(R = 5 \times 10^6, 0, 0)$. There is a change in the eccentricity of the (red) dashed curved compared to the solid line. Right panel: Comparison of the typical gravitational waveforms from the inspiral of the small BBH for the inspiraling-freefall case. The real component of $r\Psi_{4,20}$ is plotted. The red dashed (black solid) line indicates the case with (without) the third BH. The detector is located at $r = 40$. It can be seen that the gravitational potential of the third BH causes a tiny time delay in the spurious part and the effect of a time advance in the major part of the waveform.

affected by the gravitational potential of a third large BH located at $(R = 5 \times 10^6, 0, 0)$. From the plot we can see that the gravitational potential of the third large BH leads to an increase in the eccentricity of the trajectory of the BBH, compared to that of an isolated BBH. The right panel of Fig.13 shows a typical example of the difference in the gravitational waveforms between those of the BBH inspiraling under the influence of a third large BH and the inspiraling of an isolated BBH. In this figure it can be seen that the gravitational potential of the third large BH causes a time advance, a prolongation of the wavelength, and a "surprising" variation in the amplitude of the gravitational waveform.

As explained in Section 4.1, the time delay for the existence of a third large BH in a head-on collision arises from three factors: the delayed merger process, the prolonged proper distance, and the change of the coordinate time. For an inspiraling process of a BBH, there exists another effect, i.e., an increase in the eccentricity, as shown in the left panel of Fig.13. It is obvious that the increase in the eccentricity is mainly caused by the tidal effect of the third large BH. As a BBH system emits more gravitational radiation in an elliptical orbit than in a circular orbit [83, 84], the increase of eccentricity is expected to expedite the merger process of the BBH. Therefore, the time shift affecting the inspiraling process is the result of competition between the increase in eccentricity and the others. Here we look at this in more detail.

From the right panel of Fig.13 we can see that the waveform is affected by both the time delay and time advance phenomena: the spurious radiation part of the waveform is delayed, while the burst part is advanced. The beginning of the waveform, near $t = 50$, is a result of the spurious radiation hidden in the initial data. Thus, this part is somehow independent of the merging process of a BBH. However, we can still see a time delay, because of the prolonged proper distance from the BBH to the detector, at $r = 40$, due to the existence of the third BH. In the main part of the merging process, after a coordinate time of roughly $t = 100$, we can clearly see the time advance effect. This is due to the increase in eccentricity under the influence of the gravitational potential of the third large BH. Obviously the effect of the eccentricity increase overtakes other time delay factors in the competition.

Besides the effect of time advance, there is another feature shown in the waveform. If we take the highest peak in the wave-train of Ψ_4 roughly as the time of of merging [18], this divides the waveform in the gravitational wave-train into two kinds of different behaviors. Before the merger, the wave amplitude of Ψ_4 becomes larger as the gravitational potential of the third large BH becomes larger. This is because the increase in eccentricity of the BBH's trajectory is the dominant effect in speeding up the inspiral-to-merger process compared to the isolated case. This results in stronger gravitational radiation. After the merger, the wave amplitude of Ψ_4 turns out to become smaller as the gravitational potential of the third BH becomes larger. The decrease in the amplitude of the gravitational waveform mainly arises from the stronger redshift effect due to the existence of the third BH. These phenomena are illustrated in the right panels of Fig.13.

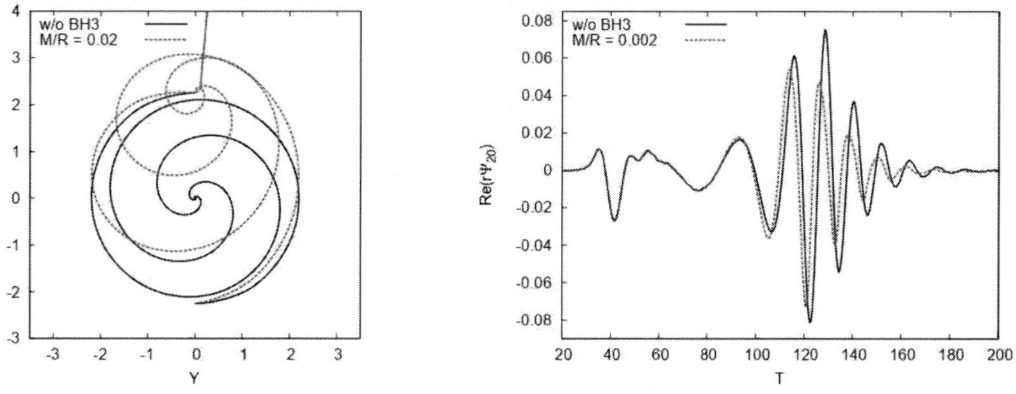

Figure 14. Left panel: Comparison of the typical inspiral trajectory of a BBH under the influence of a third large BH with one without that influence in the inspiraling-orbiting case. The (red) dashed curve indicates the trajectory of the BBH in the gravitational potential well of the third BH with $M = 5 \times 10^4$ which is located at $(R = 2.5 \times 10^7, 0, 0)$. The solid curve indicates the trajectory of an isolated BBH and shows a quasi-circular inspiral shape. Right panel: Comparison of the typical gravitational waveforms from the inspiraling of a BBH for the inspiraling-orbiting case. The real component of $r\Psi_{4,20}$ is plotted. The red-dashed (black-solid) line indicates the case with (without) the third large BH. It can be seen that, besides the decrease in the wave amplitude, the gravitational potential of the third large BH causes a tiny time delay in the spurious part and a time advance in the major part of the waveform.

Compared with the head-on collision of a BBH as discussed in Section 4.1, the higher-order mode effect due to the existence of a third large BH in the inspiraling process is more evident. The higher-order mode, $\mathrm{Re}(r\Psi_{4,33})$, shows a nonlinear dependence on the strength of the gravitational potential. This result is consistent with those detailed in the previous two subsections: This nonlinear phenomenon could be a signature distinguishing the three-BH system from the BBH system, as well as complicating the identification of the source of a gravitational wave.

4.4. Inspiraling-orbiting case
In this subsection, we investigate the inspiral-to-merger process of a BBH orbiting around a third large BH. Initially, the COM of the BBH moves around the third large BH in a quasi-circular

Journal of Physics: Conference Series **330** (2011) 012016 doi:10.1088/1742-6596/330/1/012016

orbit. Therefore, the velocity of the COM of the BBH can be approximated using Newtonian mechanics. Given the linear momentum and position parameters of the *three* BHs, we use the LORENE library to obtain the initial data by numerically solving the Hamiltonian constraint with the puncture scheme. Similar to the above subsection, we set the two BHs with irreducible mass $m_1 = m_2 \approx 0.5$ and locate them at $(0, 0, \pm\frac{D}{2})$ initially. We put the third BH at $(R, 0, 0)$. In order to get a circular orbit for the BBH around the third large BH, we use the quantity $\sqrt{M/R}$ to evaluate the velocity of the COM of the BBH with respect to the third large BH. This velocity is then added vectorially to that of each individual BH of the BBH. For the third large BH, we fix its mass parameter $M = 5 \times 10^4$ as in the previous subsection while varying its coordinate distance R.

The left panel of Fig.14 shows a typical trajectory of the BBH with the initial coordinate separation $D = 4.5$ for the current case, where the third large BH is located at $(R = 2.5 \times 10^7, 0, 0)$. It can be understood from the plot that the COM of the BBH moves roughly along a Newtonian circular orbit around the third large BH. The trajectory is in fact the combination of the orbit of the COM of the BBH and the BBH's inspiraling around the COM. Compared with the trajectory in Fig.13, it can be seen that the trajectory in Fig.14 (for the current case) is quite deformed compared to the one for an isolated BBH. This is mainly due to the BBH's orbital velocity around the third large BH.

The typical resulting gravitational waveform of a BBH in the current case is presented in the right panel of Fig.14. We found that the resultant effect of the third large BH on the waveform is stronger in the current case than in the inspiraling-freefall cases. We can see in the plot that the gravitational potential of the third large BH causes a time delay in the spurious radiation but a time *advance* in the major part of the waveform. This is similar to the previous inspiraling-freefall case. The time advance of $\mathrm{Re}(r\Psi_{4,20})$ due the existence of the third large black hole is clear and different from its spurious radiation part. This phenomenon is similar to the one in the inspiraling-freefall case. The amplitude of the waveform decreases more in the wave-train as the gravitational potential of the third BH becomes larger.

Similar to the previous case, as the initial separation of BBH increases, the phase change of the gravitational wave, due to the effect of the third large BH, transits from time-delay to time-advance. As in the previous case, a possible explanation for the change in the time difference of the inspiral-to-merger process of a BBH under the influence of a third large BH compared with the one of an isolated BBH, arises from the competition between the eccentricity effect and the gravitational redshift effect. Compared to the previous (inspiraling-freefall) case in which the $\ell = m = 3$ mode of Ψ_4 is dominant among the higher modes, the $\ell = 3$, $m = 0$ modes of Ψ_4 is dominant in the current case. The orbiting of the BBH along the x-z plane accounts for this effect. In addition, the amplitude of the $\ell = 3$, $m = 0$ mode of $\mathrm{Re}(\Psi_4)$ increases nonlinearly.

5. Summary and discussion

Motivated by the fact that all most BBH systems are located in the gravitational potential of a supermassive black hole hosted in the center of galaxy, we investigated the effect of this potential on the dynamics of the BBH. Special attention is paid to the gravitational wave form released from the binary system. To deal with this problem, we proposed a new perturbation method to take the effect of the supper-massive black hole into consideration. In our perturbation method, we ignore the back reaction of the BBH system on the supermassive black hole and look the third black hole as a background to the BBH system.

The effect of the gravitational potential includes: 1). delaying or advancing the merging process of the BBH which depends on the relative motion of the binary respect to the supermassive black hole; 2). making the gravitational wave signal propagate slower; 3). red-shifting the resulted gravitational wave form; 4). producing significantly the $\ell = m = 3$ mode of the gravitational wave form. Except the amplitude of $\ell = m = 3$ mode, all other quantities depend on the gravitational potential $\frac{M}{R}$ linearly which implies the validity of our perturbation method. And in most realistic cases the gravitational potential is much weaker than the ones considered in this paper. So our perturbation method is expected to be valid for almost realistic problems. In some sense the effect 4) is an evidence of the conjecture proposed in paper [75] in the full GR region. And it can provide valuable information for gravitational wave astronomy to distinguish if the detected BBH system is isolated or located in a potential of a supermassive black hole.

The code used in this work is an extension of our previous work. The consistence of the results obtained in this work also validate our code for BBH simulations. With this code we are ready to research other problems involved in black hole systems in the near future.

Acknowledgments

This work was supported in part by the National Science Council under the grants NSC97-2112-M-006-008 and NSC98-2112-M-006-007-MY2. Z. Cao was supported by the NSFC (No. 10731080 and No. 11005149). This work was also supported in part by the National Center of Theoretical Sciences. We are grateful to the National Center for High-performance Computing for the use of their computer time and facilities, and to the Academia Sinica Computing Center for providing computing resource.

Appendix A. Calculation of Ψ_4

We update our method of calculating Ψ_4 from [57] to a new one. Associated with the Cartesian coordinates used in our simulation, we have a spherical coordinate (r, θ, ϕ). We denote their orthonormal bases as $(\hat{r}, \hat{\theta}, \hat{\phi})$. In addition we denote the time-like unit vector normal to the constant coordinate time slice $\hat{N}$. Then the null tetrad used in the calculation of Ψ_4 read as

$$l \equiv \frac{1}{\sqrt{2}}(\hat{N} + \hat{r}), \tag{A.1}$$

$$n \equiv \frac{1}{\sqrt{2}}(\hat{N} - \hat{r}), \tag{A.2}$$

$$m \equiv \frac{1}{\sqrt{2}}(\hat{\phi} - i\hat{\theta}). \tag{A.3}$$

With

$$E_{ij} \equiv R_{ij} - K_{ik}K_j{}^k + KK_{ij}, \tag{A.4}$$

$$B_{ij} \equiv \hat{r}^k[D_k K_{ij} - D_{(i}K_{j)k}], \tag{A.5}$$

Ψ_4 can be calculated through

$$\Psi_4 = (B_{ij} - E_{ij})\bar{m}^i\bar{m}^j. \tag{A.6}$$

Appendix B. Transformation of Ψ_4

Now we consider relation of two head-on collision simulations, one is collision along z axis, the other is along y axis. These two simulations correspond to the same physical process. We use the Cartesian coordinate corresponds to the z axis collision simulation to do the analysis. Denote the corresponding spherical coordinate of z axis simulation (r, θ, ϕ), while y axis simulation (r', θ', ϕ'). Then we have

$$x = \sin\theta\cos\phi = \cos\theta', \tag{B.1}$$

$$y = \sin\theta\sin\phi = \sin\theta'\cos\phi', \tag{B.2}$$

$$z = \cos\theta = \sin\theta'\sin\phi', \tag{B.3}$$

Journal of Physics: Conference Series **330** (2011) 012016 doi:10.1088/1742-6596/330/1/012016

and

$$\begin{aligned}
\hat{\phi} &= \frac{1}{\sqrt{x^2+y^2}}\begin{pmatrix} -y \\ x \\ 0 \end{pmatrix}, \\
\hat{\theta} &= \frac{1}{\sqrt{x^2z^2+y^2z^2+(x^2+y^2)^2}}\begin{pmatrix} xz \\ yz \\ -(x^2+y^2) \end{pmatrix}, \\
\hat{\phi}' &= \frac{1}{\sqrt{y^2+z^2}}\begin{pmatrix} 0 \\ -z \\ y \end{pmatrix}, \\
\hat{\theta}' &= \frac{1}{\sqrt{x^2y^2+x^2z^2+(y^2+z^2)^2}}\begin{pmatrix} -(y^2+z^2) \\ xy \\ xz \end{pmatrix}.
\end{aligned}$$

$(\hat{\theta}, \hat{\phi})$ and $(\hat{\theta}', \hat{\phi}')$ are only up to a rotation

$$\begin{aligned}
\hat{\phi}' &= \frac{-1}{\sqrt{\sin^2\theta\sin^2\phi+\cos^2\theta}}(\cos\theta\cos\phi\hat{\phi}+\sin\phi\hat{\theta}), \\
\hat{\theta}' &= \frac{-1}{\sqrt{\sin^2\theta\sin^2\phi+\cos^2\theta}}(-\sin\phi\hat{\phi}+\cos\theta\cos\phi\hat{\theta}).
\end{aligned}$$

With notation $m \equiv \frac{1}{\sqrt{2}}(\hat{\phi} - i\hat{\theta})$ and $m' \equiv \frac{1}{\sqrt{2}}(\hat{\phi}' - i\hat{\theta}')$, we have $m' = e^{i\lambda}m$ with

$$\cos\lambda = -\frac{\cos\theta\cos\phi}{\sqrt{\sin^2\phi+\cos^2\theta\cos^2\phi}}, \tag{B.4}$$

$$\sin\lambda = -\frac{\sin\phi}{\sqrt{\sin^2\phi+\cos^2\theta\cos^2\phi}}. \tag{B.5}$$

Denoting the result of z axis simulation Ψ_4 while x axis simulation Ψ_4', we have $\Psi_4' = e^{-2i\lambda}\Psi_4$ considering the spin weight -2 property of Ψ_4. From numerical result we know Ψ is dominated by ${}_{-2}Y_{20}$ mode. so we have

$$\begin{aligned}
\Psi_4' &= e^{-2i\lambda}C(t)_{-2}Y_{20}(\theta,\phi) \\
&= C(t)[-\frac{1}{2}{}_{-2}Y_{20}(\theta',\phi') - \frac{\sqrt{6}}{4}{}_{-2}Y_{2-2}(\theta',\phi') - \frac{\sqrt{6}}{4}{}_{-2}Y_{22}(\theta',\phi')].
\end{aligned} \tag{B.6}$$

References

[1] Astone P *et al.*, Phys. Rev. D 47, 362 (1993).
[2] Mauceli E *et al.*, Phys. Rev. D 54, 1264 (1996).
[3] Blair D G *et al.*, Phys. Rev. Lett. 74, 1908 (1995).
[4] Cerdonio M *et al.*, 1st Edoardo Amaldi Conf. on Gravitational Wave Experiments (Frascati, 14???C7 June) (1997).
[5] A. Abramovici *et al.*, Science **256**, 325 (1992).
[6] B. Caron *et al.*, Class. Quant. Grav., **14**, 1461 (1997).
[7] H. Lück, Class. Quant. Grav., **14**, 1471 (1997).
[8] M. Ando *et al.*, Phys. Rev. Lett. **86**, 3950 (2001).
[9] K. Danzmann *et al.*, Pre-Phase A report, 2nd ed. (1998).

[10] Xuefei Gong, "A Feasibility Study of the Advanced LISA Mission", talk given on the 8th International LISA Symposium (2010). http://www.stanford.edu/group/lisasymposium/LISA8_GONG.pdf.
[11] X. Gong, S. Xu, S. Bai, Z. Cao, *et al.*, Class. Quant. Grav., **28**, 094012 (2011).
[12] B. Aylott *et al.*, Class. Quant. Grav., **26**, 114008 (2009); B. Aylott *et al.*, Class. Quant. Grav., 26, 165008 (2009).
[13] T. Damour and A. Nagar, Phys. Rev. D **76**, 064028 (2008); A. Buonanno *et al.*, Phys. Rev. D **76**, 104049 (2007); Y. Pan *et al.*, Phys. Rev. D **77**, 024014 (2008); P. Ajith *et al.*, Class. Quant. Grav., **24**, S689 (2007); P. Ajith *et al.*, Phys. Rev. D **77**, 104017 (2008).
[14] E. Lorenz, Trans. N. Y Acad. Sci. 25, 409 (1963); H. Poincare, Les Methodes Nouvelles de la Mecanique Celeste. 3 volumes (1892). English translation, New Methods of Celestial Mechanics, History of Modern Physics and Astronomy 13, Amer. Inst. Phys. (1993); S. Jhingan, N. Dadhich and P. Joshi, Phys. Rev. D 63, 044010 (2001).
[15] F. Pretorius, Phys. Rev. Lett. **95**, 121101 (2005).
[16] M. Campanelli, *et al.*, Phys. Rev. Lett. **96**, 111101 (2006).
[17] J.G. Baker, *et al.*, Phys. Rev. Lett. **96**, 111102 (2006).
[18] A. Buonanno, G. Cook, and F. Pretorius, Phys. Rev. D **75**, 124018 (2007).
[19] J. Baker, J. van Meter, S. McWilliams, J. Centrella, and B. Kelly, Phys. Rev. Lett. **99**, 181101 (2007).
[20] M. Boyle, D. Brown, L. Kidder, A. Mroue, H. Pferffer, M. Scheel, G. Cook, and S. Teukolsky, Phys. Rev. D **76**, 124038 (2007).
[21] B. Vaishnav, I. Hinder, F. Herrmann, and D. Shoemaker, Phys. Rev. D **76**, 084020 (2007).
[22] J. Baker, S. McWilliams, J. van Meter, J. Centrella, D. Choi, B. Kelly, and M. Koppitz, Phys. Rev. D **75**, 124024 (2007).
[23] P. Ajith *et al.*, Class. Quantum Grav. 24, S689 (2007).
[24] A. Buonanno *et al.*, Phys. Rev. D **76**, 104049 (2007).
[25] T. Baumgarte, P. Brady, J. Creighton, L. Lehner, F. Pretorius, and R. DeVoe, Phys. Rev. D **77**, 084009 (2008).
[26] Y. Pan *et al.*, Phys. Rev. D **77**, 024014 (2008).
[27] S. Husa, M. Hannam, J. Gonzalez, U. Sperhake, and B. Brugmann, Phys. Rev. D **77**, 044037 (2008).
[28] M. Hannam, S. Husa, U. Sperhake, B. Brugmann, and J. Gonzalez, Phys. Rev. D **77**, 044020 (2008).
[29] B. Krishnan, C. O. Lousto, and Y. Zlochower, Phys. Rev. D 76, 081501 (2007).
[30] J. Baker *et al.*, Astrophys. J. 653, L93 (2006).
[31] C. Sopuerta, N. Yunes, and P. Laguna, Phys. Rev. D 74, 124010 (2006).
[32] J. Gonzalez, U. Sperhake, B. Bruegmann, M. Hannam, and S. Husa, Phys. Rev. Lett. 98, 091101 (2007).
[33] C. Sopuerta, N. Yunes, and P. Laguna, Astrophys. J. 656, L9 (2007).
[34] F. Herrmann, I. Hinder, D. Shoemaker, and P. Laguna, AIP Conf. Proc. 873, 89 (2006).
[35] F. Herrmann, I. Hinder, D. Shoemaker, and P. Laguna, Class. Quantum Grav. 24, S33 (2007).
[36] M. Campanelli, C. Lousto, Y. Zlochower, and D. Merritt, Astrophys. J. 659, L5 (2007).
[37] M. Koppitz *et al.*, Phys. Rev. Lett. 99, 041102 (2007).
[38] D. Choi, B. Kelly, W. Boggs, J. Baker, J. Centrella, and J. Meter, Phys. Rev. D, **76**, 104026 (2007).
[39] J. Gonzalez, M. Hannam, U. Sperhake, B. Brugmann, and S. Husa, Phys. Rev. Lett. 98, 231101 (2007).
[40] J. Baker *et al.*, Astrophys. J. 668, 1140 (2007).
[41] M. Campanelli, C. Lousto, Y. Zlochower, and D. Merritt, Phys. Rev. Lett. 98, 231102 (2007).
[42] E. Berti *et al.*, Phys. Rev. D 76, 064034 (2007).
[43] W. Tichy and P. Marronetti, Phys. Rev. D 76, 061502 (2007).
[44] F. Herrmann, I. Hinder, D. Shoemaker, P. Laguna, and R. Matzner, Phys. Rev. D 76, 084032 (2007).
[45] M. Campanelli, C. Lousto, and Y. Zlochower, Phys. Rev. D 74, 084023 (2006); C. Lousto, and Y. Zlochower, Phys. Rev. D 76, 041502 (2007); M. Campanelli, C. Lousto, Y. Zlochower, B. Krishnan, and D. Merritt, Phys. Rev. D 75, 064030 (2007).
[46] L. Boyel, M. Kesden, and S. Nissanke, Phys. Rev. Lett., **100**, 151101 (2008); L. Boyel and M. Kesden, Phys. Rev. D, **78**, 024017 (2008).
[47] W. Tichy and P. Marronetti, Phys. Rev. D 78, 081501 (2008).
[48] P. Marronetti, W. Tichy, B. Brugmann, J. Gonzalez, and U. Sperhake, Phys. Rev. D 77, 064010 (2008).
[49] L. Rezzolla, E. Barausse,E. Dorband, D. Pollney, C. Reisswig, J. Seiler and S. Husa, Phys. Rev. D 78, 044002 (2008); L. Rezzolla, P. Diener, E. Dorband, D. Pollney, C. Reisswig, E. Schnetter, and J. Seiler, Astrophys. J. 674, L29 (2008); L. Rezzolla, Class. Quantum Grav. 26, 094023 (2009).
[50] M. Campanelli, C.O. Lousto, and Y. Zlochower, Phys. Rev. D 77, 101501 (2008).
[51] C.O. Lousto, and Y. Zlochower, Phys. Rev. D**77**, 024034 (2008).
[52] U. Sperhake, V. Cardoso, F. Pretorius, E. Berti, and J. Gonzalev, Phys. Rev. Lett., **101**, 161101 (2008); U. Sperhake, V. Cardoso, F. Pretorius, E. Berti, T. Hinderer, and N. Yunes, Phys. Rev. Lett., **103**, 131102

Journal of Physics: Conference Series **330** (2011) 012016 doi:10.1088/1742-6596/330/1/012016

(2009).
[53] B. Krishnan, C. Lousto, and Y. Zlochower, Phys. Rev. D 76, 081501 (2007); M. Campanelli, C. Lousto, and Y. Zlochower, Phys. Rev. D 79, 084012 (2009).
[54] J. van Meter *et al.*, arXiv:0908.0023.
[55] G. Lovelace *et al.*, Phys. Rev. D 82, 064031 (2010).
[56] Z. Cao, J.P. Yu, S. Bai, C.Y. Lin, and H.J. Yo, submitted to PRD.
[57] Z. Cao, H.J. Yo, and J.P. Yu, Phys. Rev. D 78, 124011 (2008).
[58] http://nsfcac.rutgers.edu/TASSL/Projects/GrACE/
[59] M. Parashar, X Li, and S. Chandra, *Advanced computational infrastructures for parallel and distributed adaptive applications*, Wiley-Interscience (2009)
[60] MPI is a message passing interface standard which can control and manage the collective calculation and data exchange over a computer cluster. See http://www.mcs.anl.gov/research/projects/mpi/
[61] OpenMP is specially for the shared memory programming which can fully utilized current multi-core architecture in our cluster. See http://openmp.org/
[62] http://www.cactuscode.org/
[63] http://www.carpetcode.org/
[64] B. Brügmann, J. Gonzallez, M. Hannam, S. Husa, U. Sperhake and W. Tichy, Phys. Rev. D **77**, 024027 (2008).
[65] T. Yamamoto, M. Shibata, and K. Taniguchi, Phys. Rev. D **78**, 064054 (2008).
[66] E. Schnetter, S. Hawley and I. Hawke, Class. Quantum Grav. 21, 1465 (2004).
[67] E. Gourgoulhon, P. Grandclement, K. Taniguchi, J. Marck, and S. Bonazzola, Phys. Rev. D 63, 064029 (2001); K. Taniguchi and E. Gourgoulhon, Phys. Rev. D 66, 104019 (2002); ibid 68, 124025 (2003).
[68] http://www.lorene.obspm.fr/
[69] M. Hannam, S. Husa, B. Brugmann, J. Gonzalez, U. Sperhake, Class. Quantum Grav. **24**, S15 (2007).
[70] G. Cook and H. Pfeiffer, Phys. Rev. D **70**, 104016 (2004).
[71] Z. Etienne, J. Faber, Y. Liu, S. Shapiro, and T. Baumgarte, Phys. Rev. D **76**, 101503(R) (2007).
[72] P. Laguna, Phys. Rev. D **69**, 104020 (2008).
[73] S. Brandt and B. Brugmann, Phys. Rev. Lett. **78**, 3606 (1997).
[74] Astone P *et al.*, Astroparticle Phys. 7, 231 (1997).
[75] Y. Torigoe, K. Hattori, and H. Asada, Phys. Rev. Lett. **102**, 251101 (2009).
[76] M. Campanelli, M. Dettwyler, M. Hannam, and C. Lousto, Phys. Rev. D 74, 087503 (2006).
[77] W. Tichy and B. Bruegmann, Phys. Rev. D 69, 024006 (2004).
[78] Y. Kozai, ApJ 67, 591 (1962); L. Wen, ApJ 598, 419 (2003).
[79] M. Washik, J. Healy, F. Herrmann, I. Hinder, D. Shoemaker, P. Laguna, and R. Matzner, Phys. Rev. Lett. 101, 061102 (2008); J. Healy, P. Laguna, R. Matzner, and D. Shoemaker, Phys. Rev. D 81, 081501(R) (2010).
[80] C.O. Lousto and H. Nakono, Class. Quantum Grav. **25**, 195019 (2008).
[81] P. Galaviz, B. Brügmann, and Z. Cao, Phys. Rev. D**82**, 024005 (2010).
[82] M. Campanelli, C. Lousto, and Y. Zlochower, Phys. Rev. D 77, 101501 (2008); C. Lousto and Y. Zlochower, Phys. Rev. D 77, 024034 (2008).
[83] P.C. Peters, Phys. Rev. **136**, B1224 (1964).
[84] C.W. Lincoln and C.M. Will, Phys. Rev. D**42**, 1123 (1990).

7th Biennial Conf. on Classical and Quantum Relativistic Dynamics of Particles and Fields IOP Publishing
Journal of Physics: Conference Series **330** (2011) 012017 doi:10.1088/1742-6596/330/1/012017

Some features and implications of exponential gravitation

N. Ben-Amots
P.O.Box 3193, Haifa 31032, Israel
E-mail: benamots@alumni.technion.ac.il

Abstract. The paper starts with a brief recapitulation of exponential gravitation concepts, and then continue by presenting the following items: a comparison of exponential gravitation with general relativity; gravitational radiation in exponential gravitation theory; supernova formation in the frame of exponential gravitation; a possible exponential gravitation explanation to the origin of jets of Active Galactic Nuclei, quasars and microquasars.

1. Introduction

This work presents and discusses features and implications of exponential gravitation. The theory of exponential gravitation can be found in refs [1-19], although under various names.

Einstein's energy equation

$$E = mc^2 \tag{1}$$

implies that in relativity, a potential energy ΔE of a body, increases its mass by a quantity

$$\Delta m = \Delta E/c^2 \tag{2}$$

In special relativity the energy E includes all sorts of energies. Yet, in Einstein's *general* relativity, gravitational energy is not included in E.

The section: "Physical basis of finiteness of self-energy when gravitation is included" of the paper by Arnowitt, Deser and Misner [20] (ADM) deals with a variation where the gravitational energy is a part of E too.

One may consider inclusion of gravitational energy in E by two approaches:

(i) E includes all sorts of energy, including the gravitational energy

$$E_g = -\frac{mMG}{r} \tag{3}$$

where r is the distance of the particle with mass m from another body with mass M, which leads to

$$m(r) = \frac{E(r)}{c^2} = m_\infty - \frac{Gm(r)M}{c^2 r} \tag{4}$$

In (4) at infinite distance r the mass m becomes m_∞ as it should.

Journal of Physics: Conference Series **330** (2011) 012017 doi:10.1088/1742-6596/330/1/012017

Solving (4) for $m(r)$ while taking the mass M as independent of r, results in the mass $m(r)$ of the particle m:

$$m(r) = \frac{m_\infty}{1 + GM/c^2 r} \tag{5}$$

and its energy E

$$E(r) = \frac{m_\infty c^2}{1 + GM/c^2 r} \tag{6}$$

Formula (5) was derived by Ben-Amots [6 ,21].
Note that the *rest mass* m turns out to be *variable*, depending on the distance to another body, with which it interacts gravitationally, and also on the mass of that body.

(ii) *Another approach* to include the gravitational energy is by finding the energy $E(r) = m(r)c^2$ whose derivative in respect with its distance r from a central much larger mass M, gives the gravitational force:

$$\frac{dE}{dr} = F = -\frac{m(r)MG}{r^2} \tag{7}$$

so that

$$\frac{d[E(r)]}{dr} = c^2 \frac{d[m(r)]}{dr} = -F(r) = \frac{Gm(r)\,M}{r^2} \tag{8}$$

The force $F(r)$ is defined as the force acting on the test particle of mass $m(r)$ in the gravitational field of the central mass M, and is equal to the derivative of the energy $E(r)$ of the test particle with respect to the distance r of the test particle from the central mass. The mass M of the central body is considered as constant in the present approximation of this theory.
Solving the equation $c^2 \frac{d}{dr}[m(r)] = Gm(r)M/r^2$ extracted from (8) for M independent of r (or equivalently solving Majerník's [14], eq. (10) below), the variation of the rest mass m with r results in:

$$m(r) = m_\infty \exp\left(\frac{-GM}{c^2 r}\right) \tag{9}$$

which is graphically represented in Figure 1.
Hereafter this second method is named *exponential gravitation*, suggested by the exponential character of the solution (9).
The corresponding gravitational force is calculated below in Eq. (20) and analyzed.
The exponential gravitation (under various names, and with different proofs) was analyzed and explained in detail by many authors, starting with Milne [1] in 1948 and followed for example by Yilmaz [2, 3], Majerník [4-5], Ben-Amots [6], Kiesslinger [7], Vankov [8-11], Turanyanin [12], Ibison [13] and recently by Majerník [14-15], Vankov [16-18] and Walker [19].
One can find further details of derivation of (8) above for example in Kiesslinger [7] or Ben-Amots [6], or equivalently in Majerník [14] who defines:

$$m_\infty = m(r) + 1/c^2 \int_\infty^r F(x)dx \tag{10}$$

where $F(x)$ is the gravitational force according to Newton's gravitation law. A different independent derivation using relativistic redshift is given by Walker [19].
While the mass of the small particle m changes in the gravitational field of the central mass M, the central mass M is also changed in the gravitational field of the small particle m. This effect gives for two *equal* variable masses $M(r) = m(r)$ in Eq. (8) the equation:

$$\frac{d[m(r)]}{d(r/2)} c^2 = \frac{G[m(r)]^2}{r^2} \tag{11}$$

Journal of Physics: Conference Series **330** (2011) 012017 doi:10.1088/1742-6596/330/1/012017

Note that r at the right hand side of (11) is the *distance* between the two masses; yet at the left hand side of (11) the *movement* of the mass is toward (or away from) the mass center, which is located in this case at *half of the distance* r between the two *equal* masses.
Solving (11) results in:

$$m(r) = \frac{m_\infty}{1 + G m_\infty / 2c^2 r} \tag{12}$$

thus showing that while the distance r approaches zero, both equal masses diminish to zero mass and finally annihilate.

Both approaches do *not* give solutions of Einstein equations. Note that in general relativity and in Newtonian theory, the rest mass *is not* variable (is not dependent on the distance r.) Definition of the force F as the derivative of energy with respect to the distance r is used in Newtonian theory, and allowed here as quasistatic approximation.
Schwarzschild's line element [22] for mass point in vacuum is:

$$ds^2 = \left(1 - \frac{2GM}{c^2 r}\right) c^2 dt^2 - \frac{dr^2}{1 - 2GM/c^2 r} - r^2 d\theta^2 - r^2 \sin^2\theta d\phi^2 \tag{13}$$

Although both the first approach of getting variable mass $m(r)$ (5) and the exponential gravitation (9) do not assume any special geometry (except the Euclidean), if, only as a means of comparison with general relativity we were to *describe* a space giving the same gravitation force as (5) and (9) by line elements in a curved space as general relativity does, Ben-Amots obtained "pseudo" line elements for the first approach and for the exponential gravitation theory as follows:

(i) For the first approach where $m(r)$ is given by (5) and $E_g = -\frac{mMG}{r}$, Ben-Amots [21] derived:

$$ds^2 = \frac{c^2 dt^2}{1 + 2GM/rc^2} - \left(1 + \frac{2GM}{rc^2}\right) dr^2 - r^2 d\theta^2 - r^2 \sin^2\theta d\phi^2 \tag{14}$$

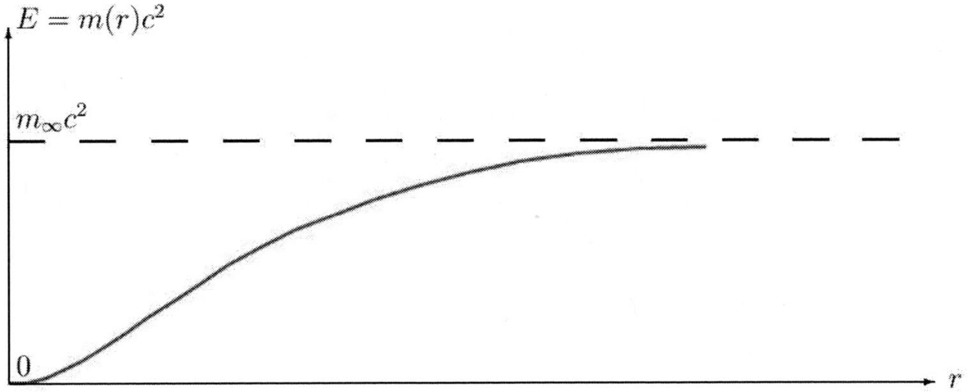

Figure 1: Energy$=m(r)c^2 = c^2 m_\infty \exp(-GM/c^2 r)$

Ben-Amots [21] also showed that (14) agrees with the present existing measurements of the three crucial tests of general relativity.

Journal of Physics: Conference Series **330** (2011) 012017 doi:10.1088/1742-6596/330/1/012017

For this first approach we derive the formula for the gravitational force of the interaction between two bodies in the *quasistatic* approximation [21]:

$$\vec{F} = -\frac{dE}{dr} = -\frac{Gm_\infty M}{(r+GM/c^2)^2}\left[\frac{\vec{r}}{r}\right] \tag{15}$$

which was obtained as the derivative of the energy E of (6) in respect to r. The difference between (15) and the Newtonian gravitational force is too small to be measured with nowadays techniques.

(ii) For exponential gravitation where $m(r)$ is given by (9), Ben-Amots [6] and Majerník [14] derived the "pseudo" line element:

$$ds^2 = \exp\left(\frac{-2GM}{c^2 r}\right)c^2dt^2 - \exp\left(\frac{2GM}{c^2 r}\right)dr^2 - r^2d\theta^2 - r^2\sin^2\theta d\phi^2 \tag{16}$$

The line element (16) was analyzed by Coleman [23] and Ben-Amots [6]. They showed also that the line element (16) agrees with the measurements of the three crucial tests of general relativity (see [23] and in the appendix of [6]).

In order to compare with observations, Schiff [24 ,25] analyzed the structure of the Schwarzschild's line element (13) by using a series expansion (written according to the notation here):

$$ds^2 = \left[1+\alpha\frac{2GM}{c^2r}+\beta\left(\frac{2GM}{c^2r}\right)^2+...\right]c^2dt^2 - \left[1+\gamma\frac{2GM}{c^2r}+\delta\left(\frac{2GM}{c^2r}\right)^2+...\right]dr^2 - r^2(d\theta^2+\sin^2\theta d\phi^2) \tag{17}$$

Schiff concluded that higher terms in (17) were not subjected yet, at the time of writing [24, 25], to experimental/observational measurements. This is still correct nowadays. See Ben-Amots [21] and refs therein.

Applying on (17) the conditions that the field is spherically symmetric and that the spacetime is asymptotically flat (see, for example, Foster and Nightingale [26] and §2 below) gives

$$ds^2 = \left[1-\frac{2GM}{c^2r}+k\left(\frac{2GM}{c^2r}\right)^2+...\right]c^2dt^2 - \frac{dr^2}{1-\frac{2GM}{c^2r}+k\left(\frac{2GM}{c^2r}\right)^2+...} - r^2d\theta^2 - r^2\sin^2\theta d\phi^2 \tag{18}$$

Expanding (16) in Taylor series of $\frac{2GM}{c^2r}$ results in

$$ds^2 = \left[1-\frac{2GM}{c^2r}+\frac{1}{2}\left(\frac{2GM}{c^2r}\right)^2-\frac{1}{6}\left(\frac{2GM}{c^2r}\right)^3+...\right]c^2dt^2$$
$$-\frac{dr^2}{1-\frac{2GM}{c^2r}+\frac{1}{2}\left(\frac{2GM}{c^2r}\right)^2-\frac{1}{6}\left(\frac{2GM}{c^2r}\right)^3+...} - r^2d\theta^2 - r^2\sin^2\theta d\phi^2 \tag{19}$$

The differences between (13), (14) and (16) when expressed in the form (19) are in k and higher orders of $\frac{2GM}{c^2r}$ of (18):

- For Schwarzschild's line element (13), the coefficient k in (18) and the higher coefficients are equal to zero.
- For the pseudo line element (14), the coefficient k in (18) is $k=1$, and higher coefficients are *not* equal to zero.

Journal of Physics: Conference Series **330** (2011) 012017 doi:10.1088/1742-6596/330/1/012017

- For the exponential gravitation (16) and the pseudo line element (19), the coefficient k in (18) is $k = 1/2$, and higher coefficients are *not* equal to zero.

The difference between the solutions of exponential theory of gravitation (16), (19) and of general relativity (13) starts with the second order of $2GM/c^2r$. The experimental observations to date are not yet sufficiently accurate to distinguish between the second order terms of the solutions, as discussed by Ben-Amots [21] and refs therein, thus from the experimental point of view the question of which theory better describes the reality remains open.

Turyshev et al. [27], Maleki et al. [28], Maleki and Prestege [29] and Ben-Amots [21] discuss in detail different suggested experiments aimed to measure k in future.

For the Schwarzschild's solution (13), when $r < 2GM/c^2$, g_{00} becomes negative and g_{11} becomes positive, which give the solution the "black hole" properties (event horizon etc.).

In (16), not as for the Schwarzschild solution (13), g_{00} and g_{11} do not change signs (for any r). So, there is no black hole associated with (16) or equivalently with (9).

Substituting $m(r)$ of (9) in (7) gives the gravitational force $F(r)$ between $m(r)$ and M as:

$$F = -\frac{GMm(r)}{r^2} = -\frac{GMm_\infty}{r^2}\exp\left(\frac{-GM}{c^2r}\right) \tag{20}$$

This force is equal to *zero* for $r = 0$, and approaches the Newtonian force $\frac{GM}{r^2}m_\infty$ as r increases to infinity (Figure 2).

For a central mass having finite dimensions in general relativity there exist Schwarzschild different exterior and interior solutions [30]. The interior solution is within a spherical body, and the different exterior solution is valid in the empty space around it. However, this exterior solution only describes the region exterior to the horizon of the "black hole."

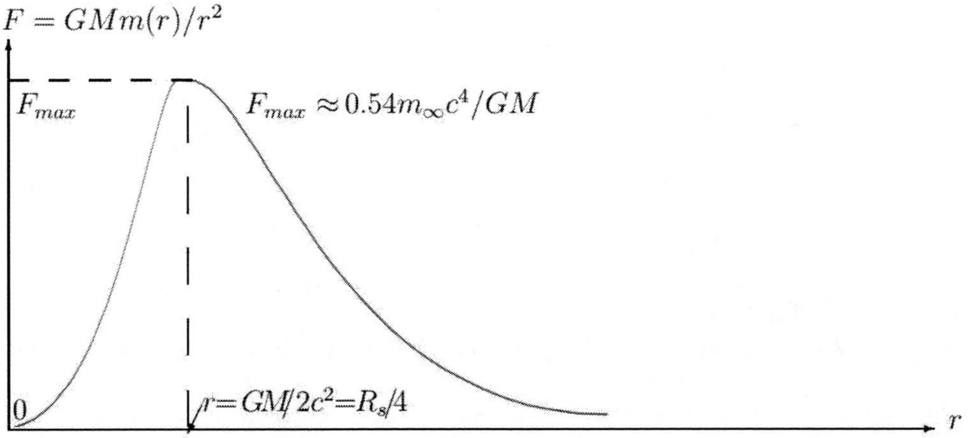

Figure 2: Force= $GMm(r)/r^2 = \frac{GM}{r^2}m_\infty \exp(-GM/c^2r)$

In Newtonian gravitation, the attraction force of a body of radius R is for the exterior case mMG/r^2 and for the interior case $mMGr/R^3$, with common value mMG/R^2 at radius $r = R$ (Figure 3). In the Newtonian solution there is a discontinuity of slope at the radius that separates between the Newtonian exterior ($r > R$) and the Newtonian interior ($r < R$) solution for the gravitational force of a spherical body (Figure 3).

Journal of Physics: Conference Series **330** (2011) 012017 doi:10.1088/1742-6596/330/1/012017

While the Newtonian force is represented by *two* equations (exterior and interior, Figure 3), the force in the exponential gravitation theory has only *one* equation (Figure 2), *without the discontinuity of slope* that exists in the Newtonian force (Figure 3).

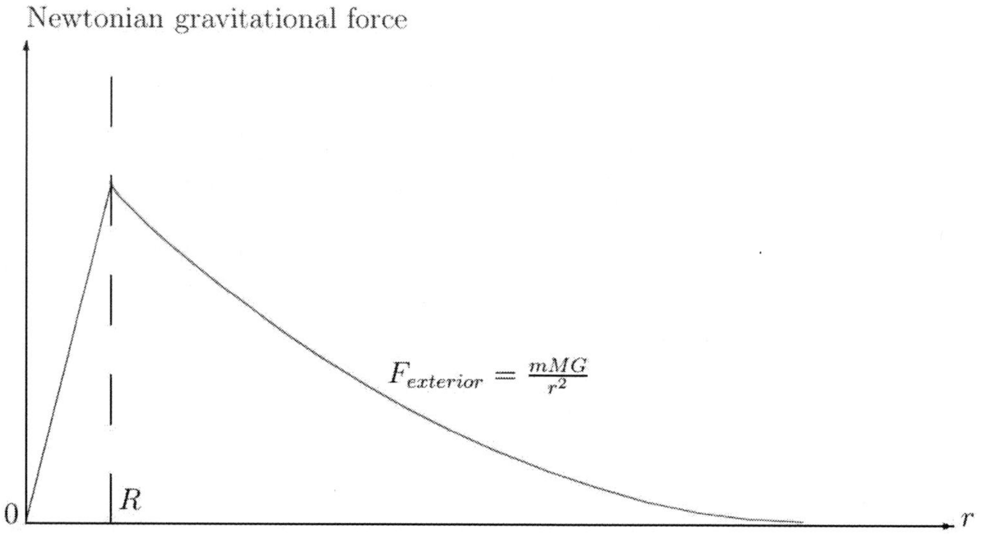

Figure 3: The Newtonian gravitational force (schematic) for a homogeneous spherical body of mass M and radius R: interior branch (left to $r = R$) and exterior branch (right). The force is zero at the center (r=0) and maximal at $r = R$

Expanding the exponential gravitation force (20) in Taylor series of $\frac{GM}{c^2r}$ results in

$$F \approx -\frac{GMm_\infty}{r^2} + \frac{GMm_\infty}{r^2}\left(\frac{GM}{c^2r}\right) - \frac{1}{2}\frac{GMm_\infty}{r^2}\left(\frac{GM}{c^2r}\right)^2 + \frac{1}{6}\frac{GMm_\infty}{r^2}\left(\frac{GM}{c^2r}\right)^3 - \ldots \tag{21}$$

where the first term is recognized as the well known Newtonian gravitational force for point mass.

It has been shown [6] that in the variable rest mass exponential gravity theory there cannot exist black holes, contrary to the possibility exhibited by Einstein's general relativity. §2 below returns to this point, while discussing Figure 2 and (23).

Ben-Amots [6, 21] detailed some implications of the first approach for $m(r)$ (where ΔE_g is $\Delta E_g = -\frac{mMG}{r}$. See above).

The present paper details other implications, of the *exponential gravitation* (the second approach).

It should be emphasized that the theoretical basis for the exponential gravitation was derived, analyzed and discussed in detail under various names by more than a dozen of papers [1-19] during more than 60 years. Although field equations were not given, yet any theory must be tested by experiments/observations. So far the two theories above and general relativity agree equally well with the experiments/observations.

2. Differences in basic assumptions between general relativity and exponential gravitation

First, this section mainly discusses Einstein's general relativity from a rather narrow point of view by focusing on the Schwarzschild [22] solution for Einstein's equations.

Quoting Foster and Nightingale [26]:

> "Schwarzschild... guiding assumptions were:
> (a) that the field is static,
> (b) that the field is spherically symmetric,
> (c) that the spacetime *is empty*,
> (d) that the spacetime is asymptotically flat.
> ...Assumption (c) means that (the solution is) to be found using the empty-spacetime [Einstein's] field equations..."

Thus, assumption (c) above means that all the components of the Ricci Tensor in general relativity are null.

Still, assumption (c) of Schwarzschild is questionable since the space around a mass comprises the gravitational field of the mass, which itself is massive. (See for example Adler [31]). Thus assumption (c) affects both Schwarzschild exterior and interior solutions. The following discussion uses the fact that a gravitational field has a mass of its own, in contrast to the assumption (c) of Schwarzschild.

Also, the present author thinks that the conservation law

$$\mathrm{div}(\text{total stress energy tensor}) = 0 \tag{22}$$

(See for example Friedman [32]) can be too limiting, if the mass of the gravitational field is included in the energy. The conservation law (22) does not allow for gravitational waves to transfer energy in the first order, so that it limits the gravitational energy transfer in lowest order to quadrupolar waves. Another less limiting condition should apply instead (22).

If the claim presented here that (22) is violated because of the gravitational field being massive is correct, then bipolar gravitational waves should exist.

In the presented version of exponential gravitation theory, the gravitational radiation carries energy not only as a second order effect, but in the first order as well. This makes the mass transfer, that is the energy transfer by gravitational radiation, significant.

As stated in his lecture in Vienna in 1913, Einstein [33, 34] rejected the idea that the *rest mass* of a body can increase when its distance from another body increases. Instead, in Einstein's theory the *rest mass is constant*, while space is curved. (In the presented theory the rest mass is variable while space is Euclidean). The assumption of constant rest mass, is a constraint that Einstein imposed (firstly in his lecture in Vienna in 1913), which was used later while deriving his equations. The presented theory is free of this constraint. So, (9), which involves a *variable mass*, is not compatible with Einstein's equations. The presented solution (9) (for (8)) is close quantitatively to the solution of the Einstein's equations for weak gravitation fields (like those in the solar system, but not for neutron stars or other dense matter). For dense matter the presented solution is different from Einstein' equations solution (to be discussed below).

The exponential theory presented here complies with conditions (a), (b), (d) of Schwarzschild. Condition (d) states that the spacetime is asymptotically flat. (In the Birkhoff theorem the field should vanish at large distances, meaning here and in general relativity that the spacetime is asymptotically flat).

According to the theory presented here, of special importance is a region of the order of GM/c^2 close to any sufficiently dense mass (see Figure 2. Features of this figure are preserved for finite radius spherical mass distribution). This region has no discontinuities, in contrast to the black hole region of Einstein's general relativity. The physical behavior near the center of a mass (interior branch of the solution, $r < GM/2c^2$) differs from the behavior far from the mass (exterior branch of the solution), yet this behavior changes gradually and continuously in exponential gravitation theory. This continuous region replaces the sharp event horizon of

Journal of Physics: Conference Series **330** (2011) 012017 doi:10.1088/1742-6596/330/1/012017

the "black hole" of Einstein's general relativity. The black hole of Einstein's general relativity is characterized by the impossibility of any mass or radiation to escape from a region confined in a sharply defined event horizon. The equivalent region within the Schwarzschild radius in the theory here has different features. Electromagnetic radiation, gravitational radiation, and massive particles may escape this region, provided they have sufficient radial velocity to overcome the gravitational potential, according to the equation:

$$v > c\left[1 - \exp\left(\frac{-GM}{c^2 r}\right)\right]^{\frac{1}{2}} \tag{23}$$

While according to general relativity as currently accepted, the neutrinos are trapped when the radius of a celestial body is smaller than Schwarzchild's radius, in the theory presented here energy may escape, including neutrinos. In the future, observations on neutrinos from objects of mass and radius which by the general relativity should be black holes may determine which theory is valid.

It is reasonable to think that an adequate theory should not contain infinities, and the theory of exponential gravitation is without any infinity while Einstein's general relativity contains infinities at Schwarzschild radius and at $r = 0$.

In a broader context, the present author thinks that the equivalence principle is only an approximation that is valid for small accelerations and weak gravitational fields only.

Einstein described in ref. [35] how general relativity should be completed, and by what Born [36] described as following: "The idea is that the field produced by a body in turn reacts on the body and thus determines its world line. This is mathematically a very complicated problem the nature of which we cannot even indicate. Einstein, Infeld and Hoffmann [37] attacked it; their first investigations were so extended that only summaries could be published. Later Infeld [38] succeeded in simplifying the calculation considerably." See also Einstein and Infeld [39].

3. Gravitational radiation

3.1. Regenerated gravitational field

Van Flandern [40] asks:

> "... are the gravitational fields for a rigid stationary source frozen, or are they continuously regenerated?"

After analysis his conclusion is:

> "We conclude that the concept of frozen gravitational fields is acausal and paradoxical. Gravitational fields must continuously regenerate, like flowing waterfall. In doing so, they must consist of entities that propagate and transfer momentum."

We ask: Why doesn't the mass of the gravitational field collapse on the central mass? As already stated, the gravitational field has mass (Adler [31]), so, like any mass, it is attracted inward towards the direction of the central mass.

The answer to this question is reasoned as following: The central attraction creates an inward flux of mass of the gravitational field. The gravitational field fall should be balanced by radiation outward by the central mass with the energetic flux of field of the same mass as the falling field.

Gravitation is thus a superposition of the inward fall of massive gravitational field, and the outward radiation flow of energetic/massive gravitational field. So, some flux transfer should be allowed instead the limitation (22).

In conclusion to the above:

Journal of Physics: Conference Series **330** (2011) 012017 doi:10.1088/1742-6596/330/1/012017

(i) A gravitational field that has mass, falls toward the gravitational source, implying that the source and this associated gravitational field emits continuously a counteracting gravitational field.

(ii) In addition, The present writer thinks that the conservation law (22) is too rigid, thus he allows the above mentioned gravitational waves of bipolar order, which may also be produced even by spherically symmetric sources of gravity.

(iii) To summarize the fundamental idea of exponential gravitation theory from which stems all the theory: The energy content of the gravitational potential that a body has, affects the mass of the body (as in Arnowitt et al. [20] (ADM)), just like the energy content of any other potential that the body has affects its mass.

3.2. Supernova formation

A supernova is usually assumed to be born in a bounceback of the inner layers of a star after their gravitational collapse (bounce usually thought to originate in neutrinos' pressure). This bounce back is supposed to meet the slower collapsing external layers of the star in a violent collision which ignites the fusion of the hydrogen in them and causes the explosion seen as supernova. Unfortunately spherically symmetric simulations are unable to reproduce this (Liebendörfer et al. [41]). These simulations lack sufficient outward pressure to produce the *explosion*; instead the simulations end with *implosion*.

The presented explanation of bounceback using exponential gravitation is as follows: The appropriate equivalent line element for (9) is [6]:

$$ds^2 = \exp\left(\frac{-2GM}{c^2r}\right)c^2dt^2 - \exp\left(\frac{2GM}{c^2r}\right)dr^2 - r^2d\theta^2 - r^2\sin^2\theta d\phi^2 \tag{16}$$

See Coleman [23] and Ben-Amots [6] for discussions of this line element.

Calculation of the stress-energy tensor component G_{00} (24), which is the *approximate* density of mass-energy gives (Figure 4):

$$G_{00} = -\frac{1}{r^2}\exp\left(\frac{-2GM}{c^2r}\right)\left[1 - \left(1 + \frac{2GM}{c^2r}\right)\exp\left(\frac{-2GM}{c^2r}\right)\right] \tag{24}$$

(while for Schwarzschild Solution $G_{00} = 0$).

G_{00} (24) is negative, which represents attraction gravitational force (See Figure 4).

Below a certain radius when the radius approaches zero, the radial force becomes smaller and approaches zero. At radius $r = 0$ the force exerted on a test particle is zero.

Figure 4: G_{00} (24) normalized: $G = c = M$=1

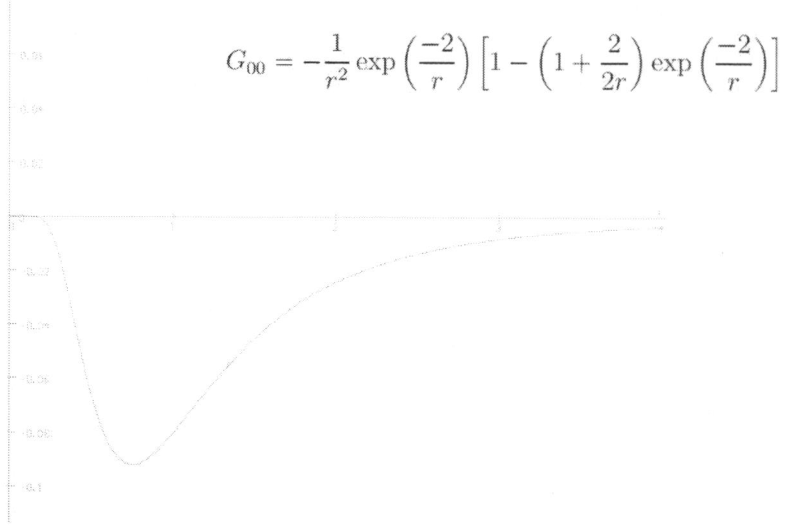

The gravitational collapse of celestial bodies compresses the matter to a very dense state. During compression the body will radiate field/mass outward with contribution to pressure given by the compressed matter. Note that at the same time the gravitational attraction decreases according to G_{00} in (24) in accord with Einstein's [42] statement that "matter cannot be concentrated arbitrarily."

This explanation based on the exponential gravitation even in the spherical symmetry context offers the above contribution to the additional pressure necessary for the bouncing outward stage which is lacking in Newtonian or general relativity models. The presented explanation will never lead to a *black hole* remnant star, because in the theory presented here such an object can never exist, and in addition a denser collapse creates an even higher outward pressure of gravitational radiation. Kiesslinger [7] arrived at this outward pressure of gravitational radiation too, but without applying to the supernova explosion.

In the presented theory, this mass/field radiation outward by the inner core during the implosion stage of a supernova, causes the core to expand outward. It stops the outer layers implosion inward, and redirects them to bounce outward. The falling outer layers collide with the expanding core and are bounced back outward. The sharply increased pressure caused by collision raises the temperature that ignites the fusion of the hydrogen and helium of the falling outer layers, and causes the final explosion of the supernova. Thus exponential gravitation theory enables a possible explanation to supernova explosions. See also §4.3 below.

Existing simulations of supernova do not consider the gravitational radiation force outward (as we suggest of being involved) so, most often the simulations of supernovas, especially spherically symmetric simulations, lead to a final implosion toward a "black hole," without bouncing outward [41], while observations imply bouncing outward[1].

General relativity or Newtonian existing model simulations fail to lead to explosion, implying that the helium and heavier elements created until the supernova formation are not dispersed in space. Introducing in the simulations the gravitational radiation force suggested by us could possibly produce the supernova explosion predicted by our model and others.

[1] Present simulations of supernovas that consider rotation either fail to reproduce observations, or fail to conserve angular momentum [41].

Journal of Physics: Conference Series **330** (2011) 012017 doi:10.1088/1742-6596/330/1/012017

3.3. A possible qualitative explanation of how jets are emitted from some astronomical bodies

The approximate potential of exponential gravitation of a *non-rotating* spherically symmetric body versus a radius r, (or of a spherical body rotating along a diameter axis of rotation z) is schematically drawn in Figure 5.

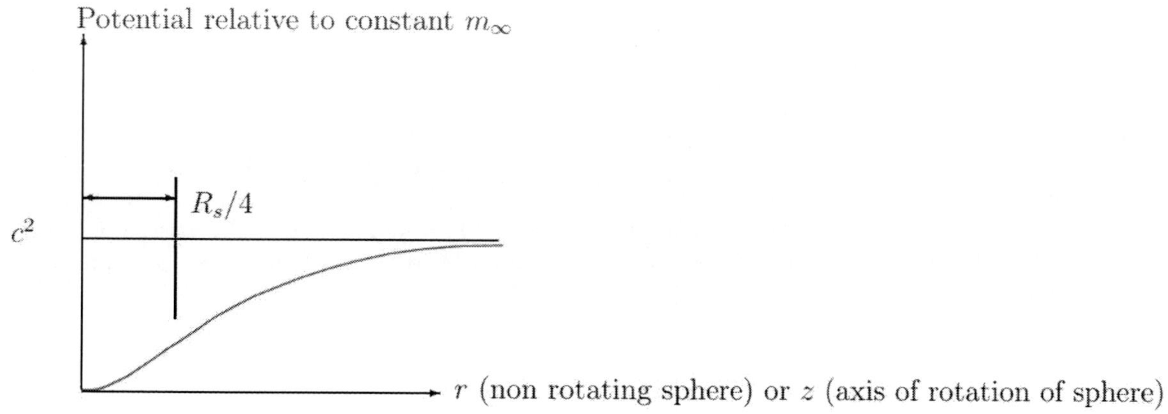

Figure 5: Schematic exponential gravitation potential along the radius r of a *non-rotating sphere*, or along the diameter z around which a spherical body rotates

The approximate potential of exponential gravitation of a *rotating* spherically symmetric body versus a radius in the equatorial plane is schematically drawn in Figure 6. This curve is for constant angular momentum along the radius. Constant angular momentum is maintained for $r > R_{max} > R_s/2$ (where R_{max} is the radius at which the potential is maximum). Continuing left of $r = R_{max}$, a constant angular momentum would bring to $v > c$.

This potential curve has a "bulge" (peaked at R_{max}) that is the contribution of centrifugal potential for exponential mass.

For very high angular velocity this "bulge" becomes a high peak that can be many orders of magnitude higher than the potential energy of a non-rotating body. Actually, this bulge is a potential barrier.

Substituting a central mass M of a typical galaxy nucleus of 200 million solar masses in G_{00} (24), the radius for the dip of G_{00} (Figure 4) results to be of the order of the distance of Mars from the Sun. Rotation causes the peak at R_{max} (Figure 6) to rise higher, and the radius for this peak to be smaller, depending on the angular velocity of the rotation. For more on this influence of rotation see below in Section 4.2.

Consider an explosive or another pressure creating event that increases the pressure in the core ($r < R_{max}$ in Figure 6) of a fast rotating very massive body. The particles very near the center lack sufficient angular momentum to cross the high potential barrier in direction perpendicular to the rotation axis, but they have two narrow outlets around the axis of rotation, in which the potential barrier is much lower and reaching a minimum of zero for the axis itself. If the particles are sufficiently energetic, they will eventually escape as two narrow jets through two narrow polar outlets.

Actually narrowly collimated jets are observed escaping along the rotation axis of active galactic nuclei and quasars on scales of hundreds of thousands of light years. There are similar smaller jets getting out of microquasars. The exponential gravitation presented here gives then this possible explanation of these jets (at this stage only a qualitative one). See also §4.2 below.

There are other suggested theories that try to explain the observed jets in other ways (for

Journal of Physics: Conference Series **330** (2011) 012017 doi:10.1088/1742-6596/330/1/012017

example, [44-48] and references therein), including explanations based on stellar magnetism and other phenomena.

4. Symmetry considerations for astronomical jets

4.1. Spherical symmetry

A test particle at the center of a spherically symmetric massive body of finite radius, is equally attracted to all directions. The resultant force of gravitational attraction acting on this particle is equal to zero. Figure 3 that shows this for Newtonian gravitational force at $r = 0$.

Figure 2 approximates the force along the radius r for exponential gravitation. This force has a left branch for $r < R_{max}$ representing the interior solution for the force. This interior force approaches zero when the radius r approaches zero. (When increasing the radius r, the force gradually approaches the Newtonian (exterior) solution $F = -mMG/r^2$ as shown in Figure 3).

For calculating the stress-energy G_{00} term for massive central body we use the mass (9) in Weinberg's [49] procedure for the line element. Slightly modifying Weinberg's procedure (Ben-Amots [6]) results in the pseudo line element (16). The stress-energy term G_{00} (24) (derived from (16)) is the normalized pressure, that is the normalized $\frac{\text{force}}{\text{area}}$, and incorporates the gravitational attraction force from *all directions*. The force diminishes gradually to zero near the center of the body. Thus this diminishing of the central force near the center just expresses spherical symmetry. (Compare to the Newtonian gravitational force for the spherically symmetric case described in Figure 3, which approaches zero at the center).

Schwarzschild exterior solution [22] is for a *mass point* surrounded by *empty space*, thus it lacks the diminishing of the attraction gravitational force toward zero, of the force in a massive body of finite radius. Tolman [30] correctly used *interior* solution (and not exterior solution) for the interior of bodies of *finite radius*. Einstein [42] concluded that "'Schwarzschild singularities' do not exist in physical reality," because "matter cannot be concentrated arbitrarily," but others [50], [51] later made manipulations on Schwarzschild exterior solution [22] to get insight on the solution near the center. See Thorne's review [52]. Yet using Schwarzschild exterior solution [22] for zero radius implies that the force at zero radius is infinite instead zero force as obtained in exponential gravitation.

Similarly, a point mass is surrounded by a gravitational field, that possesses energy and mass (Adler [31]), which is neglected in Schwarzschild solution. This mass of the surrounding gravitational field produces a zero resultant force at the center, because it attracts a test particle at the center symmetrically and equally from all directions. Schwarzschild solution ignores this, because of the assumption (c) (§2), that space is empty. So, the singularity and infinite values of Schwarzschild solution originates in *neglecting* this aspect of *spherical symmetry* (that the resultant force at the center is zero).

4.2. Examining symmetry for a rotating body

The exponential gravitation potential along any radius r of a spherical non-rotating body, or along the axis of rotation z passing through the center of a spherical rotating body, is shown in figure 5. For a rotating body, adding the potential of the centrifugal force along a radius perpendicular to the axis of rotation results in a centrifugal barrier as seen in figure 6. The faster the rotation is, the higher is the centrifugal barrier (Figure 7).

Still symmetry dictates a zero force component in the equatorial plane at the center.

Journal of Physics: Conference Series **330** (2011) 012017 doi:10.1088/1742-6596/330/1/012017

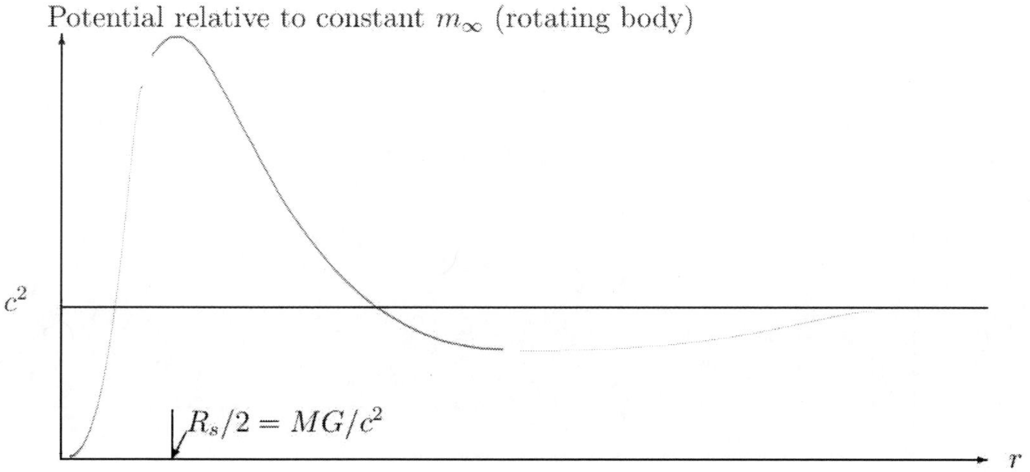

Figure 6: Schematic potential along a radius r in a section perpendicular to the axis of *rotation* passing through the center of a spherical body, where the angular momentum is constant for $r > R_{max}$. (For smaller radius the graph represents a solution for transient motion, because then there is no solution that conserves the angular momentum)

Examining the whole rotating body, centrifugal forces exist in any plane parallel to the equatorial plane, creating a potential barrier, and leaving a "tunnel" of low potential along the axis of rotation z (Figure 7).

A pressure caused by any event inside this "tunnel," cannot supply any massive particle sufficient angular momentum or the high energy necessary to overcome the high energy barrier in the radial direction. A pressure that cannot cause the explosion of the body, can be released by mass transfer only along the "tunnel" along the axis of rotation z, as two jets expelled from the poles.

In a way, the jets can be seen as caused by the symmetry of rotation, which allows for a centrifugal barrier only *around* (but not on) the rotation axis. So, this could be a possible cause that contributes to production of astrophysical jets in the rotation axis of celestial bodies.

Journal of Physics: Conference Series **330** (2011) 012017 doi:10.1088/1742-6596/330/1/012017

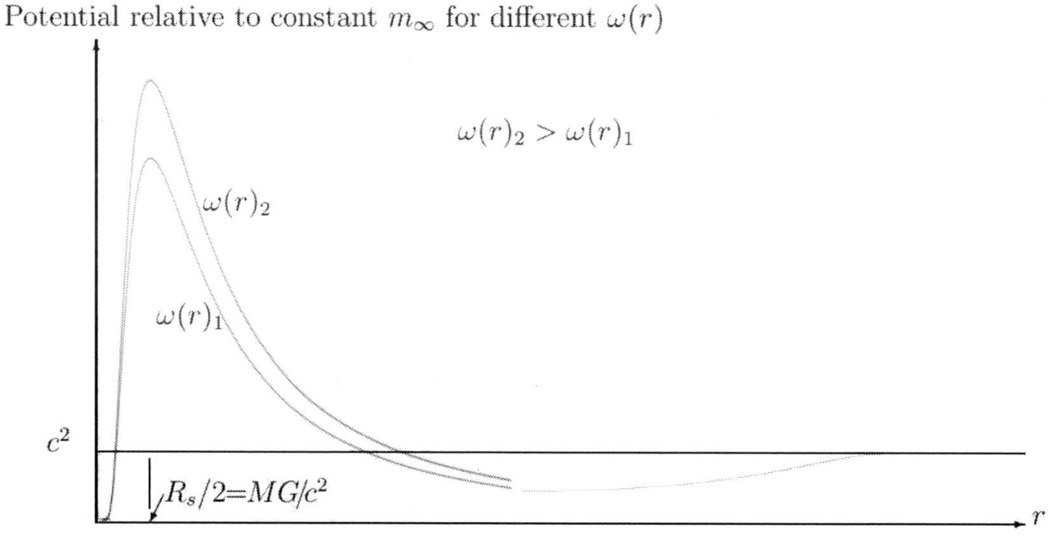

Figure 7: As in Figure 6 but the upper curve is for higher angular velocity of the body

4.3. The repelling force

The gravitational field of a spherically symmetric body has mass by itself and like any mass it is attracted inward towards the direction of the central mass. The central attraction creates an inward flux of mass of the gravitational field. Yet this cannot be the whole picture. The gravitational field fall should be balanced by radiation outward by the central mass with the energetic flux or field of the same mass as the falling field. The falling of the gravitational field, and the counteracting radiation outward by the central mass, have spherical symmetry.

Further, an implosion of a celestial spherical body of sufficiently large mass (which is a non-equilibrium event) compresses the inner core near the center, causing an increase of the gravitational radiation outward, and with it an increase of the *repelling* pressure of this radiation. The increase of the repelling pressure together with the pressure of the neutrinos, (created in the implosion stage of supernova formation), stop the implosion of upper layers and contributes to an eventual bounceback outward in the process of supernovae formation. This effect possibly contributes too to the explosion of supernovas.

5. Conclusions

The possibility to experimentally distinguish between the predictions of general relativity and exponential gravitation theories [27-29] may cause the implications of the exponential gravitation to be more significant for gravity research.

The above predictions of the exponential gravitation are solvable for two bodies, where the massive one is approximated as having constant rest mass M, and the rest mass $m(r)$ of the light body is variable and depending on the distance r between these two bodies.

Predictions based on assumption of two bodies whose both *equal* rest masses $m(r)$ are variable and depending on the distance r between these two bodies, are also obtained by the approach of exponential gravitation above, resulting in solution without exponential character, though.

Predictions based on assumption of two bodies whose both *arbitrary* rest masses $m(r)$ and $M(r)$ are variable and depending on the distance r between these two bodies, are obtained by the approach of Eq. (4) [6, 21].

Journal of Physics: Conference Series **330** (2011) 012017 doi:10.1088/1742-6596/330/1/012017

Acknowledgments
The author is indebted to Dan Igner, who helped find the appropriate wording for a clearer presentation. Thanks to IARD for allowing me to present this work in the IARD2008 conference, and to Professor Larry Horwitz for his comments on the manuscript. Thanks to Colin Walker for his comment.

References

[1] Milne E A 1951 *Kinematic Relativity* pp 203–210, 214–220. (Oxford, Uk: Clarendon Press) (1st edn 1948)

[2] Yılmaz H 1972 *Nuo. Cim. B* **10B** 79. See p 87

[3] Yılmaz H 1973 *Nuo. Cim. Lett.* **7** 337

[4] Majerník V 1985 *Ast. Space Sci.* **113** 199

[5] Richterek, L and Majerník V 1999 *Acta Univ. Palacki. Olomuc., Fac. Rer. Nat., Physica* **38** 47

[6] Ben-Amots N 2007 *Found. Phys.* **37** 773

[7] Kiesslinger R *Gravitation Theory on the Testbench* (Nürnberg, Germany)

[8] Vankov A A 2000 *Does light gravitates? (Proposal on a new test of equivalence principle)* *Preprint* arXiv/0012050

[9] Vankov A A 2001 *Proposal of experimental test of general relativity theory* *Preprint* arXiv/0105057

[10] Vankov A A 2003 *On problem of mass origin and self-energy divergence in relativistic mechanics and gravitational physics* *Preprint* arXiv:gr-qc/0311063

[11] Vankov A A 2004 *Elimination of high-energy divergence in relativistic Lagrangean formulation of gravitating particle dynamics* *Preprint* arXiv:physics/0402117

[12] Turanyanin D 2004 *On interaction of motional masses, J. Theoretics* See addendum. http://www.journaloftheoretics.com/Links/Papers/Turan.pdf

[13] Ibison M 2006 2007 *Cosmological test of the Yilmaz theory of gravity* *Class. Quant. Grav.* **23** 577 (2006) *Preprint* arXiv:0705.0080v1 (2007)

[14] Majerník V 2006 *Energy conservation at the gravitational collapse* *Preprint* arXiv:astro-ph/0609313

[15] Majerník V 2009 *Phys. Essays* **22** 413

[16] Vankov A A 2007 *Proper mass variation under gravitational and coulomb force action in relativistic mechanics of point particle* *Preprint* arXiv:0710.1119v1

[17] Vankov A A 2006 2008 *On relativistic generalization of gravitational force* *Found. Phys.* **38** 523 (2008) *Preprint* arxiv:physics/0611161v1 (2006)

[18] Vankov A A 2010 *General relativity problem of Mercury's perihelion advance revisited* *Preprint* arXiv:1008.1811v1

[19] Walker C 2010 *Calculation of gravitational potential by the method of multiple redshifts* http://sites.google.com/site/revisingnewton

[20] Arnowitt R, Deser S and Misner C W (ADM) 1962 *The dynamics of general relativity* In: Witten L 1962 *Gravitation: An Introduction to Current Research* (New York: Wiley) pp 227–265. See Section 7-7.1: Physical basis of finiteness of self-energy when gravitation is included, pp 256-257.
Reprinted in *Gen. Rel. Grav.* **40** 1997 (2008). See pp 2020-2021

[21] Ben-Amots N 2008 *A new line element derived from the variable rest mass in gravitational field*
Preprint http://arXiv.org/pdf/0808.2609

[22] Schwarzschild K 1916 *Sitzber. Deut. Akad. Wiss. Berlin, Konigl. Math.-Phys. Tech.* pp 189–196

[23] Coleman C J 1971 *J. Phys. A* **4** 611

[24] Schiff L I 1960 *Proc. Nat. Acad. Sci.* **46** 871

[25] Schiff L I 1967 *Comparison of theory and observation in general relativity* pp 105-116 (especially p 106) in: Ehlers J: *Relativity Theory and Astrophysics: I. Relativity and cosmology* (Providence, Rhode Island: Am. Math. Soc.)

[26] Foster J and Nightingale J D 1979 *A Short Course in General Relativity* (London: Longman Group). 2nd edn 1995 (New York: Springer-Verlag)

[27] Turyshev S G, Shao M and Nordtvedt K 2004 *Class. Quant. Grav.* **21** 2773

[28] Maleki L et al. 1998 *American Phys. Society APS/AAPT Joint April Meeting, Columbus, Ohio, April 18-21*

[29] Maleki L and Prestage J D 2001 *SpaceTime mission: Clock test of relativity at four solar radii* p 369 in: Lämmerzahl C, Everitt C W F and Hehl F W 2001 *Gyros, Clocks and Interferometers: Testing Relativistic Gravity in Space* (Berlin: Springer-Verlag)

[30] Tolman R C 1987 *Relativity Thermodynamics and Cosmology* (New York: Dover) pp 245–247. (1st edn 1934 Oxford)

[31] Adler R 2005 *Gravity*, Chapter 2 in: Gordon Frasir: *The New Physics for the Twenty-First Century* (Cambridge: Cambridge University Press) pp 41–68. See p 44

[32] Friedman M 1983 *Foundations of Space-Time Theories* (New Jersey: Princeton University Press) p 179

[33] Klein M J, Kox A J, Renn J and Schulmann R 1995 *The Collected Papers of Albert Einstein* Vol. 4: Doc 17 (pp. 486-503) and Doc 28 (pp. 589–597). (New Jersey: Princeton University Press)

[34] Mehra J 1974 *Einstein, Hilbert, and the Theory of Gravitation* (Dordrecht, Holland: Reidel) pp 6-8, 60-62

[35] Einstein A and Infeld L 1938 *The Evolution of Physics* (New York: Simon and Schuster) pp 256–258, 260

[36] Born M 1965 *Einstein's Theory of Relativity* (New York: Dover) p 351, 2nd edn

[37] Einstein A, Infeld L and Hoffmann B (EIH) 1938 *Ann. Math. Ser. 2* **39** 65

[38] Einstein A and Infeld L 1949 *Can. J. Math.* **1** 209

[39] Einstein A and Infeld L 1940 *Ann. Math.* **41** 455

[40] Van Flandern T 1998 *Phys. Lett. A* **250** 1

[41] Liebendörfer M et al. 2008 *J. Phys. G: Nucl. Part. Phys.* **35** 014056

[42] Einstein A 1939 *Ann. Math.* **40** 922. See p 936

[43] Soker N (13.2.2008) *Conservation of angular momentum during star evolution*, Lecture, Faculty of Physics, Technion, Haifa, Israel

[44] Blandford R D 1976 *Mon. Not. Astr. Roy. Soc. London* **176** 465

[45] Burgarella D, Livio M and O'Dea C 1993 *Astrophysical jets* (Cambridge: Cambridge University Press)

[46] Ogilvie G I and Livio M 1998 *Astroph. J.* **499** 329

[47] Livio M 1999 *Phys. Rep.* **311** 225

[48] Tsinganos K, Ray T and Stute M 2009 *Protostellar Jets in Context* (Springer-Verlag)

[49] Weinberg S 1972 *Gravitation and Cosmology* (New York: Wiley). See pp 78, 152.

[50] Thorne K S, Flammag R A and Żytkow A N 1981 *Mon. Not. R. Astr. Soc.* **194** 475

[51] Penrose R 2002 *Gen. Rel. Grav.* **34** 1141

[52] Thorne K S 1994 *Black Holes and Time Wraps* (New York: Norton)

Journal of Physics: Conference Series **330** (2011) 012018 doi:10.1088/1742-6596/330/1/012018

Relativistic radial expansion: do we need dark energy?

N. Ben-Amots
P.O.Box 3193, Haifa 31032, Israel
E-mail: benamots@alumni.technion.ac.il

Abstract. The Hubble law, for which the radial expansion velocity is linear in the distance r, implies a visible Universe limited to a horizon at $r = c/H$. No information beyond this horizon can reach us.

Here we suggest a modified Hubble law for the *measured* radial expansion velocity that does not put a limit on the radius of the visible Universe. The modified Hubble law is based on Einstein's well known addition of relativistic velocities, and differs from the (conventional linear) Hubble law in orders of Hr/c that are higher than the first order. The modified Hubble law solves the Olbers's paradox (as the linear Hubble law does), and the existence of dark energy becomes unnecessary for explaining existing astronomical data on dependence of luminosity as a function of redshift for type Ia supernovas.

1. Introduction

The linear Hubble law (in some references referred to as the Hubble-Humason law) implies a horizon. See, for example, Rindler [1] and others [2-6]. Ellis and Rothman [5] named their paper on the subject of cosmological horizons: *"Lost horizon."*

Here we present a modified Hubble's law so that no horizon exists or is necessary.

Our formulas indicate that Perlmutter's interpretation of data of luminosity versus redshift of type Ia supernovas as demanding an accelerating expanding Universe is not correct. Analyzing the data with the modified Hubble law, shows that no cosmic acceleration is present, so that invoking Dark Energy is superfluous.

2. Horizon in radial expansion

In an expanding Universe with the linear Hubble law

$$v = Hr \tag{1}$$

we get a velocity greater than the speed of light when

$$r > \frac{c}{H} \tag{2}$$

We may interpret $\frac{c}{H}$ as the horizon of the Universe, because if Hubble's law (1) is accurate, no communication is possible between us and regions beyond this limiting radius. This horizon is referred as the "Hubble radius."

Journal of Physics: Conference Series **330** (2011) 012018 doi:10.1088/1742-6596/330/1/012018

3. Integration

Special relativity gives the formula for adding relativistic velocities

$$v = \frac{v_1 + v_2}{1 + v_1 v_2/c^2} \tag{3}$$

where v_2 is the velocity of a point 2 in an inertial system moving with a velocity v_1 relative to another inertial system. Velocity v is the velocity of the point 2 relative to the other system. This formula is symmetric in v_1, v_2 and is valid also if v_2 is the velocity of a point in an inertial system moving with a velocity v_1 relative to the other system.

This formula ensures that the velocity does not exceed the speed of light c, and is derived from this assumption.

Equation (3) may be written as

$$v_2 = \frac{v - v_1}{1 - v\,v_1/c^2} \tag{4}$$

where v_2 is again the relative velocity between two systems 2 and 1 as defined above.

Using (4) and the radial expansion $v = Hr$ as in (5), the relative velocity between a point at a distance $r + \Delta r$ from us, and a point at a distance r from us on the same line of sight is:

$$H\Delta r = \frac{v(r + \Delta r) - v(r)}{1 - v(r + \Delta r)v(r)/c^2} \tag{5}$$

Going to infinitesimal $\Delta r \to dr$

$$H\,dr = \frac{dv}{1 - v^2/c^2} \tag{6}$$

that on integration leads to

$$\frac{v}{c} = \tanh\frac{Hr}{c} \tag{7}$$

or

$$v(r) = c\tanh\frac{Hr}{c} \tag{8}$$

which we suggest as a modified Hubble law for the *measured* radial expansion velocity.

Equation (8) can be written:

$$r(v) = \frac{c}{H}\text{arctanh}\frac{v}{c} \tag{9}$$

4. Radial expansion

To enable $r > c/H$, without exceeding the speed of light, we present modified non-linear possibility for the linear Hubble law:

We showed above that applying the relativistic addition of velocities (3) on the linear radial expansion (1) leads to the relativistic formula of relative velocity between distant parts of the Universe (8).

Expansion of (8) results in:

$$v = Hr\left[1 - \frac{1}{3}\left(\frac{Hr}{c}\right)^2 + \frac{2}{15}\left(\frac{Hr}{c}\right)^4 - \ldots\right] \tag{10}$$

Keeping the first term, we receive for equation (8) the linear Hubble law (1), which is valid in the view only for relatively small distances, before the correction terms in (10) become significant.

Journal of Physics: Conference Series **330** (2011) 012018 doi:10.1088/1742-6596/330/1/012018

Yet for (8) when r approaches infinity, the *measured* velocity v approaches the velocity of light c, without exceeding it.

For (8) we can write a corresponding no-horizon line element:

$$ds^2 = \frac{c^2 dt^2}{\cosh^2 \frac{Hr}{c}} - dr^2 \cosh^2 \frac{Hr}{c} - r^2(d\theta^2 + d\phi^2 \sin^2 \theta) \tag{11}$$

5. Distances and redshift

The formula for the redshift of special relativity is

$$z = \sqrt{\frac{c+v}{c-v}} - 1 \tag{12}$$

or

$$v = c\frac{z + \frac{1}{2}z^2}{1 + z + \frac{1}{2}z^2} \tag{13}$$

We see that each of the equations (12) and (13) imply that $v < c$ for any z.

Hubble's law (1) can be joined with (13) and we receive for the conventional linear Hubble law

$$r = \frac{c}{H}\frac{z + \frac{1}{2}z^2}{1 + z + \frac{1}{2}z^2} \tag{14}$$

that is, $r < \frac{c}{H}$ for any z.

We can substitute (13) in the modified Hubble's law (8) and obtain

$$r = \frac{c}{H}\operatorname{arctanh}\left(\frac{z + \frac{1}{2}z^2}{1 + z + \frac{1}{2}z^2}\right) \tag{15}$$

in which r is greater than in (14), thus for each z, each z star is further away from us compared with the distance given by the linear Hubble law (1). If the linear Hubble law (1) is an approximation of the modified Hubble law (8), then the conventional distance (14) is an approximation to our distance (15).

Substituting the linear Hubble law (1) in (12) gives:

$$z = \sqrt{\frac{c+Hr}{c-Hr}} - 1 \tag{16}$$

Substituting the modified Hubble law (8) in (12) gives:

$$z = \sqrt{\frac{1 + \tanh \frac{Hr}{c}}{1 - \tanh \frac{Hr}{c}}} - 1 \tag{17}$$

or:

$$z = \exp\left(\frac{Hr}{c}\right) - 1 \tag{18}$$

from which one gets:

$$r = \frac{c}{H}\ln(z+1) \tag{19}$$

which is a simpler expression for (15).

6. Olbers's paradox

Halley [7-8], Cheseaux [9] and Olbers [10] asked why the night sky is dark, although the light of the stars and galaxies should accumulate to a very bright sky even at night. This question is called Olbers's paradox [11-20].

Citing Weinberg [12] pp. 611-612:

> "... if the absorption is neglected, the apparent luminosity of a star of absolute luminosity B at a distance r in a naive cosmological model will be $\frac{B}{4\pi r^2}$. If the number density of such stars is a constant n, then the number of stars at distances between r and $r + dr$ is $4\pi n r^2 dr$, so the total radiant energy density due to all stars is
>
> $$\int_0^\infty \frac{B}{4\pi r^2} 4\pi n r^2 dr = Bn \int_0^\infty dr = Bnr|_0^\infty$$
>
> $$\text{Weinberg number: (16.1.1) our number:} \quad (20)$$
>
> The integral diverges, leading to an infinite energy density of starlight!"

Supposing that the universe is homogenous having a constant density n of stars of absolute luminosity B, the light reaching us from a spherical layer of the universe of thickness dr at a distance r from us is

$$4\pi r^2 n \frac{B}{4\pi r^2} dr \quad (21)$$

When integrated for an infinite universe, that is infinite r, the integral of (21) diverges, meaning infinitely bright sky, which is not observed. This is the well known Olbers's paradox.

Today two accepted answers try to solve this paradox:

A. The age of the Universe is finite, so that light from distant parts of the Universe more remote than a certain horizon has not yet reached us. Thus the light that we see is an integral of the light we see up to a certain distance.

B. The Universe is expanding according to Hubble's law, so that the light of distant galaxies is red-shifted. The integrated luminosity of the shifted light up to an infinite distance is finite.

We derive the following proof for the resolution of Olbers's paradox based on B above only. The relativistic formula for the shift of frequencies is

$$\frac{f'}{f} = \sqrt{\frac{c - v}{c + v}} \quad (22)$$

Substituting the linear Hubble law (1) in (22) we obtain:

$$\frac{f'}{f} = \sqrt{\frac{c - Hr}{c + Hr}} \quad (23)$$

Checking for Olbers's paradox integrated radiant energy density we receive the integral

$$\int_0^{\frac{c}{H}} Bn\sqrt{\frac{c - Hr}{c + Hr}} dr \quad (24)$$

which is finite.

Now we do the same for the modified Hubble law above. Substituting the modified Hubble law (8) in (22) we obtain

$$\frac{f'}{f} = \sqrt{\frac{1 - \tanh(Hr/c)}{1 + \tanh(Hr/c)}} \quad (25)$$

Journal of Physics: Conference Series **330** (2011) 012018 doi:10.1088/1742-6596/330/1/012018

or

$$\frac{f'}{f} = \exp\left(\frac{-Hr}{c}\right) \tag{26}$$

The above energy density integral for this case is:

$$\int_0^\infty Bn \exp\left(\frac{-Hr}{c}\right) dr = \frac{c}{H} Bn \tag{27}$$

The integrals (24), (27), are finite. Thus the amount of light in the sky is finite for the modified Hubble law, so the night sky is not infinitely bright, and Olbers's paradox is solved for both the linear Hubble law (1), and for the modified Hubble law (8).

7. The "dark energy" hypothesis

Observations using Perlmutter et al.'s and Riess et al.'s method to measure type Ia supernova distances (Perlmutter et al. [21], and Riess et al. [22]) were interpreted as if the expansion of the universe is accelerating. See refs. 22-26.

An acceleration in the expansion of the universe came as a big surprise. It was interpreted as if some repelling force is acting against the gravitational attraction. This assumed force was associated with a further assumed "dark energy."

Perlmutter et al. [21] estimated distance according to the observed luminosity of remote type Ia supernovas, whose absolute peak luminosity is known. The corresponding redshift was the redshift of the galaxies that host these supernovas. Remote type Ia supernovas were found fainter than expected according to the value calculated with the linear Hubble's Law (1), using the corresponding redshift, thus Perlmutter et al. and others assumed that there is an acceleration in the universe expansion.

To explain the force necessary to accelerate the expansion of space dragging all the galaxies in the Universe, a "dark energy" in the Universe causing this acceleration was assumed.

Here we give a different interpretation to the observations, showing that if the observed redshift of remote galaxies is indeed originated in Doppler effect of radial velocity of expansion, then the acceleration of the expansion of the universe does not exist and is only a misinterpretation of data.

According to the modified Hubble law (Equation (8)), the measured expansion velocity deduced from z at a certain distance r is *less* than expansion velocity Hr according to the *linear* Hubble's Law (1). So, the distance r calculated from our equation (8) is *larger* than the distance r calculated from the linear Hubble law (1). This larger distance explains the observed fainter luminosity of type Ia supernovas. Thus, the expanding universe did not accelerate, and no dark energy is necessary to explain the observations.

We conclude that the expanding universe is not accelerated, hence there is no accelerating force, and no "dark energy" is needed to explain the (non existing) acceleration of the expansion.

7. Observed luminosity and magnitude

The observed luminosity L of a celestial object is proportional to the reciprocal of its distance r from us:

$$L = \frac{kB}{r^2} \tag{28}$$

Substituting r of (14) (derived from the linear Hubble Law (1)) in Eq. (28) results in observed luminosity:

$$L = kB\left(\frac{H}{c}\right)^2 \left(\frac{1 + z + \frac{1}{2}z^2}{z + \frac{1}{2}z^2}\right)^2 \tag{29}$$

Journal of Physics: Conference Series **330** (2011) 012018 doi:10.1088/1742-6596/330/1/012018

or:

$$L = kB\left(\frac{H}{c}\right)^2\left(1+\frac{1}{z+\frac{1}{2}z^2}\right)^2 \tag{30}$$

Substituting r of (19) (derived from the modified Hubble Law (8)) in (28) results in observed luminosity:

$$L = kB\left(\frac{H}{c\ \ln(1+z)}\right)^2 \tag{31}$$

that is fainter than the luminosity according (29) (or (30)).

The defined astronomical magnitude of a celestial body is larger when its luminosity is smaller. The magnitude of a celestial body whose luminosity is smaller hundred times than the luminosity of another celestial body, is defined to be *larger* by five.

In our case the relevant measurements are all of explosions of type Ia supernovas, which have the same absolute peak luminosity, thus the observed luminosity depends solely on their distance r from us. So, for a type Ia supernova distant ten times from another type Ia supernova, the observed luminosity will be hundred times weaker, or its magnitude will be larger by five:

$$\mu = \mu_0 + 5\log_{10}\frac{r}{r_0} \tag{32}$$

where μ_0 and r_0 are the magnitude and the distance to a certain supernova of the type Ia, while μ and r are the magnitude and the distance to another supernova of type Ia.

The results of measurements are given by graphs plotting $\mu(z)$ of several type Ia supernovas.

Substituting (14) in (32) gives the expected $\mu_{linear}(z)$ for the linear Hubble law (1) as:

$$\mu_{linear} = \mu_0 + 5\log_{10}\left(\frac{c}{H\,r_0}\,\frac{z+\frac{1}{2}z^2}{1+z+\frac{1}{2}z^2}\right) \tag{33}$$

Substituting (19) in (32) gives the expected $\mu(z)$ for the modified Hubble law (8) as:

$$\mu_{modified} = \mu_0 + 5\log_{10}\left[\frac{c}{H\,r_0}\,\ln(z+1)\right] \tag{34}$$

Substituting (15) in (32) gives the expected $\mu_{modified}(z)$ for the modified Hubble law (8) as:

$$\mu_{modified} = \mu_0 + 5\log_{10}\left[\frac{c}{H\,r_0}\,\mathrm{arctanh}\left(\frac{z+\frac{1}{2}z^2}{1+z+\frac{1}{2}z^2}\right)\right] \tag{35}$$

which is equivalent to (34).

Comparing equation (35) with (33) we see that $\mu_{modified}(z)$ of (35) is larger than $\mu_{linear}(z)$ of (33) for any z >0. Similarly, the accelerating expansion graph in Figure 1 of Perlmutter et al. [21] is higher than the non-accelerating expansion graph in Figure 1 of Perlmutter et al. [21]. Both Perlmutter's μ of accelerating expansion graph in his Figure 1, and our $\mu_{modified}(z)$ in our Figure 1 here, give better fitting to data being higher than non-accelerating expansion graph in Figure 1 of Perlmutter et al. [21].

Figures 1 and 2 present the expected magnitude μ of supernovas of type Ia versus Log_{10} of the redshift z, according the linear and the modified Hubble laws. Figure 1 presents the expected magnitude μ for redshift z in the range z=0.01 to z=1 (the range covered by refs. 21-22, 26), while Figure 2 shows the expected magnitude μ in the range z=0.01 to z=10 (covered partially by observations up to $z \approx 2$ in refs. 27-29). In both figures, the lower graph is the function

$$\mu = 25 + 5\log_{10}\left(\frac{z+\frac{1}{2}z^2}{1+z+\frac{1}{2}z^2}\right) \tag{36}$$

Figure 1: Magnitude μ versus $\log_{10}$ of redshift z for z between 0.01 and 1

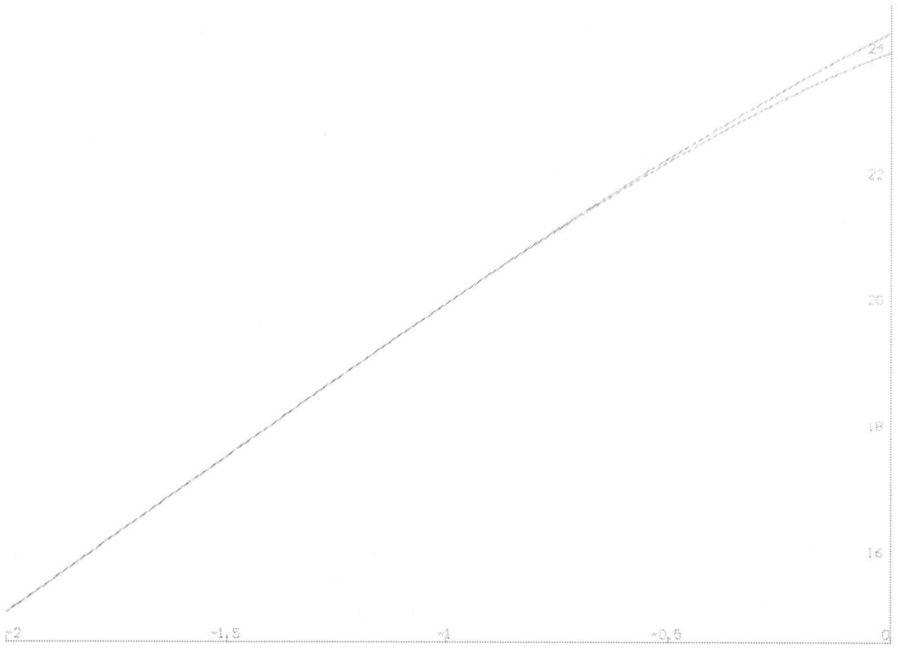

Figure 2: Magnitude μ versus $\log_{10}$ of redshift z for z between 0.01 and 10

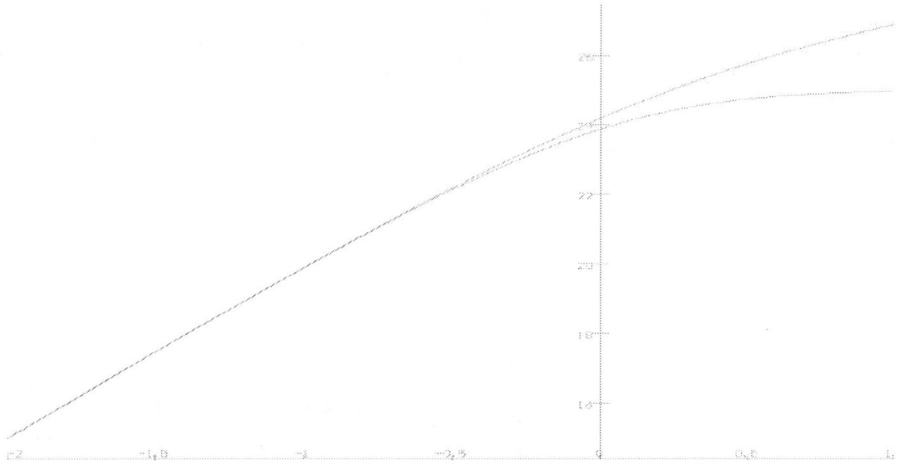

Journal of Physics: Conference Series **330** (2011) 012018 doi:10.1088/1742-6596/330/1/012018

representing the expected magnitude according to the linear Hubble law (1), while the upper graph is the function

$$\mu = 25 + 5\log_{10}\left[\ln(z+1)\right] \tag{37}$$

representing the expected magnitude according to the modified Hubble law (8).

Comparing our Figure 1 to the graphs of $\mu(z)$ in Figure 1 by Perlmutter et al. [21] in their Figure 1 shows that the difference between our upper graph (μ for modified Hubble law (8)) and the lower graph (μ for linear Hubble law (1)) in Figure 1 here, is similar to the difference between the upper (μ for accelerating expansion) and the lower (μ for expansion without acceleration) graphs in Figure 1 of Perlmutter et al. [21].

Figure 2 shows significant difference of the predicted magnitude according the modified Hubble law (8) (the upper graph), comparing to the expected magnitude according to the linear Hubble law (1) (the lower graph), in the redshift range z=0.01 to z=10.

The modified Hubble law (8) implies that measurements of redshifts of z=0.01 to z=10 of galaxies hosting type Ia supernovas will fit the upper graph of Figure 2, which significantly goes downwards for large z compared to the lower graph calculated for the case of no-acceleration linear Hubble law (1). Bertram Schwarzschild [27] and Riess et al. [28] pointed out that apparent magnitude of supernova SN1997ff whose host galaxy (HDF 4-403.0) redshift z was measured as 1.7, is significantly downwards compared to the extrapolation of the graph of accelerating universe up to z=1. See their graphs in Figure 2 of Schwarzschild [27] and figures 11 and 12 by Riess et al. [28]. Later Riess et al. [29] observed downwards trend for experimental data curves for more type Ia supernovas at larger z. They explained [27-29] the downward trend by assuming that once the Universe decelerated, then started to accelerate. Yet our Figure 2 shows that this significant downwards departure is part of the solution for both the linear Hubble law (1) and the modified Hubble law (8), without needing decelerating and accelerating Universe. The reason for this downwards departure is that according to (12) the redshift z is not linearly proportional to the expansion velocity v.

Just in case that our considerations above are not able to explain the experimental measurements sufficiently accurate, so that acceleration of the expansion is still necessary for an explanation, below in Section 9 we derive formulas that consider *constant* acceleration *and* the modified Hubble law (8).

8. Radial acceleration

As can be shown (See for example Adler and Brehme [30]), for linear motion with *constant* acceleration a:

$$\frac{v}{c} = \tanh\frac{aT}{c} \tag{38}$$

where T is the proper time of the accelerated body in linear motion.

Integration gives (Adler and Brehme [30]:

$$X(t) = \frac{c^2}{a}\left[\cosh\left(\frac{aT}{c}\right) - 1\right] \tag{39}$$

In (39) the velocity v is measured with the proper time T of *a "traveler,"* yet X represents "our" measured distance.

Applying a similar method for *constant* acceleration a and expansion Hr similarly gives

$$\frac{v}{c} = \tanh\frac{aT + Hr}{c} \tag{40}$$

that represents modified Hubble Law that also considers possible *constant* linear acceleration a in the $\vec{v}$ direction.

Formula (40) incorporates possible constant acceleration a as a free parameter.
Substituting (19) in (40) gives:

$$v(z) = c \tanh\left[\frac{aT}{c} + \ln(z+1)\right] \tag{41}$$

Formulas (38)-(41) are not for variable acceleration.

Taking advantage of these formulas that consider *constant* radial acceleration a, we may compare them to observations' results (Perlmutter et al. [21], Riess et al. [22, 28-29], Schwarzschild [27]) for finding the *constant* acceleration a that best fits the observations.

9. Summary

The linear Hubble law leads to a visible Universe limited in radius.

Here we suggested a modified Hubble law that does not limit the visible radius of the Universe, thus enabling a Universe without an horizon. The linear Hubble law may be viewed as an approximation of the modified Hubble law for small expansion velocity of relatively close galaxies.

We showed that the modified Hubble law, like the linear Hubble law, solves Olbers's paradox.

This paper shows that if the origin of Hubble's redshift of galaxies is the Doppler effect, (caused by the receding velocity of the galaxies), there exists a modified Hubble law that fits the present observations for small Hr/c – yet without exhibiting a horizon at large Hr/c. Further, we suggested explanation for Perlmutter et al.'s [21] and others' observations by the non-linear, modified Hubble's law (8), without needing dark energy.

The originally suggested Hubble law (1) is arbitrary, while the modified Hubble law (8) suggested above obeys the correct relativistic addition formula (3):

$$v = \frac{v_1 + v_2}{1 + v_1 v_2 / c^2} \tag{3}$$

Thus we derived a new formula for the Hubble law, which contains the existing as a special case for small expansion velocity of relatively close galaxies.

So, using our formula (8)

a) No horizon is needed,

b) Olbers's paradox is explained by both Hubble formulas.

c) We find that Perlmutter et al.'s and others' interpretation of accelerating expanding universe may be incorrect. We suggest that analyzing the phenomenon with the modified Hubble law, will find that no cosmic acceleration is present, so that also Dark Energy may be superfluous.

Acknowledgments

The author is indebted to Dan Igner, who helped find the appropriate wording for a clearer presentation. Thanks to Professor Larry Horwitz who read the manuscript, for his guiding questions.

References

[1] Rindler N 1956 *Mon. Not. R. Astron. Soc.* **116** 662

[2] Ellis G F R and Stoeger W 1988 *Class. Quant. Grav.* **5** 207

[3] Berry M V 1989 *Principles of cosmology and gravitation,* (Bristol: Institute of Physics) pp 107-111

[4] Harrison E 1991 *Astroph. J* **383** 60

[5] Ellis G F R and Rothman T 1993 *Am. J. Phys.* **61** 883

Journal of Physics: Conference Series **330** (2011) 012018 doi:10.1088/1742-6596/330/1/012018

[6] Allday J 1998 *Quarks, leptons and the big bang*, (Bristol: Institute of Physics Publishing) pp 257-259

[7] Halley E 1720 *Phil. Trans. Roy. Soc. (London)* **31** 22

[8] Halley E 1720 *Phil. Trans. Roy. Soc. (London)* **31(364)**

[9] de Cheseaux J P L 1744 *Traite de la comete*, (Lausanne:MM Bousquet) pp 223 ff; Reprinted in Dickson [11] Appendix II

[10] Olbers H W M 1826 *Über die Durchsichtigkeit des Welttraums*, in: *Bode's Astronomiches Jahrbuch* **111** (Berlin: CFE Spaethen) pp 110-121. First English translation: *On the transparency of space*, *Edinburgh New Phil. J.* **1** 141 (1826). Reprinted in Dickson [11] Appendix I

[11] Dickson F P 1968 *The bowl of night* (Cambridge, Mass.: M.I.T. Press) Appendixes I, II

[12] Weinberg S 1972 *Gravitation and cosmology*, (New York: Wiley) See Chapter 16.1, pp 611-613

[13] Harrison E R 1964 *Nature* **204** 271

[14] Harrison E R 1974 *Phys. Today* **27** 30

[15] Harrison E R 1977 *Am. J. Phys.* **45** 119

[16] Harrison E R 1981 *Cosmology. The science of the Universe* (Cambridge, Mass.: Cambridge Univ. Press)

[17] Harrison E 1984 *Science* **226** 941

[18] Harrison E R 1986 *Nature* **322** 417

[19] Harrison E 1987 *Darkness at night – a riddle of the Universe*, (Cambridge, Mass.: Harvard Univ. Press)

[20] Arpino M and Sardigli F 2003 *Eur. J. Phys.* **24** 39

[21] Perlmutter S et al. 1999 *Astrophys. J.* **517** 565

[22] Riess A G et al. 1998 *Astron. J.* **116(3)** 1009

[23] Glanz J 1998 *Science* **282** 2156

[24] Hogan C J et al. 1.1999 *Sci. Am.* **280** 46

[25] Krauss L M 2.1999 *Sci. Am.* **280** 52

[26] Perlmutter S 4.2003 *Phys. Today* **56(4)**

[27] Schwarzschild B 6.2001 *Phys. Today* **54(6)** 17

[28] Riess A 2001 *Astrophys. J.* **560** 49 (*Preprint* `astro-ph/0104455`)

[29] Riess A G et al. 2004 *Astrophys. J.* **607** 665 (*Preprint* `astro-ph/0402512`)

[30] Adler C G and Brehme R W 1991 *Am. J. Phys.* **59** 209. See Equation 7b.

AUTHOR INDEX

Institute of Physics
Dirac House, Temple Back
Bristol BS1 6BE UK

CURRAN ASSOCIATES INC
proceedings
.com